Practical Channel Hydraulics

Practical Channel Hydraulics

Roughness, Conveyance and Afflux

2nd Edition

Donald W. Knight

Emeritus Professor, University of Birmingham, Birmingham, UK

Caroline Hazlewood

HR Wallingford, Wallingford, UK

Rob Lamb

JBA Consulting, Skipton, UK

Paul G. Samuels

HR Wallingford, Wallingford, UK

Koji Shiono

Emeritus Professor, University of Loughborough, Loughborough, UK

CRC Press
Taylor & Francis Group
Boca Raton London New York Leiden

CRC Press is an imprint of the
Taylor & Francis Group, an **informa** business

A BALKEMA BOOK

Cover photo credit:
Rivon Avon, near Fladbury, Worcestershire, February 1979
by P.G. Samuels

Published 2018 by CRCPress/Balkema
P.O. Box 447, 2300 AK Leiden, The Netherlands
e-mail: Pub.NL@taylorandfrancis.com
www.crcpress.com – www.taylorandfrancis.com

First issued in paperback 2021

ISBN 13: 978-0-367-78142-2 (pbk)
ISBN 13: 978-1-138-06858-2 (hbk)

Visit the Taylor & Francis Web site at
http://www.taylorandfrancis.com

and the CRC Press Web site at
http://www.crcpress.com

Typeset by V Publishing Solutions Pvt Ltd., Chennai, India

Although all care is taken to ensure integrity and the
quality of this publication and the information herein, no
responsibility is assumed by the publishers nor the author for
any damage to the property or persons as a result of operation or
use of this publication and/or the information contained herein.

Library of Congress Cataloging-in-Publication Data

Names: Knight, Donald W., author.
Title: Practical channel hydraulics : roughness, conveyance and afflux/
authors, Donald W. Knight, University of Birmingham, Birmingham, UK
Caroline Hazlewood, HR Wallingford, Wallingford, UK
Rob Lamb, JBA Consulting, Skipton, UK
Paul G. Samuels, HR Wallingford, Wallingford, UK
Koji Shiono, University of Loughborough, Loughborough, UK
Description: 2nd edition. | Leiden, The Netherlands: CRC Press/Balkema,
[2018] | Includes bibliographical references and indexes.
Identifiers: LCCN 2017059717 (print) | LCCN 2017059978 (ebook) |
ISBN 9781315157771 (ebook) | ISBN 9781138068582 (hardcover : alk. paper)
Subjects: LCSH: Channels (Hydraulic engineering)
Classification: LCC TC175 (ebook) | LCC TC175 .P73 2018 (print) |
DDC 627/.042--dc23
LC record available at https://lccn.loc.gov/2017059717

Contents

Foreword 1st Edition

Forty years ago as an MSc student studying Civil Engineering hydraulics, I read the excellent '*History of Hydraulics*' by Hunter Rouse and Simon Ince. I was impressed by two common strands to many of the advances in hydraulics over the centuries. First, developments were largely driven by the needs of society. Second, these developments were usually done by practical scientists and engineers who, while building on the scientific knowledge of that day, had a good understanding of engineering application and got involved in some inspiring experimental research – often at field scale.

There are elements of all the above in *Practical channel hydraulics* and its origins. With regard to societal need, the principal driver of the research and development covered by the book has been improved flood risk management in the United Kingdom. The subject matter is of course relevant to many other applications of open channel hydraulics throughout the world.

For effective design or management of the channels and structures associated with flood defence and land drainage, practitioners must be able to predict accurately the water level at the relevant flood or storm conditions. The local increase in water level (afflux effect) caused by flow constriction at bridges and culverts is particularly important as these structures are often located close to human settlement or some other important infrastructure or environmental features that will be affected by flooding. Key requirements that have driven the tools (the *Conveyance Estimation System or CES* and the *Afflux Estimation System or AES*) associated with this book are the ability (a) to understand the sensitivity of water level to the various physical parameters that determine it, and (b) to plan 'interventions' such as cutting vegetation in channels or installing a culvert that might reduce the required performance of the flood management or land drainage system.

Most rivers or watercourses in today's world are multi-functional – as one might say 'One river, many users' – not least the flora and fauna who cannot speak for themselves! Fortunately their interests are covered in most countries by legislation, such as the Water Framework Directive in Europe, and by statutory and voluntary organisations and individuals. Channel management usually involves a dialogue with other users. A particular driver for the CES was the need to help those planning engineering works or management activities to provide evidence to, and to explore options with, others having different interests in the same channel.

Perhaps the strongest driver for the tools described in this book has been the consensus reached by the UK flood management community in 2000 that there would be significant benefit from developing new practical computer-based tools for the estimation of water level in channels. Therefore the UK authorities, led by the Environment Agency but with support from the Scottish Government and Rivers Agency Northern Ireland, have funded the programme of R&D needed to draw together the advances in the underlying science and to make this available in a practical and useful form to practitioners. This national programme involving both researchers and practitioners was deemed necessary in order to bridge the so-called 'implementation gap' between the available science and its application.

To return to the *History of Hydraulics*, Antoine Chezy (who is a senior figure in hall of fame of channel hydraulicians) made an interesting and relevant comment to the French Academy of Sciences in the 1760s when he proposed his original formula for determining the required channel cross-section to convey a given discharge (in this case the water supply to Paris). He remarked that '... *It would also be interesting to have similar observations on different brooks and streams, but it is important to note that all these observations require the greatest care, that it is difficult to make them with sufficient precision, and that one should count only on those which are made by people known for giving the most scrupulous attention*'.

Practical channel hydraulics would certainly have pleased Chezy with the extent of data that it either contains or references through the related website (*www.river-conveyance.net*). I am sure that Chezy would also have approved of the care and scrupulous attention that has been given to the supporting programmes of research and development described in Section 1.5. The programme of research in the former UK Flood Channel Facility was world class and the scientific publications resulting from it have made a significant contribution to the literature. For promoting this as well as for the initiative to produce the book, we owe particular gratitude to Professor Donald Knight. However, I must also express gratitude to the other authors each of whom has contributed an important part of the book. They are all leaders in their respective fields. Chezy would have found them all to be the type of people he could count on!

Dr. Mervyn Bramley
OBE Engineer and Environmentalist,
Formerly Head of R&D Environment Agency
Bristol,
April 2009

Foreword 2nd Edition

Just prior to writing this foreword, I was standing on the banks of the Shire River in Malawi, a short distance downstream of Lake Malawi. I was reflecting on the problems of determining discharges at key locations on this strategic river that is such an important resource for Malawi. I was also conscious that this essential source of livelihood can, too, be a destroyer of life during extreme floods, such as the catastrophic January 2015 event. The Shire River basin is a very dynamic system due to climate variability and land-use changes and many of the key river gauging sites have unstable stage-discharge relationships. The problems of capturing high flood events by gauging at key stations make it difficult to determine water level and flow on a continuous basis. My deliberations turned me to thinking about the contents of *Practical channel hydraulics*. I reflected how the tools and techniques described therein can help with the development of river basin utilisation plans, the control of watercourses including the design of hydraulic structures and the management of river systems, including issues like those experienced in the Shire Basin.

In one my main areas of professional activity, namely hydrometry, I have always felt very strongly that a sound understanding of open channel hydraulics is an essential prerequisite for achieving good streamflow determinations. Despite the advances in modern monitoring technologies, the stage-discharge (water level flow) relationship is still the main route to obtaining continuous estimates of discharge throughout the world. Even if more modern techniques such as the use of fixed velocity sensors e.g. Horizontal ADCPS, are used, it is still necessary to develop relationships between the mean velocity and the measured velocity with stage as a second parameter. It is the monitoring of extreme flows, namely floods and droughts, that is of greatest importance and yet, which can present significant difficulties. For example, flood flow monitoring, especially when out of bank flow occurs, is often not possible. Accordingly, both stage-discharge and velocity index ratings need to be extended using hydraulic concepts and models. As people who know me professionally are aware, I would never advocate the use of analytical and modelled data in preference to flow measurements. Nevertheless, it is essential that in the real river environment, measurements are supported by sound and readily applied analytical and hydraulic modelling techniques.

Throughout my career, I have undertaken a considerable amount of hydrometric training. One of the key elements that I always include is an introduction to open channel hydraulic theory with relevance to hydrometric practitioners. I address topics such as velocity distribution, conveyance and the Manning formula, while at the same time not forgetting Chezy. In this regard, I feel strongly that persons engaged in the monitoring of rivers and streams should at the very least have a rudimentary understanding of the basic concepts that are described so well in *Practical channel hydraulics*.

The 2nd edition of *Practical channel hydraulics* builds on the techniques described in the first edition, including the use of Excel and the Conveyance Estimation System (CES) and Afflux Estimation System (AES) software on the Shiono-Knight Method on which the CES-AES methods are largely based, and examples of the use of the SKM and CES methods. These are very welcome enhancements and from my own personal perspective, I greatly appreciate the new Chapter 8, which provides examples, including the estimation and use of stage-discharge relationships. The current version of the International Standard on stage-discharge relationships, *ISO 1100-2, Hydrometry— Determination of the stage-discharge relationship* (to be renumbered as ISO18320) has recently been revised. The British Standards Committee on Hydrometry, of which I am privileged to be the current chair, has taken a leading role in these revisions through a specialist working group. We have been very fortunate to have had the lead author of *Practical channel hydraulics*, Professor Donald Knight, on the working group. He has not only brought his own extensive knowledge of the subject matter to the revision process but has been able to reflect the knowledge of the co-authors. This has resulted in a wider consideration within ISO18320, of the change of river hydraulic properties with stage as well as the introduction of both the CES and SKM methods. For my part, I believe that these are very important additions. The mathematical fitting techniques which are readily available in hydrological data management software or Excel, can so easily lead to the derivation of stage-discharge relationships without due consideration of the hydraulic and other physical properties of the watercourse.

As noted in the foreword to the first edition, one of the main driving forces for the tools and techniques described therein was to meet the needs of the UK flood management community. However, it was also noted that most rivers are multi-functional and the needs of water supply, HEP and navigation authorities need to be catered for, along with the requirements of fauna and flora, and the potential impacts of the latter. The CES and AES methods can be used for a wide variety of applications and critically, are not just applicable to the UK. Worldwide applications and examples from small artificial channels to large-scale international rivers are included in the book to good effect.

Returning to my deliberations on the banks of the Shire river, I felt comforted by the fact that even though some of the contents of *Practical channel hydraulics* and the supporting software tools were UK-driven, they can truly be used by practitioners and students throughout the world. It is currently one of the most "well-thumbed" technical references on my bookshelf. I value it highly since it balances the hydraulic theory and research on which it is based with the practicalities of solving a variety of river hydraulics and engineering problems and requirements. It also provides the supporting tools to solve those problems. I am sure that my copy of the 2nd edition will be even more heavily utilised than the 1st and will sit alongside a few of the classics from younger days such as *Open Channel Hydraulics* by Ven Te Chow. I wish that this book and the supporting software tools had been available at the advent of my career. The authors are to be congratulated on producing a 'classic'.

Stewart Child
Hydrological and Hydrometric Consultant
Hydro-Logic Services International Ltd.
Current Chair of the BSI and CEN committees on Hydrometry
Former Chair of the ISO Hydrometry, Velocity-Area Methods Sub-committee
December 2017

Acknowledgements

With a book of this nature it will be only too evident that the Authors owe a considerable debt to many colleagues, research workers and practitioners who have contributed so much to its origin and present form. Several institutions have also been involved, notably the Environment Agency (EA) and the Engineering and Physical Sciences Research Council (EPSRC), both of whom funded the foundational research undertaken in the Flood Channel Facility (FCF) built at HR Wallingford, and related research carried out at various UK Universities. Other organisations include the Natural Environment Research Council (NERC), the Scottish Government, the Rivers Agency in Northern Ireland and the Association of Drainage Authorities. Especial thanks are due to Mervyn Bramley, formerly Head of R&D at the EA, who supported this collaborative work from the beginning and who has played a crucial role in getting this book to its present state.

The Authors gratefully acknowledge the EA team for their considerable efforts in bringing this book to fruition, namely Dr. Mitchell, Mr. Robinson, Dr. Baxter, Mr. Tustin and Ms. Eleanor. The views expressed in this book are, however, personal and the publication does not imply endorsement by either the EA or the Department for Environment, Food and Rural Affairs (Defra).

The Authors would like to thank various staff at HR Wallingford (Ms. Escarameia, Professor Bettess, Mr. Wills), Wallingford Software (Messrs. Fortune, Millington, Nex, Gunn and Body and Dr. Daniels) and JBA Consulting (Mr. Benn, Dr. Mantz, Dr. Atabay, Mr. Rose) who were involved in the original CES-AES development. In particular, Rob Millington has contributed substantially to the ongoing CES-AES software development as well as the documentation within this book. The CES-AES programme has also benefited from a wide range of valuable inputs outside of these organisations. Notably, the contributions of Ms. Fisher (Independent) and Drs. Dawson and O'Hare (CEH) on channel and vegetation roughness as well as Messrs. Evans, Pepper, Rickard, Spencer and Sisson, Drs. Ervine, Whitlow, Khatibi, Hamill and Riddell and Professors Pender and Wright.

The Authors would also like to acknowledge the input of many former research students at the University of Birmingham who provided much of the data referred to herein, notably Drs. Abril, Ayyoubzadeh, Brown, Chlebek, Macdonald, Mohammadi, Omran, Patel, Rezaei, Rhodes, Yuen, Professors Al-Hamid, Atabay and Lai, to name but a few. Those who led the first phase of work in the FCF at HR Wallingford, Professor Sellin, Drs. Myers and Wormleaton, together with their assistants, deserve special praise for their painstaking efforts.

Professor Shiono, of Loughborough University, whose outstanding work whilst collaborating with one of the Authors led to the development of the theoretical basis for the Shiono & Knight Method (SKM), deserves special mention. His continuing research into the nature of turbulent flow behaviour in open channels, sustained over many years, has advanced our understanding of turbulence phenomena significantly. Dr. Tang, of Birmingham University, also deserves recognition for his continued analysis and development of the SKM in recent years.

Finally, particular thanks are due to five individuals who kindly read the text and made critical and exceptionally helpful comments on it in the later stages of its preparation, namely Mr. Ramsbottom, Professor Bettess, Drs. Myers, Sterling, and Wormleaton. Without their valuable scrutiny the book would be less than it currently is. The Authors are, of course, responsible for any remaining errors and deficiencies.

Notation

A	cross-sectional area; coefficient in logarithmic velocity law, Eq. (2.49);	(m^2)
A_{ac}	Ackers' coefficient – function of Dgr, Table 5.2;	(m)
a	coefficient for Colebrook-White equation, Eq. (3.12); amplitude, Eq. (5.14);	
B	channel width; coefficient in logarithmic velocity law, Eq. (2.49); barrel span;	(m)
B_{RD}	roadway width across channel above culvert or bridge;	(m)
b	channel base width; semi breadth; pier width;	(m)
b	coefficient for Colebrook-White equation, Eq. (3.12);	
C	Chezy coefficient, Eq. (2.23);	$(m^{1/2}\ s^{-1})$
C	weir discharge coefficient, Eq. (3.60);	
C_{ac}	Ackers' coefficient – function of D_{gr}, Table 5.2;	
C_C	contraction energy loss coefficient applied upstream of bridge or culvert, Eq. (3.63);	
C_E	expansion energy loss coefficient applied downstream of bridge or culvert, Eq. (3.63);	
C_d	discharge coefficient, Eq. (3.59);	
COH	channel coherence (Ackers, 1991–93);	
C_{uv}	coefficient for sinuosity, Eqs. (3.29), (3.32) & (3.33);	
c	coefficient for Colebrook-White equation, Eq. (3.12); wave celerity, Eq. (5.18);	(ms^{-1})
D	pipe diameter; culvert diameter; culvert depth when flowing full;	(m)
D	diffusion coefficient, Eq. (5.21);	$(m^2\ s^{-1})$
D_{gr}	dimensionless particle size number, Eq. (5.3);	
D_r	relative depth $(= (H - b)/H)$, Figure (2.45) and Eq. (3.21);	
d	pipe diameter; sediment dimension;	(m)
d_{50}	median size diameter of sediment particles;	(m)
d_{90}	representative sediment size at 90% of distribution, Eq. (4.1);	(m)
E	specific energy, Eq. (2.12);	(m)
E_t	total energy, Eq. (2.39);	(m)
e	eccentricity, see Figure (2.66);	
F	force on fluid, Eq. (2.17);	(N)

F_g	force on sluice gate;	(N)
F_{gr}	particle mobility number, Eq. (5.2);	
Fr	Froude Number ($= U/(\sqrt{gA/T})$), Eq. (2.5);	
f	Darcy-Weisbach friction factor, Eqs. (2.24)–(2.27);	
f	local boundary friction factor in Eq. (2.58), adapted from Eq. (2.24) using U_d;	
f, f_z, f_b, f_t	global, zonal, local and turbulent friction factors defined variously in Eq. (2.61);	
G_{gr}	sediment transport parameter, Eq. (5.1);	
g	gravitational acceleration;	(ms^{-2})
H	local water depth, Fig. (2.30) and Eq. (3.4);	(m)
H_i	water level above datum in some bridge afflux Figures, e.g. Fig. (2.53);	(m)
HWL_{ic}	Headwater level above datum for culvert under inlet control, Table (3.10);	(m)
HWL_{oc}	Headwater level above datum for culvert under outlet control, Table (3.10);	(m)
h	local water depth;	(m)
h_c	critical depth, see Eqs. (2.6)–(2.8) & (2.10);	(m)
h_f	head loss due to friction (m), Eq. (2.24);	(m)
h_{HW}	culvert head water depth, Fig. (3.48);	(m)
h_i	culvert inlet energy loss, Fig. (3.48);	(m)
h_n	normal depth, see Eqs. (2.2)–(2.4);	(m)
h_o	culvert outlet energy loss, Fig. (3.48);	(m)
h_{TW}	culvert tail water depth, Fig. (3.48);	(m)
Δh	water level difference (m);	(m)
I	inflow, Eq. (5.22)	(m^3 s^{-1})
ILi	culvert invert level above datum at the inlet, Fig. (3.48);	(m)
ILo	culvert invert level above datum at the outlet, Fig. (3.48);	(m)
JD	blockage ratio, Eq. (3.61);	
K	cross-sectional channel conveyance, Eq. (2.41);	(m^3 s^{-1})
K_i	energy loss coefficient of ith mechanism, Eq. (3.2);	
k	roughness height (linked to ks);	(m)
k_s	roughness height after Nikuradse (1933), as in Eqs. (2.26) & (2.30);	(m)
L	backwater length (m), as in Eq. (2.46); culvert or barrel length, length of bend;	(m)
L_C	contraction length upstream of bridge or culvert;	(m)
L_E	expansion length downstream of bridge or culvert;	(m)
L_W	culvert entrance length;	(m)
l	length of pipe or channel;	(m)
$\mathbf{M}$	linear operator (FEM), as in Eq. (3.37);	
M	Momentum, as defined in Eq. (2.20); Manning Number ($= 1/n$);	(m$^{1/3}$ s^{-1})
M	bridge opening ratio ($= q/Q$), used in Section 2.6.1, see Figure (2.66);	

m_{ac}	Ackers' coefficient – function of D_{gr}, Table 5.2;	
n	Manning's resistance coefficient, Eq. (2.1);	$(sm^{-1/3})$
n_{ac}	Ackers' constant relating to grain size, Table 5.2;	
n_{clear}	Manning 'n' when the channel is clear of vegetation;	$(sm^{-1/3})$
n_{irr}	unit roughness due to channel irregularities, Eq. (3.1);	$(sm^{-1/3})$
n_l	local or unit roughness, Eqs. (3.1) & (3.15);	$(sm^{-1/3})$
n_{sur}	unit roughness due to ground material, Eq. (3.1);	$(sm^{-1/3})$
n_{veg}	unit roughness due to vegetation, Eq. (3.1);	$(sm^{-1/3})$
O	outflow, Eq. (5.22);	$(m^3\ s^{-1})$
Q	total cross-section discharge;	$(m^3\ s^{-1})$
q	discharge per unit width (= Q/B or Q/b), unit flow rate, Eq. (2.11);	$(m^2\ s^{-1})$
	unfettered flow through a gap, as if the bridge or constriction was not in place	$(m^3\ s^{-1})$
q^n	unit flow for iteration n(FEM);	$(m^2\ s^{-1})$
Δq^n	incremental change in q^n (FEM);	$(m^2\ s^{-1})$
q_s	volume of sediment transported per second per unit channel width;	$(m^2\ s^{-1})$
P	wetted channel perimeter; power as in Eq. (2.16);	$(m), (W)$
per	percentage, Eq. (3.46);	$(\%)$
R	hydraulic radius (= A/P);	(m)
Re	Reynolds Number (= UD/v for pipes and $4RU/v$ for open channels), Fig. (2.10);	
r	entrance rounding, see Figure (2.66);	(m)
S	longitudinal common uniform gradient in steady uniform flow, Eq. (2.41);	
S_e	longitudinal energy gradient, Eq. (2.40);	
$\bar{S}_f$	longitudinal friction gradient, ($S_f = h_f/l$), Eq. (2.43);	
S_f	reach-averaged friction slope, Eq. (2.44);	
S_{fx}	friction gradient in x-direction, Eq. (5.27);	
S_{fy}	friction gradient in y-direction, Eq. (5.27);	
S_o	longitudinal bed slope, see Figure (2.11);	
S_w	longitudinal water surface slope (= $\partial\eta/\partial x$), see Figure (2.11);	
S_y	local transverse bed slope, Eq. (3.6);	
s	channel side wall slope (1: s, vertical: horizontal);	
T	top width of channel at water surface;	(m)
TW	tailwater depth;	(m)
t	time;	(s)
U	average cross-section velocity (= Q/A);	(ms^{-1})
U	velocity component in the x-direction (longitudinal), Eqs. (2.57);	(ms^{-1})
U_*	shear velocity (= $\sqrt{(\tau_o/\rho)}$, Eq. (2.31);	(ms^{-1})
U_d	depth-averaged streamwise velocity (x-direction), defined by Eq. (2.52);	(ms^{-1})
U_o	average cross-section velocity (= Q/A), Eq. (3.49);	(ms^{-1})

u	local velocity component in the x-direction (horizontal) Eq. (2.47);	(ms^{-1})
u_{max}	maximum velocity, as used in Eq. (2.55);	(ms^{-1})
u_*	local shear or friction velocity, Eq. (2.47);	(ms^{-1})
V	velocity component in the y-direction (lateral), Eq. (2.57) & (3.4);	(ms^{-1})
V	section mean velocity (in bridge afflux Figures from Hamill), Fig. 2.17;	(ms^{-1})
V_d	depth-averaged velocity (y-direction), Eq. (5.24);	(ms^{-1})
W	velocity component in the z-direction (vertical), Eq. (2.57);	(ms^{-1})
X_s	mass flux of sediment per mass flux of fluid, Eq. (5.6);	(ppm)
x	horizontal direction, streamwise in plan view;	(m)
x_s	unit mass sediment flux per unit mass of flow per unit time, Eq. (5.5);	(ppm)
x	vector of all contributing functions that are independent of Δq^n(FEM);	
Y	water depth in some bridge afflux Figures, e.g. Fig. (2.62);	(m)
y	lateral direction across channel section;	(m)
y_o	tailwater depth downstream of culvert outlet;	(m)
z	vertical direction;	(m)
$\bar{z}$	distance of centre of area below water surface;	(m)
z_b	level of channel bed above datum;	(mAD)
z_o	roughness length ($= \gamma k_s$), Eq. (2.47);	(m)

Greek Alphabet

α	Coriolis or 'kinetic energy' correction coefficient, Eqs. (2.12), (2.53) & (3.54);	
β	Boussinesq or 'momentum' correction coefficient, Eqs. (2.17) & (2.54);	
Γ	secondary flow term for straight prismatic channels, Eq. (2.60);	
Γ^*_{ave}	average scaled secondary flow term, Eq. (3.24);	
$\Gamma_{mc(trans)}$	secondary flow term in the transitional main channel region, Eqs. (3.28), (6.65) & (6.66);	
Δ	increment, percentage change in;	(%)
γ	energy loss factor in Eq. (2.62); coefficient for z_o in Eq. (2.47);	
ε	eddy viscosity ($= \tau_{yx}/(\rho \partial U/\partial y)$;	(m^2 s^{-1})
ε	weighting coefficient, Eq. (5.23);	
$\bar{\varepsilon}_{yx}$	depth-averaged eddy viscosity, Eq. (2.59);	(m^2 s^{-1})
η	water level above datum ($= h + z$), Fig. (2.11);	(mAD)
κ	von Kármán's constant, Eq. (2.47);	
λ	dimensionless depth-averaged eddy viscosity, Eq. (2.59) ($= \bar{\varepsilon}_{yx}/(U^*H)$);	
μ	dynamic viscosity;	(Nsm^{-2})

v	kinematic viscosity of the fluid ($= \mu/\rho$);	(m² s⁻¹)
v_t	turbulent eddy viscosity, Eq. (3.4);	(m² s⁻¹)
ρ	fluid density;	(kgm⁻³)
σ	plan form channel sinuosity, length along river thalweg/length along valley slope;	
σ	blockage ratio (= $(B - b)/B$), Eq. (2.62), see Figure (2.66);	
τ_d, τ_{2d}	two-dimensional shear stress;	(Nm⁻²)
τ_o	boundary shear stress, Eq. (2.32);	(Nm⁻²)
τ_{yx}	Reynolds stresses on vertical plane, normal to y axis in x direction, Eq. (2.57);	(Nm⁻²)
$\bar{\tau}_{yx}$	depth-averaged value of τ_{yx}, Reynolds stresses on vertical plane, Eq. (2.59);	(Nm⁻²)
τ_{zx}	Reynolds stresses on horizontal plane, normal to z axis in x direction, Eq. (2.57);	(Nm⁻²)
ϕ	linear shape functions (FEM);	
ϕ_b	bend angle; skewness of bridge, see Figure (2.66);	(°)
χ	relaxation factor, Eq. (3.36);	
ψ	projection onto plane due to choice of Cartesian coordinate system, Eqs. (3.4) & (3.6);	

Subscripts

ave	average;
fp	floodplain;
i	ith channel element;
mc	main channel;

Acronyms

AE	Afflux Estimator
AES	Afflux Estimation System
BM	Backwater Module
CAPM	Centre for Aquatic Plant Management
CES	Conveyance Estimation System
CFD	Computational Fluid Dynamics
CIRIA	Construction Industry Research and Information Association
CIWEM	Chartered Institution of Water and Environmental Management
CG	Conveyance Generator
COH	Coherence
COHM	Coherence Method
DCM	Divided Channel Method
Defra	Department for Environment, Food and Rural Affairs
DRF	Discharge Reduction Factor
DTi-SAM	System based Analysis and Management of flood risk
EA	Environment Agency
EC	European Commission
EPSRC	Engineering and Physical Sciences Research Council
EU	European Union
FCF	Flood Channel Facility
FE	Finite Element
FEH	Flood Estimation Handbook
FEM	Finite Element Method
FHWA	Federal Highways Administration
FRMRC	Flood Risk Management Research Consortium
HA	Hydrologic Atlas
HEC	Hydrologic Engineering Center (US Army Corps of Engineers)
HEC-RAS	HEC-River Analysis System (US Army Corps of Engineers)
HW	Head Water
ICE	Institution of Civil Engineers
InfoWorks RS	Hydrodynamic modelling software (Wallingford Software)
IPCC	Intergovernmental Panel on Climate Change
ISIS	Hydrodynamic modelling software (Halcrow)
ISO	International Standards Organisation
IWRS	InfoWorks RS (see above)

LiDAR	Light Detection And Ranging
LIFE	Lotic-invertebrate Index for Flow Evaluation
MAFF	Ministry of Agriculture, Fisheries and Food
MDSF	Modelling and Decision Support Framework
MIKE11	Hydrodynamic Modelling Software (Danish Hydraulic Institute)
NERC	Natural Environment Research Council
NIRA	Northern Ireland Rivers Agency
NIWA	National Institute of Water & Atmospheric research
NRA	National Rivers Authority
PAMS	Performance based Asset Management System
PDF	Probability Density Function
PHABSIM	Physical Habitat Simulator
RA	Roughness Advisor
RANS	Reynolds Averaged Navier-Stokes equations
RASP	Risk Assessment for Strategic Planning
RDL	Road Level
ReFEH	Revised Flood Estimation Handbook
RHABSIM	River Habitat Simulator
RHS	River Habitat Survey
SCM	Single Channel Method
SEPA	Scottish Environment Protection Agency
SFL	Soffit Level
SKM	Shiono & Knight Method
SMURF	Sustainable Management of Urban Rivers and Floodplains
SPL	Springer Level
SSSI	Site of Special Scientific Interest
TW	Tail Water
UE	Uncertainty Estimator
USBPR	United States Bureau of Public Roads
USGS	United States Geological Survey
VPMC	Variable Parameter Muskingum-Cunge method
WFD	Water Framework Directive
WSPRO	Water Surface Profile analysis

Glossary of terms

Accuracy	The precision to which measurement or calculation is carried out. Potentially accuracy can be improved by better technology.
Advection	The transfer of a property of the atmosphere or water, such as heat, cold, or humidity, by the horizontal movement of an air mass or current.
Afflux	Afflux by a partial obstruction such as a bridge or culvert is defined as the maximum difference in water level, for a specified discharge, if the structure were to be removed.
Aspect ratio	Ratio of channel breadth to depth, typically B/H.
Backwater	Backwater effects occur when sub-critical flow is controlled by the down stream conditions, for example, presence of an outfall, bridge constriction, dam, etc.
Bankfull	The maximum channel discharge capacity; further discharge spreads onto the floodplains.
Berms	(i) The space left between the upper edge of a cut and toe of an embankment to break the continuity of an otherwise long slope; (ii) The sharp definitive edge of a dredged channel such as in a rock cut; (iii) Natural levee where river deposits sediment.
Bias	A statistical sampling or testing error caused by systematically favouring some outcomes over others.
Biodiversity	The variability among living organisms from all sources including, inter alia, terrestrial, marine and other aquatic ecosystems and the ecological complexes of which they are part; this includes diversity within species, between species and of ecosystems.
Boundary layer	Region of fluid influenced in behaviour by the presence of a rigid boundary. Boundary layer theory enables the velocity distribution of a fluid to be established, showing a reduction in velocity within the boundary layer.
Braided	A braided channel is a channel that is divided into two or more channels, for example, flow around an island.
Calibration	Adjustment of a model to reach an acceptable degree of accuracy.

Chaotic	Description of a dynamic system that is very sensitive to initial conditions and may evolve in wildly different ways from slightly different initial conditions.
Compound channel	A channel whose width changes markedly at the bankfull level, typically increasing from the top width of the main river channel to the width of the river and any associated floodplains.
Convection	The transfer of heat in a fluid by the circulation of flow due to temperature differences, where a change in temperature at one location affects the temperature exclusively in the flow direction such that a point only experiences effects due to changes at upstream locations.
Conveyance	Measure of the discharge carrying capacity of a channel, $K(m^3\ s^{-1})$ at a given depth and slope.
Control point	Point that controls the water surface profile in an upstream or downstream direction, typically associated with where the flow passes through the critical depth.
Critical depth	Depth at which the Froude Number equals one (i.e. $Fr = 1.0$), the energy is a minimum and where there is only one depth for a given specific energy ($h = h_c$).
Culverts	Pipes to enable the flow of water between parts of the catchment where roads and railways traverse the watershed.
Diffusive	Diffusion is the movement of a fluid from an area of higher concentration to an area of lower concentration by random fluctuations. In diffusive phenomena, such as heat conduction, a change in temperature at one location affects the temperature in more or less all directions.
Dispersive	Spreading by differential velocity field.
Discharge	The volume of water that passes through a channel section per unit time.
Emergent vegetation	Plants growing above the water but that are rooted below the surface or along the water edge.
Error	Errors are mistaken calculations or measurements with quantifiable differences.
Flow	A general term for the movement of volumes of water at a speed.
Geomorphology	The system of description and analysis of (physical) landscapes processes that change them; the study of landforms.
Habitat	An area in which a specific plant or animal naturally lives, grows and re-produces; the area that provides a plant or animal with adequate food, water, shelter and living space.
Inbank flow	Flow confined solely within the main river channel, with no flow over the floodplains.
Interpolation	Estimation of values based on a relationship within the limits of observation.
Invertebrate	An animal lacking a backbone or spinal column.

Irregularity	Channel irregularities represent variations in roughness from obstructions such as exposed boulders, trash, groynes etc or channel shape e.g. pools and riffles.
Manning's n	An engineering coefficient, n, that incorporates all losses i.e. losses due to local friction, secondary flows, form losses and lateral shear stresses (Chow, 1959). Derived for use in medium to large non-vegetated rivers with fully developed flow profiles.
Maximum depth	The vertical distance of the lowest point of a channel section from the free water surface.
Mean	The arithmetic average i.e. the sum of the data divided by the sample size.
Multi-thread	A multi-thread channel that is divided into two or more channels, for example, flow around an island. ($\approx$braided)
Normal depth	The depth of flow that will occur in a channel of constant bed slope and roughness, provided the channel is sufficiently long and the flow is undisturbed.
Offset	Distance measured laterally across the channel, measured from the left-hand side of the cross-section when looking downstream.
Overbank flow	Flow that is sufficiently large to cause the floodplains to be inundated.
Pool	A point at which the stream is relatively unenergetic due to channel widening and deepening and sediment entrainment (the counterpart of riffle).
Rating curve	Relationship between depth of water and discharge.
Regression	Mathematical analysis of applying straight line principles to an observed relationship.
Resistance	As for roughness but defined as flow-, form-, frictional or turbulent, etc.
Reynolds Number	The Reynolds Number describes the ratio of the inertia forces to the viscous forces. In open channel flow, for Re < 500 laminar flow occurs, Re > 2000, turbulent flow occurs and for 500 < Re < 2000 transitional flow occurs i.e. the flow is characterised by both laminar and turbulent effects.
Riffle	A point at which the stream is relatively energetic due to a constriction or steep gradient (the counterpart of pool).
Roughness	The effect of impeding the normal water flow of a channel by the presence of a natural or artificial body or bodies, which may be biotic for example vegetation; or abiotic/mineral for example bank and bed substrate.
Secondary flows	Flows that are typically transverse the main direction of flow, caused either by anisotropic turbulence or curvature of the mean flow (i.e. bends), manifested as slowly rotating vortices, as illustrated in Figures 2.14 & 3.12.

Shear layer	The region close to the boundary between a solid surface and the water or between two bodies of water moving at different mean velocities.
Siltation	The deposition of finely divided soil and rock particles on the bottom of streams, river beds and reservoirs.
Sinuosity	A measure of a channel's tendency to meander defined as the thalweg length over the valley length.
Stage	The vertical distance of the free surface from an arbitrary or defined datum.
Stream power	Rate of work required by a river to transport water and sediment.
Study reach	The length of a study reach or a section under investigation/measurement
Substrate	A generic term for a substance that underlies another; soil is the substrate for plants, while bedrock is the substrate for soil.
Surface material	Surface material encompasses the substrate on the bed, bank and floodplains. The roughness due to surface material includes, for example, sand, gravel, peat, rock, etc.
Thalweg	A line connecting the deepest points along a channel length.
Trash screens	Screens in front of structures (e.g. culverts) where rubbish is collected for removal.
Trend	A general tendency or inclination. To extend, incline, or veer in a specified direction.
Uncertainty	Uncertainty arises principally from lack of knowledge or of ability to measure or to calculate which gives rise to potential differences between assessment of some factor and its "true" value.
Unit roughness	Roughness due to an identifiable segment of boundary friction per unit length of channel.
Vegetation morphotype	Aquatic and marginal plant species of similar form or function in the riparian corridor.
Vena contracta	The *vena contracta* (Latin for *contracted vein*) is where the local water depth in a channel downstream of a sluice gate or similar structure is the lowest.
Vorticity	Circular motion of the fluid, often in the streamwise or planform directions; a vector measure of local rotation in a fluid flow, defined mathematically as the curl of the velocity vector.
Water course	Natural or man-made channel for the conveyance of drainage and flood water by gravity.
Wetland	Transitional habitat between dry land and deep water. Wet-lands naturally occur in river valleys where drainage is impeded either by topography or soil structure. Wetlands include marshes, swamps, peatlands (including bogs and fens), flood meadows, river and stream margins.

Chapter 1

Introduction

ABSTRACT

This chapter describes the wider issues related to flooding and flood risk management and how the concepts of conveyance, roughness and afflux are related to these issues. The context and motivation behind this book are outlined, together with a brief description of the origins of the Conveyance Estimation System (CES) and the Afflux Estimation System (AES). The CES-AES software incorporates five main components: A Roughness Advisor, a Conveyance Generator, an Uncertainty Estimator, a Backwater Module and an Afflux Estimator. These are introduced and briefly described, together with comments concerning key assumptions and some technical limitations. An outline of the structure of the book and how it might be used in conjunction with the software is also given.

1.1 CONTEXT AND MOTIVATION FOR BOOK

The origins of this book lie in a sequence of research projects and programmes commissioned within the UK since the 1980's. These focused on, and provided, increased scientific knowledge and understanding on the hydraulic factors that affect river flows and flood levels in particular. Some of these programmes brought together researchers from other countries, notably Japan and European (e.g. see Ikeda and McEwan, 2009; Knight and Shamseldin, 2006), and their work also focused on the behaviour of rivers in flood conditions.

The widespread nature of flooding at a global scale has been highlighted by Berz (2000), who has reminded the engineering community that flood disasters account for about a third of all natural disasters by number and economic losses, but are responsible for over half the deaths world-wide. Asian countries, in particular, are prone to flooding problems, with consequent high loss of life. The extent of flooding in Europe in the past decade has been highlighted by Knight and Samuels (2007), who give examples of some significant flood events, together with their causes and impacts. Figures 1.1 indicates the economic and insured losses of all types of natural disasters (floods, earthquakes, storms and others) and Figures 1.2–1.7 show some examples of fluvial flooding in various countries, and the impact that they have. Other examples of floods are given by Miller (1997), together with strategies for prevention.

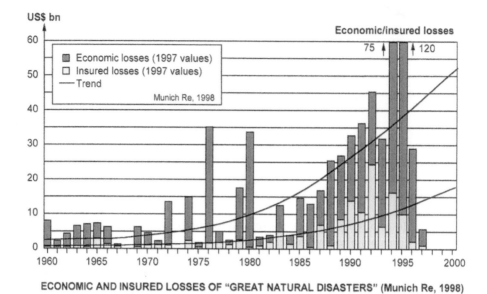

ECONOMIC AND INSURED LOSSES OF "GREAT NATURAL DISASTERS" (Munich Re, 1998)

Figure 1.1 Trends in economic and insured losses arising from all types of disaster world-wide (floods, earthquakes, storms and others, after Berz, 2000).

Figure 1.2 Initial stages of floodplain inundation – River Severn, UK (courtesy HR Wallingford).

Figure 1.3 Flood damage to a bridge in the River Oder basin, Poland, July 1997 (courtesy IIHR).

Figure 1.4 Overtopping of a river flood defence wall, China, 2006 (courtesy H Liao). [Note height of river above ground level by scale of steps & buildings].

This book deals with three key aspects of river hydraulics and is aimed at all those professionals world-wide working in the field of flood risk management, river engineering and drainage.

National and international policy and practice have now generally moved from one of flood defence to a more encompassing approach of flood risk management.

Figure 1.5 Flood damage to infrastructure and a railway line during floods in China, 2006 (courtesy H Liao).

Figure 1.6 Breach in a river embankment on the River Aire, UK, 2002 (courtesy Environment Agency).

Figure 1.7 River Severn in flood at Upton upon Severn, UK, 1990 (courtesy Environment Agency).

This recognises that flooding cannot be eliminated, but that the influence of floods on people and the environment needs to be addressed in the context of broader sustainability (Samuels *et al.*, 2006). Although the policy objective has been broadened, that has not diminished the need to understand the fundamental fluvial processes that determine the depths, velocities and extent of flooding. These factors remain essential in reaching final decisions on the design of physical flood defences and on river maintenance.

In Europe, there has been a significant development of international policy with the entry into force in late 2007 of the European Directive on the assessment and management of flood risks (European Commission, 2007). As a result of this directive, each Member State of the EU must, within their boundaries, undertake a preliminary flood risk assessment, prepare flood risk maps and develop flood risk management plans for each river basin. Clearly an understanding of the way floods propagate within a river basin is important in carrying out these responsibilities. To this end, the European Commission recently sponsored an 'Advanced Study Course' on 'River basin modelling for flood risk mitigation', which gave a broad but integrated view of flood risk management at a river basin scale (Knight and Shamseldin, 2006).

Furthermore, the WFD – the European Water Framework Directive – (European Commission, 2000) also requires an assessment of the flow conditions in rivers, although the primary environmental objectives of the WFD are in particular to achieve by 2015 "good ecological and good chemical status" for surface water bodies in general or good ecological potential for the specific case of heavily modified and artificial water bodies. The WFD also seeks that in general "no deterioration" in status of the water body and this will require the management of the quality, quantity and structure of aquatic environments. However, in the area of "hydromorphology", derogations from some of the WFD requirements are possible under Article 4 (7) for flood management operations, but only when specific criteria and conditions are met. These need a full justification through analysis of the alternatives.

In the UK, under the national policy of "Making Space for Water" there is a strong emphasis on understanding the performance of flood defence "assets", that is channels, floodplains, embankments, structures and other infrastructure to make decisions on their effective and efficient management. For river channel and floodplains this requires an assessment of their combined capacity to convey floods and how routine maintenance, upgrading of defences or construction of new works will influence this. Thus estimation of river channel conveyance and the afflux at river structures are core components of flood risk management through water level prediction and flood defence design.

As the executive agency responsible for flood risk management works on main rivers in the UK, the Environment Agency (EA) has a commitment to use sound science in undertaking its activities. The EA is assisted in these activities at a local level by internal drainage boards and local authorities, who have new statutory responsibility following the Pitt review (2008). It is usual practice to undertake assessments of river flows and flood conditions using computational models, with a process of model construction, calibration and verification, and then application to the particular issue of concern. All such assessments require the estimation of the discharge involved, estimation of the capacity of the watercourse to carry that discharge, and the degree and extent of influence of any structures on the watercourse. Discharge estimation is a core activity in hydrology and lies beyond the scope of this book, which concentrates on the capacity of a watercourse to discharge water as the water level varies (this is known as its 'conveyance') and on the water level increases caused by bridges and culverts (that is their 'afflux'). These two technical terms are defined in the glossary and dealt with in detail in Chapters 2 & 3.

All river modelling software includes one or more methods for conveyance estimation, usually based upon methods dating from research completed more than 50 years ago, with little or no account taken of more recent advances in knowledge and understanding. Hence the Environment Agency, with support from the Scottish Government and the Northern Ireland Rivers Agency, commissioned the development of new methods for the calculation of conveyance that incorporate the latest scientific knowledge on the topic. Likewise the Agency commissioned a review of methods for assessing the effects of river structures on flood levels and this work on afflux and blockage was incorporated into a single modelling tool alongside the new work on conveyance estimation.

1.2   SCOPE OF BOOK AND INTRODUCTION TO THE CES-AES

This book provides a stand-alone introduction to the principles and processes of hydrodynamics in open channel flows, set in the context of the Conveyance and Afflux Estimation System (CES-AES), developed from certain Environment Agency R&D projects outlined in Section 1.5. It aims to give the technical background to each of the components of the CES-AES, together with illustrations of their use. The five main components of the CES-AES are:

- The Roughness Advisor,
- The Conveyance Generator,
- The Uncertainty Estimator,
- The Backwater Module, and
- The Afflux Estimator.

The Roughness Advisor incorporates a database of information on the hydraulic roughness of different surfaces including descriptions, photographs and unit roughness values for a range of natural and man-made roughness types. These were sourced from over 700 references (including the UK based River Habitat Survey) and include aquatic vegetation, crops, grasses, hedges, trees, substrates, bank protection and irregularities (Defra/EA, 2003b). The Roughness Advisor also includes information on seasonal variations in vegetation roughness, the effects of weed cutting in river maintenance activities and has suggested re-growth patterns following cutting.

The Conveyance Generator estimates the channel conveyance based on the roughness information and the shape of the cross-section. The calculation considers the lateral distribution of velocity across the section and the overall flow is then determined by integrating the velocity across the river section. The outputs available include the variation with depth of important hydraulic parameters, such as: water level, flow, rating curves, velocity, area, wetted perimeter, Froude Number, Reynolds Number, etc. It is also possible to obtain lateral distributions (across a section) of velocity, boundary shear stress and shear velocity, which may be used in assessments of sediment movement and other environmental factors.

The Uncertainty Estimator provides upper and lower "credible" scenarios, providing some measure of the range of the uncertainty associated with each predicted water level. These values are derived from the upper and lower roughness estimates from the Roughness Advisor. However, they do not have any precise statistical interpretation nor do they represent any upper or lower limits on the water level for all possible conditions. They are scenarios of what can occur based on the observations in the roughness database and some expert assessments where field data were unavailable.

The Backwater Module includes a standard calculation of the backwater profile upstream of a control, i.e. a point of known stage and flow. This is based on a basic balance of work done against the river resistance to the loss of potential energy as the water flows downstream. The calculation covers only subcritical (or tranquil) flow; it works from downstream to upstream and includes the option to incorporate the variation of velocity head in non-uniform channels (i.e. changes in the bulk kinetic energy of the flow). The module can be used to simulate the change in water level between structures.

The Afflux Estimator models the influence on water level of arch and beam bridges with up to 20 openings, and pipe, box and arch culverts with up to 10 identical barrels. This code is embedded within the bridge and culvert units in the CES-AES model and, combined with the Backwater Module, produces a longitudinal water surface profile along a watercourse taking account of channel shape, roughness and hydraulic structures.

1.3   LIMITATIONS OF THE CES-AES

In any technical assessment it is important to understand the limitations of the methods being used. For the use of the CES-AES, the most important limitations are:

- Steady flow (i.e. negligible attenuation);
- Fixed bed (i.e. no scour or deposition);
- Fixed roughness (i.e. uninfluenced by velocity);
- Small to medium sized channels and rivers (say 0.5 to 500 m width, 0.2 to 30 m depth, gradients 10^{-2} to 10^{-4} approximately), low to moderate sinuosity;
- Unobstructed bridges and culverts (i.e. not partially blocked by debris or sediments).

The assumption of steady flow arises in the Backwater Module and is appropriate for many practical purposes. The main implication is that the Backwater Module, through its simple model of the watercourse, must not be used for the assessment of the effects of changes in flood storage. It may be used with confidence to investigate the sizing of river and drainage channels, the dimensioning of structures and the assessment of maintenance works through vegetation cutting regimes and dredging. The calculation of conveyance as a characteristic property of the channel cross-section is not affected by the assumption of steady flow; the conveyance calculations can be used in unsteady flow simulations.

The assumption of a fixed bed means that the variation of the shape of the section is not taken into account in any of the components of the CES-AES. This restriction is important in any watercourse that is morphologically active, especially where changes in section dimensions occur over the course of a flood. In such cases the conveyance should be estimated for the initial and final dimensions of the sections or a full sediment transport model should be employed. This limitation is likely to be of importance in flash flood cases, such as the flood in Boscastle (Cornwall, UK) that occurred in August 2004.

The assumption of fixed roughness can be significant in several circumstances. For channels where the bed material is fine or is easily transported, the effective roughness may depend upon the flow velocity as ripples or dunes appear on the river bed or are washed out. Such alluvial friction processes currently lie outside the scope of the CES-AES and are not covered in this book. The limitation probably only affects the use of the CES-AES on the lower reaches of larger rivers and these are more often in the tidal zone anyway (Tagg and Samuels, 1989).

The limitation to small and medium sized channels reflects the range of cases used in the development and validation of the CES. Mc Gahey *et al.* (2008) report that the new methods included in the CES are appropriate for all except the largest or the steepest rivers in the international data assembled. It is possible that for the largest rivers the assumption (as above) of fixed roughness independent of velocity is not appropriate, and in steep mountain rivers the roughness from boulders (which are large in comparison with the flow depth) is not well-represented by the current resistance equations in the code, although an alternative is suggested.

The Afflux Estimation System component of the CES-AES applies only to fixed dimensions of the hydraulic structures, and there is no transient allowance for the changes in effective shape and area through blockages that may form during a flood (Mantz and Benn, 2009). There are a range of opening types available in the CES-AES for bridges and culverts, with some flexibility to define the shapes of the openings; however more complex types such as multiple culvert openings with different shapes and invert levels or bridges with relief culverts cannot be represented directly.

1.4   OUTLINE OF BOOK

The book has been written so that the various topics are introduced initially at a simplified level, and then again progressively at more complex levels in later chapters. In this way it is hoped that the reader can learn about certain practical and theoretical aspects of river engineering early on, without recourse to all the details normally required when actually modelling a particular river in earnest. Thus, Chapter 2 deals with some basic theoretical concepts in open channel flow, illustrated via simple examples, followed by certain practical issues illustrated via case studies. These are largely based on the Shiono and Knight (1988 & 1991) method of analysis (SKM), a pre-cursor to the CES methodology described in detail in Chapter 3. Chapter 2 deals with inbank and overbank flows in simple engineered channels, flows in natural channels, afflux at bridges and culverts, roughness, hydraulic resistance and data sources. In Chapter 3 further details of the scientific issues related to flow structures, their mathematical representation, governing equations and assumptions are provided, expanding on what was initially presented in Chapter 2. Having now fully established these concepts, full details of the CES-AES methods related to roughness, conveyance and afflux are given, together with an introduction to uncertainty and the CES-AES software. Chapter 4 then deals with the application of the CES-AES software, covering many practical issues such as estimating and using stage-discharge relationships, assessing lateral distribution of velocities and boundary shear stress, estimating afflux at bridges and dealing with vegetation. These issues are again illustrated via case studies and examples, and serve to demonstrate the wide range of application of the CES-AES software. Chapter 5 considers how the CES and AES methodologies can be used more broadly in assessments covering ecological issues, habitats, sediment and geomorphology and some more generic issues in river modelling.

Chapter 6 gives full details of the Shiono & Knight Method (SKM), upon which the CES-AES is based, including the theoretical background to the governing differential equation and its physical basis. The SKM provides an alternative approach to solving the governing equation, either by analytical means, as shown in Appendices 4 & 5, or by numerical means using Excel. A simple Fortran program is given in Appendix 6, which provides another tool for undertaking flow analysis and for checking results. Particular attention is paid in Chapter 6 to the determination of boundary shear stress distributions in channels of different shape, and to appropriate formulations for resistance and drag forces, including those on trees. Chapter 7 has many worked examples for a wide range of channel types and discharges, using Excel to obtain both analytical and numerical solutions. These examples range from small scale laboratory flumes ($Q = 2.0$ ls^{-1}) to European rivers ($\sim$2,000 m^{3}s^{-1}), and even large-scale world rivers (>23,000 m^{3}s^{-1}), a $\sim$10^7 range in discharge. Chapter 8 provides further worked examples for a range of channel types (size, shape, cover, sinuosity), based on the CES, illustrating its capabilities and providing comparative results with those in Chapters 6 & 7, based on the SKM. Sites from rivers in the UK, France, China, New Zealand and Ecuador are considered. In this respect, it should be noted that International Standards are now beginning to cite commercially available software for use in evaluating stage-discharge relationships, as for example ISO 18320 (2018), which refers to both the CES and SKM as one method of achieving this objective. For advanced readers who wish to improve their understanding of the science behind current river engineering practice, Chapter 4 might provide a suitable entry point, only referring back to Chapters 1–3 where necessary.

The CES-AES stand-alone software is available to all as a free download at the CES-AES Website: *www.river-conveyance.net*. It is also available to the Environment Agency of England and Wales staff through their internal Corporate Information Services. The CES conveyance calculation and roughness information has also been made available through the 1-D hydrodynamic modelling software InfoWorks RS (Version 6.5 and onwards) and ISIS Flow (Version 2.3 and onwards), providing an alternative to the Manning equation for evaluating cross-section conveyance. Information and documentation from the original CES and AES development projects may be found at: *www.river-conveyance.net/ ces* and *www.river-conveyance.net/aes* respectively. For up-to-date information on the CES-AES software including documentation, training, Frequently Asked

Questions and contact details for provision of feedback, please visit the CES-AES Website.

1.5   ORIGIN OF THE CES-AES

The scientific basis of the Conveyance Estimation System and Afflux Estimation System (CES-AES) was formulated in UK funded research that was undertaken in the decade 1985–1995. It was focused on the conveyance capacity of river and flood-plain systems, and involved a programme of large-scale experiments on the convey-ance of compound channels in the Flood Channel Facility (FCF) at HR Wallingford. The experimental programme, which included straight, skewed and sinuous cases with a variety of geometries, roughness conditions and flow conditions, is sum-marised by various authors, notably Knight and Sellin (1987), Knight and Shiono (1990), Sellin *et al.* (1993) and Knight (2008a). Figures 1.8–1.11 illustrate some of the fixed bed experimental geometries that were used in Phases A & B of this work. In a later phase (Phase C), a loose boundary was employed and sediment re-circulated in both straight and meandering channels. The observations explored systematically

Figure 1.8 Straight channel with two symmetric smooth floodplains in FCF (courtesy HR Wallingford).

Figure 1.9 Skewed channel in FCF with smooth floodplains (courtesy HR Wallingford).

Figure 1.10 Meandering natural channel in FCF with sinuosity of 2.043 (courtesy HR Wallingford).

Figure 1.11 Dispersion experiments in FCF with a trapezoidal meandering channel with a sinuosity of 1.374 (courtesy HR Wallingford).

discharge capacity, boundary shear stresses and turbulence stresses in the body of the flow. The data and further details of the FCF experimental programme may be found at *www.flowdata.bham.ac.uk*. These data produced greater physical insight into the fluid mechanics of these flows and various research papers indicated that traditional methods of analysing these flows could have large uncertainties of the order of 30% or more. However, this improved knowledge of the processes of flood propagation was not taken up readily into practice.

The former National Rivers Authority in England and Wales (1994) commissioned the derivation of hand calculation methods for assessing the conveyance of straight and compound channels based upon the research undertaken in the first two phases of the experimental programme of the Flood Channel Facility (Ackers, 1992b and James and Wark, 1992). Although these methods performed well in tests against experimental measurements and those from real rivers, these methods remained largely unused. Potential reasons for this slow take-up of the research were deemed to be:

- changes in the profession away from hand calculation to much greater reliance on standard computational models;

- the lack of a pathway into use for research in internal procedures from the principal end user.

A key issue addressed by the joint research programme into Flood and Coastal Defence established by the former Ministry of Agriculture, Fisheries and Food (MAFF) and the Environment Agency in 2000 was the improved dissemination and implementation of the outputs of research. See, for example, Leggett and Elliott (2002) and Townend *et al.* (2007).

In 2000 the Engineering and Physical Sciences Research Council (EPSRC), together with the Government Department for Food and Rural Affairs (Defra) and the Environment Agency (EA) established a joint research network on river and floodplain conveyance in the UK, with over 60 academics and practicing river engineers participating. Following the extensive flooding in the UK in the autumn and winter of 2000 to 2001, the Institution of Civil Engineers Presidential Commission on flood management issues identified the importance of improving the application of technical knowledge on the hydraulics of flood level calculation and the supporting models in practical use (ICE, 2001; Fleming, 2002). An early action resulting from the joint EPSRC/Defra/EA network was the commissioning by the Environment Agency of a Scoping Study on reducing uncertainty in river flood conveyance calculation. The setting up of the CES project was supported by the Natural Environment Research Council (NERC), the Rivers Agency in Northern Ireland and the Scottish Government. The Environment Agency was particularly concerned that the Scoping Study should address the whole of flood risk management as the end objective, not just the design function (as had been the case in the earlier work of the National Rivers Authority). In addition, the Scoping Study needed to reconcile the needs for both research openness and the supply of commercial software systems for simulation of river flow.

The objective of the scoping study was to identify a programme of work to synthesise current knowledge and how to fill in the gaps, in order to improve the tools available for river management practice. The project involved directly, and through consultation, recognised academic experts and researchers, operational staff and consulting engineering practices. The Scoping Study included a questionnaire targeted at specific individuals and distributed more broadly within the UK and internationally. The industry and academic communities were involved through workshops on the user needs and to validate the draft of the research recommendations.

Figure 1.12 Elliptical arch bridge with river level close to the soffit (courtesy HR Wallingford).

Figure 1.13 Multiple arch bridge with inundation of approach road (courtesy HR Wallingford).

Figure 1.14 Multiple arch bridge with debris beginning to collect around central pier (courtesy University of Birmingham).

Figure 1.15 Arch bridge with a parapet destroyed by flood debris.

The study also included the preparation of a series of expert review papers on flow measurements (field and experimental), effects of vegetation, conveyance calculation methods in 1-D models, implications for 2-D and 3-D models and the use of remote sensing. These expert papers were prepared by members of the EPSRC network and the texts are contained in an annex to the final report (Defra/EA, 2001). This documentation on the state-of-the-art underpinned many of the scoping study research recommendations. The subsequent research involved a partnership between academic researchers, experts and end-users. Particular emphasis was placed on issues of concern, such as: the effects of riverine vegetation, the influence of natural shaped (and re-naturalised) channels and the interaction between river channels and floodplain flows. The output from this programme was the novel Conveyance Estimation System (CES) and its inclusion in several modelling packages. The CES also includes a Roughness Advisor (RA) to facilitate access to the wide body of knowledge on the estimation of fluvial resistance. These are described in detail in Chapters 3 & 4.

At the same time, the Environment Agency acted on the need to provide similar guidance and methods for estimating the effects of river structures on flood conditions. Figures 1.12–1.15 illustrate some of the effects of flood flows on bridges. First of all, a study collected information on current knowledge and practice (Defra/EA, 2003a) and this was followed by the development of the Afflux Estimation System (AES) to operate with the CES and the Afflux Advisor (originally begun as just an MS Excel spreadsheet application). The AES project was also supported by others, including the Scottish Government. A particular impetus to the CES-AES project was provided by the report on 'Sediment transport and alluvial resistance in rivers' by Bettess (2001), that helped to demonstrate to practitioners and funders the relevance of drawing together existing knowledge for an improved tool on conveyance.

The main purpose of developing the AES was to improve the tools available in engineering practice to simulate the effects of bridges and culverts on flood water levels at high flows, particularly for the representation of afflux (the increase in upstream water levels attributable to the structure). The project researched a range of methods of calculating afflux, to determine their applicability to the various types of bridges and culverts found in UK channels and incorporated the most appropriate methods and algorithms, following detailed testing, into the AES software (Defra/EA, 2004a&b). The development again involved close collaboration between academic researchers, experts and practitioners.

The CES and the AES have now been integrated into a single software module so that they can operate separately, or together where needed, to estimate water levels along an open channel reach containing bridge or culvert crossings. For completeness, the software includes backwater calculations along a watercourse as a basic model of open channel flow.

This 2nd Edition has been greatly expanded through the addition of Chapters 6–8, which now supply full details of the Shiono and Knight Method (SKM), upon which the CES-AES is largely based, and provide further examples of the use of the CES. The background to the SKM is given by Shiono & Knight (1988 & 1991) and reproduced in Appendix 4. How it might be used to determine stage-discharge relationships is described by Knight *et al* (1989 & 2012), Knight & Abril (1996) and Abril & Knight (2004).

Chapter 2

Practical and theoretical issues in channel hydraulics

ABSTRACT

This chapter is an introduction to the many practical and theoretical issues that need to be considered when attempting to model the flow of water in open channels and rivers. These issues are illustrated through several worked examples, as well as a number of case studies on actual rivers. Consideration is also given to several related topics, such as: how to schematise a river cross-section or reach, the difficulty of estimating hydraulic roughness and resistance, the hydraulic characteristics of inbank and overbank flows, and flows through bridges and culverts. The data requirements for effective modelling are also considered.

2.1 GETTING STARTED WITH SOME PRACTICAL EXAMPLES ON CALCULATING FLOWS IN WATERCOURSES

The flow of water in a river, water course or drainage system is governed by physical laws related to the conservation of mass, energy and momentum. Their application to open channel flow is the subject of this book. Although these physical laws and principles are sometimes difficult to apply to all types of fluid flow in river engineering, as well as other areas in engineering and science, mainly due to the difficulty of accounting for energy degradation, some physical characteristics of open channel flow may be readily understood and are surprising. In order to appreciate some key features of open channel flow, the following 5 examples are given to get started on some practical problems. They also help to introduce and define certain terms and parameters, as well as illustrating certain issues addressed in later sections. Examples 1 to 3 are concerned with a freely flowing small-sized river or drainage channel, and Examples 4 & 5 are concerned with flow in a reach where the channel has some hydraulic controls. Although these examples are straightforward for those familiar with hydraulic principles and text books (e.g. Chadwick *et al.*, 2004; Chanson, 1999; Chow, 1959; Cunge *et al.* 1980; Henderson, 1966), they should be worked through none-the-less as they serve to illustrate some basic topics and issues in modelling mentioned later. They introduce parameters such as specific energy, momentum, roughness, resistance to flow and energy losses, as well as the notation used in this book.

Example 1

A river has a simple trapezoidal cross section with a bed width, b, of 15.0 m, side slopes, s, of 1:1 (vertical : horizontal), a Manning's $n = 0.030$ and a longitudinal bed slope, S_o, of 3.0×10^{-4}. Determine the discharge in the channel when the depth of flow, h, is 1.5 m, using Manning's equation (1891), given as Eq. (2.1). Figures 2.1(a) & (b) illustrate typical river channels that might be considered to be approximately trapezoidal in shape, but with different roughnesses.

$$Q = AR^{2/3}S_o^{1/2}/n \tag{2.1}$$

(a) low Manning n value (~0.03)   (b) high Manning n value (~0.30)

Figure 2.1 Two trapezoidal channels with different Manning n roughness values.

Answer & comment

Area, $A = (b + sh)h = (15 + 1.0 \times 1.5) \times 1.5 = 24.75$ m².
Wetted perimeter, $P = b + 2h\sqrt{(1 + s^2)} = 15 + 2 \times 1.5 \sqrt{(1 + 1^2)} = 15 + 3\sqrt{2} = 19.243$ m.
Hydraulic radius, $R = A/P = 24.75/19.243 = 1.286$ m.
Discharge, using Eq. (2.1), $Q = 24.75 \times 1.286^{2/3} \times 0.0003^{1/2}/0.030 = 16.90$ m³s⁻¹.

Note that the discharge is given explicitly by the Manning equation, because the depth, h, is specified, and the whole cross section is used. For a long reach, with a constant slope and roughness, this depth is referred to as the 'normal depth' and often given the symbol h_n. Since the whole section is used, this is essentially a 'single channel' method of analysis, employing a simple 'one-dimensional' (1-D) approach that gives the discharge, but no other information about the velocity distribution within the channel (i.e. across the width or over the depth). For further details on the background to the Manning equation, see Yen (1991).

Example 2

If the discharge in Example 1 is now reduced to 10 m³s⁻¹, determine the new depth of flow (normal depth), h_n, and the 'critical depth', h_c. Also determine the ratio of normal to critical depth, h_n/h_c, the Froude Number and whether the flow is 'subcritical' or 'supercritical'.

Answer & comment

Equation (2.1) may be re-arranged with all known parameters on one side and all the geometric parameters involving depth on the other side. Noting that $R = A/P$, then:

$$Qn/S_o^{1/2} = A^{5/3}/P^{2/3} \tag{2.2}$$

Since h occurs both in the area and wetted perimeter, whose constituent equations depend on the shape of the cross section, it is generally now not possible to obtain an answer for the depth in explicit form using Manning's equation. Pivoting around the highest power of h, a numerical method, such as fixed-point iteration, may be shown to give a new value for the normal depth, h_{j+1}, in terms of a previous value, h_j, as:

$$h_{j+1} = \left[\frac{Qn}{S_o^{1/2}}\right]^{3/5} \frac{\left[b + 2h_j\sqrt{1+s^2}\right]^{2/5}}{(b + sh_j)} \tag{2.3}$$

Equation (2.3) may be solved numerically for h_{j+1}, by assuming initially that $h_j = 0$, and then by repeated iteration until h_{j+1} is sufficiently close to h_j, giving ultimately $h_n = 1.095$ m. The normal depth of flow for the discharge of 10 m³s⁻¹ is therefore 1.095 m.

Equation (2.3) may also be formulated into a dimensionless form, convenient for other purposes, as

$$\frac{h_{j+1}}{b} = \left[\frac{Qn}{b^{8/3}S_o^{1/2}}\right]^{3/5} \frac{\left[1 + (2h_j/b)\sqrt{1+s^2}\right]^{2/5}}{(1 + sh_j/b)} \tag{2.4}$$

Figure 2.2 shows the solution for h for flows in trapezoidal channels and circular culverts or pipes running part full, to illustrate the influence of each dimensionless group. These types of flow are commonly found in engineering design and feature in this book in relation to simple shaped rivers, culverts and drainage conduits.

The critical depth, h_c, is given when $Fr = 1.0$, where the Froude Number, Fr, is defined by

$$Fr = \frac{U}{\sqrt{gA/T}} \tag{2.5}$$

in which U = section mean velocity (= Q/A) and T = top width of channel (i.e. at the free surface). At the critical depth and 'critical flow', $Fr = 1$, and the energy is also

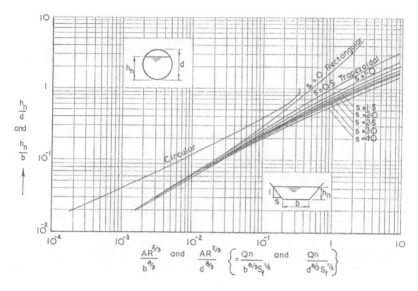

Figure 2.2 Uniform depth relationships based on Eq. (2.4), and equivalent equation for circular culverts running part full, showing influence of individual parameter groups.

a minimum, as may be deduced later from Eq. (2.12). The condition necessary for critical flow then follows from Eq. (2.5) and may be written in the general form for channels of any cross sectional shape, as

$$\frac{Q^2 T}{g A^3} = 1 \tag{2.6}$$

In order to calculate h_c, a similar numerical procedure to the normal depth calculation method is required. In this particular example, with a trapezoidal channel, $T = b + 2sh$, which is a function of depth and therefore initially unknown. Re-arranging Eq. (2.6) for fixed-point iteration yields

$$h_{j+1} = \left[\frac{Q^2}{g}\right]^{1/3} \frac{[b + 2sh_j]^{1/3}}{(b + sh_j)} \tag{2.7}$$

This may also be put into a dimensionless form again, convenient for other purposes, as:

$$\frac{h_{j+1}}{b} = \left[\frac{Q}{g^{1/2}b^5}\right]^{2/3} \frac{[1 + 2sh_j/b]^{1/3}}{(1 + sh_j/b)} \tag{2.8}$$

Using Eq. (2.7) iteratively, assuming initially again that $h_j = 0$, gives $h_c = 0.354$ m. Thus $h_n/h_c = 1.0952/0.3537 = 3.096$ (i.e. $h_n > h_c$, therefore subcritical flow. See Fig. 2.12)

At the new normal depth, $A = (b + sh)h = (15 + 1.0 \times 1.0952) \times 1.0952 = 17.628$ m².
Top width, $T = b + 2sh = 15 + 2 \times 1.0 \times 1.0952 = 17.190$ m.
$U = Q/A = 10.0/17.6275 = 0.567$ ms⁻¹
$Fr = 0.5673/(9.807 \times 17.6275/17.1904)^{1/2} = 0.179$.
Hence, since $Fr < 1.0$ the flow is confirmed as being subcritical.

Note that an iterative approach is required to solve Eq. (2.1) because this is the inverse problem to Example 1, in which the discharge is given and not the depth. Since $h_n > h_c$, this means that the flow is subcritical, a fact confirmed by the Froude Number being less than 1.0. The general form of the Froude Number, Eq. (2.5), should always be used in open channel work, and distinguished from the particular form adopted in channels with a rectangular cross section. In that particular case, since $A = bh$ and $T = b$, the Froude Number is given by

$$Fr = \frac{U}{\sqrt{gh}} \quad \text{(rectangular channel only)} \tag{2.9}$$

and the critical depth may be readily obtained from Eq. (2.6) in the form

$$h_c = \sqrt[3]{q^2/g} \quad \text{(rectangular channel only)} \tag{2.10}$$

where the discharge per unit width, q, is defined by

$$q = Q/b \tag{2.11}$$

The discharge per unit width is a useful concept, used widely in designing channel transitions, and also in a depth-averaged form of the Reynolds Averaged Navier-Stokes equations (RANS) which form the basis of the CES system described more fully in Chapter 3.

Example 3

If the river in Example 1 is re-graded, maintaining the same bed width and side slopes, but with a Manning's $n = 0.025$, what bed slope is required to convey 30 m³s⁻¹ at a depth of 1.5 m? What will be the increase in water level for $Q = 30$ m³s⁻¹, if subsequently the roughness of the river reverts back to its original Manning's n value, with S_o remaining as calculated?

Answer & comment

$A = 24.75$ m²; $P = 19.243$ m; $R = 1.286$ m as before, then solving for slope:
Bed slope, using Eq. (2.1), $S_o = [30 \times 0.025/(24.75 \times 1.2862^{2/3})]^2 = 6.565 \times 10^{-4}$.
For original $n = 0.030$, depth will increase. Using Eq. (2.3) gives $h_n = 1.673$ m.
Increase in water level = 0.173 m.

Note that with reduced roughness, but the same depth, the bed slope is increased because the specified discharge has increased from 16.9 m³s⁻¹ to 30 m³s⁻¹. With the original roughness, depth, new discharge but reduced bed slope, the depth is increased. This illustrates how useful the Manning equation is in relating the 4 basic parameters $\{Q, h, S_o, n\}$ that commonly occur in many design problems in river engineering. Furthermore, Eq. (2.1) also shows how easy it is to calculate changes in any one parameter that are governed by the other three.

Example 4

This example is based on the physical situation shown in Figure 2.3, and the corresponding diagrams in Figure 2.4. A horizontal channel with a rectangular cross section is 4.5 m wide and conveys 13.5 m³s⁻¹. A vertical sluice gate is incorporated in the channel and the gate is regulated until the depth of flow just upstream is 3.0 m. Assuming no energy loss during the passage of water under the gate, find the depth of flow just downstream of the gate. Figure 2.3 shows a picture of this type of flow in the laboratory, and Figure 2.4 shows the corresponding depth versus specific energy (h v E) and depth versus specific momentum (h v M) diagrams. The depth at section 2 is known as the *vena contracta* (Latin for *contracted vein*), i.e. where the depth in the channel is lowest. The flow with a hydraulic jump is covered in Example 5.

Answer & comment

This example requires one to understand the concept of specific energy, E, defined by

$$E = h + \alpha \frac{U^2}{2g} \tag{2.12}$$

Figure 2.3 Flow under a laboratory vertical sluice gate with a hydraulic jump downstream.

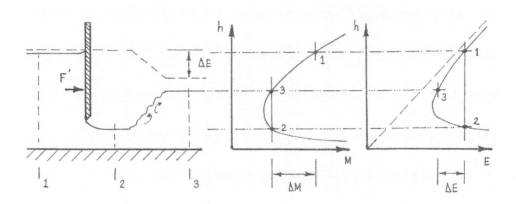

Figure 2.4 Schematic diagrams of specific energy and momentum for flow under a vertical sluice gate with a hydraulic jump downstream.

where E = specific energy per unit weight, α = kinetic energy correction coefficient, taken here as 1.0 in order to simplify the problem. The technical definition of α, and the corresponding momentum correction coefficient, β, are dealt with more fully in Section 2.2.5.

Equation (2.12) may be developed for flow in a rectangular channel of breadth, b, with $\alpha = 1.0$, as a cubic equation in h

$$E = h + \frac{q^2}{2gh^2} \tag{2.13}$$

Thus with $q = Q/b = 13.5/4.5 = 3.0$ m^{3}s^{-1}m^{-1}, then $E_1 = 3.0 + 3.0^2/(2 \times 9.807 \times 3.0^2) = 3.051$ m.

Assuming no energy loss across the sluice gate, as shown in Figure 2.4, then $E_1 = E_2$ and the depth downstream of the gate, h_2, may be obtained from

$$E_2 = h_2 + \frac{q^2}{2gh_2^2}$$

Solving this for $E_2 = 3.051$ m gives $h_2 = 0.417$ m. Hence $A_2 = 4.5 \times 0.4174 = 1.878$ m^2 and $U_2 = 13.5/1.8783 = 7.187$ ms^{-1}. Also, from Eq. (2.10), $h_c = 0.972$ m, indicating that $h_2 < h_c$.

Checking that the value of h_2 gives the same specific energy, the kinetic head at section $2 = U_2^2/2g = 7.1874^2/(2 \times 9.807) = 2.6337$ m and hence $E_2 = 0.4174 + 2.6337 = 3.0511$ m ($\approx E_1$).

Note that for a given specific energy there are generally two possible depths of flow, h_1 and h_2, known as 'alternate depths', one subcritical ($Fr < 1$) and the other supercritical ($Fr > 1$). These are denoted in Figure 2.4 by the two points 1 & 2 in the h v E diagram, where a vertical line for the given value of E intersects the curve for a given q. Note the value of $U_2^2/2g$, and how the sluice gate changes the flow from subcritical to supercritical flow. The point on the h v E curve where the energy is a minimum, and where there is only one depth for a given energy, is where critical flow occurs (i.e. $Fr = 1.0$). The depth there is known as the critical depth, h_c, as already defined by Eq. (2.6), and is associated usually with the transition from subcritical to supercritical flow. It should also be noted that the downstream depth, h_2, is seen to be much smaller than the upstream depth h_1 and is defined by h_1 and q, not the gate opening, h_g. The flow downstream is also confirmed to be supercritical since $h_2 < h_c$. The contraction coefficient, C_c ($= h_2/h_g$), may be calculated for the sluice gate if h_g is known. Typically, C_c is around 0.61 for such sluice gates. Alternatively, if h_g, C_c and q are known, then the upstream depth, h_1, may be calculated.

Example 5

If a hydraulic jump occurs downstream of the sluice gate in Example 4, as shown in Figures 2.3 & 2.4, determine the maximum tail water level which could be tolerated before the sluice gate becomes 'drowned out', i.e. when the downstream depth begins to affect the depth of flow upstream of the gate. Also determine the height of the jump and the energy and the power lost in the jump.

Answer & comment

The standard equation for a hydraulic jump in a rectangular channel (Chow, 1959) is as follows:

$$\frac{h_2}{h_1} = \frac{1}{2}\left[\sqrt{(1 + 8Fr_1^2)} - 1\right] \tag{2.14}$$

where h_1 and h_2 are the depths before and after the jump, known as 'conjugate depths'. These should not be confused with the two 'alternate depths' defined in Example 4 in relation to specific energy. Applying Eq. (2.14) to sections 2 & 3, where section 2 is upstream of the jump and section 3 downstream, then:

From Eq. (2.9), $Fr_2 = 7.1874/(9.807 \times 0.4174)^{1/2} = 3.552$.
From Eq. (2.14), with appropriate notation changes, $h_3/h_2 = 1/2$ [$(1 + 8 \times 3.5524^2)^{1/2} - 1$] = 4.549.
Hence $h_3 = 4.5487 \times 0.4174 = 1.899$ m.
Thus the height of the jump = $h_3 - h_2 = 1.898 - 0.417 = 1.481$ m.
The maximum tail water level to prevent drowning of the sluice gate is thus around 1.90 m.

In order to determine the energy loss, one needs to calculate E_3.
Downstream of the jump, $U_3 = Q/A_3 = 13.5/(1.8986 \times 4.5) = 1.580$ ms^{-1}.

Hence $E_3 = 1.8986 + 1.5801^2/(2 \times 9.807) = 2.026$ m.

The specific energy lost in the jump $= E_2 - E_3 = 3.051 - 2.026 = 1.025$ m.

This may also be determined from the standard relationship for a jump in a rectangular channel (see Chow, 1959), where again h_1 and h_2 are the depths before and after the jump

$$\Delta E = \left(h_2 - h_1\right)^3/\left(4h_1h_2\right) \tag{2.15}$$

Thus between sections 2 & 3, $\Delta E = (1.8986 - 0.4174)^3/(4 \times 0.4174 \times 1.8986) = 1.024$ m.

The power lost is given by

$$P = \rho g Q \Delta E \tag{2.16}$$

Hence power loss $= 1000 \times 9.807 \times 13.5 \times 1.0251 = 135713$ Watts {or 135.7 kW}.

Note how the specific energy and momentum vary between sections 1, 2 & 3. The value of h_3 in Figure 2.4 is seen to be much less than h_1, indicating an energy loss somewhere between 1 & 3. Most of this loss is known to occur as a result of the intense mixing and turbulence within the hydraulic jump. Between 1 & 2, across the sluice gate, it was therefore assumed that $\Delta E = 0$ and $\Delta M \neq 0$, whereas across the hydraulic jump, between 2 & 3, it was assumed that $\Delta E \neq 0$ and $\Delta M = 0$. Since both energy and momentum feature later with regard to hydraulic structures, the momentum function, M, is now introduced. Applying Newton's 2nd law for flow in an open channel

$$F_1 - F_2 + F = \beta \rho Q \left(U_2 - U_1\right) \tag{2.17}$$

where F_1 & F_2 are the hydrostatic forces acting on upstream and downstream cross sections ($= \rho g A \bar{z}$), $\bar{z}$ is the distance below the water surface to the centre of area of the cross section, F is any additional force acting on the fluid in the direction of flow and $\beta =$ momentum correction coefficient. Then in general for 1-D flow

$$\rho g A_1 \bar{z}_1 - \rho g A_2 \bar{z}_2 + F = \beta \rho Q \left(U_2 - U_1\right) \tag{2.18}$$

Re-arranging terms in Eq. (2.18), dividing by ρg and assuming here for simplicity that $\beta = 1$, then

$$\left\{\frac{Q^2}{gA_1} + A_1 \bar{z}_1\right\} - \left\{\frac{Q^2}{gA_2} + A_2 \bar{z}_2\right\} + \frac{F}{\rho g} = 0 \tag{2.19}$$

Defining the momentum function as

$$M = \frac{Q^2}{gA} + A\bar{z} \tag{2.20}$$

then the force acting on the fluid in the flow direction between any two sections is

$$\frac{F}{\rho g} = M_2 - M_1 \qquad (2.21)$$

For the special case of a rectangular channel, with $\bar{z} = h/2$, Eq. (2.20) shows that the momentum function is related to the specific force (= F/b, i.e. force per unit width) by

$$M = b\left[\frac{q^2}{gh} + \frac{h^2}{2}\right] \qquad (2.22)$$

Equation (2.22) has certain similarities with the Eq. (2.13) as already shown in Figure 2.4. Thus in Example 4, the force acting on the sluice gate, F_g, (= −F) may be calculated from Eq. (2.21), giving

M_1 = 4.5 × [3^2/(9.807 × 3.0) + 3^2/2] = 4.5 × [0.3059 + 4.5] = 21.627 m^3.
M_2 = 4.5 × [3^2/(9.807 × 0.4174) + 0.4174^2/2] = 4.5 × [2.1986 + 0.0871] = 10.286 m^3.
F = 1000 × 9.807 × (10.2858 − 21.6266) = − 111218.98 N or − 111.22 kN.
F_g = −F = +111.22 kN (i.e. force is positive and therefore acts in the direction of flow).

It should be also noted that in the derivation of Eq. (2.14) it is assumed that the frictional force, F_f, on bed under the short length of the jump is negligibly small. This is equivalent to stating that in Eq. (2.17) the force $F = -F_f = 0$. However, in straight channels where there may be blocks, or forward and rearward facing steps present (as shown in Figure 2.22), the relevant forces have to be specified in Eq. (2.17) in order to produce an equation equivalent to Eq. (2.14). Where supercritical flow occurs in channels with bends, care must be taken in dealing with the high super-elevation that might occur and any cross waves that may develop as a result of a hydraulic jump forming in the channel. See Chow (1959) for further details on these topics.

2.2   COMMON DIFFICULTIES IN MODELLING FLOW IN RIVERS AND WATERCOURSES

The five examples in Section 2.1 were selected to illustrate simple flow behaviour in channels where the geometry of the cross section is prismatic, i.e. a constant cross sectional shape in the flow direction, the roughness is uniformly distributed around the wetted perimeter and characterised by the Manning equation, and there are no complex hydraulic structures, vegetation or sediment within the channel. In reality, in most rivers and watercourses, all of these issues are generally of concern and must be modelled appropriately.

The previous pages and some of the examples have highlighted the three physical laws related to the conservation of mass, energy and momentum. In each example the flow was considered to be steady, with no additional discharge being introduced with distance along the channel, thus always ensuring mass conservation ($Q_1 = Q_2$). The conservation of energy was assumed in the analysis of the vertical sluice gate

$(E_1 = E_2)$, allowing the depths either side to be determined, whereas the force acting on the fluid (or the reaction on the gate) was shown to be related to the change in momentum (i.e. $M_1 \neq M_2$). In the case of the hydraulic jump, an energy loss occurred (i.e. $E_2 \neq E_3$), but momentum was conserved ($M_2 = M_3$). Dealing with energy losses is, as shown later, one of the key issues to be considered in modelling flows in rivers.

Although the same three physical principles outlined in Section 2.1 apply to flow in natural watercourses, the flow structure is typically three-dimensional and unsteady. The geometry of the river itself is also generally more complex and highly variable in the streamwise direction. All of this makes the modelling of flow in rivers and watercourses more difficult. Furthermore, there is often little or no data concerning channel roughness or resistance, which are typically flow and season dependent. Many of the hydraulic structures that have been built in rivers are often outflanked by the flow in flood conditions, or built to non-standard conditions, thus creating uncertainties in the application or interpretation of various national or international codes of practice. There are also scale issues to consider when modelling at basin, reach, or cross-section scales, and even the schematisation of a short stretch of river requires careful consideration. Hydraulic controls may also vary with flow conditions, due to changing channel resistance, the behaviour of particular hydraulic structures or even human input. For these reasons, some of the common difficulties and issues in modelling actual river flows are now considered.

2.2.1 Modelling flow in rivers and watercourses

The use of models in investigating and solving practical river engineering problems is now so commonplace, that it easy sometimes to overlook their origin, nature and purpose. There are different types of model that are now routinely used to analyse both general flow behaviour patterns and also particular hydraulic issues, such as the extent of flooding, sediment deposition or erosion patterns, pollution incidents, vegetation control and habitat biodiversity. A wide range of hydrodynamic tools are also frequently used in developing optimum design strategies for dealing with specific practical problems. For example, these tools might include models for flood risk mapping, or models that simulate dispersion phenomena, sediment movement, bank erosion or general flow behaviour. Such tools are typically based on a mathematical representation of the physical processes (sometimes incorrectly referred to as a 'model'), contain advanced numerical procedures, and employ complex computer techniques to carry out not only all the routine calculations, but also to present the results in a convenient form to the end-user, typically in graphical as well as numerical format.

The origin of modern, computer-based, river modelling began in the 1960's and the background to this is described by Abbot (1991), Cunge *et al.* (1980) and Nakato & Ettema (1996). Some of the key steps involved in modelling flow behaviour may be summarized as follows:

- Identify the nature of the practical problem requiring solution;
- Identify the physical processes involved;
- Develop some mathematical representation of the physical processes;
- Understand any limitations to the mathematical representation;

- Replace the (simplified) mathematical governing equations with a set of alternative equations based, for example, on finite difference, finite element or finite volume equations;
- Define the 'domain' of the river to be studied and represent it by a suitable mesh or grid. This process is sometimes known as the 'schematization';
- Define the boundary conditions to the domain, e.g. flows, velocities or water levels at specified boundaries;
- Solve the set of equations using appropriate numerical techniques;
- Check the answers by solving any 'benchmark' test cases to similar problems and compare the solutions;
- Calibrate the model for the particular problem in hand, optimizing key features such as water levels, velocity distributions, flow capacity, etc. by appropriate selection of any calibration coefficients;
- Check on the values and range of any calibration coefficients used, preferably by direct measurement or by comparison with other values in similar circumstances, with reference to their physical meaning;
- Optimize the design or analysis by repeated use of the model.

The word 'model' is now used frequently in a broader sense, not just for the mathematical equations governing the physical processes, but to describe the whole modelling process, as described above, often encapsulated in a particular software 'package'. A good example of the development from 'physical processes' to a conceptual 'theoretical model' is shown in the work of the ASCE Task Force on River Width Adjustment (Thorne *et al.*, 1998; Darby *et al.*, 1998). Another example showing the development from initial data on overbank flow, through to conceptual thinking, model development and calibration issues is described by Knight (2008a&b).

Modelling thus involves not only selecting the right 'tool' for the right 'job', but also being aware of the various steps involved, particularly concerning the limitations and uncertainties in the chosen methodology. It is quite possible that, for example, the numerical procedures may vary between models even of the same type. Furthermore, most models use empirical coefficients that have to be selected carefully at the calibration stage, often without adequate field measurements on the particular river in question. Turbulent parameters are also often based on inadequate full-scale measurements, sometimes even on flows other than in rivers, and in some cases on flows not even amenable to experimental scrutiny at all. See Nezu & Nakagawa (1993) and Ikeda & McEwan (2009) for further details.

The use of resistance coefficients to account for roughness effects is particularly problematic, as indicated by Yen (1991) and Morvan *et al.* (2008) and described later in more detail in Section 2.2.3 and in Chapter 3. The engineer is thus faced with some considerable difficulty in blending theoretical and empirical knowledge together in a pragmatic manner in order to simulate river processes and develop any type of 'model' at all. See Ikeda & McEwan (2009), Knight (1996 & 2008a&b) and Wang (2005) for further details.

2.2.2 Schematisation of channel geometry

The schematisation of river geometry may be relatively straightforward where the cross section is of a simple shape and prismatic. Figure 2.5 illustrates a trapezoidal

Figure 2.5 River Main, with compound trapezoidal cross section (i.e. a prismatic channel).

Figure 2.6 River Severn in flood at Upton upon Severn, February 1990 (courtesy Environment Agency).

compound section, an 'engineered' channel, which forms part of a specially con-structed flood relief scheme for the River Main in Northern Ireland. Between 1982 and 1986, an 800 m reach of the channel was reconstructed and realigned to form a double trapezoidal channel from Lisnafillan Weir to the junction with the Braid River. The cross-section comprises a main channel with a bankfull depth of ~0.9–1.0 m, a top width of 14 m, two berms covered with heavy weed growth, making a total width, inclusive of floodplains, of 27.3–30.4 m. The floodplains slope towards the main channel with a 1:25 gradient. The reach-averaged longitudinal bed slope is 0.00297. The river bed consists of coarse gravel, with a d_{50} of the order 10–20 mm and the main channel side slopes consist of quarry stone (0.5 tonne weight, 100–200 mm size). This reach has been closely monitored (Myers & Lyness, 1989) and subsequently modelled extensively, as discussed later.

Many natural rivers, particularly in times of flood, exhibit much more variabil-ity in both cross section shape and plan form geometry than engineered channels. Figure 2.6 shows the lower reaches of the River Severn during extensive flooding in February 1990, with the town of Upton upon Severn at the centre of inundation. In this case the river has overflowed its banks and occupied part of its floodplain. Figure 2.6 thus illustrates the need to include sufficient floodplain area, together with

Figure 2.7 River Severn in flood at Worcester, February 1990. Note land use management.

embankments, bridges and urban areas in any numerical model for effective flood risk mapping and management. Figure 2.7 shows an example of why it is important to include flood plain storage elements into any model. The flooded racecourse in the foreground will make a positive contribution towards reducing flooding, as it provides additional short-term storage whenever particular floods occur. This illustrates a good example of integrated policy on land use. The reclaimed land in the centre, on the other hand, may indicate poor land use policy, since it demonstrably restricts the flood plain width and hence exacerbates flooding locally.

The severe economic damage that such flooding causes to infrastructure, the loss of life and the potential impact of climate change, have forced many politicians to re-consider land use policy. The global scale of the impact of flooding is described by Knight (2006a) and the impact of climate change is dealt with by Bronstert (2006a&b), Green *et al.* (2009) and Oshikawa *et al.* (2008). Examples of the impact of recent flooding in 7 European countries are described by Knight & Samuels (2007). The European Parliament ratified the Flooding Directive in 2007 (European Commission, 2007) that requires the following for every river basin district:

- Preliminary flood risk assessment to identify areas for subsequent investigation;
- Flood risk maps;
- Flood risk management plans.

The role of numerical models is therefore likely to play a significant part in all of these activities. As described elsewhere by Knight (2006c), the schematisation of a river, including its floodplain, is an important element in constructing a successful mathematical simulation model. It should not be treated merely as an exercise in digitising numerous survey data, but rather as an art: that of blending the geometry and the hydraulic flow features together in parallel. General rules for the location of cross-sections and the data requirements for 1-D models are given in Samuels (1990 & 1995), Cunge *et al.* (1980), Castellarin *et al.* (2009) and Defalque *et al.* (1993), summarized below:

- At model limits (boundary conditions)
- Either side of structures (for afflux or energy loss calculations)

- At all flow and level measuring stations (for calibration purposes)
- At all sites of prime interest to the client
- Representative of the channel geometry
- About 20 B apart (as a first estimate only), where B = bankfull width
- A maximum of 0.2 H/S_o apart
- A maximum of $L/30$ apart where L is the length scale of the physically important wave (flood or tide)
- The area for successive sections lies between 2/3 and 3/2 of the previous section.
- The conveyance for successive section lies between 4/5 and 5/4 of the previous sections.

The overall plan form geometry of a river, and its associated floodplain boundaries, as well as other large topographical features, will inevitably steer the flow into certain patterns and produce large scale or macro flow structures. For example, dense bankside vegetation or clumps of trees/hedges on the floodplain will create plan form vorticity that will govern lateral momentum transfers and mixing processes. The variation in the shape of the main river channel around a meander bend likewise causes significant variation in the longitudinal flow pattern. Convective accelerations and decelerations along the channel will not only affect turbulence levels and resistance, but the balance of terms in the 1-D equation also. Likewise, non-prismatic floodplains will produce transverse fluxes of mass and momentum which need to be considered (see for example Bousmar, 2002; Bousmar & Zech, 1999; Bousmar *et al.*, 2004; Chlebek & Knight, 2008; Chlebek, 2009; Elliott & Sellin, 1990; Rezaei, 2006; Sellin, 1995). The degree of sinuosity, as well as the depth of flow and the aspect ratio of the river channel, will affect the position of the filament of maximum velocity, and thus influence where bank erosion might take place (Okada & Fukuoka, 2002 & 2003). Such knowledge is clearly required before designing bank protection works, targeting maintenance work or estimating flood wave speed (Tang *et al.*, 2001).

Although survey data are collected routinely at particular cross-sections, it should be remembered that the hydraulic equations are generally applied to a reach (Laurensen, 1985; Samuels, 1990) and that energy losses are quantified over a specific distance. Although care may be taken in identifying reaches that are approximately prismatic, there will inevitably be some longitudinal variation in shape and energy gradient that will cause changes in hydraulic parameters in the streamwise direction. An example of the use of 1-D models to determine the appropriate 'bankfull discharge' along a series of river reaches is given by Navratil *et al.* (2004).

2.2.3   Roughness and resistance

Although it is relatively straightforward to obtain digital terrain data, it is much harder to obtain the corresponding roughness data to go with the terrain data to the same level of detail (Defra/EA, 2003b; Hicks & Mason, 1998). For this reason the CES includes a Roughness Advisor that allows the modeller to select appropriate roughness value for various substrates, types of vegetation and seasonal growth patterns. Figures 2.1(a) & (b) have already shown the variation that can occur in lowland and upland rivers. To illustrate seasonal growth of instream and bankside vegetation,

Figures 2.8 & 2.9 show two reaches of the River Blackwater in Hampshire, near Coleford bridge. The effect of this vegetation on the roughness coefficient and conveyance capacity of the river throughout the year is documented by Sellin & van Beesten (2004).

A distinction should be made between 'roughness', which denotes a surface textural property and 'resistance', which is the cumulative effect of the roughness resisting any hydraulic flow over or through a particular type of roughness. The meaning of the term 'roughness' in the field of fluvial hydraulics, and how it is often formulated as a 'resistance to flow' term in 1-D, 2-D & 3-D numerical models, is described elsewhere more fully by Morvan et al. (2008). The Manning formula (1891), which was introduced in Section 2.1, is but one of a number of equations that attempt to quantify hydraulic resistance to flow. Although it is well founded and popular with river engineers (Yen, 1991), it should be recognized that there are a number of alternative formulations that describe resistance. A brief review of these and related topics is now given.

One of the earliest resistance equations was that by Chezy (1768), who expressed the discharge in an open channel in terms of a roughness 'constant', C, where

Figure 2.8 River Blackwater, showing seasonal growth (Winter to Summer) in a narrow reach.

Figure 2.9 River Blackwater, showing seasonal growth in a meandering reach.

$$Q = CAR^{1/2}S_o^{1/2} \tag{2.23}$$

This was followed by the Darcy-Weisbach equation (1857), which gave the head loss, h_f, over a length, l, in either a circular pipe or open channel as

$$h_f = \frac{f\ell U^2}{2gd} = \frac{f\ell U^2}{8gR} \tag{2.24}$$

where h_f = head loss, f = friction factor, l = length of pipe or channel, d = pipe diameter, R = hydraulic radius (= $A/P = d/4$ for pipe running full), U = section mean velocity (= Q/A), and S_f = friction slope (= h_f/l). When the friction slope, S_f, is equal to the bed slope, S_o, (see Section 2.2.4 for further details), then Eq. (2.24) may be expressed in a similar way to Eq. (2.23) as

$$Q = \left(\frac{8g}{f}\right)^{1/2} AR^{1/2}S_o^{1/2} \tag{2.25}$$

The Colebrook-White equation (Colebrook, 1939; Colebrook and White, 1937) clarifies how f varies in smooth and rough pipes, conduits running part full and open channels. The variation of f is usually expressed in terms of two parameters, the relative roughness, k_s/d and the Reynolds Number, Re (= $4UR/v$). For clarity, two forms of the Colebrook & White equation are given, one for pipes and another for channels, but these are essentially the same, since $d = 4R$:

$$\frac{1}{\sqrt{f}} = -2.0\log\left[\frac{k_s}{3.71d} + \frac{2.51}{(ud/v)\sqrt{f}}\right] \text{(pipes \& culverts running full)} \tag{2.26}$$

$$\frac{1}{\sqrt{f}} = -2.0\log_{10}\left[\frac{k_s}{14.8R} + \frac{2.51}{(4UR/v)\sqrt{f}}\right] \text{(open channels and pipes part full)} \tag{2.27}$$

The variation of f with k_s/d and Re is often shown plotted in the form of the so-called 'Moody diagram' (1944), as illustrated in Figure 2.10. The roughness of any surface is now characterised by k_s, the so-called Nikuradse equivalent sand roughness size (see Nikuradse, 1933), a measure of the size of excrescences, k, on a flat surface that would yield the same resistance as that in a circular pipe roughened with uniform grains of sand. The Colebrook-White equation is physically well founded, since it tends towards two theoretically limiting cases, described elsewhere by Schlichting (1979). For example, for very smooth surfaces, as $k_s \rightarrow 0$, then Eq. (2.27) becomes the Prandtl 'smooth law', in which f depends solely on Reynolds Number, Re, giving:

$$\frac{1}{\sqrt{f}} = 2.0\log_{10}\left(Re\sqrt{f}\right) - 0.80 \tag{2.28}$$

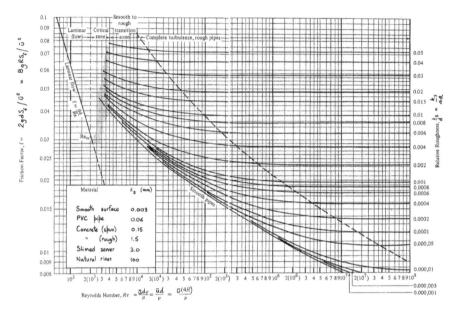

Figure 2.10 Variation of friction factor, *f*, versus Re and k_s/d or $k_s/4R$ (Moody, 1944).

On the other hand, as $Re \to \infty$, Eq. (2.27) becomes the fully 'rough law', in which *f* is independent of *Re* and depends solely on the ratio of surface roughness, k_s, to hydraulic radius, *R*, i.e. $k_s/4R$, giving:

$$\frac{1}{\sqrt{f}} = -2.0 \log_{10} \left[\frac{k_s}{14.8R} \right]$$

(2.29)

This is particularly important in river engineering as it may be shown that this links Manning's *n* in Eq. (2.1) with k_s, in Eq. (2.27), by the relationship:

$$n = k_s^{1/6} \left[\frac{(R/k_s)^{1/6}}{18 \log(11R/k_s)} \right]$$

(2.30)

The Colebrook-White equation essentially expresses the transition region from 'hydraulically smooth' to 'hydraulically rough' flow. Technically these terms are defined by the shear Reynolds Number, $(U_* k_s/v)$, as

Hydraulically smooth $(U_* k_s/v) < 5$
Transitional flow $5 < (U_* k_s/v) < 70$
Hydraulically rough $(U_* k_s/v) > 70$

where U_* is known as the shear velocity, defined by

$$U_* = \sqrt{\tau_o/\rho}$$

(2.31)

with τ_o = boundary shear stress and ρ = fluid density. The shear stress and shear velocity are important parameters in open channel flow as they frequently used when dealing with sediment transport and erosion issues (Chang, 1988). The boundary shear stress may be shown to be related to geometric parameters as follows.

Consider a short length of a prismatic channel, Δx in length, with a cross-sectional area, A, then in uniform flow, there will be a balance between the weight force acting down the channel bed slope (= $mg\sin\theta$, where m = mass, θ = channel bed slope and $\tan\theta = S_o$), with the resisting force acting on the wetted area (= $\tau_o P\Delta x$, where τ_o = mean boundary shear stress, and P = wetted perimeter). For small channel bed slopes (< ~0.1), since $\theta \approx \sin\theta \approx \tan\theta = S_o$, then for uniform flow $\rho g A \Delta x S_o = \tau_o P\Delta x$, noting that $R = A/P$, the mean boundary shear stress is given by

$$\tau_o = \rho g R S_o \tag{2.32}$$

This is essentially the momentum principle applied to this particular case. For practical purposes, and in order to provide a link between shear stress and velocity (or discharge), the energy principle needs to be invoked. If the shear stress, τ_o, is assumed to be proportional to the square of the velocity (for dimensional reasons), then this suggests that

$$\tau_o \propto \rho U^2 \tag{2.33}$$

The constant of proportionality must then be $f/8$, where f is the friction factor, in order to satisfy Eq. (2.24). Hence

$$\tau_o = \frac{f}{8}\rho U^2 \tag{2.34}$$

It may then be observed that by equating Eqs. (2.31) & (2.34), the Darcy-Weisbach friction factor, f, is simply the ratio of two velocities, U_* and U, since

$$f = 8\left(\frac{U_*}{U}\right)^2 \tag{2.35}$$

2.2.4   Energy and friction slopes

So far, two longitudinal slopes have been introduced, the bed slope, S_o, and the friction slope, S_f (= h_f/l) via Eqs. (2.1) or (2.23) and (2.24) respectively. These only have the same value when the depth and velocity do not change with distance, x, along the channel, i.e. uniform flow exists and $h = h_n$. Although this is an important type of flow, often used when calculating conveyance capacity, as shown in the worked Examples 1–3, natural river flows are frequently affected by upstream or downstream controls, making the flow non-uniform, that is to say, the depth and velocity now vary with distance x along the channel, as well as possibly with time, t, if the flow is also unsteady. See Figure 2.11 for general non-uniform flow behaviour and notation.

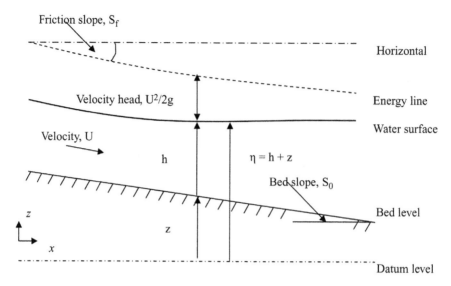

Figure 2.11 General non-uniform flow behaviour, with notation.

Under these circumstances, the friction slope is given by the following 1-D equation

$$S_f = S_o - \left[\frac{\partial h}{\partial x} - \frac{U}{g}\frac{\partial U}{\partial x} + \frac{1}{g}\frac{\partial U}{\partial t} \right] \qquad (2.36)$$

in which $\{h \ \& \ U\}$ vary as functions of distance and time $\{x \ \& \ t\}$. Since the first two terms on the right hand side are equivalent to the water surface slope, S_w, $(= \partial \eta / \partial x)$, then Eq. (2.36) may be simplified and expressed in a form useful for field measurements, as

$$S_f = \frac{\partial \eta}{\partial x} - \frac{U}{g}\frac{\partial U}{\partial x} - \frac{1}{g}\frac{\partial U}{\partial t} \qquad (2.37)$$

Equation (2.36) is important in extending rating curves (h v Q relationships), since it may be combined with Manning's equation, in which S_f replaces S_o in Eq. (2.1) for non-uniform flow. The discharge in unsteady non-uniform flow, Q, in terms of the discharge for steady flow at the same depth, based on the same Manning roughness coefficient, Q_n, is then given by

$$Q = Q_n \left[1 - \frac{1}{S_o}\frac{\partial h}{\partial x} - \frac{U}{gS_o}\frac{\partial U}{\partial x} - \frac{1}{gS_o}\frac{\partial U}{\partial t} \right]^{1/2} \qquad (2.38)$$

The concept of specific energy that was introduced via Eq. (2.12) relates the depth and kinetic energy of the flow relative to the bed (invert) of the channel. If the level of the channel bed (invert) above a horizontal reference frame is given by z_b, then the total energy, E_t relative to this reference frame is

$$E_t = z_b + h + \alpha \frac{U^2}{2g} = z_b + E \tag{2.39}$$

The gradient of total energy in the downstream direction, S_e, is then given by

$$\frac{\partial(E_t)}{\partial x} = \frac{\partial(z_b + E)}{\partial x} = -S_e \tag{2.40}$$

in which the bed slope $S_o = -\partial z_b/\partial x$. In steady uniform flow, all three slopes, S_o, S_f & S_e have the same value, S, and normal flow occurs ($h = h_n$). Under these circumstances, the conveyance of a channel, K, is defined as

$$Q = KS^{1/2} \tag{2.41}$$

Conveyance is thus a quantitative measure of the discharge capacity of a watercourse, relating the total discharge, Q (m^3s^{-1}) to a measure of the gradient or slope of the channel. By comparing Eqs. (2.1) & (2.41), it may be seen that K (m^3s^{-1}) is essentially related to the geometry and roughness of the channel. For example, when Manning's equation is used

$$K = A^{5/3}/(nP^{2/3}) \tag{2.42}$$

For steady non-uniform flow, Eq. (2.36) may be reduced to an ordinary differential equation

$$\frac{dh}{dx} = \frac{S_o - S_f}{1 - Fr^2} \tag{2.43}$$

Equation (2.43) may be solved to give the variation of depth, h, with distance, x, along a channel. The local depth is seen to depend on the longitudinal variation of 3 parameters, the bed slope, S_o, the friction slope, S_f and the Froude Number, Fr. For a reach of length, Δx, then the average friction slope, $\overline{S}_f$, is required (Laurensen, 1985), and Eq. (2.43) is commonly written in the form

$$\Delta x = \frac{\Delta E}{S_o - \overline{S}_f} \tag{2.44}$$

Equation (2.43) is usually known as the 'gradually varied flow' (GVF) equation, and leads to a way of classifying longitudinal water surface profiles in open channels. For a wide rectangular channel, such that $b >> h$, then $R \approx h$ and Eq. (2.43) may be written as

$$\frac{dh}{dx} = \frac{S_o[1 - (h_n/h)^{10/3}]}{1 - (h_c/h)^3} \tag{2.45}$$

This shows that water surface profiles depend upon values of S_o, h_n and h_c. Traditionally, these profiles have been divided into 5 classes, depending on the bed slope, and each class is further sub-divided depending on the depth of water, h, relative to the normal and critical depths, h_n and h_c. For a given shape of cross section, discharge, bed slope and roughness, the values of h_n and h_c may be determined, as shown previously, and these will effectively divide the flow space into 3 depth zones. The water surface, being disturbed from its equilibrium normal depth, may lie in any of these 3 depth zones. Flow profiles are then classified by denoting the type of channel bed slope by a letter and the depth zone by a number. For example, an M3 profile would imply that the channel bed slope is mild (M) and that the flow was in zone 3, i.e. supercritical. Since there are 5 classes of channel slope: horizontal (H), mild (M), critical (C), steep (S) and adverse (A), one might expect 15 possible profiles. In fact 2 are inadmissible (H1 and A1) and C2 is technically uniform flow with $h_n = h_c$, so there are only 12 distinct gradually varied flow profiles.

Of most interest to river engineers are those on mild or horizontal slopes, illustrated in Figures 2.12 & 2.13 respectively. In Figure 2.12, the left hand side indicates the 3 depth zones, with schematic M1, M2 and M3 water surface profiles sketched in relation to imaginary h_n and h_c lines drawn parallel to the bed. On the right hand side of the Figure, the smaller diagrams (a) to (f) represent some examples of where such profiles might typically occur in practice. For mild slopes ($h_n > h_c$), the backwater profile (M1) is frequently encountered, particularly in relation to bridges and control structures, as shown later in Section 2.6. An estimate of the length of a backwater profile has been given by Samuels (1989b), as

$$L = 0.7h/S_o \qquad\qquad (2.46)$$

where L = the distance upstream over which the water depths are disturbed by more than 10% from their equilibrium condition ($h = h_n$). The actual water surface profile may be determined by numerical solution of the GVF equation, using the 'standard step' method for non-prismatic channels, or the 'direct step' method for prismatic channels. See Chow (1959) and Henderson (1966) for further details of these so called

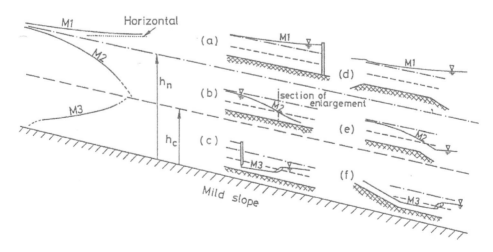

Figure 2.12 Gradually varied flow profiles on a mild slope.

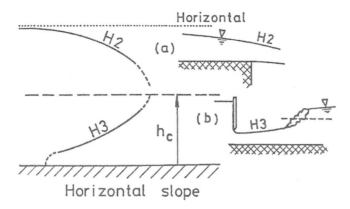

Figure 2.13 Gradually varied flow profiles on a horizontal slope.

'step' methods. Since there is a widespread need to calculate backwater lengths or profiles, a backwater routine is included within the CES and further details are given in Chapters 3 & 4 (Sections 3.2.4 & 4.3 respectively).

In addition to the bed slope and friction slope, the energy slope, S_e, is also important. Since this involves the kinetic energy correction coefficient, to account for the 3-D nature of most flow fields in rivers, the next section considers a number of issues related to the velocity distributions within channels and their implications for 1-D modelling.

2.2.5 Velocity distributions and implications for 1-D modelling

Up until now, the flow has been considered as one-dimensional (1-D), with the section mean velocity, U ($= Q/A$) being the only one used in all resistance equations and water surface profile relationships. As mentioned earlier, this has certain implications with respect to energy, kinetic head, and approximating 3-D effects into a lateral distribution model such as is used in the CES. Figure 2.14 shows a typical velocity field for inbank flow in a simple trapezoidal channel, taken from the work of Yuen (1989). Figure 2.15 shows how the flow field becomes even more complex when overbank flow occurs, partly due to the strong interaction that may develop between the flows in the main river channel and on the floodplain(s). Since 3-D concepts and flow structures are described in more detail in Section 3.1, only those that are relevant to the velocity field are introduced now, hopefully enabling the reader to understand more about modelling flows in rivers.

The first issue to appreciate is the 3-D nature of the flow, even in straight channels. Figure 2.14 shows a typical isovel pattern, secondary flow cells (Einstein & Li, 1958; Nezu & Nakagawa, 1993) and the resulting distribution of boundary shear stress around the wetted perimeter (Knight *et al.* 1994). The maximum velocity may be at the free surface for a wide river, as shown, but very often is just below the surface, particularly in channels with a low aspect ratio (*b/h*). Boundary layers (Schlichting, 1979) develop from the bed and sides of the channel, with the local velocity, *u*, being zero at every solid/liquid interface around the wetted perimeter (i.e. a boundary condition of

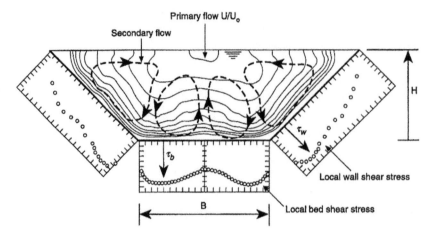

Figure 2.14 Isovels and boundary shear stresses for inbank flow in a trapezoidal channel (after Knight et al., 1994).

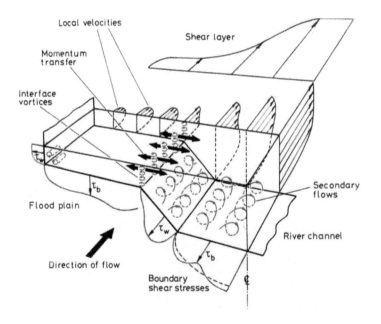

Figure 2.15 Flow structures in a straight two-stage channel (after Knight & Shiono, 1996).

$u = 0$). If the channel is wide (large aspect ratio, b/h), then the velocity distribution over a vertical at the centre-line position may be assumed to be approximately logarithmic, described by Prandtl's logarithmic velocity law for a rough surface as

$$\frac{u}{u_*} = \frac{1}{\kappa} \ln\left(\frac{z}{z_0}\right) \tag{2.47}$$

where u = velocity a distance z above the bed, u_* is the local shear velocity, κ is von Karman's constant, $z_0 = \gamma k_s$, γ = constant and k_s is the Nikuradse equivalent sand roughness size. Thus z_0 is assumed to be proportional to the size of the roughening excrescences, k_s, as for smooth surfaces. Early experiments gave values of $\gamma = 1/30$ and $\kappa = 0.40$ (now taken as ~0.41). Hence the velocity distribution law for rough surfaces is often assumed to be governed by an equation of the form

$$\frac{u}{u_*} = 2.5\ln\left(\frac{z}{k_s}\right) + 8.5 \quad \text{or} \quad \frac{u}{u_*} = 5.75\log_{10}\left(\frac{z}{k_s}\right) + 8.5 \tag{2.48}$$

The corresponding equation for velocity distribution over a smooth surface is of the general form

$$\frac{u}{u_*} = A\ln\left(\frac{z}{k_s}\right) + B \tag{2.49}$$

where A & B are constants. Experiments show that B is a function of (u_*k_s/ν), given by

$$B = 2.5\log_{10}\left(\frac{u_*k_s}{\nu}\right) + 5.5 \text{ (only valid for } (u_*k_s/\nu) < 5.0) \tag{2.50}$$

Combining Eqs. (2.49) & (2.50), and using Prandtl's original constants, gives the corresponding velocity distribution over a smooth surface as

$$\frac{u}{u_*} = 2.5\ln\left(\frac{u_*z}{\nu}\right) + 5.5 \quad \text{or} \quad \frac{u}{u_*} = 5.75\log_{10}\left(\frac{u_*z}{\nu}\right) + 5.5 \tag{2.51}$$

The second issue is to consider the implications of using a depth-averaged model. The depth-averaged velocity, U_d, is defined as

$$U_d = \frac{1}{h}\int_0^h u\,dz \tag{2.52}$$

Typical lateral distributions of U_d across a channel are shown in Figure 2.16. As might be expected, $U_d = 0$ at the two edges and increases towards the central region. The presence of secondary flows may mean that the maximum value of U_d does not actually occur at the centerline of the channel.

The third issue is to consider how a 3-D velocity field might be included in a 1-D formulation of energy or momentum. As noted earlier in Eqs. (2.12) & (2.17), this is usually achieved by use of the kinetic energy and momentum correction coefficients, α and β, here formally defined as

$$\alpha = \frac{1}{U^3 A}\int_A u^3 dA \tag{2.53}$$

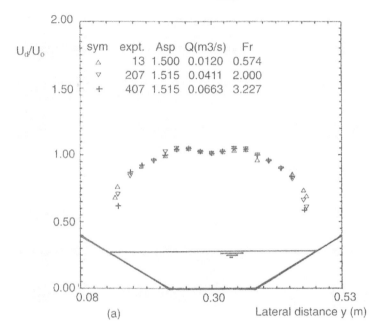

Figure 2.16 Variation of U_d with lateral distance across a trapezoidal channel (Knight *et al.*, 1994).

Table 2.1 Typical values of kinetic energy and momentum correction coefficients.

Correction coefficient	Prismatic channels	Rivers	Using the seventh power law, Eq. (2.55)
α	1.15	1.30 (and higher)	1.045
β	1.05	1.10 (and higher)	1.016

$$\beta = \frac{1}{U^2 A} \int_A u^2 dA \qquad (2.54)$$

These coefficients are introduced to ensure that the energy flux ($=1/2\ \rho AU^3$) or momentum flux ($=\rho AU^2$) based on the area, A, and section mean velocity, U, agree with the actual energy and momentum fluxes ($= \int_A \frac{1}{2}\rho u^3 dA$ and $= \int_A \rho u^2 dA$) obtained by summing over the whole area, using local velocities, u, applied to each individual element of area, dA. This effectively allows one to use the section mean velocity, U, rather than knowledge of the complete velocity field ($u = f\{y, z\}$, which can be obtained from measurement or 3-D computational fluid dynamics (CFD) and integrating over the areal surface. The use of these two correction coefficients has great practical merit in that only section mean velocities are used, and these are generally known. Typical values of α and β are shown in Table 2.1.

As indicated in Table 2.1, a seventh power law is sometimes used to represent the distribution of velocity over a vertical in turbulent open channel flow, rather than the more complex logarithmic laws, expressed earlier as Eqs. (2.48) & (2.51), on account of its simplicity and closeness of fit to logarithmic distributions. This distribution may be written as

$$\frac{u}{u_{max}} = \left(\frac{z}{h}\right)^{1/7}$$ (2.55)

When this is inserted into Eq. (2.52), to obtain the depth-averaged value, it gives

$$U_d = \frac{7}{8}u_{max}$$ (2.56)

Although Eq. (2.55) has the advantage that it is a simpler relationship to use than Eqs. (2.48) & (2.51), it has the disadvantage from an engineering perspective that it needs the maximum velocity, u_{max} to be known, as well as its position within the cross section. If it is assumed that u_{max} occurs at the free surface, then it may be shown that U_d occurs at approximately $z = 0.4\,h$, regardless of which equation, Eq. (2.48), (2.51) or (2.55), is used. This is helpful, since it forms the basis of river gauging in open channels.

The fourth issue following on from the third, is the application of these α and β coefficients to flows in natural channels. As already indicated in Eq. (2.12), the kinetic head term based on the section mean velocity, $U^2/2g$, needs to be multiplied in every energy relationship by α. In a similar way, wherever the momentum equation is used, the term ρQU should be multiplied by β, as indicated in Eq. (2.18), to give the correct force-momentum relationship. The energy line (or surface) may be conceived as an imaginary line drawn a distance $\alpha U^2/2g$, above the water surface, as already indicated by the dotted line in Figure 2.4. This is now shown more clearly in Figure 2.17 for flow in a section of open channel in which there is a constriction, such as arising from

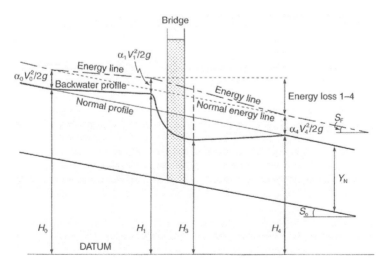

Figure 2.17 Illustration of variation of kinetic heads and energy losses through a bridge (after Hamill, 1999).

bridge piers or a measuring flume. Upstream of the bridge, the velocities are lower and so the energy line is closer to the water surface than downstream where the velocities are larger. Note that the afflux is the increase in water level upstream of a bridge, above the water level that would occur in the absence of the bridge, and should not be confused with the energy loss across the whole bridge structure.

In Figure 2.17 water surface dips as it passes through the section with a reduced width, on account of the increase in velocity from considerations of continuity ($Q = U_1 A_1 = U_2 A_2$). Since the frictional losses are proportional to the square of the velocity, the frictional losses in the narrow section will be larger than those in the channel upstream. Because of the change in cross section, there may also be additional energy losses in the contraction and expansion regions, with the losses in any expansion region generally being larger than those in a contracting region. The overall losses are then the sum of the frictional and additional energy losses, with the overall energy line indicating the total budget of energy and how it is degraded. A distinction should therefore be made between the energy and friction gradients, S_e and S_f.

2.2.6 Hydraulic structures and controls

A particular difficulty in modelling water surface profiles in rivers arises from the wide variety of hydraulic structures that are commonly used in river engineering. These range from weirs and flumes (Ackers *et al.*, 1978; Bos *et al.*, 1984), to bridges and culverts (Neill, 1973; Ramsbottom *et al.*, 1997), gates and sluices (Franke & Valetin, 1969; Kolkman, 1994; Larock, 1969; Rajaratnam & Subramanya, 1969; Rajaratnam & Humpries, 1982), siphon spillways (Novak *et al.*, 1996) and free overfalls (Chow, 1959). Each structure will cause a change in water level across the structure and should be modelled accordingly. For bridges, the 'head loss' may be different from the 'afflux', as considered further in Section 2.6.

Wherever the flow changes from subcritical to supercritical there exists a unique relationship between h & Q, and this may form one or two 'control points'. For example, in the case of the sluice gate, illustrated in Figures 2.3 & 2.4 (Example 4 in Section 2.1), the control point on the upstream side of the gate controls all the water levels in the upstream direction, i.e. in the subcritical part of the flow. This is referred

Figure 2.18 Bridge arches in flooding River Avon (courtesy University of Birmingham).

Figure 2.19 Scour in channel with bridge (courtesy HR Wallingford).

Figure 2.20 Flow downstream of the Holme sluices, River Trent (courtesy University of Birmingham).

to as a backwater profile and, if the sluice gate were sited in a mildly sloping channel, it would be an M1 profile (see Figures 2.12(a) & (c)). On the downstream side of the gate, at the *vena contracta*, the control point controls all the water levels in the downstream direction, i.e. in the supercritical part of the flow, creating the H3 profile in Example 4.1 or an M3 as shown in Figure 2.12(c) if the channel has a mild slope. Knowledge about where control points occur, or rules about where they form, are a pre-requisite for deciding which of the 12 gradually varied flow profiles will occur and for modelling any water surface profiles in open channels. Some examples of flow through bridges and some typical control structures are shown in Figures 2.18–2.22.

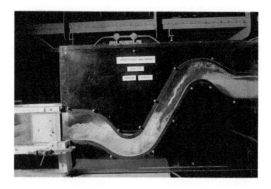

Figure 2.21 Model of air regulated siphon Figure (model scale 1:10).

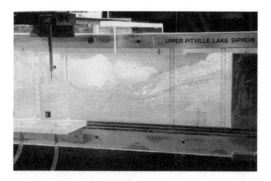

Figure 2.22 Siphon outlet, showing flow over energy reducing blocks.

2.2.7 Calibration data for river models

One of the main difficulties frequently experienced in the development of a mathematical model of a river is the lack of suitable field data for calibration, especially that
related to channel roughness, water surface behaviour and turbulence parameters under
extreme flow conditions. Despite hydrometric survey work being routinely carried out
in the UK, there are generally never enough gauging stations in the right position, or
indeed any monitoring of other key hydraulic parameters at all. This has two main consequences: that of making the model less useful than it might have been and, secondly,
the model results will have larger uncertainties associated with them than perhaps originally hoped for. This issue has been discussed by several authors (e.g. Anastasiadou-
Partheniou & Samuels, 1998; Knight, 2008a&b; Seed *et al.*, 1993; Vidal *et al.*, 2007)
many of whom stress the need for a proportion of any modelling budget to be spent on
acquiring appropriate data. Well focussed physical model studies, such as those undertaken in the UK Flood Channel Facility (Knight & Sellin, 1987; website *www.flow-
data.bham.ac.uk*), are sometimes the only source of data at sufficiently comprehensive
spatial and temporal scales to be used in the effective calibration of numerical models
of river flow. These, and other sources of data, will be explored further in Section 2.7.

In addition to the hydraulic variables, there are many meteorological, hydrological and hydro-geological factors that affect the discharge in a river and also influence its variability. Such factors might be related to rainfall-runoff issues, precipitation and snow melt issues, groundwater issues, the underlying geology of the river basin, the number of tributaries and the behaviour of individual sub-catchments. The Flood Estimation Handbook (FEH, 1995) and the revised version (ReFEH, 2007) contain much of this information that should be consulted by all modellers requiring further details.

2.3   FLOW IN SIMPLE ENGINEERED CHANNELS

This section, and subsequent sections in Chapter 2, now explores briefly the more complex flow conditions that can occur in actual rivers and watercourses and how they might be modelled successfully. A full explanation of the scientific basis behind the CES-AES software is given in Chapter 3, and more examples illustrating further practical issues are given in Chapter 4.

The modelling strategy used within the CES is largely based on the Shiono & Knight Method (SKM) of analysis, which gives the lateral distribution of depth-averaged velocity, U_d across channels of any prismatic shape. It is applicable to inbank and overbank flows in straight channels, and may be extended to flows in meandering channels by adding an additional term. The method captures certain 3-D flow effects and embodies them into a simpler 1-D approach. A fuller explanation of the methodology is given in Chapter 3, and further details of the model may be found in Shiono & Knight (1988 & 1991), Knight & Abril (1996), Abril & Knight (2004), Mc Gahey (2006), Knight, Omran & Tang (2007), Knight & Tang (2008), Mc Gahey et al. (2006 & 2009) and Tang & Knight (2009). Chapters 6 & 7 provide full details of the SKM and now only a brief introduction to the SKM approach is given, using H = water depth to make the notation appropriate to the CES method (see Figure 2.23).

The equations which govern the behaviour of fluids in motion are known as the Reynolds Averaged Navier-Stokes (RANS) equations. See Schlichting (1979) and Drazin and Riley (2006) for further details on the RANS equations. The governing Reynolds Averaged Navier-Stokes equation for the streamwise motion of a fluid element in an open channel, with a plane bed inclined in the streamwise direction, may be combined with the continuity equation to give

$$\rho\left[\frac{\partial}{\partial y}(UV)+\frac{\partial}{\partial z}(UW)\right]=\rho g S_o+\frac{\partial \tau_{yx}}{\partial y}+\frac{\partial \tau_{zx}}{\partial z} \qquad (2.57)$$

where $\{UVW\}$ = velocity components in the $\{xyz\}$ directions, x-streamwise parallel to the channel bed, y-lateral and z-normal to the bed, ρ = fluid density, g = gravitational acceleration, S_o = channel bed slope, and $\{\tau_{yx},\tau_{zx}\}$ = Reynolds stresses in the streamwise direction on planes perpendicular to the y and z directions respectively. Figure 2.23 indicates these stresses, the notation and some of the key terms used. A physical interpretation of Eq. (2.57) would be: Secondary flows (streamwise and planform) = Weight force + Reynolds stresses (lateral + vertical).

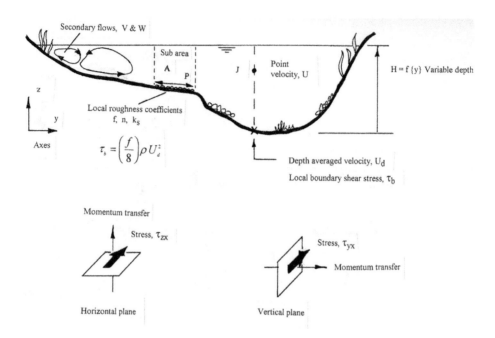

Figure 2.23 Flow in a natural channel and the various terms used in SKM and the CES.

Shiono and Knight (1991) obtained a depth-averaged velocity equation by integrating (2.57) over the flow depth (H), to give

$$\rho g H S_o - \rho \frac{f}{8} U_d^2 \left(1 + \frac{1}{s^2}\right)^{1/2} + \frac{\partial}{\partial y}\left\{\rho \lambda H^2 \left(\frac{f}{8}\right)^{1/2} U_d \frac{\partial U_d}{\partial y}\right\} = \frac{\partial}{\partial y}[H(\rho UV)_d] \quad (2.58)$$

where f = Darcy-Weisbach friction factor; λ = dimensionless eddy viscosity; s = the channel side slope of the banks (1:s, vertical: horizontal) and U_d = the depth-averaged velocity. Solving Eq. (2.58) yields U_d as a function of y, the lateral co-ordinate. Other terms are defined by:

$$U_d = \frac{1}{H}\int_0^H U \, dz; \quad \tau_b = \left(\frac{f}{8}\right)\rho \, U_d^2; \quad \overline{\tau}_{yx} = \rho \overline{\varepsilon}_{yx} \frac{\partial U_d}{\partial y}; \quad \overline{\varepsilon}_{yx} = \lambda U_* H \quad (2.59)$$

where $U_* = (\tau_b/\rho)^{1/2}$ = shear velocity. Eq. (2.58) thus includes the effect of secondary flows (streamwise vorticity about horizontal axes) and planform vorticity (vorticity about vertical axes). Both these types of vortex structure are important in many open channel problems, and particularly so in overbank flow, as shown schematically in Figure 2.15. Based on experimental results, the term $(\rho UV)_d$ is assumed to vary approximately linearly, and therefore of the form

$$\frac{\partial}{\partial y}[H(\rho UV)_d] = \Gamma \tag{2.60}$$

where Γ is the lateral gradient of the advective term. In essence the SKM method allows values of the three calibration parameters (f, λ and Γ) to be assumed for each part of the flow and then Eq. (2.58) to be solved either numerically or analytically. In order to illustrate its use, some simple examples are now given, beginning with flow in straight channels.

Since the two simplest types of cross section that commonly occur, or are used in modelling, are rectangular and trapezoidal, these are used first to demonstrate the modelling strategy. See Figure 2.24 for examples of natural channels, and Figures 2.25–2.29 for examples of laboratory channels used in refined calibration studies (Chlebek & Knight, 2006; Chlebek, 2009; Knight *et al.*, 2007; Liao & Knight, 2007a&b; Omran *et al.*, 2008; Sharifi *et al.*, 2009; Tominaga *et al.*, 1989).

Figure 2.24 Inbank flows in approximately rectangular and trapezoidal river channels.

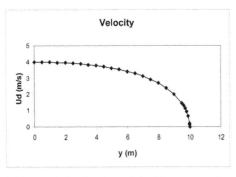

(a) Lateral distribution of depth-averaged velocity, U_d

(b) Lateral distribution of bed shear stress, τ_b

Figure 2.25 Velocity and bed shear stress results for flow in one half of a 20 m wide rectangular channel (single panel results, with $y = 0$ at centreline).

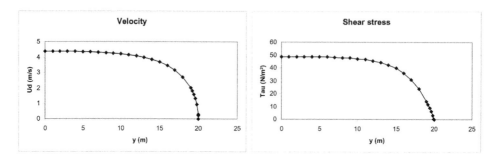

(a) Lateral distribution of depth-averaged velocity, U_d

(b) Lateral distribution of bed shear stress, τ_b

Figure 2.26 Velocity and bed shear stress results for flow in one half of a 40 m wide rectangular channel (single panel results, with $y = 0$ at centreline).

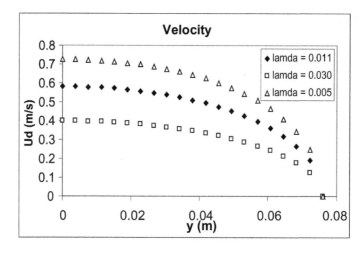

Figure 2.27 Effect of λ on lateral distribution of velocity in a rectangular channel.

The cross section of a rectangular channel may be treated singly or sub-divided into any number of panels, each of equal depth, but with varying flow parameters (f, λ and Γ) set for each panel. A trapezoidal channel may likewise be sub-divided, but this time the panels will be of different depths if they are on the sloping side region, and constant in the flat bed region. In this case, the least number of panels to model the whole channel would be 2, one for each region, as shown in Figure 2.30 later. The choice concerning the number of panels is important and may affect the results in a number of ways. If the flow and cross section are symmetric, as maybe sometimes for certain flows in straight channels with rectangular and trapezoidal cross sections, then only half the channel requires analysis.

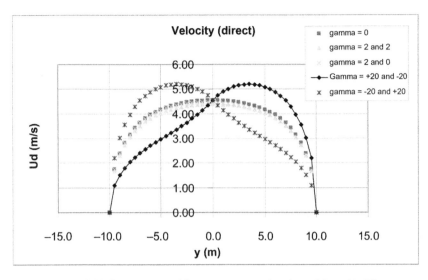

Figure 2.28 Simulated bend flow in a rectangular channel (variable Γ).

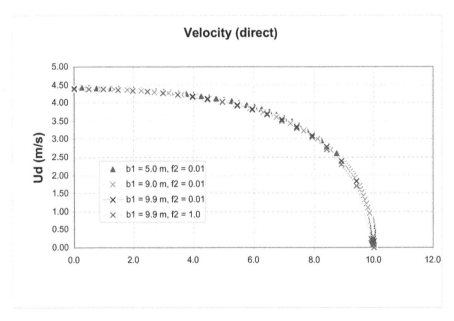

Figure 2.29 Flow in a 20 m wide rectangular channel with variable panel widths (2 panels, with y = 0 at centreline).

The philosophy adopted within SKM is to choose the minimum number of panels, commensurate with obtaining the maximum useful output from the results. Any greater accuracy is unwarranted since it should be remembered that the method itself is only an approximate one. However, there are a number of principles that should be

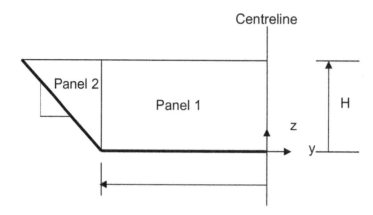

Figure 2.30 Cross-section of trapezoidal channel, schematized with two panels.

followed concerning panel selection that are described in detail in Chapter 3. For the present purposes, the following examples are restricted to single or, at most, a small number of panels.

2.3.1   Flows in rectangular channels

Example 6

Determine the discharge and velocity distribution in a 20 m wide rectangular channel that flows at a depth of 5.0 m at a bed slope of 0.001, assuming a Darcy f of 0.02, and constant values for λ and Γ of 0.07 and 0.15 respectively.

Answer & comment

Consider only half the channel so one panel has a width of 10 m. The results are as shown in Figure 2.25 at 0.5 m intervals for $y = 1$–9 m, then at 0.1 m intervals to 9.9 m and finally at closer intervals near the wall. It can be seen that $U_d = f\{y\}$ for the one panel, even though the three calibration parameters (f, λ and Γ) are constant. The distribution of U_d is physically realistic, in that U_d = maximum at the centreline ($y = 0$) and zero at the wall ($y = 10$ m). The method thus mimics the anticipated shear layer, and also gives the bed shear stress distribution since, once U_d is known, values of τ_b may be calculated from Eq. (2.59). Both distributions are hyperbolic, and this illustrates how some details of the flow have now been calculated, in a way that was not possible using the simple 1-D Manning resistance equation, Eq. (2.1).

Having determined the lateral distribution of U_d, the individual values may be multiplied by the appropriate area of each sub-panel and the discharge obtained. This gives $Q = 330$ m³s⁻¹, which is somewhat different from the value of 362 m³s⁻¹ calculated below, on the basis of using the same value of f in Eq. (2.25).

From Eq. (2.25):

$U = (8gRS_o/f)^{1/2} = (8 \times 9.807 \times 3.3333 \times 0.001/0.02)^{1/2} = 3.616$ ms⁻¹.
$Q = UA = 3.616 \times 20 \times 5 = 361.6$ m³s⁻¹.

The same value of Q should not in fact be expected, despite Eq. (2.25) being essentially a single channel method (SCM), since the influences of the other two parameters (λ and Γ) have not yet been considered. For example, halving the value of λ (i.e. less lateral shear) and putting $\Gamma = 0$ (no secondary flow) in Eqs. (2.58) & (2.59) gives $Q = 364$ m³s⁻¹ which is much closer to the value of 362 m³s⁻¹ calculated using Eq. (2.25).

The lateral distribution of velocity in Figure 2.25(a) is seen to give $U_{max} = 4.0$ ms⁻¹ at the centreline and a mean value of $U \approx 3.2$ ms⁻¹. The mean value is again slightly different from the value calculated from Eq. (2.25), $U = 3.616$ ms⁻¹, for the reasons given above. The centre-line value of bed shear stress in Figure 2.25(b) is seen to be approximately $\tau_b \approx 40.0$ Nm⁻², which is lower than the mean bed value based on assuming two-dimensional flow, $(\tau_{b2d} = \rho gHS_o)$ and higher than the mean boundary shear stress for the whole cross-section $(\overline{\tau}_o = \rho gRS_o)$. This is to be expected, since:

$\tau_{b2d} = 1000 \times 9.807 \times 5.0 \times 0.001 = 49.03$ Nm⁻².
$\overline{\tau}_o = 1000 \times 9.807 \times 3.3333 \times 0.001 = 32.69$ Nm⁻².

The value of maximum bed shear stress will approach the two-dimensional value $(\tau_b = \rho gHS_o)$ as the channel is widened. For example if the width is now doubled to 40 m, and assuming the same values for the parameters $(f, \lambda$ and $\Gamma)$ as before, gives the distributions shown in Figure 2.26. In this case, the centre-line bed shear stress, τ_b, is now much closer to the two-dimensional value of 49.03 Nm⁻², since the aspect ratio of the channel (width/depth) has risen to 40/5 = 8. This is still not technically a 'wide' channel as the simulation results here are some way from those for a wide channel (see Figure 9 of Knight & MacDonald (1979), Knight et al. (1992 & 1994) and Al-Hamid (1991) for further details). For practical purposes the aspect ratio needs to be at least >10 to get within 95% of τ_{b2d} and generally >25.

Figure 2.26(a) shows $U_d = f\{y\}$ for the 40 m wide channel, varying from $U_{max} = 4.38$ms⁻¹ at centerline to zero at the wall. Integrating across the width gives $Q = 774.9$m³s⁻¹. This is less than the value of $Q = 792.24$ m³s⁻¹ below, based on Eq. (2.25):

$U = (8gRS_o/f)^{1/2} = (8 \times 9.807 \times 4.0 \times 0.001/0.02)^{1/2} = 3.9612$ ms⁻¹.

$Q = UA = 3.9612 \times 40 \times 5 = 792.24$ m³s⁻¹.

Lowering λ to 0.035 causes the values of U_d to rise, especially nearer the walls, giving $Q = 810.3$ m³s⁻¹, now above the value of 792.3 m³s⁻¹. This example has shown that the individual values of the parameters λ and Γ, as well as f, will also influence the discharge, and so this is explored further in the next example.

Example 7

Using only one panel to represent half of a rectangular channel, explore the combined influence of all 3 parameters (f, λ and Γ).

Answer & comment

In order to test the methodology further, data are needed on channels where both velocity and boundary shear stress distributions have been measured with known accuracy. Since such data are rare at full scale in large channels, recourse has to be made to data from small scale laboratory channels where it had been possible to measure both parameters accurately. One set of data that is suitable is highlighted by Chlebek & Knight (2006), taken from experiments in a series of small rectangular channels. Further discussion on data and availability are given in Section 2.7.

The general effect on Q of varying all 3 parameters and the percentage of shear force on the channel walls, $\%SF_w$, is shown in Table 2.2, based on numerical simulations of all the experimental results from the various channels, covering a wide range of aspect ratios. The $\%SF_w$ values were obtained by integrating the bed shear stresses over the semi width of the channel, determining the overall shear force on the bed, SF_b, and then subtracting it from the overall mean shear force for the whole channel, $(\tau_o \times P)$ to give SF_w, which is then expressed as a percentage of the total (bed and both walls).

Figure 2.27 shows the effect of varying λ on the lateral distribution of U_d, while holding f and Γ constant for one particular simulation. In this case, the experimental channel had a width of 152 mm, a depth of 200 mm, a Q of 13.78 l/s with a value of $\%SF_w$ of 73.08%. The $\%SF_w$ value is understandably high, given the narrowness of the channel (aspect ratio of only 0.76). In general, it was found that for a given depth, when λ was increased, with fixed f & Γ, it caused Q to decrease and $\%SF_w$ to increase. The reverse occurred when λ was decreased. Tables 2.2 & 2.3 illustrate these changes quantitatively, based on this one depth ($H = 200$ mm) with reference values of $Q = 13.78$ ls^{-1} and $\%SF_w = 73.08\%$ for which $f = 0.018$, $\lambda = 0.011$ and $\Gamma = 0.047$ gave the closest approximation to the experimental data, using just one panel.

The results in rows 2 & 3 of Table 2.3 indicate that the effect of varying f from 0.040 to 0.005 (from optimum of 0.018) is to change Q from 13.78 ls^{-1} to 10.46 and

Table 2.2 Effect of varying one calibration parameter, holding the other two constant.

Variable parameter	Fixed parameters	+/−	Q	%SF$_w$
f	λ & Γ	increase	decreases	decreases
	λ & Γ	decrease	increases	increases
λ	f & Γ	increase	decreases	increases
	f & Γ	decrease	increases	decreases
Γ	f & λ	increase	decreases	increases
	f & λ	decrease	increases	decreases

Table 2.3 Effect of varying parameters f, λ and Γ on values of Q and $\%SF_w$ (Highlighted row shows closest simulation to the experimental data, i.e. optimum values).

f	λ	Γ	Q (litres sec^{-1})	$\%SF_w$ (%)
0.018	**0.011**	**0.047**	**13.78**	**73.08**
0.040	0.011	0.047	10.46	65.23
0.005	0.011	0.047	20.52	83.08
0.018	0.030	0.047	9.36	87.54
0.018	0.005	0.047	17.35	57.44
0.018	0.011	0.500	11.97	79.68
0.018	0.011	0.000	13.95	72.93

20.52 ls^{-1} respectively, and to change $\%SF_w$ from 73.08% to 65.23% to 83.08%. The effect of varying λ from 0.030 to 0.005 (from optimum value of 0.011) is shown in rows 4 & 5, and the effect of varying Γ from 0.500 to 0.000 (from optimum value of 0.047) is shown in rows 6 & 7. This example highlights the need to consider the combined effect of all 3 calibration parameters on two objective functions (discharge and a boundary shear force, i.e. Q and $\%SF_w$). Before considering this in detail, and using substantive sets of data outlined in Section 2.7, the influence of the secondary flow term, Γ, as well as the appropriate number of panels to be used to represent the channel cross section, also need to be investigated. These are now considered together in Example 8.

Example 8

A rectangular channel is 20 m wide, 3.0 m deep and has a bed slope of 0.001. Using two panels to represent half of this channel, investigate the influence of the secondary flow term, Γ, and panel selection on the flow results, assuming $f = 0.01$ and $\lambda = 0.10$ for all panels.

Answer & comment

In the absence of suitable full scale data, particularly those in which Γ has been measured, this exercise on secondary flow effects and the number of panels that should be used has to be a purely numerical one. Detailed numerical simulation of multi-panels is reserved for an example in Section 2.3.2, using laboratory data for both U_d and $\%SF_w$ and examples in Chapter 3 using river data, for which data exists on U_d but not on boundary shear stress distributions.

Figure 2.28 shows the results of these simulations for a 20 m wide channel in which two panels of equal width ($b = 10$ m) make up the total width of the rectangular channel. This is equivalent to modelling half the channel with just a single panel 10 m wide. The results show that for $\Gamma = 0$, the distribution of U_d is symmetric, as expected. For $\Gamma = 2$, in all panels the velocities are diminished slightly. When Γ is changed in sign for each panel either side of the centerline, i.e. $\Gamma_1 = +20$ on one side and $\Gamma_2 = -20$ on the other side, then the velocity distribution is no longer symmetric, but biased towards one side or another. Such a distribution is that which might occur in a left-handed or right-handed bend, with a higher velocity on the outside of the bend. Not surprisingly this is seen to be produced here by altering the secondary flow term, Γ. As shown later in Chapter 3, there are other ways of attempting to simulate flow in meandering channels.

Figure 2.29 shows the effect of splitting a single panel into a number of sub-panels. In this case, a single panel of width b, equivalent to half of the channel width, is divided further into 2 sub panels, of widths b_1 and b_2, to make up half of a rectangular channel of semi-width, b, thus making $b_2 = b - b_1$. The width of one panel, b_1, is altered progressively from 5.0 m to 9.9 m in order to see if the smaller panel adjacent to the wall captures the very high shear in that region. As can be seen, with $b_1 = 9.9$ m and $b_2 = 0.1$ m the results compare quite well with the simulation undertaken with two equal sized panels. The effect of increasing the roughness in the smaller panel adjacent to the wall by a factor of 100 is seen to reduce the velocity in that region, thus indicating a possible way of refining any calibration to fit data. This is dealt with more thoroughly in the next section that deals with flows in trapezoidal channels.

Example 9

Attempt some of these exercises using the CES software, which is based on the SKM, and compare your answers with those given in this Chapter. All the results in Examples 6–8 were obtained using spreadsheets to evaluate the analytical solutions to Eq. (2.58) directly.

2.3.2   Flows in trapezoidal channels

The previous examples have indicated some general features of the SKM for flows in rectangular channels. The use of multi-panels is now explored more thoroughly, using data from laboratory experiments undertaken by Yuen (1989) and Tominaga *et al.* (1989). Figure 2.30 shows a typical schematization for a trapezoidal channel, based initially on just two panels. Figure 2.31 shows the results of a simulation based on using such a schematization, together with some experimental data. Although Figure 2.31(a) shows that the lateral distribution of U_d is reasonably well simulated, Figure 2.31(b) shows that the corresponding lateral distribution of boundary shear stress, τ_b, is not simulated as well. A much better simulation for the latter is obtained when each of the previously used 2 panels are sub divided further, making 4 panels in all, as shown by the results in Figures 2.32(a) & (b). It was found by Omran (2005) that 4 panels are normally quite adequate for practical purposes when modelling most types of flow in trapezoidal channels.

In particularly demanding cases, where even greater precision is required, such as when attempting to simulate the boundary shear stress distribution very accurately in corner regions, then 6 panels may be required. Under these

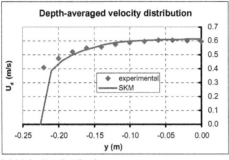 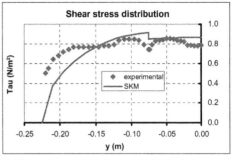

(a) Velocity distribution   (b) Boundary shear stress distribution

Figure 2.31 Results of velocity and boundary shear stress simulations of flow in a trapezoidal chan-nel (2 panels). [Exp. 16, Yuen, 1989, with $f_1 = 0.0185$, $f_2 = 0.020$, $\lambda_1 = \lambda_2 = 0.07$, $\Gamma_1 = 0.5$, $\Gamma_2 = -0.2$].

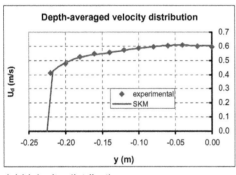 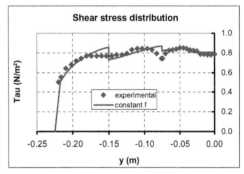

(a) Velocity distribution   (b) Boundary shear stress distribution

Figure 2.32 Results of velocity and boundary shear stress simulations of flow in a trapezoidal channel (4 panels, Exp. 16, Yuen), after Knight, Omran & Tang (2007).

circumstances, the number, pattern and strength of the secondary flow cells should be taken into account, as shown in Figure 2.33 by Tominaga *et al.* (1989). The results of simulating the same experiment shown in Figures 2.31 & 2.32 are shown in Figure 2.34. It can be seen that by using 6 panels the sharp decrease in boundary shear stress in the corner region ($y = 0.075$ m) is much better simu-lated. Another feature that is important when dealing with boundary shear stress with multiple panels of different roughness is to include some lateral smoothing into the friction factor, f, in order to overcome any discontinuity in τ_b that arises from use of a depth-averaged approach based on Eqs. (2.34) & (2.52). This is shown in Figure 2.34, where lateral distributions of boundary shear stress, τ_b, are shown for constant f values (saw tooth plot, 4 panels), linearly varying f values (4 panels) and modified linearly varying f values (6 panels). Further details of

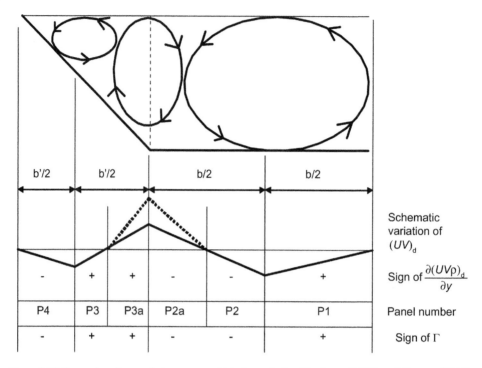

Figure 2.33 Secondary flow cells in a trapezoidal channel, after Tominaga (1989) and Omran (2005).

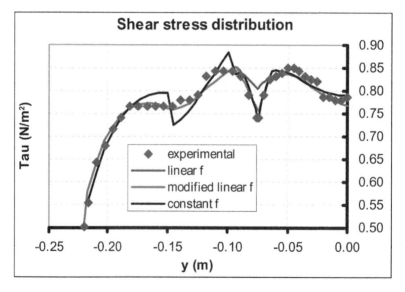

Figure 2.34 Results of boundary shear stress simulations of flow in a trapezoidal channel (6 panels, with variable λ, Exp. 16, Yuen), after Omran (2005).

this and related issues may be found in Omran (2005), Omran & Knight (2006), Knight *et al.* (2007) and Omran *et al.* (2008). Figures 2.31–2.34 thus indicate that the main issues that need to be considered are the number of panels and the values of each parameter within individual panels.

2.4 INBANK FLOW IN NATURAL RIVERS

The previous section has shown some of the background to modelling flows using SKM/CES approaches, applied to simple prismatic channels, and how useful some laboratory data might be for validation purposes. However, river engineers are usually concerned with flows in natural rivers, where the cross-section is generally irregular in shape, not necessarily prismatic, the distribution of roughness more varied around the wetted perimeter, and other practical issues such as vegetation, culverts and bridges need to be considered. This section and subsequent ones now address all of these concerns and illustrates them via several case studies.

Figure 2.23 shows that the flow in a natural channel is more complex than those cases considered in Section 2.3. Firstly the irregular cross section has to be schematised using survey data to create multiple panels, and should be undertaken with some knowledge of the flow characteristics. Secondly, because there is often a lack of knowledge concerning the velocity distributions from suitable field data, care needs to be taken over the choice of the 3 calibration parameters (f, λ and Γ) to be used in each panel. Arising from this, a measure of uncertainty in the predicted results also needs to be established. These features are now illustrated using the CES applied to an actual river, the River Ngunguru in New Zealand.

The example of the River Ngunguru at Drugmores Rock has been deliberately chosen to illustrate inbank flow in a natural channel that has many of the difficult features cited above. It will also serve to highlight further modelling issues, and is totally unlike any of the previous examples, based on simple prismatic laboratory channels. Furthermore, one particular reach has been subject to ongoing measurements since 1969 (Hicks & Mason, 1998). The Ngunguru River drains a catchment area of 12.5 km², has a mean annual flood discharge of 61 m³s⁻¹ and an average flow rate of 0.41 m³s⁻¹. Water level and discharge measurements were taken at three cross-sections along an 80 m reach, with a gentle bend in the downstream portion. The sections were approximately 12 m wide and 2 m deep. The water surface slope varied from 0.0037 at low depths to 0.0064 at large depths. The river bed consists of gravel and cobbles and the banks are lined with grazed grass and scattered brush, as shown in Figure 2.35. The observed Manning n values vary with depth in the range 0.051 to 0.160, with the values increasing substantially at lower depths. The cross section is shown in Figure 2.36.

The cross section was divided into 100 panels or elements and values of the 3 calibration parameters (f, λ and Γ) chosen for each panel, based on the rules outlined in Chapter 3. The CES software was then used at a specified depth to solve for the lateral distribution U_d v y, which was subsequently integrated laterally to give the discharge at that same value of depth. Repeated application over a range of depths led to the stage-discharge relationship (H v Q) shown in Figure 2.37. As can be seen in Figure 2.37, although the CES stage-discharge prediction and data agree well for the

(a) looking upstream   (b) looking downstream

Figure 2.35 River Ngunguru at Drugmores Rock looking upstream and downstream from the mid-reach section (Hicks & Mason, 1998).

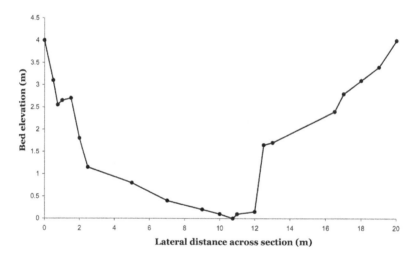

Figure 2.36 Surveyed cross-section geometry for the River Ngunguru at Drugmores Rock (Hicks & Mason, 1998).

majority of flows, there is some discrepancy at shallow depths. This is related to the physical representation of the large roughness in the river (Manning's $n = 0.1$ to 0.16), a topic which is discussed further in Chapter 3 when dealing with boulder roughness. Figure 2.38 shows the actual measured Manning n values, together with those back calculated from the CES. The back calculated values agree well with the data for depths in excess of 0.9 m. This Figure also demonstrates how Manning n often varies with depth in rivers, decreasing from a high value at low depths. This is a key feature, that indicates that roughness coefficients (n, C, f or k_s) should always be plotted in a similar manner and checked when calibrating a numerical model. Other examples are given by many authors (e.g. Chow, 1959; Hicks & Mason, 1998; Knight, 1981; and

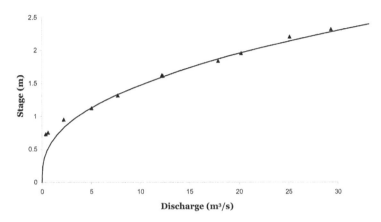

Figure 2.37 CES stage-discharge prediction and data for the River Ngunguru at Drugmores Rock (Hicks & Mason, 1998; Mc Gahey, 2006).

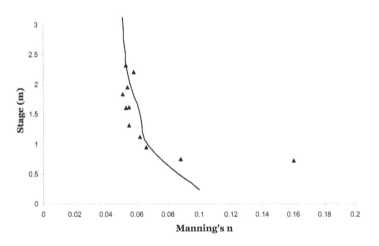

Figure 2.38 Back-calculated CES Manning *n* values and measured Manning *n* values from the River Ngunguru at Drugmores Rock (Hicks & Mason, 1998; Mc Gahey 2006).

Yen, 1991). Following consideration of resistance in overbank flow (see for example Fig. 2.47), roughness is considered further in Secions 3.2.1 & 3.4.1.

2.5 OVERBANK FLOW IN NATURAL AND ENGINEERED RIVERS

Overbank flow is somewhat different in a number of respects from inbank flow. The key difference is that more complex three-dimensional (3-D) flow structures may be present (Ashworth *et al.*, 1996). When overbank flow occurs, and floods occupy the floodplains, there are a number of issues that require special consideration, dealt with more fully by Knight & Shiono (1996). These are summarised here as follows:

- use of hydraulic radius, *R*, in calculations (abrupt change at bankfull stage);
- interaction between main river and floodplain flows (strong lateral shear);
- proportion of flow between sub-areas is important (e.g. river and floodplains);
- heterogeneous roughness (differences between river and floodplains);
- vegetation zone (effect on flow, plan form vortices);
- unusual resistance parameters (global, zonal and local);
- significant variation of resistance parameters (with depth and flow regime);
- distribution of boundary shear stresses (affects sediment, mixing and erosion);
- sediment transport (rate, equilibrium shape, deposition, etc.);
- flood routing parameters (wave speed and attenuation parameters are affected);
- critical flow (definition and control points);
- valley and channel slopes and sinuosities (may be different for meandering channels);
- greater plan form variations in geometry (due to meandering patterns).

Many hydraulic equations, such as Eq. (2.1), were developed primarily for inbank flows and are based on a mean velocity, *U*, applied to the whole of the cross-section area, *A*. Manning's equation is therefore classed as a single channel method (SCM). Other methods divide the area into zones in which the roughness and flow might vary within the cross-section, giving rise to the so-called divided channel methods (DCM). There are several categories of DCMs, as shown by many authors, e.g. Huthoff *et al.* (2008); Knight & Demetriou (1983); Knight (2006b&c); Lambert & Myers (1978); Myers & Lyness (1997); Wormleaton (1988 & 1996); Wormleaton & Merrett (1990); Wormleaton *et al.* (1982, 1985) and Yen (1991). A variant on this theme is the 'coherence method' (COHM) of Ackers (1991, 1992a&b & 1993a&b), which deals with heterogeneous roughness and shape effects. The 'coherence', COH, is defined as the ratio of the basic conveyance, calculated by treating the channel as a single unit with perimeter weighting of the friction factor, to that calculated by summing the basic conveyances of the separate zones. All of these simple sub-division methods, and the composite roughness methods given in Chow (1959), are now known to be inappropriate for simulating overbank flow accurately (Wormleaton & Merrett, 1990). The CES, on the other hand, was designed specifically with overbank flows in mind, as shown by the following examples.

2.5.1 Overbank flow in an engineered river

A good example of an engineered channel is that of the River Main, already shown as Figure 2.5, details of which are also given in Section 2.2.2. The main channel cross section is approximately trapezoidal, and so links in with the previous laboratory examples, and the floodplains are prismatic. Figure 2.39 shows the survey data at Bridge End bridge.

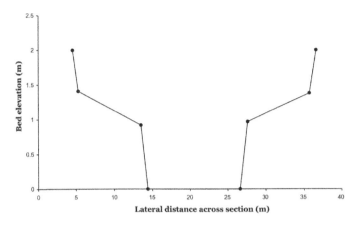

Figure 2.39 Surveyed cross-section geometry for the River Main at Bridge End bridge (Myers & Lyness, 1989).

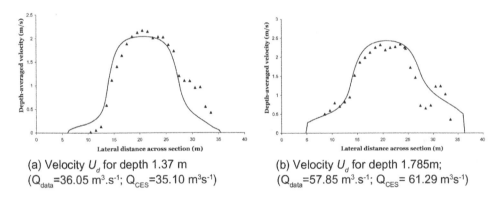

(a) Velocity U_d for depth 1.37 m
(Q_{data}=36.05 m³.s⁻¹; Q_{CES}=35.10 m³s⁻¹)

(b) Velocity U_d for depth 1.785m;
(Q_{data}=57.85 m³.s⁻¹; Q_{CES}= 61.29 m³s⁻¹)

Figure 2.40 Comparison of lateral distributions of velocity by CES with data (Myers & Lyness, 1989; Mc Gahey, 2006).

The CES was used to simulate the lateral distribution of U_d for a number of flow depths and discharges. Two of these are shown in Figure 2.40 for flow depths of 1.37 m and 1.79 m, together with the field data (Myers & Lyness, 1989). The depth-averaged velocity profiles are reasonably well simulated, but there appears to be a small degree of skew in the velocity data at low depths. This may result from the effects of the upstream bend circulations being transported downstream. Since the channel bed material is gravel, with no vegetation present, these higher velocities are plausible but are not captured by the CES where the effect of the slightly higher calibrated bank side roughness is present. Subsequent integration of these, and all the data sets, produced the predicted stage-discharge curve shown in Figure 2.41. The *H* v *Q* relationship is well simulated, including the bankfull variations around the depth of ~0.95 m. The differences between CES predicted and all data flow values are shown in Figure 2.42, giving an average difference of 5.6% and a standard deviation value

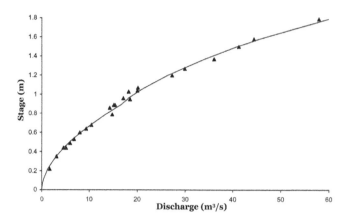

Figure 2.41 CES stage-discharge prediction and data for the River Main at Bridge End Bridge (Myers & Lyness, 1989; Mc Gahey, 2006).

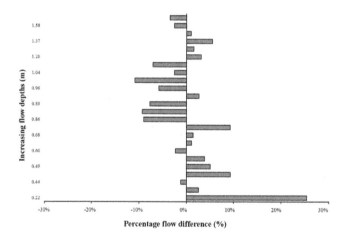

Figure 2.42 Differences between CES predicted and data flow values for the River Main at Bridge End Bridge (Mc Gahey, 2006).

of 7.6%. These differences obviously depend on the technical capability of the model, the accuracy of the measured data and the calibration parameters used in the numerical model, especially on the unit roughness which is the most significant parameter. These roughness values and equivalent roughness heights, k_s, are shown in Table 2.4, together with the Manning's n values suggested by previous authors. It is instructive to note that the various author estimates vary substantially. The large CES unit roughness, n_p (defined by Eq. (3.15) in Chapter 3) on the left floodplain is explained by the dense vegetation immediately below the bridge, where the measurements were taken. The CES back-calculated Manning's n values are shown in Figure 2.43, illustrating

Table 2.4 Calibrated roughness values for the River Main at Bridge End Bridge (Mc Gahey, 2006).

Roughness parameter	Left floodplain	Left Bank	Main channel	Right bank	Right floodplain
n_l	0.0750	0.0350	0.0300	0.0350	0.0450
k_s (m)	2.225	0.316	0.172	0.316	0.713
Author			Manning's *n* values		
Myers et al (1999) based on measurement	0.0380	0.0420	0.0380		
Ackers (1991)	0.0200	0.0300	0.0200		
Wark (1993)	0.0400	0.0278	0.0400		

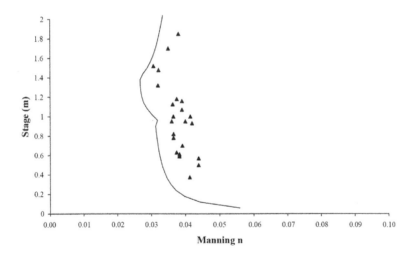

Figure 2.43 Back-calculated CES Manning n values and measured Manning n values from for the River Main at Bridge End Bridge (Myers & Lyness, 1989; Mc Gahey, 2006).

a reasonable distribution with depth, albeit ~15% lower than the measured data. Further details of this simulation are given in Mc Gahey (2006).

This example serves to show how sensitive the solution is to roughness, even apart from other parameters, and how compromises sometimes have to be made with regard to fitting both individual data sets (such as U_d v y) and overall relationships (H v Q or H v n). It is for these reasons that an Uncertainty Estimator (UE) has been incorporated into the CES to assist modellers at the calibration stage. An example showing the sensitivity to roughness is given after the next example which illustrates some other important technical issues associated with overbank flow.

Figure 2.44 The River Severn at Montford Bridge (a) looking upstream from right bank, (b) looking downstream from the cableway and (c) the cable way at the bridge (Courtesy HR Wallingford).

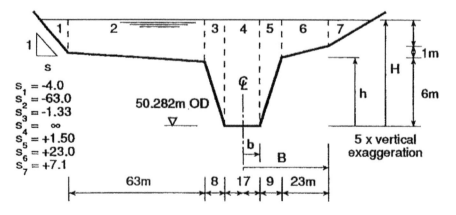

Figure 2.45 Cross-section of River Severn at Montford Bridge (after Knight et al., 1989).

2.5.2 Overbank flow in a natural river

A reach of the River Severn at Montford Bridge (Figure 2.44) has been extensively monitored for practical hydrometry and research purposes, providing a large body of accurate current metering data. It is a natural cross-section with a cableway extending

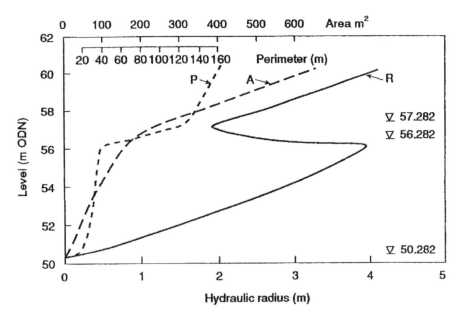

Figure 2.46 Variation of hydraulic parameters with level, River Severn at Montford Bridge (after Knight *et al.*, 1989).

over the full width, including the floodplains. The bankfull width and depth are 40 m and 6.3 m respectively. The total width, including the floodplains, is approximately 120 m and the reach-averaged longitudinal bed slope is 0.000195. The floodplains are grass-covered and the gauged station is on a straight section of the river. The measured data includes water surface slope, water levels, velocity and discharge.

The cross section, shown schematically in Figure 2.45, indicates a bankfull depth of around 6.0 m and two flood plains of differing widths. This section was modelled by Knight *et al.* (1989) prior to the development of the CES, and undertaken analytically using the 7 panels shown to represent the geometry and the secondary flow term ignored. It was later modelled numerically by Abril & Knight (2004) with all the terms included. The variation of area, wetted perimeter and hydraulic radius with depth are shown in Figure 2.46, which serves to illustrate a key problem when dealing with 1-D modelling of overbank flow and the use of standard 1-D resistance laws based on the hydraulic radius, R. When the flow rises just above the bankfull level (56.282 m OD), the hydraulic radius ($R = A/P$) changes markedly, since for a small change in depth there is little change in A but a significant increase in P. As a result, R decreases sharply, unlike inbank flow in which R generally increases with depth monotonically. This can have certain implications when applying resistance laws to sub-regions of the flow, as indicated in Figure 2.47.

Figure 2.47 shows that if the river is treated as a single channel, then the overall or composite n value deceases with increasing depth from 0.036 to 0.030 at the bankfull level, as might be expected from the trends shown in Figures 2.38 & 2.43 for inbank flows. However, with further increase in water level above the bankfull level,

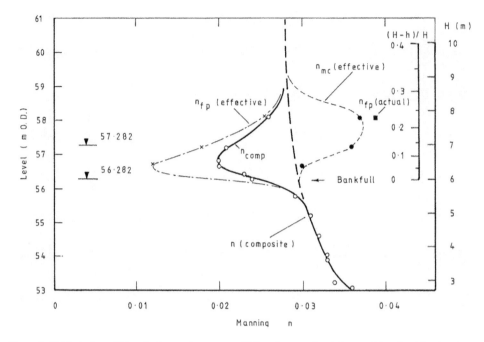

Figure 2.47 Variation of Manning's *n* resistance coefficient for overbank flow at Montford Bridge, River Severn (after Knight *et al.* 1989).

n_{comp} appears to decrease sharply to around 0.020, before increasing again to join the characteristic smooth *H* v *n* relationship at *n* ~ 0.029. This apparent reduction in overall resistance coefficient is entirely fictitious, and due to the changes in *R* noted in Figure 2.46. In fact the overall resistance of the river actually increases when overbank flow occurs, due to the effects of bankside and floodplain vegetation. Using measured water surface slope and velocity data, Knight *et al.* (1989) showed that when the channel was divided into 3 zones, the LH floodplain (panels 1 & 2), the main channel region (panels 3–5) and the RH floodplain (panels 6 & 7), then the effective *n* value for the main channel n_{mc} (effective), based on the main channel area, appeared to increase from 0.030 to around 0.038, and the effective *n* value for the floodplains n_{fp} (effective), based on the inundated floodplain areas, appeared to decrease from 0.030 to 0.012. This is despite the actual floodplain roughness being estimated as ~0.04, as shown in Figure 2.47.

The physical explanation for this seeming anomaly is straightforward. By dividing the channel into separate zones, typically using vertical lines that form panel boundaries, then using measured U_{mc}, A_{mc} and S_f values, the resulting zonal Q_{mc} takes no account of the retarding effects of the shear layers between the slower moving floodplain flows and the faster moving main channel flow. Consequently, to match the observed Q_{mc} a larger 'resistance' is apparently needed. The reverse is true of the zonal floodplain flow, Q_{fp}, where it appears that an impossibly low resistance value is required to account for the accelerating effects of the main channel flow on the floodplain flow (0.012 compared with the actual value of ~0.040). This highlights one of the main difficulties that all the divided channel methods (DCMs), mentioned in the introduction, have had to deal with. Many of them attempted to insert the apparent shear forces between zones or panels, based on arguments about transverse shear and secondary flows. The CES approach considers this issue from first principles and is based on extensive experimental data and observations.

When using resistance coefficients, care needs to be taken to distinguish between the section-mean velocity, U_A, the zonal velocity U_z, the depth-mean velocity, U_d, and any local near bed velocity, u, used in the 'law of the wall' in a turbulence model. It is also important to distinguish between 'global', 'zonal' and 'local' friction factors used in 1-D, 2-D & 3-D river models, where:

$$\tau_o = \left(\frac{f}{8}\right)\rho U_A^2; \quad \tau_z = \left(\frac{f_z}{8}\right)\rho U_z^2; \quad \tau_b = \left(\frac{f_b}{8}\right)\rho U_d^2; \quad \tau_b = \left(\frac{f_t}{8}\right)\rho u^2$$

(global)   (zonal/sub-area)   (local/depth-averaged)   (local/turbulence)

1-D model   quasi 1-D model   2-D model   3-D model

$$(2.61)$$

It should be noted here that, within the CES, the third option is used, with the shear stress on the bed assumed to be in the same streamwise direction as U_d. The topic of how resistance is represented in numerical models is addressed more fully by Morvan *et al.* (2008). Some of the errors involved in using DCMs are highlighted by Knight & Demetriou (1983), Knight (2006b&c), Lambert & Myers (1998), Myers (1978) and Wormleaton (1996).

The same reach of the River Severn, as well as other rivers, were modelled by Abril & Knight (2004), who also put forward a novel link between boundary shear and the secondary flow parameter, Γ. As a result, certain rules were formulated for the variation of the 3 calibration parameters (f, λ and Γ) in different zones (main channel and

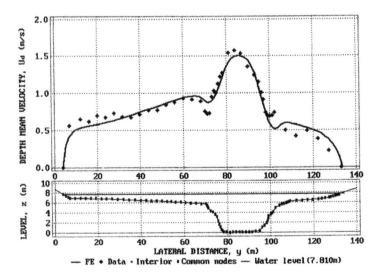

Figure 2.48 Finite element simulation of the lateral distribution of depth-averaged velocity in the River Severn at Montford Bridge (after Abril & Knight, 2004).

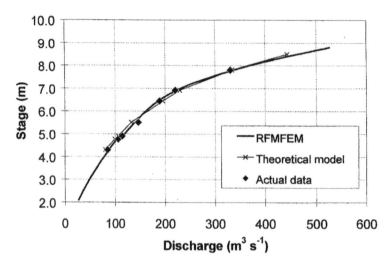

Figure 2.49 Stage-discharge prediction for the River Severn at Montford Bridge (after Abril & Knight, 2004).

Table 2.5 Percentage change in flow rate as a result of change in unit roughness for the River Main and River Severn at Montford Bridge (adapted from Mc Gahey, 2006).

Percentage change in n_l	Depth (m)	Percentage change in flow rate (%)						
		−50%	−30%	−10%	0%	+10%	+30%	+50%
River Main	1.785	+92	+40	+10	0	−8	−21	−31
River Severn	7.805	+69	+30	+8	0	−6	−16	−23

floodplains). These are described and utilized in Chapters 3 & 4 in relation to several other rivers. The results of one simulation of the River Severn at Montford bridge at a depth of 7.81 m are shown in Figure 2.48. Lateral integration of these U_d values with the corresponding elemental areas gave the total discharge for this particular depth. Repeated simulations over a range of depths gave the complete stage-discharge relationship, shown in Figure 2.49. The H v Q relationship may also be obtained analytically for steady inbank and overbank flow for certain shapes of open channel, as shown by Liao & Knight (2007a&b). See also the relevant International Standards, ISO 1100-1 (1996) and ISO 1100-2 (1998). Unsteady flows require special treatment, as indicated by Schmidt & Yen (2008), and also by Eq. (2.38).

Since the outputs from any model are especially sensitive to the input roughness values used, two examples are now given to illustrate the sensitivity of discharge to roughness value. These are based on the River Main and River Severn, which have been illustrated already in Figure 2.41 (River Main) and Figure 2.49 (River Severn). Table 2.45 illustrates the percentage change in flow rate as a result of changes in n_l of ±10, 30 and 50% at the two particular depths, 1.785 and 7.805 m, used previously. The variation in discharge is seen to be from +92% to −31% for the River Main and from +89% to −23% for the River Severn. This highlights the importance of estimating the roughness as accurately as possible when modelling flows in rivers. As noted from Eq. (2.61), the resistance coefficients have to be carefully defined for overbank flows, and may differ considerably depending upon whether global, zonal or local situations are being considered. For further details on resistance of compound channels see Myers & Brennan (1990) and Myers *et al.* (1999).

2.6   FLOWS THROUGH BRIDGES AND CULVERTS

In addition to the energy losses caused by channel resistance, energy losses also occur at river control structures such as bridges, culverts and weirs. Since these may affect water levels in any simulation model, the Afflux Estimator (AE) is included within the CES, together with a backwater module. For detailed information on flow through bridges, culverts, flumes and weirs, reference should be made to papers, textbooks, manuals and international standards, such as Ackers *et al.* (1978), Benn & Bonner (2001), Biery & Delleur (1962), Biglari & Sturm (1998), Bos *et al.* (1984), Brown (1985 & 1988), Chow (1959), Hamill (1999), Henderson (1966), Katz & James (1997), Knight & Samuels (1999), Mantz & Benn (2009), May *et al.* (2002), Ramsbottom *et al.* (1997), Seckin *et al.* (1998, 2007 & 2008a&b), ISO 4374 (1990), ISO 4377 (2002) and ISO 1438 (2008). The AE concentrates particularly on flows through bridges and culverts, since these are often more difficult to model accurately than flows through weirs and flumes and they are frequently encountered.

2.6.1   Flows through bridges and contractions

The flow of water through bridges and contractions is illustrated in Figure 2.50 & 2.51. The cross-sections 1.4 have been positioned at those places that are frequently adopted in the analysis of flow through constricted waterways. Section 1 is traditionally taken upstream of the bridge, prior to the commencement of any contraction of the streamlines due to the bridge. Section 2 is generally taken on the upstream face of the bridge, with Section 3 either on the downstream face or at the position of the *vena contracta*. Section 4 is taken some distance further downstream, typically at the end of the expansion of the streamlines, where the flow returns to normal depth conditions in the river channel. Note the differences between afflux, head loss and energy loss. The afflux is defined as the increase in water level upstream of a bridge above the water depth that would occur in the absence of the bridge, here indicated by the normal depth at section 1, and should not be confused with the energy loss across the whole bridge structure, i.e. between sections 1 & 4. See Mantz & Benn (2009) for further details of computation of afflux ratings and water surface profiles through bridges and culverts.

Figure 2.52 shows a typical variation in water surface profile through a laboratory scale model bridge, taken from Atabay & Knight (2002). Figures 2.53 & 2.54 illustrate how such experiments were conducted in a compound channel at the University of Birmingham, and Figures 2.55–2.58 illustrate some actual flows through bridges with piers and arches of various shapes. Figures 2.18 & 2.19 have already shown some actual bridges during or after a flood event.

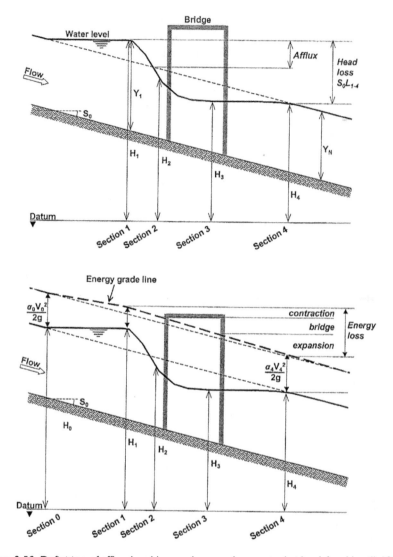

Figure 2.50 Definition of afflux, head loss and energy losses at a bridge (after Hamill, 1999).

There are a number of different methods for determining afflux, the main ones being:

- Energy equation method (various authors);
- Momentum equation method (various authors);
- USGS method;
- USBPR method;
- FHWA WSPRO method;
- HR method (arch bridges);

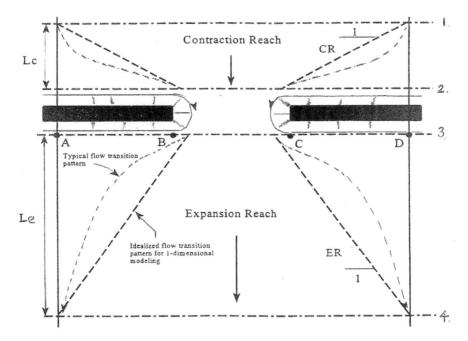

Figure 2.51 Plan view of flow though a bridge or constriction (as in HEC-RAS R&D, 1995, but with reversed notation to make it compatible with standard practice and the USBPR method).

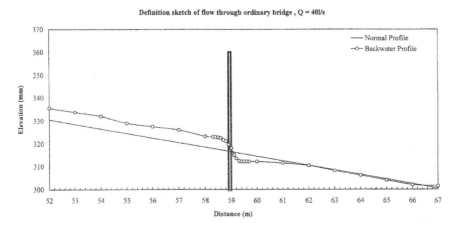

Figure 2.52 Observed water surface profile through a model bridge in the laboratory (Atabay & Knight, 2002).

- Biery & Delleur method (arch bridges);
- Pier loss methods – Yarnell, Nagler and d'Aubuisson.

Although it is not appropriate to reproduce details of these methods here, since they are well documented elsewhere (Defra/EA, 2004), it should be appreciated that

Figure 2.53 Flume with model bridge set in a compound channel with smooth floodplains (University of Bimingham).

there are several conceptual differences between them, so that the affluxes computed by the various methods will differ according to the method used, the coefficients adopted, and the expertise of the user in applying the method to both standard and non-standard situations. In the same way, methods for analyzing flows through culverts have also been well documented, see for example, Bradley (1978), Chow (1959), CIWEM (1989), Hamill (1999), HEC-RAS (1996 & 2008), Henderson (1966), Miller (1994) and Ramsbottom *et al.* (1997).

Figure 2.59 illustrates schematically the various types of flow that may occur through a bridge, and how a distinction needs to be drawn between free, submerged and drowned flow. At low discharges, with the channel controlling the flow, the flow is referred to as an open channel type flow, as shown for a subcritical state in Figure 2.59(a). When the discharge and upstream water level reach a value at which submergence of the upstream face occurs, then the hydraulic behaviour switches from a free flow state to a sluice gate-like flow state, as shown in Figure 2.59(b).

If re-attachment occurs on the underside of the bridge soffit, then the contraction effect within the bridge opening will be affected, typically causing the whole area at the outlet to become a flow area and thereby changing the longitudinal

Figure 2.54 Model double arch bridge in a flume with roughened floodplains (University of Birmingham).

Figure 2.55 Flow through a model bridge with multiple piers.

pressure gradient, as shown in Figure 2.59(c), Type 2. The flow is now referred to as a drowned orifice type flow, with both the structure and the channel controlling the flow. Depending on the tailwater level, the downstream face of the structure might also become submerged, as shown in Figure 2.59(c), Type 1. At very large discharges,

Figure 2.56 Flow through a model bridge with single elliptical arch.

Figure 2.57 Flow through a model double arch bridge (viewed from upstream).

some flow may also occur over the bridge parapet or roadway, leading to overtopping or bypassing of the structure. In this case the flow is part orifice-like, through the bridge opening, and part weir-like, over the bridge parapet and any adjoining embankments. The simulation of these different types of flow during a flood event must, therefore, be reproduced in any numerical model, using different algorithms, and the transition from one type of flow to another clearly linked to key water levels relative to the structure. Another way of categorising flows through bridges, different to that shown in Figure 2.59, is given in the USBPR method, based on the work of Bradley (1978). Since this is one of the most widely used methods for determining the afflux through a constricted waterway, this is summarised as follows:

Type I   subcritical flow throughout a constriction;
Type II A   critical at throat only, i.e. choking at throat;
Type II B   critical for a short distance downstream, then a jump and subcritical flow further downstream;
Type III   choking at throat, then supercritical flow downstream.

Should supercritical flow occur, then bridges have to be treated particularly carefully as the depth of flow will increase in the constriction (i.e. the opposite to that

Figure 2.58 Flow through a model double arch bridge (viewed from downstream).

shown in Figure 2.59(a)). Under these circumstances the water level may reach the soffit, and a hydraulic jump forms on the upstream side of the bridge, thereby creating a new control point. Where bridges are sited on bends with supercritical flow, then splitter walls may be required to reduce the super-elevation so that the water surface does not reach the level of the bridge soffit. Since bridges are often sited in two-stage or compound channels, with the embankments crossing part or whole of the flood-plain, then particular care needs to be taken over the definition of sub and supercriti-cal flow in such channels. It is sometimes possible for rivers with floodplains to have sub and supercritical flow co-existing within the same cross section, and therefore the Froude Number, Fr, as defined by Eq. (2.5), is no longer appropriate for defining critical flow in such channels with a complex geometry. Further information on this is given by Blalock & Sturm (1981) and Yuen & Knight (1990).

Since flow conditions upstream of bridges are sometimes not dissimilar from the flow conditions upstream of other types of hydraulic structure (e.g. vertical sluice gates, Tainter gates and other similar structures), then technical knowledge concerning the drowning characteristics of these other types of structure may also be relevant (e.g. Escaramia *et al.* (1993), Knight & Samuels (1999) and Rajaratnam (1997)). In the same manner, technical information on contraction and expansion losses in other types of fluid flow are also relevant and should not be overlooked (e.g. Cebeci & Bradshaw (1977), Chow (1959), Henderson (1966), Reynolds (1974) and Schlichting (1979)).

The flow through a constriction may be dominated by 'choking', a phenomenon that may be readily explained by reference to specific energy, E, as defined in Eq. (2.12). Since the specific energy, depth and discharge per unit width relationship is theoreti-cally fixed for a particular shape of cross section, the flow within a constriction may become critical if the gap width is too small. Further reduction in the gap width will only maintain the hydraulic control at that point, but cause the water level upstream to rise. Standard hydraulic theory applied to flow in a rectangular channel of breadth B, in which a single pier of width b is placed, will lead to the following equation for choking

$$\sigma^2 = \frac{27 Fr_1^2}{\gamma^3 (2 + Fr_1^2)^3} = M_L^2 \qquad (2.62)$$

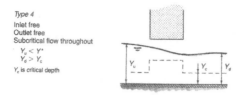

(a) Open channel type flow - channel control

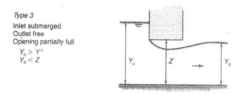

(b) Sluice gate type flow - structure control

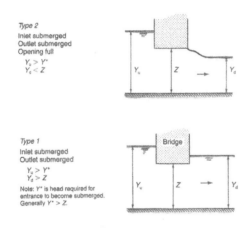

(c) Drowned orifice type flow - structure/channel control

Figure 2.59 Types of flow through a bridge crossing (after Hamill, 1999).

where $\sigma = (B - b)/B$, Fr_1 is the upstream Froude Number and γ is a factor to account for any energy loss between the upstream section (1) and the throat (2), such that $\gamma E_1 = E_2$, where $\gamma < 1$ and E refers to the specific energy. It should be noted that despite the use of b as defined here for the pier width, and not the gap width, σ becomes the same as the bridge opening ratio, M. The bridge opening ratio, M, is defined as a ratio of discharges, $M = q/Q$, where here q = unfettered flow through the gap as if the bridge or constriction was not in place and Q = total flow in the river. Thus for known B, Fr, and γ, the value of σ, and hence the width of the bridge pier, b, or limiting opening ratio, M_L, that will cause choking may be determined. Eq. (2.62)

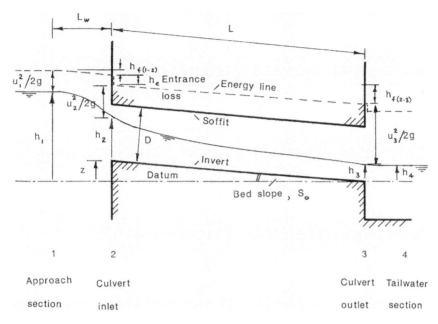

Figure 2.60 General characteristics of flow through a culvert, with notation.

is similar to that presented by Hamill (1999), but with the energy loss term included. Many alternative equations may be produced, based on similar principles (e.g. Wu & Guo, 2002; Wu & Molinas, 2001), but these are more complex to use than Eq. (2.62).

2.6.2   Flows through culverts

The flow of water through a culvert is somewhat similar to that under a bridge. It depends on whether the hydraulic control is at the inlet or outlet and also on upstream and downstream water levels. The general features of flow through a culvert are shown in Figure 2.60, together with notation. It should be noted that bridges are defined as structures with a small length to the opening size ratio ($L/D < 1$), whereas culverts are structures with typically much larger L/D values. The various types of flow through culverts are shown in Figures 2.61 & 2.62.

The head-discharge, and hence afflux, relationships depend on the type of flow. These are dealt with in Section 2.6.3 and the various geometrical parameters that affect the flow are dealt with in Section 2.6.4. Further details are reserved until later Chapters, where worked examples are also given, along with a description of the AES.

2.6.3   Head-discharge and afflux relationships

The head-discharge relationships for all the types of flow, identified in Figures 2.59–2.62 for flow through bridges and culverts, are inevitably complex (see Chapters 3 & 4). The afflux relationships, together with worked examples, are likewise given later.

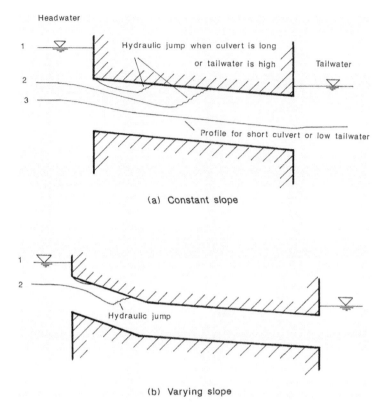

Figure 2.61 Several modes of flow in a partially filled culvert (Type 6 flow, as in Figure 2.62).

In general terms it may be noted that when a bridge is flowing 'free', the discharge – head relationship is given by $Q \sim \Delta h^2$, where Δh = water level difference between upstream and downstream sections, but when flowing like an orifice, it changes to $Q \sim \Delta h^{1/2}$. The transition from one type of flow to another is an important feature to model correctly and governed by many factors, further details of which may be found in Ramsbottom et al. (1997) and Hamill (1999). In order to understand the impact that certain geometric parameters have on the flow through bridges and contractions, the following section describes their influence and the related issues that need to be considered by the modeller.

2.6.4 Geometrical parameters affecting flow through bridges

The principal parameters that govern the flow of water through bridges and constrictions are illustrated in Figure 2.63. These parameters have traditionally been grouped under two major headings, geometric and flow, although as will be shown later, this distinction is somewhat blurred. They are, however, presented and discussed briefly here under a number of different headings.

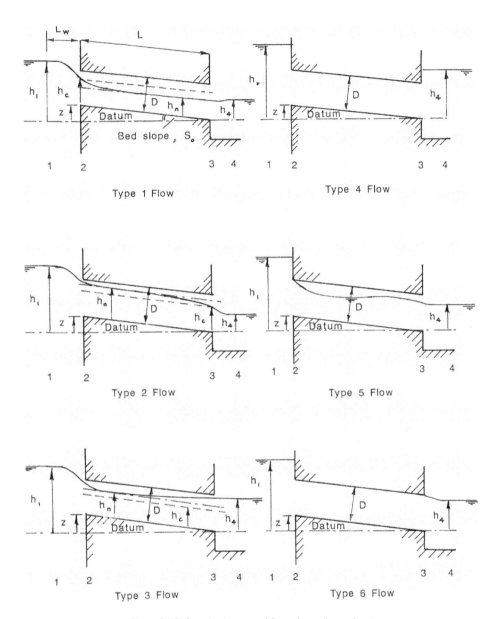

Figure 2.62 Standard types of flow through a culvert.

2.6.4.1   Bridge opening ratio, M

The bridge opening ratio, M, is defined as a ratio of discharges, $M = q/Q$, where q = unfettered flow through the gap as if the bridge or constriction was not in place and Q = total flow in the river. Thus $M = 1.0$ when there is no effect arising from a constriction, and $M < 1.0$ when there is an effect, see Figure 2.63(a). The bridge opening ratio

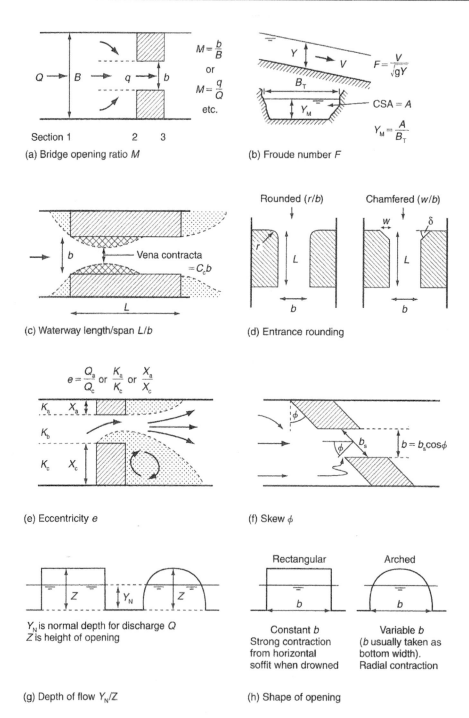

Figure 2.63 Principal hydraulic variables affecting flow through bridges (after Hamill, 1999). (a) bridge opening ratio, M; (b) Froude Number, Fr; (c) waterway length span, L/b; (d) entrance rounding; (e) eccentricity, e; (f) skew, ϕ; (g) depth of flow, h/Z; (h) shape of opening.

should be distinguished from the blockage ratio, m, commonly defined as $(1 - q/Q)$, giving $m = 0$ when there is no blockage, $m = 1.0$ for full blockage. It should be noted here that in evaluating q, the depth is usually taken to be the normal depth in the river, h_n, and the width, b, is associated with that particular portion of the river occupied by the bridge opening. The value of M is, therefore, strictly dependent on the flow distribution within the river channel, which in turn is generally influenced by the geometric shape of the river channel, the roughness distribution and the upstream flow conditions. Thus the proportion of total flow occurring in a given width, located at an arbitrary lateral position within a river cross-section, is governed by the 3-D nature of the flow, as illustrated earlier in this Chapter. Technically M should therefore be regarded as a flow parameter, and not a purely geometric one.

Most 'standard' textbooks dealing with the issue of afflux do not consider the 3-D nature of the flow, and assume that in a simple rectangular channel, M is simply equivalent to the width ratio, b/B, (i.e. $M = b/B$), where b = waterway opening and B = total channel width. It should be noted that even for a rectangular channel, this is never strictly true because of the boundary layers that develop from both sides and the bed of the channel, as shown in Figure 2.25. The value of M would vary from a minimum when the opening b is located near the walls, to a maximum when b is located at the channel centreline. The numerical values of M would also vary with depth (i.e. discharge), since as the aspect ratio of the channel changes from a large value at low depths to a lower value at high depths (discharges), the proportion of flow in any sub-region or zone would also vary. In the case of a compound channel M might correspond to the percentage of the total flow occurring in the main river channel, $\%Q_{mc}$, i.e. excluding the flow on the floodplains, which again has to be known.

2.6.4.2 Eccentricity, e

The alignment of the flow with the bridge/channel boundary is also important, as shown in Figure 2.63(e). Some consideration must be given to the macro flow field, as well the local flow field immediately adjacent to the bridge or culvert. The effect of the macro flow field is traditionally simulated somewhat crudely by two parameters, the eccentricity, e, and the skewness, ϕ, shown in Figures 2.63(e)&(f) respectively. The very local or micro flow field is traditionally simulated by a rounding parameter, r, or chamfering parameter, w, shown in Figure 2.63(d).

To some extent the eccentricity parameter, e, which equals the ratio of the bridge abutment lengths intruding from either side of the channel, X_a/X_c, is one traditional attempt at accounting for some of the 3-D effects of flow distribution on the coefficient of discharge. The eccentricity, e, was originally regarded as a purely geometric parameter (e.g. Kindsvater et al., 1953), but later assumed by some (e.g. Matthai, 1967) to be equal to the ratio of flow discharge occurring either side of the bridge opening, Q_d/Q_c, in direct proportion to the specified lengths, in a somewhat similar way to the parameter M. However, from the comments made earlier, it should be clear that the link between discharge and length is not a simple linear one.

Experimental work in rectangular shaped channels by Kindsvater & Carter (1955), Matthai (1967) and Bradley (1978) indicate that when $e > 0.12 - 0.2$, or when one embankment length is less than 5 to 8 times the other, there is little reduction in discharge performance. For $e = 0$, the USGS method states that the discharge reduction is only around 5%. However, it should be noted that the eccentricity parameter, e, does not strictly deal with all possible geometries and flow conditions. For example, it is possible to have $X_a = 0$ in Figure 2.63(e), but different values of X_c, all giving $e = 0$. In the case of $X_a = 0$, i.e. when there is no embankment on one side of the channel, the *vena contracta* effect would be suppressed on that side, leading to a smaller contraction laterally and therefore an increase in discharge, provided X_c is not too large. However, when $X_a = 0$ and X_c is large, or even for some very small values of e, the flow from one side towards the gap would cause a very large contraction, reduce the discharge capacity, increase the afflux, and might even be sufficient to produce supercritical flow locally. It would therefore seem sensible to redefine the eccentricity in such a way that introduces the ratio of gap width to embankment length, b/X_c or b/X_a in addition to the ratio X_a/X_c. It should be noted that e is usually defined in such a way that it never exceeds unity, the longer embankment length always being placed in the denominator.

In the case of flow in prismatic channels with two floodplains, the link between proportionate discharge and floodplain width will be especially complex, due to the lateral exchange of momentum between the main river channel and its associated floodplains. At flow constrictions, where typically embankments extend across the full width of both floodplains, the momentum exchange is further complicated by the flow spilling off the floodplain upstream of the bridge and onto the floodplain downstream of the bridge. These flow exchanges, which also occur in compound channels with converging or diverging floodplains, will directly affect the momentum – force balance and introduce additional energy losses, as shown by Bousmar (2002), Rezaei (2006), Chlebek (2009), Chlebek & Knight (2008) and Sellin (1995). The theoretical determination of e, based on flow proportioning, is therefore especially difficult for overbank flow in natural channels with floodplains.

2.6.4.3   Skewness, ϕ

For cases where the bridge crossing is not normal to the predominant flow direction, an angle of skew, ϕ, is traditionally introduced, as shown in Figure 2.63(f), where $\phi = 0°$ when the embankments are perpendicular to the flow. As illustrated in Figure 2.64, the gap width normal to the mean direction of river flow is then given by $b = b_s \cos \phi$, where b_s is the span width of the opening between the abutments of a skewed bridge, as measured along the highway centreline. It should be noted that the width b will have an impact on the determination of q (and hence M), and again highlights the confusion between geometric and flow parameters in traditional approaches. The alignments of the ends of the abutments also become important, as illustrated in Figure 2.64 for skewed opening Types 1, 2 & 3.

In the case of a skewed opening of type 1A, the water level will be higher on the left-hand side of the channel (looking downstream), since a dead zone will develop there. The lower velocities in that region will give a smaller velocity head term, causing the water surface to be closer to the energy line or energy surface in this case. A second dead zone, larger than the upstream one, will occur on the right-hand side of the channel immediately downstream of the right-hand embankment. Since the water level in the upstream dead zone on the left-hand side will generally be higher than the average water level in the main river channel upstream, due to

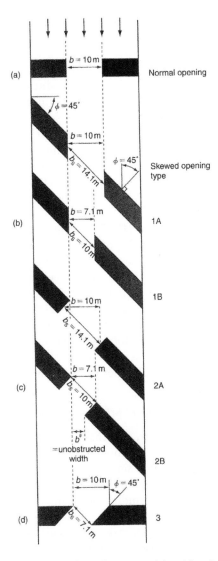

Figure 2.64 Comparison of normal and skewed crossing (after Hamill, 1999). (a) normal crossing; (b) type 1 skew with abutments parallel to the flow; (c) type 2 skew with abutments at 90? to approach abutments; (d) type 3 skew with abutments at an angle ϕ to both flow and approach abutments.

the recovery of the velocity head, it may, therefore, have consequences for the onset of submergence, as this generally occurs on the side that has the higher upstream water level.

The hydraulic efficiency of the waterway crossing is affected by skewness in a number of ways. Although the span width, b_s, increases with ϕ, the flow now has to turn through an angle in order to enter the gap. Furthermore, the orientation of the ends of the abutments may cause flow separation resulting in additional eddying, as well as contraction effects downstream, if the alignments of the end faces are unsuitable. See Figure 2.64, in which $b = 10$ m for the normal opening ($\phi = 0°$), as well as in 1A and 2A, giving $b_s = 14.1$ m for $\phi = 45°$ in both 1A and 2A. The flow through skewed openings is, therefore, complex. In 1A and 2A the afflux will be less relative to the afflux through the normal opening, but not in direct proportion to b_s/b, arising from the effects of changing the direction of the flow normal to the waterway opening, b_s, as well as the effects of separation due to possible differences in the alignment of the ends of the abutments. In 1B and 2B there will be a corresponding increase in afflux relative to the afflux through the normal opening, but again not in proportion to b/b_s. It should be remembered that it is not uncommon for a river course to change its direction during a flood, and therefore the alignment of the bridge abutments and mean skew angle of the river flow to the embankments will vary and will need to be considered as well.

Experimental results by Matthai (1967) indicate that the effect of skewness is most significant when ϕ is around 45°, producing a possible 30% reduction in discharge for the same upstream water level. Once a crossing is fully submerged, the influence of skewness is, however, reduced, as the opening then behaves more like an orifice, with the flow separating strongly from the underside of the bridge (soffit or deck) causing the flow to contract strongly in a vertical plane rather than the horizontal plane. In these circumstances the discharge capacity is then related to b_s, the width of the opening between the abutments, and is less dependent on ϕ. It thus appears that the 'effective discharge width' of a skewed crossing is complex, and care should be taken in evaluating the 3-D flow field under free or submerged conditions.

2.6.4.4   Rounding of piers, entrances to bridges or culverts

The introduction of a corner radius, r, or a chamfer length, w, into a waterway crossing is illustrated in Figure 2.63(d). It is known that the upstream rounding of piers and entrances to both bridges and culverts significantly influences the flow characteristics of such structures, by preventing or delaying flow separation at the changes in geometry. Small local changes, such as rounding or chamfering, will, therefore, generally reduce the contraction effect, increase the discharge capacity, and this increase is usually related to the parameter M. The increase is particularly large when the structure is operating under submerged conditions, i.e. orifice flow. The bridge opening ratio, M, is usually calculated for simple geometrical shapes, such as arch bridges, by assuming that the flow is wholly normal to the structure and that the flow area is identical to the geometric area. On this basis equations for the variation of M with depth for various types of arch bridge may be produced (e.g. Hamill, 1999).

The influence of r/b and w/b on the discharge coefficient, and hence the discharge capacity, is given by Kindsvater et al. (1953), Kindsvater & Carter (1955) and Hamill (1997). The USGS results indicate that the greatest increase in discharge capacity through entrance rounding will occur in narrow waterways (e.g. discharge increase of 20% for $M = 0.2$ & $r/b = 0.14$). As in all hydraulic flows, the streamlining of the macro and micro flow fields is important for enhancing discharge capacity and reducing energy losses. In the context of bridge crossings, streamlining is especially difficult to ensure throughout the full depth range, as the direction of the approaching flow may change significantly as the floodplains become inundated upstream of the bridge.

2.6.4.5 Length to breadth ratio, L/b

The length to breadth ratio, L/b, of a waterway crossing is shown in Figure 2.63(c), where L is the length between the upstream and downstream faces of the constriction, or waterway length, and b is the width of the bridge opening or a characteristic height dimension of the culvert. The length to breadth ratio, L/b, is important for distinguishing between bridges and culverts, and common practice dictates that a culvert is defined by having an integral invert and $L/b > 1$, so that the flow conditions become similar to those in a pipe with 6 main categories of flow (Chow, 1959; CIWEM, 1989; Ramsbottom *et al.*, 1997). The length is important in determining where re-attachment of the flow to the sides or soffit of a bridge takes place, which generally enhances the discharge capacity. The USGS results indicate that the increase in discharge is strongly related to L/b and M values and may be as high as 30%. A large culvert is taken here to mean $b > 1.5$ m (or an equivalent height or diameter), and provided $L/b < 1$ it will behave in a similar manner to a bridge.

2.6.4.6 Channel shape

For a single waterway opening, the cross-section shape of the river channel is important as it will affect the length of piers subjected to drag resistance, the choking behaviour, the kinetic energy correction coefficient values, as well as the determination of Q, the conveyance capacity, q, the zonal conveyance capacity corresponding to the width b without the bridge present, the normal and critical depths of flow, etc. Although these considerations are important, it should be remembered that in many cases, even without an integral invert, the shape of the waterway opening is often configured into a relatively simple shape by the structure. The customary convergence and acceleration of the flow towards a constriction generally helps to re-distribute the velocity and reduce the turbulence. Arch bridges will generally affect the flow more than those with rectangular openings, due to the separation and contraction effects produced by the upper surface once the water level is above the springing of the arch.

The traditional method for dealing with the channel shape and 3-D influences in standard 1-D open channel theory is to account for it by including kinetic energy and momentum correction coefficients. However, these are unlikely to be appropriate when dealing with complex flows through bridges. Since most of the experimental work on bridges and culverts has been conducted in channels of simple prismatic shape, generally rectangular or trapezoidal, there is a need to break away from this traditional approach and to investigate more realistic cross-sectional shapes. The most obvious one is to consider bridges in natural channels with floodplains. A considerable number of bridges in the UK fall into this category (e.g. see Pool bridge, River Wharfe, given as Figure 1 by Knight & Samuels, 1999), in which the general river shape is a compound channel and the embankments are perched higher, giving multiple openings at different invert levels. Apart from the experiments at the University of Birmingham (Atabay & Knight, 2002), there is a dearth of investigative work on bridge crossings in compound channels.

2.6.4.7   Multiple openings

For a waterway crossing with multiple openings, the division of flow has to be determined or assumed prior to any analysis. Where there are multiple openings with different invert and drowning characteristics, this will be problematic. As already commented upon, the answer lies in estimating the flow field, by using 2-D depth averaged models that give both the magnitude and the direction of the approaching flow, or by using full 3-D CFD modelling. An example of a single opening culvert with a square headwall with no rounding of the corners is shown in Figure 2.65. A multiple opening culvert, with inverts at different levels is shown in Figure 2.66. Note the tendency for silt to deposit and block the entrance to the right hand side opening.

2.6.4.8   Blockage

One remaining geometric issue concerning the shape of the opening is that of blockage. However this is conceptualised in any modelling process, the discharge characteristics and any additional head loss will be affected by the debris. Blockage may occur through floating debris trapped against the deck or soffit, trash against piers, ice accumulation, siltation or other types of trash, such as shopping trolleys, cars and even sometimes caravans. Modelling these by simply deceasing the waterway area is not always appropriate, and other methods should be attempted such as those that proportion the area to the floating debris (i.e. varying the proportion of the flow area that is blocked depending on the distance from the water surface downwards), those that include vertical or horizontal strips of blockage, those that consider random blocked area patches and those that mimic siltation by reducing the area from the invert upwards. Care needs to be taken in recognising the impact that these different geometric changes might have on the hydraulic behaviour and functioning of the structure. Some attention should also be paid to collecting evidence of where bridges have failed, due to debris accumulation, impact forces (overturning moments) and scour. There is very little experimental data on measured forces on bridges with which to check any momentum-force balance

Figure 2.65 Pre-cast concrete box culvert with one cast in-situ headwall (Ramsbottom *et al.*, 1997).

Figure 2.66 Cast in-situ twin box culvert with invert lower to carry normal flow (Note deposition on right side entrance).

approach, and what are available often relates to specific case studies, such as required in low cost bridge crossings in India or Australia (e.g. Roberts *et al.*, 1983).

2.7 DATA SOURCES USED IN THIS BOOK

2.7.1 Conveyance data

This Chapter began by emphasising the physical principles of fluid flow, theoretical concepts and how these ideas are encapsulated into mathematical models of river flow and the CES-AES in particular. The various case studies have shown the wide range of issues and some of the difficulties in developing, validating and calibrating any numerical model, particularly for overbank flow. As a result, empirical evidence from different sources was introduced, both from the field and laboratory. The laboratory data were found to play an essential part in elucidating flow mechanisms and turbulence parameters, albeit at small scale, that are virtually impossible to obtain at full scale at appropriate spatial and temporal scales on the grounds of cost and control. Figure 2.67 shows some views of the experimental arrangements used in the Phase A & B studies in the Flood Channel Facility (FCF). See Knight & Sellin (1987) and Knight & Shiono (1990 & discussion in 1991) for further details of these studies. The FCF was built at HR Wallingford for the Engineering and Physical Sciences Research Council (EPSRC) for joint collaborative work between 10 UK universities. These data, as well as many other data sets related to overbank flow studies and bridge flow may be found at *www.flowdata.bham.ac.uk*. Recent work by HR Wallingford is given at *www.river-conveyance.net* and should be consulted for further details.

The recent UK Foresight project, the recent European Commission (EC) Flood-site and Peseta projects on climate issues all indicate the strength of activity in flood research (see Table 2.6 for links to the various websites). The EC has in fact funded

Figure 2.67 FCF flumes showing Phase A with (a) smooth and (b) roughened floodplains and (c) the skewed channel layout and (d) Phase B with a sinuosity of 1.374 (Courtesy HR Wallingford).

over 100 research projects on flooding since the early 1980s, and these are itemised in a report by the University of Birmingham (see actif-ec.net website in Table 2.6).

2.7.2   Bridge afflux data

Table 2.7 summarises the most significant data sets for bridge afflux and their suitability (and availability) for use in development of the methods described in Chapter 3 of this book. Much of the early twentieth century data is inaccessible and often specific to certain types of bridge found in the USA, but not common in the UK.

After careful review, it was concluded (Mantz *et al.*, 2007, Lamb *et al.*, 2006) that the most relevant measurements for development of the AES were contained within two data sets, the HEC (1995) field scale numerical model results and the recent University of Birmingham laboratory measurements of Atabay and Knight (2002) and Seckin *et al.* (2007 & 2008a&b). Both data sets provide detailed water surface elevation data for some distance upstream and downstream of a bridge. Additionally, the USGS (1978) field survey data were used in testing the AES. These measurements are for wide (~100 to ~1000 m), rough, vegetated floodplains with low bed slopes, are very detailed and remain the most comprehensive field measurements available for bridges on wide vegetated floodplains.

Table 2.6 Links to useful flood studies websites.

Project	Link
CCRA	*www.hrwallingford.com/projects/CCRA* (Climate Change Risk Assessment)
CES	*www.river-conveyance.net* (Conveyance Estimation System)
CES	*www.hrwallingford.com/projects/CES* (Conveyance Estimation System)
CES/AES	*www.hrwallingford.com/software* (software for CES/AES)
Data	*http://www.flowdata.bham.ac.uk* (University of Birmingham data)
EU Directive	*http://ec.europa.eu/commission/index* (The EU Floods Directive)
Floods	*www.eprints.hrwallingford.co.uk/floods*
FLOODSITE	*www.floodsite.net*
FORESIGHT	*www.webarchive.nationalarchives.gov.uk/foresight/future flooding*
FRMRC	*gtr.rcuk.ac.uk* (Flood Risk Management Research Consortium)
InfoWorksRS	*www.innovyze.com/products/InfoWorksRS*
PESETA	*www.ec.europa.eu/jrc/climatechange* (Projection of Economic impacts of climate change in Sectors of the EU based on bottom-up Analysis—river floods)
Pitt review	*www.webarchive.nationalarchives.gov.uk/foresight/The Pitt Review*
RASP	*www.gov.uk/RASP* (Risk Assessment of flood and coastal defence systems for Strategic Planning)
RIBAMOD	*eprints.hrwallingford.com/projects* (River Basin Modelling/Floods)
SUDS	*www.hrwallingford.com/projects/newsudsmanual* (SuDS Manual)

The University of Birmingham data were measured in a laboratory flume, as illustrated in Figures 2.53–2.59. Subject to appropriate scaling they provide the detailed information needed for a rigorous analysis of afflux. A total of 325 afflux measurements were made for arch bridge types of single and multiple openings, and beam bridges with or without piers. An important feature is that the tests were made for a compound channel and for bridges at various skew angles. Multiple bridge and channel roughness configurations were tested in a fixed bed, compound channel, as follows:

- 15 tests with an elliptical arch bridge (designated ASOE) using three different roughness conditions;
- 20 tests with a semi-circular arch bridge (designated ASOSC) using four different roughness conditions;
- 15 tests with a twin semi-circular arch bridge (designated AMOSC) using three different roughness conditions;
- 45 tests with a single span beam bridge (designated BEAM) using three different roughness conditions and three different bridge spans;
- 50 tests with a beam bridge having three piers (designated PBEAM) using four different roughness conditions and three different bridge spans;
- 180 afflux measurements, were made for bridges located at an angle to the flow direction ('skew bridges').

In 1978–79 the USGS published detailed field data for bridge afflux analysis in the 'Hydrologic Investigations Atlas' series numbers HA-590 to HA-611. These surveys were conducted for 35 floods at 22 bridge sites in southern USA, and they have

Table 2.7 Sources of data for bridge afflux.

Source	Date	Details	Available	Notes
Rehbock	1922	Over 2000 laboratory experiments		
Nagler	1918	256 laboratory experiments on 34 different bridge models		
Yarnell	1934	2600 laboratory experiments with pier bridges		
Kindsvater, Carter and Tracy	1953	Laboratory data from the Georgia Institute of Technology		
Biery and Delleur	1962	Laboratory study on arched bridges		
Matthai	1967	Verified the Kindsvater et al. (1953) method with data from 30 field sites		
Bradley	1978	Laboratory experiments, Colorado State University	No	
USGS	1978, 1979	Hydrologic Investigations Atlases. Observed water surface for 35 flood events, 22 field sites	Paper maps.	1
HR Wallingford	1988	Laboratory study, arched bridges, 203 tests (see Brown, 1988)		
HR Wallingford	1985	Field data from bridges in the UK, 66 data sets (see Brown).	Yes, (partial records)	2
Hamill	1993	Field observations from Canns Mill bridge, Devon (see Hamill, 1999)	Yes	3
HEC	1995	Two dimensional numerical model results for 5 real and 76 idealised cases	Yes	1
Atabay and Knight	2002	University of Birmingham laboratory tests in compound channels, 145 measurements, bridges normal to flow direction		
Seckin, Knight, Atabay and Seckin	2004 2008	Extension of University of Birmingham lab tests in compound channels, 225 measurements, skewed bridges	Yes	4

Notes
 1 Digitised and used in development of AES.
 2 Archives reviewed for AES development but not suitable for use.
 3 Provided subjective checks for AES but not calibration data.
 4 Used in development of AES.

been described in detail by Kaatz and James (1997). Of the surveyed floods, 17 events at 13 bridge sites have previously been satisfactorily modelled using HEC-RAS (HEC, 1995) although some bridge sites were eliminated because of inadequate water level data near to the bridge. The modelling has been repeated more recently at JBA Consulting in development of the Afflux Estimation System using the HEC-RAS software. In the AES development, 15 events at 11 bridge sites could be satisfactorily modelled. The rivers and bridges from the USGS (1978) surveys that were used in development of AES are listed in Table 2.8 and their locations shown in Figure 2.68.

The US Army Corps of Engineers Hydrologic Engineering Center (USACE HEC) used data from three of the USGS Hydrologic Atlas field sites in an investigation of flow transitions in bridge backwater analysis (HEC, 1995) based on two dimensional (2-D) numerical modeling. To extend the scope of the study, HEC (1995) also created 2-D models of an idealized bridge crossing based on a uniform compound channel of 305 m width with 76 different combinations of opening width, discharge, slope and roughness.

Table 2.8 Bridges in the USGS (1978) Hydrologic Atlas (HA) series used with AES.

HA series no.	River	Bridge span (m)	Floodplain width (m)
591	Bogue Chitto	230	1360
596	Okatoma Creek	65	560
600	Alexander Creek	75	285
601	Beaver Creek	80	560
603	Cypress Creek	40	240
604	Flagon Bayou	70	480
606	Tenmile Creek	160	640
607	Buckhorn Creek	80	280
608	Pea Creek	78	340
609	Poley Creek	62	400
610	Yellow River	78	400

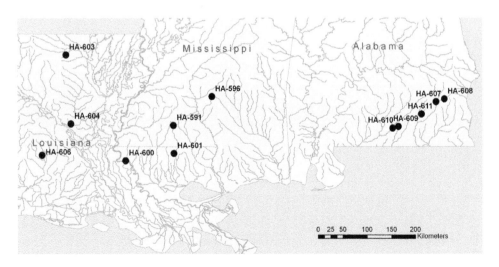

Figure 2.68 USGS Hydrologic Atlas (HA) field surveys used with the AES.

The HEC (1995) numerical model data, USGS (1978) field surveys and Birmingham University laboratory data provide a basis for analysis over a range of scales from ~1 m to ~1000 m, and a wide range of roughness conditions. There is a lack of measured field data for a scale of ~100 m. Field measurements from UK rivers were used in the HR Wallingford (Brown, 1988) work on arch bridges. However, the field data proves very difficult to use in practice because it gives estimates of head loss rather than the water surface profiles and channel survey data are needed to recover afflux estimates. Additionally, there is scant information to determine the exact locations of the measurements or the flow conditions. The bridges were not originally instrumented for the study of afflux and so these limitations are not surprising. Only one detailed set of measurements exists for afflux analysis at this scale, which is the data of Hamill (1999) at Canns Mill in Devon. Paper records were available to make subjective checks in the development of AES methods. Whilst the laboratory measurements made under careful scaling principles are very useful in calibrating and testing afflux calculation methods, there remains a gap in the data at the field scale.

Chapter 3

Understanding roughness, conveyance and afflux

ABSTRACT

This chapter is an introduction to the roughness, conveyance and afflux concepts and governing equations as adopted in the CES-AES in 2005/06. The topics also include the backwater calculation, the method assumptions and limitations and the numerical solution techniques. Where the science has subsequently advanced, references are given which may be used for future evolutions of the CES-AES. The issue of uncertainty is introduced and a pragmatic approach is set-out for dealing with this when estimating water levels in practice. The CES-AES software components are presented with a description of the overall modular interaction and example screen shots.

3.1 FLOW STRUCTURES IN OPEN CHANNEL FLOW

Identifying and understanding the various flow features is critical to estimating the overall channel flow capacity. The flow structure influences the energy dissipation, boundary shear stresses, resistance to flow and hence the overall conveyance (Knight, 2001). The flow features are defined here as 'energy transfer mechanisms', which typically convert the energy from one form to another through the development of vortex structures on a variety of length scales. Once vorticity is created, its rotational energy cascades down in length scale into increased turbulence intensity until it dissipates as heat through viscosity. The streamwise translational kinetic energy may, for example, be transferred in part to rotational kinetic energy, which no longer contributes to the streamwise channel conveyance. It is possible to identify different situations in which vorticity is created in open channel flow.

3.1.1 Boundary shear

Vorticity may be generated from boundary shear due to surface roughness (Nezu & Nakagawa, 1993), the bursting phenomenon (Reynolds, 1974; Grass *et al.*, 1991) as described below or a combination of these two mechanisms.

In uniform open channel flow, the boundary layer is fully developed and extends from the channel boundary throughout the flow depth. It includes a viscous laminar sub-layer adjacent to the boundary, a transitional zone characterised by both viscous and inertia effects and a turbulent zone where the inertia forces dominate. The effect of boundary resistance is felt throughout this layer, and is closely linked to the channel velocity distribution (Stokes, 1845). A hydraulically smooth boundary condition (Figure 3.1a) occurs where the boundary roughness elements are submerged within the viscous sub-layer. The streamlines adjacent to the bed are therefore parallel to the bed, and separation is unlikely to occur. A hydraulically rough boundary condition occurs where the roughness elements (height k_s) penetrate through the viscous sub-layer (Figure 3.1b). The boundary velocities adjacent to the bed tend to accelerate over the crest of the roughness elements and come to an abrupt standstill between them, substantially reducing the velocity and resulting in flow separation. Hydraulically rough boundary conditions are present in most rivers.

Bursting may also occur on smooth and rough beds. Bursting is a randomly occurring event comprising gradual local lift-up of the fluid in the area of low velocity parallel to the channel bed, causing sudden oscillations and ejection of fluid away from the boundary (Kline et al., 1967). This is usually stimulated by an inrush or "sweep" event (Figure 3.1), where high velocities move in close to the boundary, intensifying near wall vorticity by lateral spanwise stretching and generating new vorticity which is subsequently transported away by the ejections. Theodorsen (1952) first proposed a simple horseshoe vortex model (Figure 3.2) as the central element of the turbulence generation in shear flows.

Lighthill's (1963) explanation for the bursting phenomenon is that it represents the only means by which the very large gradients in mean vorticity present at the wall in turbulent boundary layers can be maintained against the corresponding outward diffusion and transport of vorticity by viscous and ejective action. This argument applies independently of wall roughness. More recently, Grass et al. (1991) undertook experiments to establish the full three-dimensional form of these turbulence structures over rough boundaries, and found that the horseshoe vortex structures are a central feature and play an important role in rough wall flows.

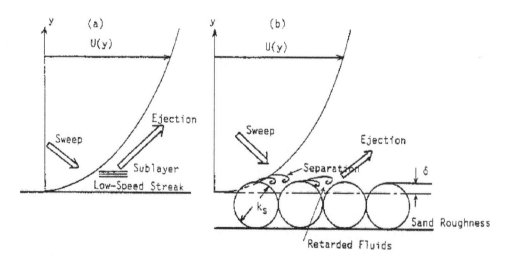

Figure 3.1 Turbulent flow over (a) smooth and (b) rough beds (Nezu & Nakagawa, 1993).

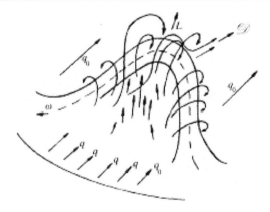

Figure 3.2 Horseshoe vortex model first proposed by Theodorsen (1952).

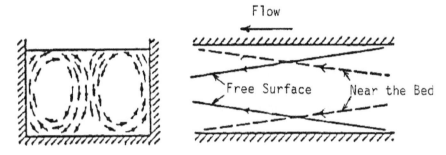

Figure 3.3 Schematic flow pattern of secondary currents in a narrow open channel (Gibson, 1909).

3.1.2 Vertical interfacial shear

Small-scale vorticity is generated from vertical interfacial shear in straight channels due to the steep velocity gradients at the main channel floodplain interface. These eddy structures tend to be smallest at the floodplain bed (i.e. at a relative depth of zero), and they expand gradually through to the water surface. These interfacial vortices are shown in Figure 2.15, Section 2.2.5.

3.1.3 Transverse currents

More than a century ago, river engineers (e.g. Stearns, 1883) first inferred the presence of secondary or "transverse" currents, as the maximum streamwise velocity occurred well below the free surface in open channel flow. Gibson (1909) first envisaged the flow pattern depicted in Figure 3.3 and estimated that the transverse velocities would be about 5% of the mainstream velocities. This simple pattern was later suggested to repeat itself across wide channels (Vanoni, 1946; Karcz, 1966), supported by the cyclical variations in spanwise sediment concentration.

Prandtl (1925) categorised the two types of secondary flows into flows of the first kind, derived from mean flow skewing, and those of the second kind, caused by inhomogeneity of anisotropic wall turbulence.

- Flows of the first type are driven by the channel geometry, which may affect non-uniform flow in the streamwise direction and hence the generation of streamwise vorticity through vortex stretching. In curved or meandering channels, the centrifugal driving force results in secondary currents in both laminar and turbulent flows with magnitudes typically 20–30% of the mainstream velocity (Rozovskii, 1957; Nezu & Nakagawa, 1993).

- Flows of the second type are generally smaller in magnitude and arise in straight channels due to the transverse gradients of the Reynolds stresses and anisotropy between the fluctuating velocity component (Prandtl, 1925; Brundrett & Baines, 1964; Perkins, 1970; Hinze, 1973). The production and suppression mechanisms for this vorticity are perhaps best understood through detailed examination of the terms in the vorticity equations, derived from the cross-differentiation of the Reynolds-Averaged Navier-Stokes equations in the y- and z-directions, to eliminate the pressure term (Perkins, 1970; Nezu & Nakagawa, 1993; Shiono & Muto, 1998). Detailed measurements of the various vorticity production terms (Perkins, 1970) suggest that the normal stresses do play an essential role (Figure 3.4).

Regions of flow confined by two perpendicular boundaries, for example, the corner of a rectangular (Figure 3.3) or trapezoidal (Figure 2.14, Section 2.2.5) channel are characterised by spontaneously occurring secondary flows of Prandtl's second kind. These circulations are only present in non-circular sections in turbulent flows, and the circulation is oriented towards the corners along the bisector line. In open channel flow, this secondary flow patterns alters slightly, in that a larger "free vortex" is formed near the surface and a smaller "bottom vortex" is formed at the channel bed, both still oriented along the bisector line (Nezu & Nakagawa, 1993). These transverse flows are small in comparison to the primary flow, in the order of 1%. They affect

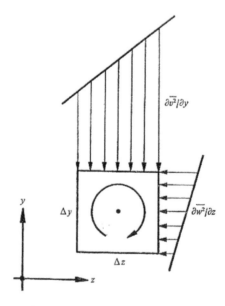

Figure 3.4 Mechanism of vorticity production by the direct stresses (Perkins, 1970).

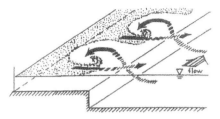

Figure 3.5 Transverse currents in a straight two-stage channel (Fukuoka & Fujita, 1989).

the isovel and boundary shear stress distributions significantly (Knight *et al.*, 1994; Tominaga & Knight, 2004), resulting in the maximum shear stress and velocity no-longer occurring at the channel centre line and free surface (Figure 2.14, Section 2.2.5). The number and orientation of these secondary flows have been related to the peaks and troughs in the lateral shear stress and mean velocity profiles (Imamoto & Ishigaki, 1989; Omran, 2005).

Post 1960, researchers still attribute the position of the maximum velocity filament to these secondary currents (e.g. Rajaratnam & Muralidhar, 1969). The cause of this velocity dip may be related to the aspect ratio (Nezu & Rodi, 1985), giving two channel types:

i   narrow channels with small aspect ratios ($b/h < 6$) where the velocity dip is caused by the free surface effect, which dampens the vertical velocity fluctuations, resulting in a different turbulent anisotropy from that in closed channels; and

ii   wide channels with large aspect ratios ($b/h > 6$) where the side-wall effects are not felt so much in the channel centre, and a series of secondary circulations occurs across the channel width. This results in low- and high-speed streaks, where the maximum velocity filaments in the high-speed streaks are below the free surface, but not always located at the centre of the channel.

Secondary currents of Prandtl's second kind can also be observed (Fukuoka & Fujita, 1989) along the flood plain main channel interface in compound channels (Figure 3.5). These transverse currents vary with relative depth, changing in orientation as the flow moves from inbank to low overbank and then to high overbank flow.

In skewed or meandering channels, secondary flows of the first kind are largely dominated by the additional contribution of the primary shear stresses. Toebes and Sooky (1967) observed that these secondary flow rotations change direction before and after inundation at the bend apex, which is supported by the detailed three-dimensional velocity measurements of Muto (1997), Shiono and Muto (1998) and Shiono *et al.* (1999b). Here, the influence of the secondary flows on eddy viscosity was found to be significant.

3.1.4 Coherent structures

Large plan form eddies or coherent structures are generated by shear instability (Ikeda & Kuga, 1997) in regions of high velocity gradient, for example, at the

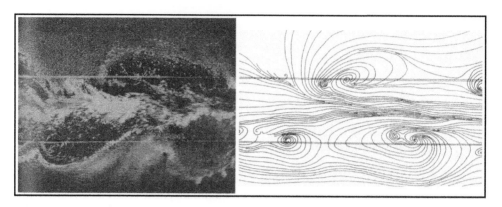

Figure 3.6 Surface velocity patterns and streamlines relative to a moving camera (Sellin, 1964).

main channel flood plain interface. Sellin (1964) identified these flow features in a two-stage experimental channel by scattering aluminium powder on the free water surface (Figure 3.6). The 4.6 m long channel was designed to maximise the flood plain main channel interaction through having vertical main channel side slopes. A strong vortex structure in the region of contact between the flood plain flow and that in the main channel was revealed (Figure 3.6a), with the surface streamlines providing further evidence of these large eddy structures (Figure 3.6b). Ikeda *et al.* (2002) carried out laboratory experiments for two-stage channels with varying relative depths (D_r). The presence of these coherent structures at low relative depths (e.g. 0.180, Figure 3.7a) was identified, however at higher relative depths (e.g. 0.344, Figure 3.7b) the periodic plan form vortices disappeared and active intermittent boils were observed. Nezu *et al.* (1997, 1999) observed that these boils actually grew stronger with increasing relative depth.

A vortex street may develop along the interfacial zone, and in compound channels the effect of these tends to vary if the parallel vortex streets are in or out of synchronisation with one-another, tending to increase the energy transfers for the in synchronisation case. Where adverse streamwise pressure gradients are present, this may result in flow separation. These patterns are further influenced by the channel aspect ratio. A large aspect ratio results in more lateral freedom for these eddies, allowing increased momentum transfer to the outer banks. Figure 3.8 provides an example of these coherent structures as observed in 1981 during a flood in the River Tone, Japan (Kitagawa, 1998).

Ikeda *et al.* (2002) undertook experiments in a two-stage curvilinear channel. Here, large-scale coherent structures were observed on both the inside and outside of the bend (e.g. Figure 3.9, $D_r = 0.5$).

Van Prooijen (2003 & 2004) considered the plan form rotation structure, highlighting the additional energy transfers due to the expansion and contraction of the flow as it rotates into and out of the main channel (Figure 3.10). In meandering channels, this rotational structure is further complicated, since the co-existence of plan form vortices with opposite rotations has been observed near the inner bank of a two-stage

Figure 3.7 Visualisation of the free surface in a straight two-stage channel with $D_r = 0.180$ and $D_r = 0.344$ (Ikeda *et al.*, 2002).

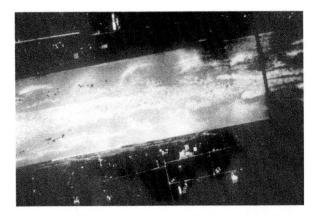

Figure 3.8 Horizontal vortices observed in the River Tone, Japan, during the 1981 flood. The flow is from left to right. (Courtesy of the Ministry of Construction of Japan).

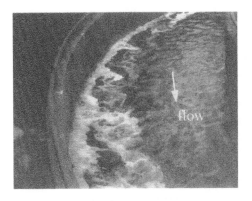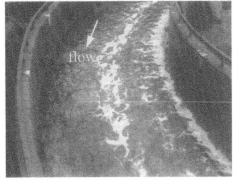

Figure 3.9 Flow visualisation of the free surface in a curvilinear two-stage channel ($D_r = 0.5$) at the outer bank and inner bank (Ikeda *et al.*, 2002).

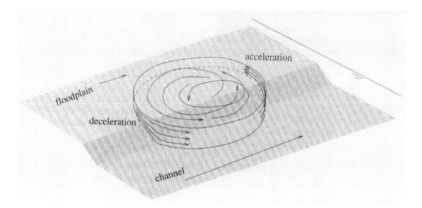

Figure 3.10 Large-scale coherent structure at the main channel floodplain interface in a two-stage channel (Van Prooijen, 2003).

curvilinear channel (Figure 3.9). In doubly meandering channels, the vortex path may be further influenced by the phasing of the main channel sinuosity relative to that of the flood plain levees. In rivers, these are rarely in phase. Fukuoka and Fujita (1989) show that the effect of this levee alignment on flow characteristics, i.e. on the position of maximum velocity filament and velocity distribution, is small compared to other influences and the main channel sinuosity has a dominant impact.

The lateral transport of fluid momentum associated with horizontal vortices becomes relatively greater, compared with that by secondary flows, as the water depth becomes shallower. If the channel curvature increases, the secondary flows transferring momentum to the outer banks increases, and this influences the strength of the horizontal vortices to induce still more momentum transport to the outer bank. The result is that the horizontal vortex structures are generally larger at the main channel floodplain interface on the outer bend relative to those at the inner bend.

3.1.5 Horizontal interfacial shear

Vorticity may be generated from horizontal interfacial shear due to steep velocity gradients at the main channel to floodplain interface. Here, the two water bodies may be oriented with different bulk flow directions, typically observed in meandering channels (Elliott & Sellin, 1990; Shiono & Muto, 1998). Figure 3.11 illustrates this with a flow visualisation technique. At low relative depths (e.g. $D_r = 0.15$), the out-of-bank main channel flow tends to follow the main channel flow direction. At high relative depths (e.g. $D_r = 0.25$), the out-of-bank flow direction is parallel to the flood plains. Water moving from the flood plain into the main channel shears over the main channel flow, and the relative magnitude and direction of this interfacial shear may enhance or retard the main channel rotations. Typically, the main channel rotational structures expand into the bend in a helical shape, and gradually dissipate beyond the bend where the shearing is no longer present. The orientation of the main channel rotations or secondary currents changes before and after the bend (Figure 3.12), since the angle of the floodplain relative to the main channel changes the magnitude and

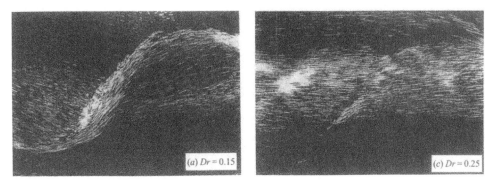

Figure 3.11 Flow visualisation using sawdust showing (a) flow in the streamwise direction ($D_r = 0.15$) and (b) parallel to the floodplain ($D_r = 0.25$) (Shiono & Muto, 1998).

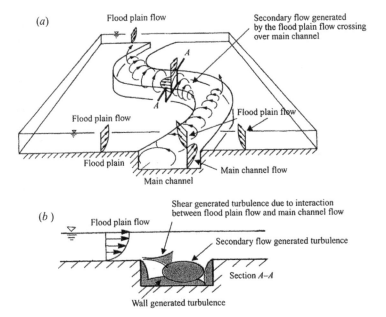

Figure 3.12 Flow mechanisms in a compound meandering channel (Shiono & Muto, 1998).

direction of the horizontal interfacial shear. These examples are taken from idealised laboratory flow where the floodplains are relatively smooth. In nature, the floodplains are typically more rough, offering greater resistance, and thus the interfacial shearing is less intense and the centrifugal accelerations are not always carried through to the next downstream apex.

If the floodplain flow entering the main channel region is slowed, for example, through roughness, it may result in less or no vorticity as the relative velocity is low.

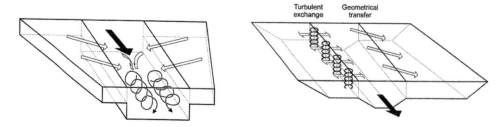

Figure 3.13 Non-prismatic channels with (a) converging and (b) diverging flow (Bousmar, 2002).

This may result in the in channel circulation occurring along the centre of the main channel, rather than the channel edges (Elliott & Sellin, 1990).

Horizontal interfacial shearing may occur in converging (narrowing floodplains) and diverging (enlarging floodplains) channel geometries (Bousmar, 2002; Rezaei, 2006; Chlebek, 2009):

- For converging flow, the floodplain flow moves over the main channel flow, causing a horizontal shearing action and main channel eddies that rotate with a different orientation to straight prismatic channels. The flow plunges to the channel bed along the main channel centreline, and then rotates back out towards the main channel edges, forming a rotating helical structure in the streamwise direction (Figure 3.13a).
- For diverging flow, there is a transfer of momentum from the main channel to the flood plain flow (Figure 3.13b). This is similar to the skewed compound channel, where the water on the enlarging floodplain is accelerated due to the faster main channel velocities passing onto the floodplain, through a smaller depth of flow, resulting in increased lateral momentum transfer relative to that in a straight channel.

3.1.6 Vorticity in rivers

The various instances in which vorticity may occur are closely related to the nature of the channel geometry i.e. a simple or compound, straight or meandering, depth and nature of flow etc. Experimental set-ups tend to idealise flow conditions to restrict the number of unknown parameters. Purpose-made river-measurements are often based on sections with idealised shapes e.g. River Main (Section 2.2.2). Vorticity in rivers, with non-idealised section shapes, may differ from those outlined thus far, for example:

- *Asymmetry of floodplains or berms* – Asymmetric channels (e.g. one floodplain or berm present) will have different in channel circulations as the flood plain flow will only influence one side of the main channel.
- *Multiple berms* – A main channel cross-section with multiple berms will affect the secondary flow structure and orientation with flow depth.

- *Braiding* – A braided channel with common floodplains may result in the flood-plain flow structures (e.g. coherent vortices) overlapping and hence interacting. This may influence the stability of the vortex pattern, where a decreased stability (i.e. the vortex patterns of the two channels are out of synchronization with one-another) is characterized by flow separation.

- *Shape* – Natural rivers have large width to depth ratios, resulting in non-homogenous turbulence. The flow is affected by bottom generated small-scale isotropic "wall turbulence", scaled by depth, and "free turbulence" from the large-scale horizontal eddy structures which have lateral freedom (Ikeda, 1999). These large-scale vortices constitute more than 75% of the lateral fluid momentum transport in rivers with large aspect ratios.

- *Bank vegetation* – Main channel banks may have thick vegetation. This may reduce or eliminate the vertical interfacial turbulent exchanges, acting as a stream-wise vertical barrier between the main channel and flood plain flow. This may reduce the overall conveyance as the vegetation provides increased flow resistance or enhance the conveyance as less lateral momentum exchange takes place.

- *Floodplain features* – Clumps of trees or urban form may generate additional streamwise vorticity through stream drag or alternatively suppress the large-scale horizontal eddy structures.

- *Structures* – Hydraulic structures such as bridges or weirs may initiate a vortex street, the affect of which may be felt some way downstream (Section 3.1.7).

- *Pools and riffles* – Meandering channels in nature invariably have a pool and riffle sequence. These undulations result in localised head losses, which are larger in low overbank flows. Riffles occur at the cross-over point between consecutive bends. The pools generally occur at the bend apex in regions of maximum bend velocities. It is possible to determine the position of these maximum velocity filaments, through a flow classification based on sinuosity and relative depth (Okada & Fukuoka, 2002). The secondary currents in these pool and riffle sequences are convergent and divergent respectively, providing deposition or erosive properties.

It is possible to formulate some general rules for the occurrence and development of secondary cells. The classic rectangular channel case is effectively half a square duct, hence the symmetry in the circulations. The cells are always stronger in the corners, at the bisector line. Weaker cells develop across the channel width, with the cell orientation changing for each adjacent cell, resulting in pairs of circulation with a radius similar to half the depth of flow. Trapezoidal channels have additional cells in the top corners. For two-stage channels, the flow in the main channel is similar to inbank flow and additional cells develop at the sharp floodplain edges. These are generally weaker in rivers whereas the plan form vortices have a far greater effect than in laboratory channels, for example, in the River Tone the planform eddies are up to 300 m wide (Kitagawa, 1998).

3.1.7 Special features near structures

The nature and orientation of vortices changes near hydraulics structures such as weirs, bridges, groynes and culverts. In-line weirs (e.g. broad crested, vertical, V-notch) obstruct the main flow direction, forcing the flow through a smaller cross-sectional area above the structure. This results in the higher velocities initiating a slowly rotating vortex upstream of the structure, with a horizontal axis (Figure 3.14).

Flow structures around bridge abutments and piers are closely related to the size, shape and orientation of the structure. For example, as flow approaches circular bridge piers, a bow wave forms at the free surface on the upstream side. Along the sides of the pier, the free surface is drawn down as the flow accelerates around it. Wake vortices (with a vertical axis of rotation) are formed and shed downstream of the pier as the flow about either side of the pier separated from the structure (Figure 3.15).

Bridge abutments typically occupy more cross-sectional area than bridge piers and commonly result in a contraction of the flow through the bridge (converging streamlines) and flow expansion further downstream. Abutments occur in variety of shapes, see, for example, Hoffmans and Verheji (1997). Consider a simple vertical-wall abutment. As the flow comes into contact with the bridge abutment, the velocity is reduced to zero and the resulting pressure gradient causes a downflow on the upstream face of the abutment. Downstream of the abutment, the main flow is separated from a large slowly rotating vortex in the lee of the abutment by a vortex-street. The flow

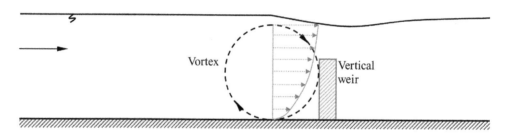

Figure 3.14 Vortex prior to a vertical weir (long-section view).

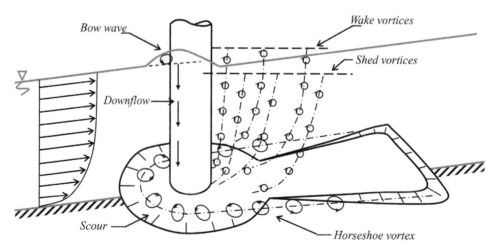

Figure 3.15 Flow patterns and vortex structures at a cylindrical pier (adapted from Raudkivi, 1998).

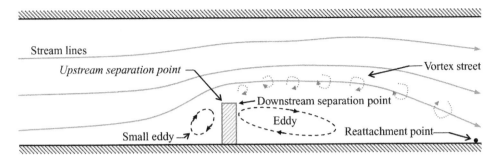

Figure 3.16 Flow patterns around a vertical-wall abutment (plan view).

decelerates over a significant distance downstream and normal depth is re-established further downstream beyond the reattachment point (Figure 3.16). Where the bridge has two similar abutments on opposite river banks, this pattern is symmetrical.

For further reading on vorticity, flow structures and other issues that need consideration, see Ashworth *et al.* (1996), Nezu & Nakagawa, (1993), Knight (1996) and Anderson *et al.* (1996).

3.2 GOVERNING EQUATIONS

3.2.1 Roughness Advisor methods

Roughness characterisation of channels and their associated floodplains is essential to water level estimation. Roughness reduces the discharge capacity through energy expenditure from boundary generated turbulence (Section 3.1.1) as well as physical blockages due to vegetation.

Traditionally, flow resistance datasets are based on average values of Manning's *n*, Chezy C or Darcy *f* for whole river sections (Chapter 2). These include the influence of bed material, vegetation and large-scale topographic effects. Table 3.1 provides a summary of some of the approaches for estimating *n*, C and *f* from the literature and used in practice today. These include methods for establishing lateral distributions of equivalent 'local' values for non-homogenous roughness across a channel section. Important assumptions of these approaches are steady, uniform flow conditions and a fixed channel bed.

The flow of water in a river, water course or drainage system is described by certain physical laws related to the conservation of mass, energy and momentum. The CES conveyance calculation requires a lateral and depth distribution of the friction factor *f* (Section 3.2.2). Due to the widespread use of Manning's equation, most resistance advice in the literature is expressed in terms of Manning *n* (Table 3.1). The Roughness Advisor (RA) is based therefore, on an input unit roughness n_l with similar units ($sm^{-1/3}$) to Manning *n*, rather than the friction factor *f*, to maintain user familiarity and confidence. This unit roughness describes identifiable segments of boundary friction only and the additional turbulence generation due to cross-section and plan

Table 3.1 Approaches for estimating Manning *n*, Chezy *C* and Darcy *f*.

No	Approach	Description
Manning *n*		
1	Direct solution of the Manning (1889) equation from measurements (Eq. 2.1)	River survey to measure flow, depth, cross-section topography and longitudinal water surface (or bed) slope.
2	Photographic sources	The most widely used reference is Chow (1959). Others include Barnes (1967) and Hicks & Mason (1998), which are based on American and New Zealand channels respectively.
3	Tabular sources	An extensive list of typically used Manning's *n* values is available in Chow (1959) and Yen (1991).
4	Formulae for evaluating composite Manning *n* for non-homogenous roughness (e.g. main channel, banks & floodplain)	These include, for example, the methods of: – Pavlovskii (1931) – Horton (1933) and Einstein (1934) – Lotter (1933) – Krishnamurthy and Christensen (1972) – Einstein and Banks (1950) which vary in their assumptions relating to local shear force, velocity and hydraulic radius.
5	Formulae for evaluating composite Manning *n* due to multiple resistance sources (e.g. vegetation, meandering)	Cowan's (1956) additive approach evaluates the total Manning's *n* resistance value from multiple sources. The standard Manning *n*-value (e.g. Chow, 1959) is inflated by the additional resistance due to surface irregularities, channel shape and size, obstructions and vegetation and the total is then multiplied by a factor for channel curvature.
6	Formulae for evaluating Manning *n* where large boundary elements are present (e.g. sediments, cobbles)	– Strickler (1923) relates Manning's *n* to the average sediment diameter *d* (ft) for flow in channels with cobbled beds – the Manning-Strickler formula relates Manning's *n* to Nikuradse's size k_s and the hydraulic radius *R*. – Ackers (1958) provides a simple approximation for the Manning Strickler in the range $5 < R/k_s < 500$.
Chezy *C*		
7	Direct solution of the Chezy (1768) equation from measurements (Eq. 2.23)	River survey to measure flow, depth, cross-section topography and longitudinal water surface (or bed) slope.
8	Formulae	A number of formulae exist for determining this coefficient, see for example Ganguillett & Kutter (1869), Bazin (1897) and Powell (1950).
Darcy *f*		
9	Direct solution of Darcy (1857) – Weisbach (1845) equation from measurement (Eq. 2.24)	This requires estimates of flow, hydraulic radius and longitudinal water surface (or bed) slope. This is typically used in open channel flow by setting the flow depth equal to 4 times the radius (Chow, 1959).

(Continued)

Table 3.1 (Continued).

No	Approach	Description
10	Blasius (1913) equation	Applicable for smooth turbulent pipe flows where the flow conditions are dominated by the viscous forces, typically used for experimental flumes (4000 < Re < 105).
11	Kármán-Prandtl (1925) equation	For use at higher Reynolds Numbers, where the Blasius equation under-predicts f. See Schlichting (1979).
12	Hazen-Williams equation	Derived for pipe flow.
13	Iterative solution of the Colebrook-White (1937) equation from measurements (Eq. 2.26)	This requires estimates of flow, roughness size, hydraulic radius and approximate cross-section shape. Many studies have been undertaken to establish shape coefficients for open channel flow (Table 3.3, Section 3.2.2).
14	Moody (1944) diagram (Figure 2.10)	A drawback in application of the Colebrook-White equation is that it is implicit in f. Moody sought to overcome this through developing a diagram for pipe resistance (Figure 2.10), where the pipe diameter can be replaced by four times the hydraulic radius R (m) for open channel flow. The diagram relates f to the relative roughness R/k_s and the Reynolds' Number for laminar, transitional and fully turbulent flow through a family of curves.
15	Direct solution of Barr (1975 & 1979) and Yen's (1991) form of the Colebrook-White equation from measurements	Barr (1975 & 1979) provides an alternative explicit equation to the Colebrook-White formula for pipe flow, which Yen (1991) has provisionally adapted for wide open channels where Re > 30,000 and R/k_s > 20. This requires estimates of flow, roughness size, hydraulic radius and approximate cross-section shape.
16	Approaches for establishing the lateral variation of f	Power Law approach (Shiono & Knight, 1991; Rhodes & Knight, 1995). Relates the ratio of Darcy f on the floodplain f_{fp} and main channel f_{mc}, to the relative depth by a power of 3/7. Calibrated relationships (Abril, 1995; Knight & Abril, 1996). Relates the main channel resistance f_{mc} to the floodplain resistance f_{fp} by an empirically derived factor R_r.

form shape is explicitly represented in the conveyance calculation (Section 3.2.2). The unit roughness n_l is then converted to an equivalent turbulence length scale k_s and thence to a friction factor f by an iterative procedure (due to velocity dependence) for input to the conveyance calculation (Section 3.2.2).

The unit roughness values are notionally associated with a depth of 1.0 m, selected as a 'representative' depth of flow for UK channels and one at which roughness elements (e.g. sand, pebbles, weeds) are normally small relative to flow depth i.e. the roughness does not vary rapidly with depth. This then excludes cases where the roughness is comparable with the flow depth (e.g. boulders, emergent vegetation), for which special rules may be introduced.

The unit roughness is comprised of three component values: surface material n_{sur} (e.g. sand, bedrock), vegetation n_{veg} (e.g. reeds, crops) and irregularities n_{irr} (e.g. tree roots, urban trash) and is evaluated from:

$$n_l = \sqrt{n_{veg}^2 + n_{sur}^2 + n_{irr}^2} \tag{3.1}$$

The "root sum of the squares" approach is adopted as the roughness is squared before being combined since the energy loss is related to the square of the local velocity. This approach highlights the contribution of the largest component roughness, for example, a tree would have more influence than a grain of sand.

The RA provides advice on the component unit roughness values, including minimum, maximum and expected values. This is based on an extensive literature review of over 700 references (Defra/EA, 2003b), and where possible, gives descriptions accompanied by photographs. The RA values, descriptions and advice are used together with information from the site to evaluate n_{sur}, n_{veg} and n_{irr} for a given cross-section or portion of cross-section (e.g. bank, channel bed). Where known, the seasonal variations in plan form aquatic vegetation coverage and n_l are provided as well as advice on expected regrowth patterns following cutting. These are derived from

Table 3.2 Unit roughness values for twelve vegetation morphotypes (Defra/EA, 2003b).

| Aquatic vegetation by RHS vegetation type | | Unit roughness n_l | | |
		Min	Mean	Max
1	None – if no vegetation visible			
2	Free-floating plants typically algae or duckweeds, medium deep drainage channel [depth = 1.1 to 2.5 m, velocity = 0.1 to 0.6 ms⁻¹]	0.010	0.030	0.040
3	Filamentous algae attached shallow nutrient rich waters [depth = 0.05 to 0.5 m, low velocities]	0.000	0.015	0.050
4	Mosses attached to bed or banks [depth ≈ 1 m]	0.000	0.015	0.030
5	Trailing bank-side plants [depth = 0.5 to 3+ m]	0.000	0.050	0.100
6	Emergent reeds, rushes, flag and large grasses	0.020	0.150	0.200
7	Floating-leaved typically water lilies in deeper slower waters	0.030	0.100	0.140
8	Emergent broad-leaved rooted plants, water cress or water parsnip	0.050	0.150	0.50+
9	Submerged broad-leaved, pondweeds [depth = 0.06 to 1.2 m, velocity 0.2 to 0.9 ms⁻¹]	0.020	0.100	0.200
10	Submerged fine-leaved, shallow rivers, chalk streams, water crowfoot, pondweeds [depth = 0.2 to 0.6 m, velocity = 0.2 to 0.7 ms⁻¹]	0.020	0.300	0.45+
11	Submerged fine-leaved, medium depth rivers, regular management [depth = 0.6 to 1.2 m, velocity = 0.3 to 0.8 ms⁻¹]	0.021	0.100	0.249
12	Submerged fine-leaved, medium to deep rivers, some management [depth > 1.2 m, velocity = 0.3 to 1.0 ms⁻¹]	0.010	0.080	0.120

previous measurements and expert knowledge, giving a series of embedded rules for regrowth pending vegetation type (and hence biomass), time of year and percentage cut (Defra/EA, 2003b).

In the absence of any survey data or channel description, the RA provides advice using data obtained through the UK national River Habitat Survey (RHS) (Raven *et al.*, 1998). Sample locations were sere selected within a 10 by 10 km square grid. At each location, observations were made at 10 cross-sections spaced at 50 m intervals. This enables provision of advice on expected in-channel and bank-side aquatic vegetation based on the UK grid reference for a study reach.

The Roughness Advisor includes values for:

- natural surface materials e.g. bedrock, cobbles, gravel, sand, silt, peat, earth, bare ploughed soil;
- man-made surface materials e.g. sheet/wood piling, stone block, hazel hurdles, concrete, rip-rap;

Figure 3.17 Example vegetation morphotypes including (top left) emergent broad-leaved brooklime *Veronica beccabunga*, (top right) submerged fine-leaved pond water crowfoot *Ranunculus peltatus*, (bottom left) willow moss *Fontinalis antipyretica* and (bottom right) trailing bank-side sweet grass *Glyceria*.

- natural vegetation covering 12 morphotypes (Table 3.2, Figure 3.17) which are closely linked to the RHS categories;
- vegetation subject to human intervention e.g. grass, hedges, trees, shrubs and crops such as maize;
- natural irregularities e.g. groynes, exposed boulders, undulations on the flood-plain and obstructions such as debris deposits, exposed stumps, logs and isolated boulders;
- man induced irregularities e.g. urban trash.

A more complete description of the information sources for the RA is provided in the Roughness Review (Defra/EA, 2003b). See also related work by Baptist (2005).

3.2.2 Conveyance Estimation System methods

Conveyance (Chapter 2, Section 2.2.4) relates total discharge to a measure of the gradient or slope of the channel under normal flow conditions, giving

$$Q = KS^{1/2} \qquad\qquad (2.41)$$

where K (m^{3}s^{-1}) is the conveyance, Q (m^{3}s^{-1}) is the discharge and S is the common uniform gradient. For normal flow there is no ambiguity about this slope since the water surface slope S_w, the bed slope S_o and the so-called "energy" or "friction" slope S_f all coincide. In practice, this is approximated from the water surface slope (e.g. in numerical models) or bed slope.

Conceptual and chronological development of conveyance

The approaches to quantifying conveyance have varied throughout the late nineteenth and twentieth century. These may be broadly divided into five categories, reflecting the basis of their derivation (Figure 3.18):

1 *Hand calculation methods* – commonly termed Single (SCMs) and Divided Channel Methods (DCMs) where the channel cross-section is treated as a single unit or divided into more than one flow zone, respectively. These vary in terms of the number and nature of the sub-divisions (Figure 3.19) and the assumptions

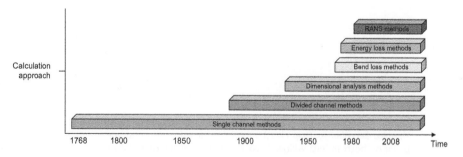

Figure 3.18 An approximate timeline of conveyance estimation approaches (Mc Gahey *et al.*, 2008).

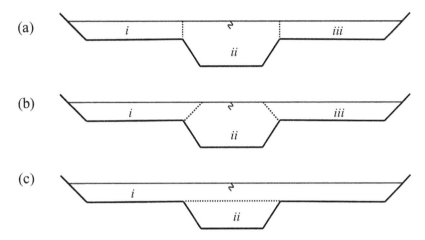

Figure 3.19 Example of (a) vertical, (b) diagonal and (c) horizontal division of a compound channel (Mc Gahey, 2006).

regarding velocity, shear stress and momentum transfer – including the so-called apparent shear force methods. These include, for example, the methods of Chezy (1768), Manning (1889), Lotter (1933), Yen and Overton (1973), Ervine and Baird (1982), Knight and Demetriou (1983), Knight, Demetriou & Hamed (1984), Wormleaton and Merrett (1990), Ackers (1991), Lambert and Myers (1998) and Knight and Tang (2008).

2 *Dimensional analysis approaches* – where channel properties likely to affect channel conveyance are identified and grouped into dimensionless combinations such as relative roughness, Froude Number and Reynolds Number. A regression analysis is applied to ascertain the value of any coefficients and the relationships are evaluated for scale models and assumed similar for the prototype e.g. Ackers (1991), Shiono *et al.* (1999a), Rameshwaran and Willetts (1999).

3 *Additional energy losses due to bends* – methods which account for the additional losses around channel bends, for example, the Soil Conservation Service and Linearised Soil Conservation Service approach and the methods of Rozovskii (1957), Leopold *et al.* (1960), Toebes and Sooky (1967), Argawal *et al.* (1984), Chang (1983 & 1984) and James (1994).

4 *Energy 'loss' methods* – which are based on an extension of the early DCMs, with consideration for the true flow mechanisms present within each sub-area. These assume that the energy transfers within each flow zone (e.g. main channel or floodplain region) are proportional to the square of the local flow velocity and they are mutually independent and hence the principle of superposition can be applied:

$$\sum_{i=1}^{n} K_i \frac{U_i^2}{2g} = S \text{ (straight)}; \quad \sum_{i=1}^{n} K_i \frac{U_i^2}{2g} = \frac{S_o}{\sigma} \text{ (meandering)} \qquad (3.2)$$

where K_i represents the energy loss coefficient. Well known methods include those of Ervine and Ellis (1987), Greenhill (1992), James and Wark (1992), Shiono *et al.* (1999b), Bousmar and Zech (1999) and van Prooijen (2003 & 2004).

5   *Reynolds Averaged Navier-Stokes (RANS) approaches* – which are based on the depth-integration of the RANS equations for flow in the streamwise direction. These equations are derived from the equations of fluid flow and are therefore physically based and theoretically sound. The unit flow rate q (m²s⁻¹) or depth-averaged velocity U_d (ms⁻¹) is solved for at discrete locations across the channel, and then a lateral integration provides the overall flow rate Q (m³s⁻¹) i.e.

$$Q = \int_0^B q\,dy = \int_0^B U_d\,dA \qquad (3.3)$$

where y is the lateral distance across the channel cross-section (m) and B is the total channel width (m). The various approaches differ in their assumptions, emphasis on different terms, the processes that are modelled rather than directly evaluated, the calibration coefficients and the turbulence model selected for closure. Significant research contributions include Shiono and Knight (1988), Wark (1993), Abril and Knight (2004), Spooner and Shiono (2003); Bousmar and Zech (2004), van Prooijen *et al.* (2005) and Tang and Knight (2009).

Methods adopted in practice

Industry practice in Europe and the United States still adopt variations of the early hand-calculation methods such as the Manning and Chezy equations. These equations are attractive due to their simplicity, limited data requirements, transparency of outputs and the ease with which they can be embedded in modelling systems. For example:

- *Flood Modeller Pro (formerly ISIS Flow)* – developed by CH2M incorporates the Manning equation with the Lotter (1933) approximation for non-homogenous roughness.
- *InfoWorks RS/ICM* – developed by Innovyze incorporates the Manning equation with the Lotter (1933) approximation for non-homogenous roughness.
- *HEC-RAS* – developed by US Army Corps of Engineers incorporates the Manning equation with the Horton (1933) and Einstein (1934) approximation for non-homogenous roughness.
- *MIKE11* – developed by Danish Hydraulic Institute incorporates the Manning and Chezy equations.

Today, with new research substantially improving our understanding of the flow behaviour, the advent of computing power allowing for more sophisticated solution techniques and the availability of flow measurements on a range of scales providing a sound basis for method testing (e.g. the EPSRC Flood Channel Facility at HR Wallingford, purpose-made river measurements) – it is feasible to consider alternatives to these early methods. For example, the CES conveyance calculation is now incorporated in the UK modelling software Flood Modeller Pro and InfoWorks RS, providing an alternative to the more traditional Manning equation.

The CES conveyance calculation

The CES conveyance calculation is derived from the depth-integration of the Reynolds-Averaged Navier-Stokes (RANS) equations at specified sections along the water course. See Schlichting (1979) and Drazin and Riley (2006) for further details of the RANS equations. The underpinning theory stems from the original Shiono Knight method (1988, 1990–91), and subsequent evolutions of this, for example, Abril (1997, 2001), Abril and Knight (2004), Knight and Abril, (1996), Ervine *et al.* (2000), Defra/EA (2003a) and Mc Gahey (2006). A complete description of the derivation from first principles is available in Mc Gahey (2006). Here, the method is described from the depth-averaged form of the RANS equation for flow in the *x*-direction, with *x* orientated streamwise in the *x-y* plane (or plane with normal *z*, Figure 3.20):

$$\rho g H \left(\frac{\partial H}{\partial x} - S_o \right) - \psi \tau_o + \frac{\partial}{\partial y}\left(\rho H \nu_t \frac{\partial U_d}{\partial y} \right) - \frac{\partial}{\partial y} \rho \big[H(UV)_d \big] = 0 \qquad (3.4)$$

$$\quad\;\;\text{(i)}\quad\;\;\text{(ii)}\qquad\quad\text{(iii)}\qquad\quad\;\text{(iv)}$$

where ρ (kgm^{-3}) is the fluid density, ψ is the projection onto plane due to choice of Cartesian coordinate system, τ_o (Nm^{-2}) is the boundary shear stress, ν_t (m^{2}s^{-1}) is the turbulent eddy viscosity and H implies $H(y)$ i.e. the local depth at any point within the cross-section. The remainder of this section describes the derivation of the four terms in Eq. 3.4:

i   Hydrostatic pressure term
ii   Representation of turbulence due to boundary friction
iii   Representation of turbulence due to transverse currents effecting vertical interfacial shear
iv   Representation of turbulence due to secondary circulations

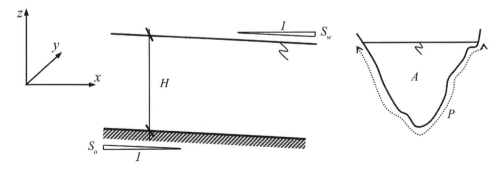

Figure 3.20 Definition of the co-ordinate system.

culminating in a system of equations that represent the complete CES conveyance methodology. An important theme is the extension of theoretically or experimentally derived models and coefficients for application in the field. This involves consideration of effects of scale, transferability of application extents, applicability of parameter values for field studies and, in particular, the potential of a depth-averaged model for predicting meandering channel flow conditions.

Term (i) – Hydrostatic pressure term

The first term in Eq. 3.4 represents the variation in hydrostatic pressure along the reach. As the flow depth H is assumed uniform along the representative reach, dH/dx is considered negligible, giving the expression for the hydrostatic pressure as

$$\rho g H S_o \tag{3.5}$$

The reach-averaged longitudinal bed slope, S_o, is a parameter based on the available reach information. Where water level measurements are present, it may be approximated by the longitudinal water surface slope. This latter option is a significant improvement as it is influenced by both bed slope and changes in water level and it will thus reduce any implications of the steady, uniform flow assumption.

Term (ii) – Representation of turbulence due to boundary friction

The second term in Eq. 3.4 represents the turbulence generated due to boundary friction and it comprises the primary balance of momentum for Term (i). The slope projection parameter ψ may be evaluated directly from (Wark, 1993)

$$\psi = \sqrt{1 + S_y^2 + S_o^2} \tag{3.6}$$

where the estimated lateral distribution of S_y is improved with increased density of channel survey data, particularly in areas of rapidly varying cross-section shape. For vertical channel banks, ψ tends to infinity as S_y tends to infinity (albeit it at a slower rate). The nature of the quasi-2D discretisation (see solution technique), whereby the surveyed data points must increase laterally, limits S_y to be a finite large number, the size of which is dependent on the grid resolution. Ideally, the force on the vertical bank should be evaluated.

The boundary shear stress τ_o (Nm^{-2}), can be related to the shear velocity u_* (ms^{-1}) (Prandtl, 1925), by (re-arrange Eq. 2.31)

$$\tau_o = \rho u_*^2 \tag{3.7}$$

The shear velocity is commonly related to the depth-averaged velocity by a dimensionless local friction factor f, giving (re-arrange Eq. 2.35)

$$u_* = \left(\frac{f}{8}\right)^{\frac{1}{2}} U_d \tag{3.8}$$

While this provides a useful expression for evaluating the shear velocity for cases of low slope, this relation originates from equating the shear stress in a uniform wide channel (Eq. 2.32) i.e.

$$\tau_o = \rho g R S \tag{3.9}$$

with the Darcy-Weisbach equation (re-arrange Eq. 2.24)

$$S_f = f U^2 / 8 g R \tag{3.10}$$

for total channel flow, with cross-section all-encompassing resistance parameters. These parameters take account of, for example, shape effects and shear generated turbulence, rather than a local friction factor f. The wide channel assumption implies that the lateral velocity distribution is uniform and hence the lateral shear affects are not considered. Knight (2001) advocates this approach, on the basis that the local friction parameter is being related to a local depth-averaged velocity and a local boundary shear stress. Term (ii) can thus be expressed as

$$-\psi \rho \frac{f}{8} U_d^2 \tag{3.11}$$

It is worth noting that equating Terms (i) and (ii) in Eq. 3.4 results in an equation similar to the Darcy-Weisbach equation (Eq. 3.10).

Evaluating the local friction factor f

The local friction factor is essentially a parameter that relates the local boundary shear stress to the depth-averaged velocity, and as such, it should consider the full three-dimensional flow effects throughout the depth of flow. This concept is incorporated through the use of the Colebrook-White equation,

$$\frac{1}{\sqrt{f}} = -c \log_{10} \left[\frac{k_s}{aR} + \frac{b}{Re\sqrt{f}} \right] \tag{3.12}$$

which evaluates a lateral distribution of f, at each depth of flow, based on an iterative estimate of the local depth-averaged velocity. Here, Re is the Reynolds Number and R (m) is the hydraulic radius (= A/P). This equation is preferred in contrast to, for example, the Chezy or Manning equation, as it covers smooth-turbulent, transitional and rough-turbulent flows; it has a strong physical basis in that it is derived from the logarithmic velocity profile together with the channel geometry; and it incorporates the variation of roughness with depth. It is therefore believed that this model will cover the experimental and prototype cases.

Table 3.3 provides a summary of the advised Colebrook-White coefficients (a, b, c) which are based on channel shape, experimental findings and experience. The CES conveyance methodology has adopted values of 12.27, 3.09 and 2.03 for a, b and c respectively. The value of c is based on a value of 0.41 for the von Kármán constant, κ. The values of a and b are based on c and assuming a trapezoidal cross-section

Table 3.3 Values of constants for Colebrook-White type formula in steady uniform flow with rigid impervious boundary (after Yen, 1991).

Researcher	c	a	b	Description
Colebrook & White (1937)	2.00	14.83	2.52	Full circular pipe
Zegzhda (1938)	2.00	11.55	0.00	Rectangular with dense sand
Keulegan (1938)	2.03	12.27	3.41	Wide & smooth flow channel
Keulegan (1938)	2.00	12.62	2.98	Wide & fully rough channel
Keulegan (1938)	2.03	12.27	3.09	Smooth trapezoidal channel
Keulegan (1938)	2.00	13.99	2.27	Rough trapezoidal channel
Rouse (1946)	2.03	10.95	1.70	Wide channels
Thijsse (1949)	2.03	12.20	3.03	Wide channels
Sayre & Albertson (1961)	2.14	8.89	7.17	Wide channels
Reinius (1961)	2.00	12.40	3.40	Wide channels
Reinius (1961)	2.00	14.40	2.90	Rectangular – width/depth = 4
Reinius (1961)	2.00	14.80	2.80	Rectangular – width/depth = 2
Henderson (1966)	2.00	12.00	2.50	Wide channels
Graf (1971)	2.00	12.90	2.77	Wide channels

shape. In reality, river cross-section shapes are irregular except for a small percentage of heavily urbanised waterways. A trapezoidal shape was adopted as one more closely representative of an irregular shaped river section with sloping side banks than, for example, a narrow or wide rectangular section with vertical banks. The implication of this assumption is considered minor in that all three coefficients (*a*, *b* and *c*) are close to the average (12.40, 3.17 and 2.02) of the available values (Table 3.3) which cover a range of possible shapes.

The values in Table 3.3 were derived based on treating the cross-section as a whole whereas the intended application is to derive a localised value of *f*. The predicted values of *f* were compared therefore to localised *f*-values back-calculated from high quality boundary shear stress and velocity measurements in a two-stage trapezoidal channel from the Flood Channel Facility (*www.flowdata.bham.ac.uk*). This indicated good agreement over a range of depths, including both inbank and out-of-bank flow (Defra/EA, 2003a). Ideally further testing should be undertaken for a range of natural channel shapes, for which both local depth-averaged velocity and boundary shear stress measurements are available.

Conversion of n_l to k_s

Eq. 3.12 contains a further unknown, the absolute roughness size k_s, a parameter not widely used in river engineering practice. It is therefore necessary to relate the unit roughness n_l (Section 3.2.1) to the roughness size k originally defined in terms of the sand grain dimension (Nikuradse, 1933). Here, k_s is interpreted as an 'equivalent' turbulence length scale i.e. a measure of the turbulent eddy size. This length scale may be greater than the water depth, for example, willow trees and large boulders, as it represents the horizontal mixing action, whereby eddies may be larger in plan than the local water depth. An upper limit on k_s may be defined as the full channel width, where the large-scale roughness essentially represents a different physical process, with very low velocities or blockage.

Knight and MacDonald (1979) classified six categories of flow patterns over strip roughness, and illustrated the importance of the roughness element's shape, size and spacing on the overall resistance. Here, all three are wrapped into the k_s dimension. The equivalent k_s value can be determined from (variation of Strickler, 1923; Ackers, 1958),

$$n = 0.0342\,k_s^{1/6}\,(ft) = 0.0417\,k_s^{1/6}\,(m) \tag{3.13}$$

which is applicable to a limited range of R/k_s ($5 < R/k_s < 500$) with an accuracy range of 10%. This conversion is suitable for use in flow conditions under which the equation was derived, in particular, application to a whole cross-section or reasonably smooth channels. For application to natural channels with large-scale roughness elements, two issues arise:

- A solution is required locally at each point across the section, and hence, for a given location, the local roughness height may be equivalent to a large fraction of the depth or in some cases, exceed it, for example, emergent floodplain vegetation.
- Large typically-used Manning n values, and the corresponding unit roughness n_l – values, result in highly inflated k_s values. For example, should the bed material be characterised by large boulders, with an equivalent n_l value of 0.5, this would produce a k_s value of over 2,000 km. Values of this magnitude are meaningless in terms of flow depth, river width and reach length.

The CES conveyance method, therefore, incorporates an alternative conversion approach for natural channels. This is based on the rough-turbulent component of the Colebrook-White law, given by the first term in Eq. 3.12,

$$\frac{1}{\sqrt{f}} = -2.03\log_{10}\left[\frac{k_s}{12.27H}\right] \tag{3.14}$$

And the expression for f given by relating the Darcy-Weisbach and Manning equations:

$$f = \frac{8gn_l^2}{H^{1/3}} \tag{3.15}$$

Hence, at a depth of 1.0 m, the depth associated with the unit roughness n_l values, k_s is given by

$$k_s = 12.27H10^{\left[\frac{H^{1/6}}{\sqrt{8gn_l}(-2.03)}\right]} = 12.27\times10^{\left[\frac{1}{\sqrt{8gn_l}(-2.03)}\right]} \tag{3.16}$$

An alternative approach for relating n_l to f would be the direct use of Eq. 3.15. This would eliminate the use of the Colebrook-White rough law to convert n_l to k_s and the iterative solution of the full Colebrook-White law in evaluating f at each of depth of flow. This direct approach has not been adopted for two reasons:

- Eq. 3.15 is derived from relating the Manning and Darcy-Weisbach resistance equations, which are intended for analysis of the total cross-section resistance rather than the localised boundary friction factor.

- Use of the full Colebrook-White equation allows for smooth and rough turbulent flow conditions i.e. the influence of the Reynolds Number on the flow; solution of the lateral and depth-varying distribution of f; and the use of equation coefficients which capture the cross-section shape effects. To use this approach, the rough turbulent conversion (Eq. 3.16) is required to evaluate k_s.

Although Eq. 3.15 is not used directly, it is still embedded within Eq. 3.16, and it will therefore still have an influence on the results, albeit reduced from the influence of a more direct application.

The n_l predictions based on Eq. 3.13 and 3.16 are plotted in Figure 3.21 for a typical range of n_l values. As n_l approaches infinity, the bracketed term in Eq. 3.16 approaches zero, and k_s is asymptotic to 12.27, the coefficient in the Colebrook-White law. In the Ackers' approach, as n_l increases to ~0.1, k_s tends to infinity. For the region $0.01 < n_l < 0.025$, both laws predict a similar relationship. Neither approach is strictly correct for experimental flumes where the smooth-turbulent law dominates. For example, the values of k_s for $n_l \leq 0.01$ decrease rapidly. In the absence of an alternative approach, the CES conveyance methodology incorporates Eq. 3.13 for this conversion in experimental flumes. Experimental conditions are generally characterised by smooth-turbulent flow conditions, and thus a lower limit for k_s is set, $k_s \geq 0.1$ mm, which corresponds to a minimum n_l of 0.009. This is reasonable since the final f distribution is based on solution of the full Colebrook-White law (Eq. 3.12), within which the rough-turbulent term will have a virtually insignificant contribution, and the distribution of f will be largely based on the viscous effects. For natural channels, the CES conversion is based on Eq. 3.16.

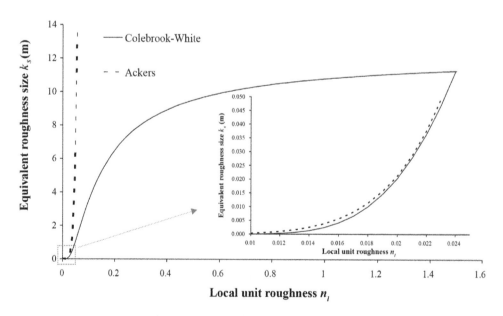

Figure 3.21 Conversion or n_l to k_s for a typical range of n_l values (Mc Gahey et al., 2008).

Conversion of k_s to f

The full Colebrook-White law (Eq. 3.12) is used to evaluate the lateral distribution of local values of f at each depth of flow, which assumes k_s is constant with depth. Eq. 3.12 was originally derived for pipe flow with small-scale roughness features, and when applied elsewhere, it can provide unrealistically large values of f if the k_s/H ratio is large, for example, in natural channels where the roughness size may be comparable with the depth, as with boulders.

Thus, at high k_s/H ratios a power law (Samuels, 1985) is introduced as the mathematical formulation of the velocity distribution in the overlap region of the boundary layer (i.e. between regions dominated by either viscous or inertia forces) is typically of the logarithmic or power law form. The former (the basis for Eq. 3.12) is generally favoured as it covers a wider range of Reynolds Numbers. The CES power law coefficients are derived from ensuring both laws are continuous in f and the derivative of f with respect to depth, df/dH, giving (Mc Gahey & Samuels, 2004):

$$f = \frac{8}{41.3015} \frac{k_s}{H} \qquad (3.17)$$

At $k_s/H = 1.66$, these laws coincide. An upper limit of f is set at $k_s/H = 10$, which corresponds to an f-value of 1.937. This upper limit restricts the value of f from approaching infinity, as H approaches zero for shallow flows. The maximum predicted f in Nikuradse's (1933) experiments is approximately 0.125 in the laminar region and the entire turbulent region falls well below a value of 0.100. In practice, this large value of f would occur in regions with extremely shallow water relative to the roughness elements, which are expected to have a negligible contribution to the overall conveyance. Thus, the exact upper limit of f carries little practical significance.

Figure 3.22 illustrates the combined friction laws for an n_l value of 0.05, with the k_s/H ratio determining the applicable friction law. The Colebrook-White equation gives a rapidly increasing value of f below a stage of 0.5 m, i.e. the region where the k_s/H ratio approaches 12.27. The power law replaces the Colebrook-White law in this region, and the predicted value is restricted to 1.937 for $k_s/H > 10$.

The complete CES methodology for estimating the lateral distribution of f for a given depth of flow, based on the local n_l values provided by the Roughness Advisor is shown in Figure 3.23.

Term (iii) – Representation of turbulence due to transverse currents

The third term in Eq. 3.4 represents the turbulence generated due to transverse currents effecting vertical interfacial shearing. This typically occurs in regions characterised by steep velocity gradients such as the main channel floodplain interface and adjacent to river banks. To evaluate this term, a turbulence closure model is required for the eddy viscosity $\nu_t (m^2 s^{-1})$. The CES methodology incorporates the approach of Cunge et al. (1980),

$$\nu_t = \lambda u_* H \qquad (3.18)$$

where λ is the dimensionless eddy viscosity and the depth $H(m)$ represents the characteristic length scale for the mixing layer. Ideally this model should separate out the

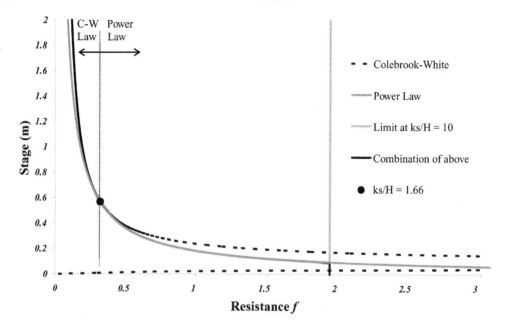

Figure 3.22 Graphical representation of the combined friction laws for $n_l = 0.05$ (Mc Gahey & Samuels, 2004).

boundary and shear generated effects (Wormleaton, 1988). This is not practicable in that the model is intended for a variety of rivers, for which shear layer information may not be readily inferred (Section 3.4). Substituting Eq. 3.18 and 3.8 into Term (iii) of Eq. 3.4 gives,

$$\frac{\partial}{\partial y}\left(\rho \lambda H^2 U_d \left(\frac{f}{8}\right)^{\frac{1}{2}} \frac{\partial U_d}{\partial y}\right) \tag{3.19}$$

Evaluating the dimensionless eddy viscosity λ

The CES methodology incorporates a recently proposed model (Abril, 2001; Abril and Knight, 2004) for estimating the lateral distribution of λ on the floodplain,

$$\lambda = \lambda_{mc}\left(-0.2 + 1.2D_r^{-1.44}\right) \tag{3.20}$$

Here, the relative depth D_r is defined as,

$$D_r = \frac{H}{H_{max}} \tag{3.21}$$

to enable application to inbank flow and D_r tends to 1.0 for the deepest part of the inbank section. The main channel dimensionless eddy viscosity λ_{mc} is 0.07 (Ikeda, 1981) for experimental channels and 0.24 (Elder, 1959; Cunge *et al.*, 1980) for natural

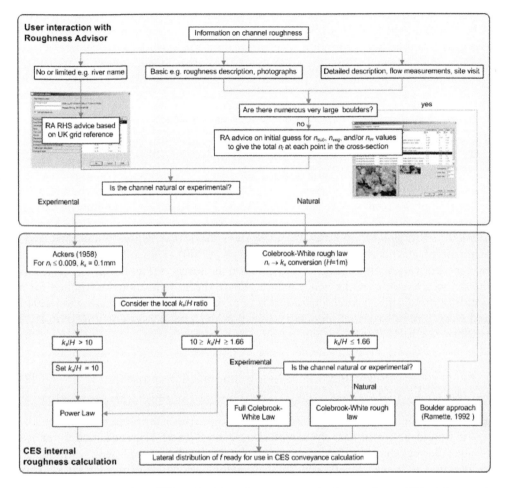

Figure 3.23 Summary of the CES methodology for estimating the friction factor *f* (Mc Gahey *et al.*, 2009).

channels. The 0.07 value is supported by experimental measurements of the viscosity distribution with depth (Nezu & Nakagawa, 1993), where the depth-averaged value is ~0.07. The measured wind tunnel dispersion coefficient value is ~0.13 (Rhodes & Knight, 1995). The values in natural channels are generally larger as the rougher banks and longitudinal irregularities result in more boundary generated turbulence and stronger lateral shearing than in laboratory channels. Elder (1959) suggests a value of ~0.23 for uniform flow in a wide open channel. Main channel and floodplain values as high as 0.5 and 3.0 respectively, have previously been adopted (Knight *et al.*, 1989; James & Wark, 1992).

The original derivation of Eq. 3.20 was based on Flood Channel Facility experimental measurements (*www.flowdata.bham.ac.uk*) for a two-stage channel, with two distinct flow depths for the main channel and floodplain. The area of variable depth between the main channel and floodplain was not considered and Eq. 3.21 would therefore provide two distinct values, of which one was always $D_r = 1.0$ (main channel). In practice, Eq. 3.21 will predict a range of relative depth values across the channel, and thus Eq. 3.20 requires an upper limit for λ, since at the channel edges, as D_r approaches zero, λ approaches infinity. These large λ values result in unrealistic boundary layer effects and the CES approach has adopted an upper limit on λ of 5.

Term (iv) – Representation of turbulence due to secondary circulations

The fourth term in Eq. 3.4 represents the energy transfers generated through transverse currents, coherent structures and horizontal interfacial shear (Knight *et al.*, 2007; Omran *et al.*, 2008; Tang and Knight, 2009). As this term contains an additional unknown, the lateral velocity V, a model is required for closure. The CES evaluation of this term is based on a hypothesised approach for combining two models, one for transverse currents in straight prismatic channels (Shiono & Knight, 1990; Abril & Knight, 2004) and one for the helical-type transverse currents characteristic of meandering channels (Ervine *et al.*, 2000). The contributions of these two models vary with the degree of plan form sinuosity σ (Figure 3.24), defined as the thalweg length over the valley length, where the straight model is applied for a sinuosity of 1.0, the meandering model is applied for a sinuosity greater than 1.015 and a combination of both models is applied in the low sinuosity or transition zone $(1.000 < \sigma < 1.015)$ i.e.

$$\frac{(1.015 - \sigma)}{0.015}[\text{straight approach}] + \frac{(\sigma - 1)}{0.015}[\text{meandering aproach}] \qquad (3.22)$$

This summation method is an arbitrary choice to provide a working model.

It should be noted that predictions based on a depth-averaged model applied to meandering channel flow conditions should be interpreted with caution. In meandering channels, the flow is fully three-dimensional, and the lateral velocity component V has a substantially larger role than in a straight channel. For the CES, x is defined as the horizontal flow direction i.e. in the horizontal x-y plane. In the main channel, the flow direction and hence orientation of x varies throughout the vertical, from being parallel to the main channel banks for inbank flow to being parallel to the valley direction for out-of-bank flow. The depth-averaging assumption is thus

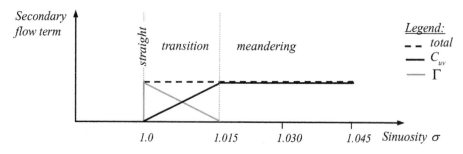

Figure 3.24 Contributions from the two models with increasing sinuosity (Defra/EA, 2003a).

not strictly valid, and the validity decreases with increasing sinuosity. Despite this, previous application of the proposed model has provided reasonable stage-discharge and depth-averaged velocity predictions for a range of sinuosity values (Ervine *et al.*, 2000; Guan *et al.*, 2002; Spooner & Shiono, 2003, Liu *et al.*, 2016). Thus, in the absence of an alternative model within the depth-averaging framework, this model has been adopted for the current (2004) version of the CES.

Secondary flow model for straight prismatic channels

The secondary flow model for straight channels is based on that of Shiono and Knight (1990) in which a secondary flow term Γ is defined as:

$$\Gamma = \frac{\partial}{\partial y}[H(UV)_d] \qquad (3.23)$$

This is a calibration parameter that varies laterally across the section and with relative depth. The laterally averaged boundary shear stress τ_{ave} (Nm^{-2}), and the two-dimensional shear stress τ_d ($= \rho g H S_o$) (Nm^{-2}) can be related for areas of constant depth, and the average scaled secondary flow term, Γ^*_{ave}, is thus (Abril & Knight, 2004),

$$\Gamma^*_{ave} = \rho g S_o (1 - k_\Gamma) \qquad (3.24)$$

a novel approach for linking boundary shear stress and the secondary flow term. For overbank flow (Abril & Knight; 2004),

$$\Gamma_{mc} = 0.15 \rho g H_{mc} S_o \qquad (3.25)$$
$$\Gamma_{fp} = -0.25 \rho g H_{fp} S_o \qquad (3.26)$$

and for inbank flow,

$$\Gamma_{mc} = 0.05 \rho g H_{mc} S_o \qquad (3.27)$$

The coefficients 0.15, −0.25 and 0.05 are calibrated values based on observed distributions derived from the Flood Channel Facility data (Shiono & Knight, 1991) for different relative depths in a straight compound channel. The applicability of these models for natural channels has not been widely tested, in part, as a result of limited availability of suitable river data. More recently (Omran, 2005; Omran *et al.*, 2008), the variation of Γ has been linked to the local nature and orientation of the secondary circulations within the channel section.

For practical application, a key difficulty is that natural channels are not readily divided into distinct regions with a main channel portion and two floodplains with identical bed elevations. For this reason, the CES requires the user to provide information on the estimated top-of-bank positions i.e. where the floodplain sections commence. This information is used to divide the flow domain into distinct flow regions (Figure 3.25), equivalent to those originally defined in Abril and Knight's model. Where the two floodplain bed elevations differ in level or for asymmetrical

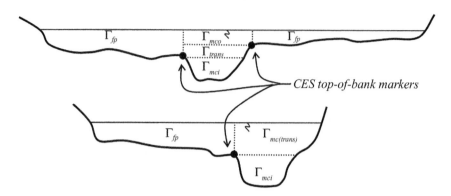

Figure 3.25 Example of natural channels with transitional secondary flow regions (Mc Gahey, 2006).

channels (i.e. only one floodplain present) a linear transition for Γ is applied in the main channel flow region. This allows for an averaging of the secondary flow effects based on the area of influence of the floodplain main channel vertical interface, which drives the magnitude and orientation of the transverse currents. The transitional main channel secondary flow term, $\Gamma_{mc(trans)}$, is given by (Mc Gahey, 2006),

$$\Gamma_{mc(\text{trans})} = \frac{(\Gamma_{mco} - \Gamma_{mci})}{(Z_H - Z_L)}(h_l - Z_L) + \Gamma_{mci} \tag{3.28}$$

where the suffixes i and o denote inbank and overbank flow, respectively, Z_H and Z_L (mAD) are the bed elevations of the higher and lower floodplain top-of-bank marker, respectively, and h_l (mAD) is the water surface level for the given depth of flow. The, $\Gamma_{mc(\text{trans})}$ region may be compared to Regions 1 and 2 in the Coherence Method (COHM – Ackers, 1991–93), where the discharge initially drops (due to the interaction and resistance offered by the floodplains) and then increases above bankfull. Here, however, the non-symmetrical floodplain affects are averaged within the main channel portion and the flow over the single floodplain is treated independently, whereas the COHM is based on a symmetrical compound channel. An alternative approach would be the use of vertically layered models i.e. to incorporate two (symmetrical case) or three (non-symmetrical) layer models.

The extent of the lateral shear layer where these secondary circulations are present is important. In practice, these shear layers extend across the entire main channel region (Samuels, 1988) and over a portion of the flood plain. The flood plain extent of Γ_{fp} is, therefore, approximated in the CES based on the lesser of the main channel width and ten times the average bankfull depth.

Secondary flow model for meandering channels

The meandering secondary flow model is based on the work of Ervine *et al.* (2000), where the secondary currents are related to the depth-averaged velocity by

$$(UV)_d = C_{uv}U_d^2 \tag{3.29}$$

Substituting this expression into Term (iv) of Eq. 3.4 gives

$$-\frac{\partial}{\partial y}[\rho H C_{uv} U_d^2] \qquad (3.30)$$

Ervine *et al.* (2000) applied this model within the main channel portion of the cross-section, with the C_{uv} coefficient attaining a single value related to sinuosity, relative depth and relative roughness. The effect is to introduce a non-symmetrical lateral profile for the depth-averaged velocity U_d in the main channel region and a corresponding decrease in the overall flow capacity.

Expanding Eq. 3.30, assuming C_{uv} is constant across the main channel, gives:

$$-\rho\, C_{uv}\frac{\partial}{\partial y}\left(HU_d^2\right) = -\rho C_{uv} U_d\left[2H\frac{\partial U_d}{\partial y} + U_d\frac{\partial H}{\partial y}\right] \qquad (3.31)$$

The expected distributions of the various bracketed terms for a simple trapezoidal channel are plotted in Figure 3.26, illustrating in particular the various non-symmetrical influences. Thus, evaluation of the right hand side of Eq. 3.31 produces a skewed main channel velocity profile, damping the velocity on falling lateral slopes (i.e. increasing lateral depths of flow $dH/dy > 0$), having little or no effect for horizontal channel beds and enhancing the velocity on rising slopes (i.e. decreasing lateral depths of flow $dH/dy < 0$).

The magnitude and sign of C_{uv} influences the size and orientation of the resulting skew, respectively. Figure 3.27 illustrates the effect of increasing positive C_{uv} values on the depth-averaged velocity profile, for a trapezoidal channel. On the laterally rising channel side slope, the aforementioned velocity enhancement causes an unrealistic increase or 'spike' in the depth-averaged velocity. However, on the laterally falling side slope, the main channel 'skew' effect is similar to that observed in real meandering channel data. Based on this observation, the CES methodology only incorporates the Ervine *et al.* (2000) model in main channel areas where the lateral slopes are less than or equal to zero. More recently, some limitations of this approach have been shown (Tang and Knight, 2009) which will be considered for future versions of the CES.

Evaluating the meandering channel coefficient C_{uv}

The CES evaluation of C_{uv} is based on the plan form sinuosity σ. Previous research (Ervine *et al.*, 2000) has identified an additional dependence of C_{uv} on relative roughness (i.e. main channel versus floodplain roughness) and relative depth. In natural channels, the relative roughness is not easily determined as it requires determination of the two reference locations (e.g. main channel and floodplain), which may not be obvious for an arbitrary shaped channel with heterogeneous roughness. Since there is uncertainty with respect to use of the meandering model within the depth-averaged framework, and coupled with this there is limited available calibration data for testing these dependencies, the CES approach for estimating C_{uv} is based on sinuosity alone.

A simple regression exercise based on 12 cross-sections taken at the bend apex for a small range of channel sinuosities (1.0; 1.0038; 1.012; 1.1; 1.18; 1.374; 2.04)

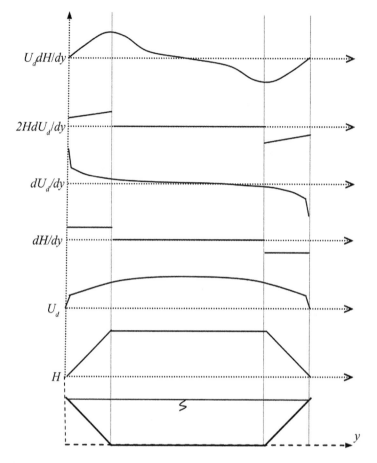

Figure 3.26 Example of meandering flow term lateral distributions in a trapezoidal channel (Mc Gahey, 2006).

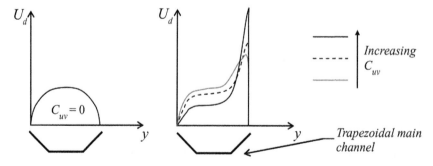

Figure 3.27 Effect of the meandering secondary flow term on the depth-averaged velocity profile in a simple trapezoidal channel (Mc Gahey & Samuels, 2004).

and best-fitting the overall flow rate and depth-averaged velocity profile (Defra/EA, 2003a) for inbank flow gives,

$$C_{uv} = 4.33\sigma^2 - 7.87\sigma + 3.540 \, [1.0 < \sigma \le 2.5 \text{ i.e. } 0\% < C_{uv} \le 11\%] \qquad (3.32)$$

with a regression fit of $R^2 = 0.99$, and for overbank flow

$$C_{uv} = 7.17\sigma - 6.626 \, [1.0 < \sigma \le 2.5 \text{ i.e. } 0.5\% < C_{uv} \le 11\%] \qquad (3.33)$$

with a regression fit of $R^2 = 0.9755$. Here, inbank flow is defined as flow below the average elevation of the top-of-bank markers, i.e. $< (Z_L + Z_H)/2$. An upper limit of 2.5 is provided for the sinuosity as due to the regression approach, the provided equations should only be used within the range for which they were derived. This is not a significant restriction as a channel with a sinuosity greater than 2.5 is very uncommon in nature.

Figure 3.28 illustrates typically observed depth-averaged velocity profiles in meandering channels for inbank and out-of-bank flow conditions. It is apparent that the velocity 'skew' changes orientation from inbank to out-of-bank flow, and for this reason, the sign of C_{uv} is reversed above bankfull flow. This has a marginal impact

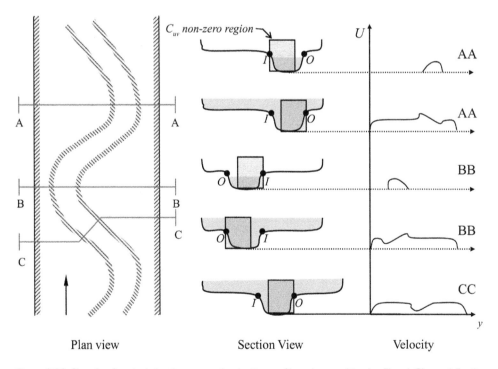

Plan view   Section View   Velocity

Figure 3.28 Sketch of typical depth-averaged velocity profiles observed in the Flood Channel Facility in a meandering two-stage channel for inbank and overbank flows illustrating inside "*I*" and outside "*O*" of bend top-of-bank markers and corresponding non-zero C_{uv} regions (Defra/EA, 2003a).

on the overall flow rate, except in cases where the main channel geometry is highly non-symmetrical, with bank slopes that differ substantially i.e. a large impact on the *dH/dy* term.

Practical considerations for evaluating C_{uv}

The CES requires additional user input to aid evaluation of C_{uv}. This includes the sinuosity magnitude, a directly measurable parameter, and the orientation of the meander bend or channel skew. This information is inferred through labeling the top-of-bank markers as "inside" or "outside" of bend (Figure 3.28). In instances where the cross-section is taken in a cross-over reach, these labels should assume the orientation of the previous bend, as the flow effects from the upstream bend are typically still present.

Practical considerations for secondary flow models in braided channels

In river modelling, a common application for conveyance estimation is in braided channels. The CES methodology calculates flow in braided channels through evaluating the conveyance within each individual channel at low depths, and then applying a simple summation (Figure 3.29).

For the case of overbank flows, where common floodplains or inundated islands are present, the island is treated as an additional floodplain when evaluating the secondary flow term Γ (Figure 3.30).

For the case of overbank flows in channels with varying sinuosity (Figure 3.31), the sinuosity and bend orientation for each channel is provided. For low sinuosities in the transitional range $1.0 < \sigma < 1.015$, the linear summation for the secondary flow models (Eq. 3.22) is applied, and the contributions from each channel in the common floodplain is averaged.

Practical modelling of low velocity regions

Rivers and their floodplains may be characterised by regions of low or zero velocity. Examples of low velocity regions include slowly rotating plan form eddies upstream

Figure 3.29 Addition of flow in braided channels (Defra/EA, 2003a).

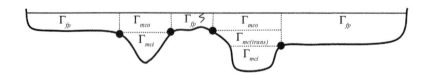

Figure 3.30 Example of straight secondary flow model Γ in a braided channel (Defra/EA, 2004b).

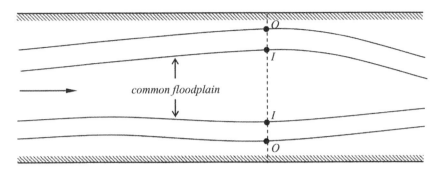

Figure 3.31   Plan view example of a braided channel with low sinuosity, a common floodplain and the appropriate top-of-bank marker labels (Defra/EA, 2004b).

or immediately downstream of narrow flow constrictions and areas within the floodplain such as water trapped between hedges or garden fences perpendicular to the bulk flow direction, off-line natural reservoirs, buildings where the water enters through doors and windows etc. In standard 1-D modelling software such as HEC-RAS or InfoWorks RS, these regions are typically termed "non-conveyance regions"; however, this term is avoided to prevent confusion.

The CES methodology incorporates a simple approach for modelling the low velocity regions within a cross-section. As with current 1-D modelling practice, the user identifies apparent low velocity regions, for example, a portion of floodplain area where the flow is trapped behind a low wall perpendicular to the bulk flow direction. The CES assigns an extremely high roughness value to this region, which significantly reduces the conveyance associated with this region. The equations are then solved at each point in the section, however, when integrating the unit flow rate across the section to determine the overall flow rate (Eq. 3.3), only the conveyance portions are included. The result is that low velocity areas are effectively storage volumes that exert some resistance on the principal flow, but do not contribute to the overall cross-section conveyance. This approach assumes low velocity regions comprise static rotations or storage and does not take account of instances where separation occurs, resulting in the generation of vortices which are swept further downstream.

3.2.3   Summary of CES methods, outputs and solution technique

Complete system of CES equations

The previous sections have detailed the CES methodology for evaluating Terms (i) to (iv) in Eq. 3.4.

$$\rho g H\left(\frac{\partial H}{\partial x} - S_o\right) - \psi\tau_o + \frac{\partial}{\partial y}\left(\rho H \nu_t \frac{\partial U_d}{\partial y}\right) - \frac{\partial}{\partial y}\rho[H(UV)_d] = 0 \qquad (3.4)$$

$$\quad\ \ \text{(i)}\qquad\quad\text{(ii)}\qquad\quad\text{(iii)}\qquad\qquad\quad\text{(iv)}$$

Previous research (Samuels, 1989a; Knight *et al.*, 2004) advocates the solution of the unit flow rate, q (ms^{-2}), over the depth-averaged velocity U_d, due to the strong continuity properties of q with variations in depth, for example, across a vertical face or 'step' in an engineered channel cross-section. The unit flow rate is defined here as,

$$q = \int_{0}^{H} u\, dz = HU_d \tag{3.34}$$

Thus, back-substituting the derived expressions for Terms (i) to (iv) in Eq. 3.4 and expressing the depth-averaged velocity in terms of the unit flow rate q, gives the final equations adopted within the CES conveyance methodology as:

Final CES conveyance equations

$$gHS_o - \psi \frac{f}{8} \frac{q^2}{H^2} + \frac{\partial}{\partial y} \left(\lambda \left(\frac{f}{8} \right)^{1/2} Hq \frac{\partial}{\partial y} \left(\frac{q}{H} \right) \right) - \left(\frac{(1.015 - \sigma)}{0.015} \right) \Gamma$$

$$+ \frac{(\sigma - 1.0)}{0.015} C_{uv} \frac{\partial}{\partial y} \left(\frac{q^2}{H} \right) \Biggr) = 0$$

$\sigma = 1.0$ inbank; $1.0 \le \sigma \le 1.015$ overbank

$$gHS_o - \psi \frac{f}{8} \frac{q^2}{H^2} + \frac{\partial}{\partial y} \left(\lambda \left(\frac{f}{8} \right)^{1/2} Hq \frac{\partial}{\partial y} \left(\frac{q}{H} \right) \right) - C_{uv} \frac{\partial}{\partial y} \left(\frac{q^2}{H} \right) = 0$$

$\sigma > 1.0$ inbank; $\sigma > 1.015$ overbank   (3.35a&b)

Eq. 3.35a&b are evaluated to determine the lateral distribution of the unit flow rate q and the total flow rate Q (m^{3}s^{-1}) is evaluated from:

$$Q = \int_{0}^{B} q\, dy = \int_{0}^{B} U_d\, dA \tag{3.3}$$

Table 3.4 provides a summary of the expressions for evaluating the unknown input parameters in Eq. 3.35a&b. Figure 3.32 provides an example of typical lateral distributions for f, λ, Γ and C_{uv} in a two-stage trapezoidal channel.

The CES method assumptions

The CES method derivation includes various assumptions. Table 3.5 provides a hierarchical summary of the assumptions made prior to developing the depth-averaged RANS equations (Eq. 3.4), identifying those which may have a significant impact on

Table 3.4 Summary of parameter evaluation for the final CES equations (Eq. 3.35a&b).

Parameter	Evaluation technique	Eq. No.
Gravitational acceleration g	9.81 ms^{-2}	
Longitudinal bed slope S_o	User input taken from best available measurements e.g. OS maps, water levels measurements, model outputs	
Lateral bed slope S_y	Evaluated from the surveyed cross-section	
Cartesian projection parameter ψ	$\psi = \sqrt{1 + S_y^2 + S_o^2}$	3.6
Friction factor f	$\dfrac{1}{\sqrt{f}} = -2.03\log_{10}\left[\dfrac{k_s}{12.27R} + \dfrac{3.09}{Re\sqrt{f}}\right] k_s/H < 1.66$	3.12
	$f = \dfrac{8}{41.3015}\dfrac{k_s}{H} \; k_s/H > 1.66$	3.17
	Upper limit $f = 1.937$; Lower limit $k_s = 0.0001$ m	
Dimensionless eddy viscosity λ	$\lambda = \lambda_{mc}\left(-0.2 + 1.2D_r^{-1.44}\right)$	3.20
	λ_{mc} experimental $= 0.07$; λ_{mc} rivers $= 0.24$	
Straight secondary flow term Γ	$\Gamma_{mci} = 0.15\rho g H S_o$	3.27
	$\Gamma_{mco} = 0.15\rho g H S_o$	3.25
	$\Gamma_{fp} = -0.25\rho g H S_o$	3.26
	$\Gamma_{mc(trans)} = \dfrac{\left(\Gamma_{mco} - \Gamma_{mci}\right)}{\left(Z_H - Z_L\right)}\left(h_i - Z_L\right) + \Gamma_{mci}$	3.28
Sinuosity σ	User input. Measured from OS map.	
Meandering secondary flow term C_{uv}	$C_{uv} = 4.3274\sigma^2 - 7.8669\sigma + 3.5395$ inbank	3.32
	$C_{uv} = 7.1659\sigma - 6.6257$ out-of-bank	3.33

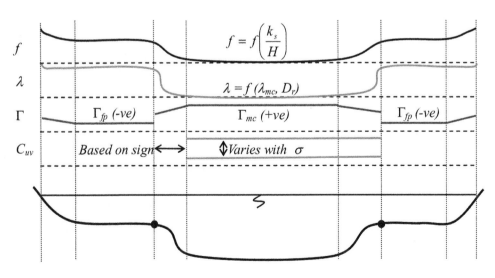

Figure 3.32 Example of lateral distributions for f, λ, Γ and C_{uv} in a typical two-stage channel.

Table 3.5 Summary of derivation assumptions, the implications and their significance.

Assumption	Implication	Significance to CES applications
Standard global		
The density is constant throughout the flow domain	i Air entrainment may vary the density and/or depth of flow.	Low
	ii Sediment concentration may affect density.	Low
	iii Tidal water is saline e.g. estuaries.	Low
	iv Heat dissipation in rivers at power stations.	Low
The flow is incompressible	Density is constant.	Low
Hydrostatic pressure distribution	i Angle of longitudinal bedslope with horizontal is sufficiently small such that the vertical depth of flow is not significantly different from the depth measured perpendicular to the channel bed.	Low
	ii The longitudinal bedslope is not curved in a convex/concave manner.	Low
	iii Flow does not pass over any obstructions or areas of bedslope sufficiently steep for air entrainment to occur.	Medium
The flow is steady	Cannot model event hydrograph, capture the change in water slope for the rising and falling hydrograph limbs or incorporate flood attenuation.	Medium
Coriolis forces resulting from the rotation of the earth about its own axis are considered small relative to other body forces	i <12 mm change in water level in UK rivers.	Low
	ii Up to 3.5 m in the tidal range between the Irish and Welsh coastline.	High
Gravity is the only body force acting on the fluid		Low
Surface forces due to wind shear	Significant force may occur at gale force winds. These depend on the 'fetch' length in, for example, tidal areas and straight drains. In rivers, these my cause set-up and affect pump drainage.	Medium
Centrifugal forces due to streamline curvature in meander bends are small relative to other body forces		Medium
Isothermal flow	Temperature variations in open channel flow are small relative to other influences.	Low
Velocity adjacent to the channel bed is zero	Velocity at bed should average to zero over time i.e. may be instantaneously positive or negative.	Low
The fluid is isotropic		Low

(Continued)

Table 3.5 (Continued).

Assumption	Implication	Significance to CES applications
3-D to 1-D simplification		
Streamlines are approximately parallel		Low
The flow is uniform	The cross-section is representative of a reach length, where the section shape (and water surface slope) is the same throughout the reach. In reality, non-engineered river channels vary in section shape.	Medium
Eddy viscosity	The 3-D eddy viscosity is depth-averaged in the vertical, reach-averaged in the longitudinal and only assumed to vary transversely across the channel.	Medium
The momentum equation is applied in the streamwise direction only	Ignores any large transverse or vertical currents i.e. lateral velocity movements. [Here, an energy balance approach would have included the various meandering effects.]	High
Vertical accelerations are negligible		Low
Detailed method specific		
Rigid bed	Most UK rivers are not alluvial in nature, however the cross-section shape may change over time.	Low-medium
Depth-averaging	i Is not stratified ii Does not change direction in the vertical (strong meanders) iii Is well-mixed in the vertical.	Medium
Turbulence modelled through RANS approach	Complete modelling of the physical properties	Medium

the predicted output. Based on these, the CES conveyance calculation is not strictly applicable for:

- flow in alluvial channels with large bedforms;
- fully three-dimensional flows such as major confluences, strongly meandering channels and adjacent to structures such as bridges, weirs and culverts (See Section 3.2.5 on afflux estimation); and
- for determining the stage through major flood hydrographs where the water surface slope is unknown. This may be handled in 1-D hydrodynamic software such as InfoWorks RS, where the changing surface water slope is calculated with time and the conveyance is based on the embedded CES calculation.

There is further work in progress addressing some of the more important assumptions.

Additional CES conveyance calculation outputs

The CES conveyance calculation is primarily focused on evaluating the stage-conveyance and stage-flow relationship. Additional outputs are also evaluated (Table 3.6) for each depth of flow for the whole cross-section, including:

- Average velocity U (ms^{-1})
- Cross-sectional area A (m^2)
- Perimeter P (m)
- Width W (m)
- Back-calculated, all-encompassing Manning n
- Froude Number Fr
- Reynolds Number Re

Table 3.6 Summary of additional CES conveyance outputs and their evaluation.

Output	Equation	Comment
U	$U = Q/A$	
A	$A = \sum_{i}^{n} A_i$	This area A_i of each integration slice is summed. These areas are a trapezoidal or rectangular shape other than the triangular end slices.
P	$P = \sum_{i}^{n} P_i$	The boundary perimeter P_i of each integration slice is summed. This is the section linking successive points.
W		Top width i.e. width at the water level being evaluated
n	$n = \dfrac{AP^{2/3}S^{1/2}}{Q}$	This is based on the overall cross-sectional area i.e. floodplains and main channel and the user entered slope. The 'n' value is equivalent to the all-encompassing Manning n value, which takes account of plan form and cross-section shape and turbulence due to lateral shear and secondary currents.
Fr	$Fr = \dfrac{U}{\sqrt{g(A/W)}}$	Here, A/W represents the hydraulic mean depth.
Re	$Re = \dfrac{U(A/P)}{\nu}$	Here, A/P represents the hydraulic radius.
B_o	$B_o = \dfrac{\sum_{i}^{n} U_{di}A_i^2}{UA^2}$	The Boussinesq or momentum coefficient provides a measure of the local velocity variations across a section compared to the average values. B_o is never less than 1.0, is equal to 1.0 if the flow is uniform across the section and rarely exceeds 1.05.
C_o	$C_o = \dfrac{\sum_{i}^{n} U_{di}A_i^3}{UA^3}$	The Coriolis or energy coefficient provides a measure of the local velocity variations across a section compared to the average values. C_o is never less than 1.0, is equal to 1.0 if the flow is uniform across the section, rarely exceeds 1.15 in simple channels and is ~2.0 for two-stage channels.
U_d	$U_d = \dfrac{1}{h}\int_0^h u\,dz$ (Eq. 2.47)	Depth-averaged velocity derived from the unit flow rate.
τ_o	$t_o = \dfrac{f}{8} - \rho U_d^2$ (Eq. 2.34)	Boundary shear stress derived from the depth-averaged velocity and friction factor.
u_*	$U_* = \sqrt{\tau_o/\rho}$ (Eq. 2.31)	Shear velocity derived from the boundary shear stress.

- Boussinesq coefficient B_o
- Coriolis coefficient C_o

An advantage of the chosen conveyance approach is that it also determines the lateral variation across the section of the following flow parameters:

- Depth-averaged velocity U_d (ms^{-1})
- Boundary shear stress τ_o (Nm^{-2})
- Shear velocity u_* (ms^{-1})

These parameters are important for the consideration of wider issues such as geomorphology and ecology (Chapter 5).

The CES conveyance calculation solution technique

Eq. 3.35a&b form a set of non-linear, elliptic, second order partial differential equations. The CES employs a numerical solution based on the Finite Element Method (FEM) (Zienkiewick, 1977; Davies, 1980), which is well suited to the solution of elliptic equations. The cross-sectional area of flow represents the solution domain, which is discretised laterally into a number of elements (Figure 3.33). The variable q is replaced with piecewise linear approximations or 'shape functions'. The result is a system of discrete equations, which can be linearised, assigned boundary values and solved iteratively.

Linearising the system of equations

Eq. 3.35a&b is linearised by expressing the unit flow rate 'q' for iteration $n + 1$ in terms of the unit flow rate of the previous iteration n,

$$q^{n+1} = q^n + \chi \Delta q^n \tag{3.36}$$

where χ is the relaxation factor and Δq^n is the incremental change in q^n to improve the previous estimate. The initial estimate for q^n is based on the Manning equation applied locally to each fluid element, and thereafter q^n is a known quantity based on the previous iteration. Substituting Eq. 3.36 into Eq. 3.35a&b and expressing the result as a system of linear equations for the solution of Δq^n, yields

$$\mathbf{M}[\Delta q^n] = \mathbf{x} \tag{3.37}$$

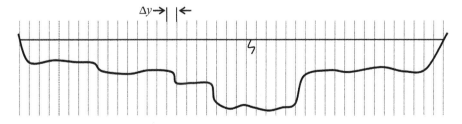

Figure 3.33 Example of the cross-section discretisation (Mc Gahey, 2006).

where $\mathbf{M}$ is the linear operator defined by

$$\mathbf{M}[\Delta q^n] = A\frac{\partial^2 \Delta q^n}{\partial y} + g\left(\frac{\partial \Delta q^n}{\partial y}, \Delta q^n, y\right) \tag{3.38}$$

and $\mathbf{x}$ is the vector of all contributing functions that are independent of Δq^n.

Incorporating piecewise functions

The FEM provides an approximate solution for $\Delta q^n(y)$ through piecewise functions in a total of E_e elements. Thus for a general element i, an approximation is sought in such a manner that outside i (Davies, 1980),

$$\Delta q_i^n(y) = 0 \quad i = 1,...,E_e \tag{3.39}$$

It follows that

$$\Delta q^n(y) = \sum_{E}^{i=1} \Delta q_i^n(y) \tag{3.40}$$

Linear shape functions ϕ are introduced where,

$$\Delta q_i^n(y) = \sum \Delta q_i^n \phi_i(y) \tag{3.41}$$

and ϕ is linear in y. Linear shape functions are the lowest order polynomials used in FEMs. A FEM using linear shape functions is similar to second order finite difference schemes. Higher order polynomials provide the option for fewer elements as they describe the parameter distribution more closely within each element. However, for the case of river channels, there is value in having more elements to capture the local variations in the input unit roughness and cross-section geometry. Thus, the CES is set to a default 100 elements and the linear shape functions are considered sufficient to capture the parameter variations across each of these elements.

Similar approximations can be introduced to describe the f, H and ψ distributions. Eq. 3.37 is integrated over the flow domain R_d,

$$\int_{R_d} \mathbf{M}[\Delta q^n]\phi(y)dy - \int_{R_d} x\phi(y)dy = 0 \tag{3.42}$$

Substituting the piecewise linear functions into Eq. 3.38 yields,

$$\sum_{E_e} \mathbf{M}\left(\sum \Delta q_i^n \phi_i(y)\right)\phi_j(y) - \sum_{E_e}\left(\sum x_i \phi_i(y)\right)\phi_j(y) = 0 \tag{3.43}$$

The numerical integration is based on Gauss-Legendre quadrature (Zienkiewick, 1977) with two sampling points on each flow element. Δq^n is evaluated for each element, q^n is then updated and the procedure is repeated until a specified tolerance on Δq^n is satisfied.

Boundary conditions

The Dirichlet boundary condition, $q = 0$, is prescribed at the boundary nodes at the edges of the flow domain. The nature of the numerical solution ensures continuity of q, i.e.

$$q_i = q_{i+1} \tag{3.44}$$

at the element boundaries. Knight *et al.* (2004) considered the boundary conditions applied at panels for the analytical solution of the SKM approach. As with Samuels (1989a), the continuity of q rather than U_d is advocated where the channel geometry has a vertical interface or 'step'. The numerical FEM solution intrinsically implies continuity of force at each element interface, as the cross-section discretisation results in any vertical interfaces being approximated through a near vertical wall (i.e. almost 90 degrees). For the analytical solution, the boundary conditions between elements or panels are prescribed and thus vertical interfaces should be treated with care. Knight *et al.* (2004) illustrate the importance of including the shear force on the vertical wall in the force balance to achieve continuity in the unit flow rate.

Iteration strategy

The iterative procedure for determining q^{n+1} is based on Eq. 3.36 with a default relaxation of 1.0. The solution is designed to converge nearly quadratically. It is therefore unlikely to take more than 5–7 iterations to reach the iteration tolerance of $\Delta q \le 0.001$ m²s⁻¹. In rare instances, for example, sections with a steep lateral roughness gradient, the solution may take longer to converge. Thus, to reduce the iteration requirement in achieving the solution tolerance, an approximate function for f at iteration $n + 1$,

$$f^{n+1} = f^n + \frac{\partial f^n}{\partial q^n} \Delta q^n \tag{3.45}$$

is substituted into Eq. 3.35a&b. The $\partial f / \partial q$ term is approximated through numerical differentiation (Newton's Method, 1664),

$$\left(\frac{df}{dq} \right)^n \approx \frac{f((1 + per) \cdot q) - f(q)}{per \cdot q} \tag{3.46}$$

where the percentage '*per*' is in the range $0 < per \le 0.5\%$.

Relaxation strategy

As outlined above, the iterative procedure for determining q^{n+1} is based on Eq. 3.36. If the following condition occurs,

$$\Delta q^n < 0 \quad \text{and} \quad |\Delta q^n| \ge q^n \tag{3.47}$$

then q^{n+1} will be negative. The interpretation of $q^{n+1} \le 0$ (or flow leaving the system) is that the previous 'q^n' estimate was too high and thus the solution is overcorrecting.

This typically occurs at low flow depths, for example, floodplain depths just above bankfull. The implication for the CES methodology is that in the Colebrook-White equation (Eq. 3.12), where $Re = f(q^n)$, the logarithm (base 10) term is zero or a negative number leading to an infinite number. Two possible solutions are identified: (i) set q^n as a small positive number or (ii) alter the value of the relaxation parameter χ. The first is not easily applied, as it requires interpretation of what 'small' is. The CES relaxation strategy is therefore based on the second.

The final system of equations is solved directly with a tri-diagonal matrix solver (Press *et al.*, 1996) appropriate for the resulting diagonally dominant and well-conditioned matrix coefficients. This solves linear simultaneous equations using Gaussian elimination and back-substitution. The equations and their finite element approximations are provided in Appendix 1.

3.2.4   Backwater calculation methods

Backwater profiles occur upstream of structures such as reservoirs, weirs, bridges and free outfalls where the upstream flow is controlled by the downstream structure. The backwater extends from where the control or disturbance to normal depth occurs to the upstream location where normal depth is re-established. These lengths may be tens of kilometres long (e.g. Table 3.7), having a substantial impact on flow depths far upstream of the control point (e.g. a new housing development or a flow gauging weir). A range of methods exist for determining backwater profiles using direct, graphical or numerical integration, for example, the direct or standard step methods (see Chow, 1959; Henderson, 1966). The CES adopts a simple iterative (as the velocity head is incorporated) energy balance approach to estimate the backwater profile.

The backwater calculation starts at the downstream control of known depth and flow rate and moves upstream, applying energy balances at each consecutive cross-section. The backwater effect is present until ~99% of normal depth is achieved at some backwater length L (m) upstream of this downstream control (Figure 3.34). Where either the flow rate or the flow depth is known, the CES backwater approach

Table 3.7 Backwater lengths for some UK Rivers (Samuels, 1989b).

River	Location	Depth (m)	Slope S_o	Backwater length (km)
Avon	Pershore	4.2	0.46	6.90
Cleddau	Haverfordwest	2.5	1.70	1.00
Lagan	Dunmurry	2.5	1.20	1.40
Lagan	Lisburn	3.0	0.29	7.20
Lagan	Banoge	2.0	1.30	1.10
Little Ouse	Lakenheath	2.5	0.05	35.00
Rhymney	Llanedeyrn	2.0	2.20	0.64
Severn	Shrewsbury	4.5	0.50	6.90
Severn	Tewkesbury	8.5	0.10	59.50
Soar	Loughborough	3.5	0.25	9.80
Stour	Christchurch	4.5	1.0	3.20
Taff	Cardiff	6.0	2.5	1.70
Wharfe	Otley	2.0	1.8	0.76

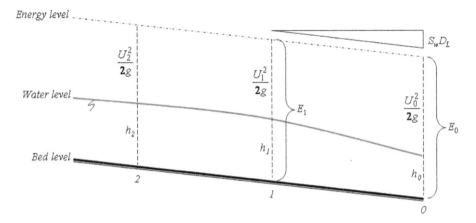

Figure 3.34 Backwater calculation method from downstream (cross-section 0) to upstream (cross-section 2).

may be used to calculate the corresponding normal depth of flow or discharge, respectively. This is then used as the downstream control.

At the control point, the flow Q (m³s⁻¹) and corresponding depth h_0 (m) is known. The conveyance for this depth of flow, K_0 (m³s⁻¹), is estimated for the downstream cross-section with the CES conveyance calculation (Section 3.2.2). The water surface slope S_w is approximated from:

$$S_w = \left(\frac{Q}{K_0}\right)^2 \tag{3.48}$$

The section average velocity U_0 is calculated from:

$$U_0 = \frac{Q}{A_0} \tag{3.49}$$

The total downstream energy level E_0 is determined from:

$$E_0 = h_0 + \frac{U_0^2}{2g} \tag{3.50}$$

The surface water slope is extrapolated to predict the energy level E_1 for the upstream cross-section i.e.

$$E_1 = E_0 + S_w D_L \tag{3.51}$$

where D_L is the streamwise chainage between the two cross-sections. The upstream depth of flow h_1 is then solved iteratively using a Newton-Raphson Solver i.e.

- set h_1 as E_1 for the initial estimate
- calculate the cross-section area A_1 (by-product of the CES calculation)

- calculate the velocity U_1 (Eq. 3.49)
- calculate the energy level E_1 from (Eq. 3.51)
- if E_1 (step above) is equal to E_1 (Eq. 3.51) within a tolerance of $\pm$ 0.001 m stop, if not update $h_1 = h_1 - E_o/E_1$ and repeat. Note, if $E_1 < 0.0$, then update $h_1 = 1.1 * E_1$.

The conveyance K_1 is then estimated for this upstream cross-section and the procedure is repeated moving progressively further upstream (Figure 3.34).

The CES backwater approach assumes:

- subcritical flow (for subcritical flow where the channel bed is steep, the calculation may result in a negative upstream energy head, and hence a negative depth, being calculated);
- single reaches i.e. no branches, confluences, junctions, loops or spills into floodplain storage areas;
- steady flow conditions i.e. the flow rate is constant with time.

3.2.5   Afflux Estimation System methods

The AES methods for bridge and culvert structures are designed to model

- discharge through the structure,
- upstream and downstream transition reaches,
- free surface flow through the structure,
- pressurised flow through the structure when it is flowing full,
- overtopping flow when the structure is drowned.

The AES methods are based on a one dimensional analysis, which is a simplification of the physics observed in reality, as discussed in Chapter 2. The analysis is based on estimation of energy losses as water flows through the structure. This simplification of the physics requires the use of empirical or semi-empirical approximations, many of which are derived from previous work reviewed in Chapter 2 and references cited therein. The AES methods incorporate both re-analysis of some existing data sets (HEC, 1995) and also new data (Atabay and Knight, 2002, Seckin *et al.,* 2008b) to provide the best available empirical support. Figure 3.35 illustrates the longitudinal schematisation adopted in the AES (note that the order of numbering for cross sections differs from some other presentations). Figure 3.35 includes a normal depth profile for the 'unobstructed stream', representing uniform flow conditions in the channel that would exist if the structure were not present. Section 1 is interpreted as the location where the water surface profile downstream of the bridge would depart from the normal depth.

There are many methods available for calculation of afflux or head loss at bridges. One of the practical difficulties can be deciding which calculation approach is most suitable. In many cases, there will be several formulae that are relevant, especially when there is a need to analyse a wide range of flows in a channel from low flow through to extreme flood conditions. The AES software combines calculations for flows through and, where necessary, over a structure.

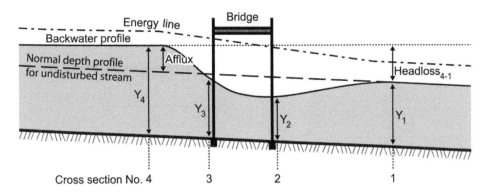

Figure 3.35 Long section profile schematic for the Afflux Estimation System bridge analysis.

Bridge flow modes represented in the AES

At least nine modes of flow may occur through a bridge structure, depending on the flow depth and the upstream and downstream conditions. Figure 3.36 illustrates these modes of flow. The flow modes are generally subdivided as sub-soffit flows and super-soffit flows. For the case where a bridge deck has an underlying girder, the soffit is taken as the elevation at the bottom of this girder. And for the case where a bridge deck has an overlying parapet, the road elevation is taken as that at the top of the parapet. (These overtopping elevations may also sometimes be called the 'relief elevation'). The following modes are illustrated in order of increasing water level at the bridge structure:

Mode 1 – Supercritical flow occurs when the flow is shallow relative to the normal depth and is associated with a Froude Number (*Fr*) greater than unity. For this flow, the afflux is theoretically zero since no backwater can exist. There will, however, be a local bridge energy loss due to form drag. If this mode is encountered in the AES, then the flow is set to critical depth by default.

Modes 2 and 3 – When critical flow occurs at the structure, the flow is said to be choked. This is because critical flow is a condition of minimum energy at constant discharge, and any increase in potential energy upstream will increase the upstream water level but not increase the critical flow level. This may occur, for example, when the bridge entry becomes partially blocked by debris. The longitudinal water surface profile may change gradually (Mode 2) or rapidly (Mode 3 as a hydraulic jump) downstream. Where supercritical flow is detected between the upstream and downstream faces of the bridge, the flow defaults to critical depth in the AES.

Mode 4 – Flow throughout the structure is subcritical and is modelled using a backwater calculation.

Mode 5 – Sluice gate flow occurs when the upstream water level is greater than the soffit level and the downstream soffit level is not submerged. The upstream soffit is effectively acting as a sluice gate, and associated sluice gate equations are used. In common with the Water Surface Profile Program (WSPRO, 1986), sluice gate flow is not modelled in the AES as the mode only occurs over a small region of the discharge rating.

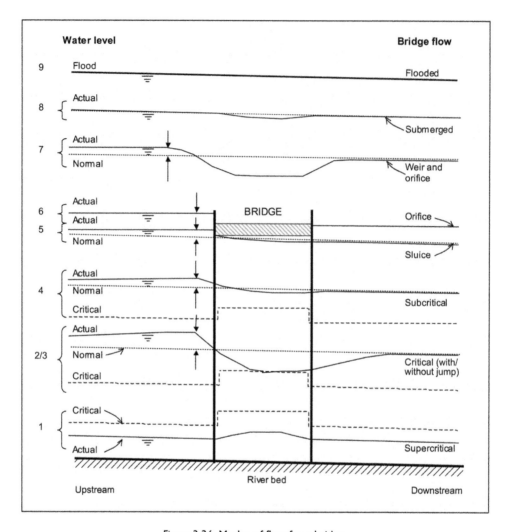

Figure 3.36 Modes of flow for a bridge.

Mode 6 – When the downstream soffit level is submerged, the flow under the bridge becomes fully pressurised. Since the bridge length is rarely much greater than the flow depth, the pressurised flow acts more like a submerged orifice than a closed conduit. The flow is therefore described by orifice flow equations in the AES.

Mode 7 – Once the upstream water level is above the road or parapet, a weir flow will ensue. If the weir flow submergence (downstream water level above road level divided by upstream water level above road level) is less than about 85%, then the weir flow is modular, and its discharge coefficient is determinable (Ackers *et al.*, 1978). Note that pressure flow is still occurring beneath the bridge (provided it is not totally obstructed). The simultaneous weir flow and pressure flow are calculated in the AES using an iterative procedure.

Mode 8 – As the downstream flow level increases further, the weir becomes submerged and the flow becomes a non-modular weir flow. This occurs for a submergence of between 85% and 95%, and the discharge coefficient is accordingly reduced. As for Mode 7, the simultaneous weir flow and pressure flow are calculated in the AES using an iterative procedure.

Mode 9 – When a submergence of about 95% is reached, the weir is drowned and the structure is assumed to act in effect as a feature of the river bed profile. As for Mode 7, the flow through the structure continues to be calculated.

In summary, the AES computes solutions for the most important flow modes, that is Mode 4 (subcritical), Mode 6 (orifice) and Modes 7, 8 and 9 (modular, submerged and drowned weir respectively). Supercritical flow (Modes 1, 2 and 3) defaults to critical depth. The detail of the computational limits used for each mode is given in Table 3.8.

Culvert flow modes represented in the AES

The AES incorporates a culvert calculation module that automates methods set out for hand calculation in the CIRIA Culvert Design Guide (CIRIA, 1997) and the Federal Highway Administration (FHWA, 2001) manuals. As for bridges, the AES computes overtopping weir flows for culverts to allow the user to obtain a full upstream rating curve.

Culvert flow may be controlled by high resistance to flow at the inlet structure or at the outlet due to a long barrel structure. If the inlet structure allows the least discharge for the same upstream energy as applied to the outlet, then the culvert is under inlet control and the flow is usually supercritical. If the culvert outlet allows the least discharge for the same upstream energy as applied to the inlet, then the culvert is under outlet control and the flow is usually subcritical. Since the flow control determines the computation of a water surface profile, it must first be determined for both free and full flow conditions.

Table 3.8 Computational limits for bridge modes in the AES.

Flow mode	Bridge flow	Start condition	Finish condition	Constraints
1	Supercritical	Fr > 1 upstream	Fr = 1 upstream	Flow defaults to critical depth
2–3	Critical	Fr = 1 within bridge		Flow defaults to critical depth
4	Subcritical	Minimum river level	Section 3 level > soffit level	Supercritical flow defaulted to critical
6	Orifice flow	Section 3 level > soffit level < road level	Section 3 level > road level	
7	Weir and orifice	Section 3 level > road level	Submergence > 0.85	Defaults to critical depth at downstream face
8	Submerged weir and orifice	Submergence > 0.85	Submergence > 0.95	Defaults to critical depth at downstream face
9	Drowned weir and orifice	Submergence > 0.95	Maximum river level	

The AES initially calculates the inlet control energy using the FHWA (2001) methods for either an unsubmerged or submerged inlet condition, depending on the flow conditions. If both the inlet and outlet are submerged and the culvert is flowing under inlet control, a hydraulic jump and possible instability in the water surface profile may occur in the culvert. This is not a flow mode that would be designed for and is ignored in the AES for simplicity. The AES calculates the outlet control energy (for the same discharge) using the standard step backwater method to model the water surface profile through the culvert barrel. The higher of the two energies determines the control of the culvert flow for any given combination of discharge and tailwater depth. When the culvert outlet is nearly submerged under outlet control, the culvert flow becomes equivalent to that of full conduit flow. When the culvert inlet water level reaches the road level, modular weir flow, submerged and drowned weir flow can occur sequentially with increased flow elevation.

The AES therefore computes a total of eight culvert flow modes, as illustrated in Figure 3.37 and described below.

Modes 1 and 2 – Inlet control condition (submerged or free inlet flow). The high friction at the culvert entry causes a high energy loss and water level fall. The FHWA (2001) formulae are used to calculate the energy loss at the inlet.

Mode 3 – Drawdown outlet control. The outlet water level is below the culvert normal depth, and thus a drawdown profile is developed within the culvert, returning to normal depth upstream for a long culvert. In the AES, the water surface profiles are computed using the standard step method.

Mode 4 – Backwater outlet control. The outlet water level is above the culvert normal depth, and thus a backwater profile develops within the culvert, which may revert to normal depth upstream for a long culvert.

Mode 5 – Submerged inlet, full flow, outlet control. It has been empirically shown by FHWA (2001) that when the outlet water level is greater than $(d_c + D)/2$, where d_c is the culvert critical depth and D is the barrel depth (see Figure 3.48), then the culvert friction loss is the same as if it were in full flow. This criterion is therefore used as an outlet condition in the AES if the associated outlet depth is less than D and $(d_c + D)/2$.

Mode 6 – Inlet water level above road level, weir flow. If the weir flow submergence (downstream water level above road level divided by upstream water level above road level) is less than about 85%, then the weir flow is modular, and its discharge coefficient is determinable (as for the bridge flow). Pressure flow is still occurring within the culvert. The simultaneous weir and pressure flow are calculated for each discharge rating in the AES using an iterative procedure.

Mode 7 – Non-modular weir flow. This occurs for submergence between 85% and 95%, and the discharge coefficient is accordingly reduced (as for the bridge flow). The AES continues to solve for the pressure flow through the culvert.

Mode 8 – Drowned weir. For submergence of 95%, the weir is drowned. See bridge Mode 9 above.

The AES assumes that the culvert is of uniform slope, dimensions, material and condition. Under these assumptions, culverts of any length can be computed. A minimum culvert rise or span of 0.45 m is recommended and the culvert invert slope is assumed to be equal to the river channel slope.

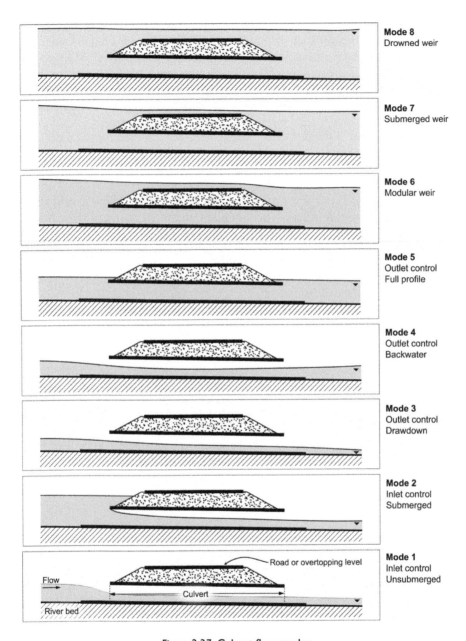

Figure 3.37 Culvert flow modes.

Channel conveyance

In the 1-D view adopted by the AES, afflux is related to the energy losses that occur when a channel is obstructed by a bridge or culvert. To determine the afflux from measured data, it is necessary to analyse both the flow conditions for the obstructed channel and those that would exist without the structure. Hence it is necessary to model the conveyance through the channel with and without the structure.

The AES uses the divided channel method to calculate channel conveyance. Figure 3.38 illustrates a channel cross-section similar to the River Main in Northern Ireland (Section 2.2.2). Eight offset-elevation points have been used to draw the geometry, and the cross-section may be vertically subdivided into seven areas (A1-7) for any of the incremental depths from zero to 5 m, which is the maximum side boundary elevation.

The cross section is divided into three panels, the left floodplain, main channel (MC) and right floodplain, separated by location markers at left overbank (LOB) and right overbank (ROB) positions. The wetted perimeter (P) and top width (T) can also be calculated for each of the sub-areas.

The AES sums A, P and T for each of the sub areas at each flow depth, and computes A, P and T for each of the left floodplain, main channel and right floodplain panels. Manning's equation is then used to compute the conveyance (K) for each flow panel as follows:

$$K = [A(A/P^{2/3})/n] \tag{3.52}$$

where n is the Manning friction coefficient for each panel. The channel conveyance curve is derived as the relationship between the sum of the panel conveyances and the water surface elevation. Additionally, the kinetic energy coefficient (α) is computed as:

$$\alpha = \sum (K_i^3/A_i^2)/(K^3/A^2) \tag{3.53}$$

where K_i and A_i are the panel conveyance and flow area respectively, and K and A are the total cross section conveyance and flow area.

The use of Manning's equation rather than the CES method is expedient for three reasons. Firstly the CES and AES developments were parallel projects and so final outputs of the CES research were not available from the outset in developing the AES. The second reason is that although the CES method could in principle be applied for

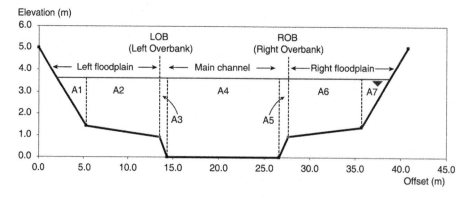

Figure 3.38 Divided channel method for the River Main.

channel conveyance, it does not yet handle the range of cross section geometries needed for the AES such as arches and pipes (see Section 6.2 on further developments). The third reason is that the AES development required numerous 1-D hydraulic models to be built to enable analysis of laboratory and field data for calibration. The USACE software HEC-RAS was used for this purpose, in part for consistency with earlier analysis of field scale data. The conveyance calculations in HEC-RAS are based on Manning's equation.

Bridge conveyance

The AES includes five bridge opening types. There are three arch bridge types, with parabolic, elliptical or semi-circular arch geometry, which are specified by the user in terms of the springer and soffit positions (offset and elevation). There is a 'custom' arch geometry, which allows a user to specify an arbitrary symmetrical arch profile. There is also a beam bridge type, which represents a beam bridge with continuous piers that are assumed to extend from upstream to downstream for the complete bridge length and to be of constant cross section. In all cases, the AES allows for multiple openings, which may have different shapes and sizes. The bridge conveyance calculation requires a Manning friction coefficient, which is assumed to be the same for all openings for a bridge with multiple spans. The recommended values have been adapted from the CIRIA (1997) recommendations for culvert surfaces.

Channel conveyance adjacent to the faces of an arch bridge is calculated within four vertical 'panels' defined in terms of the water level as follows:

*   *Minimum elevation (Zmin) to springer level (SPL)*: Conveyance is computed for the obstructed channel cross section (including bridge abutments/piers and using a roughness coefficient for the combined structure and channel)

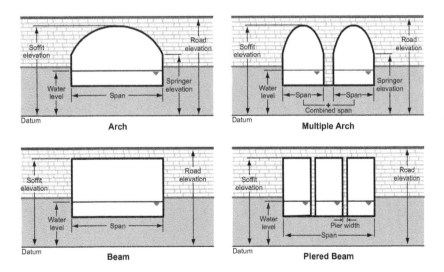

Figure 3.39 Generic bridge opening types in the AES showing variables used to define the bridge geometry.

- *Springer level (SPL) to soffit level (SFL)*: Flow within the arch. As the water level rises within the arch, the rate of increase in flow area with water level reduces due to the shape of the arch. At the same time, the wetted perimeter increases and so the conveyance can decrease as the water level rises towards the soffit, resulting in an inflection in the conveyance rating curve.
- *Soffit level (SFL) to road level (RDL)*: Orifice flow. This flow begins when the water surface reaches the soffit level.
- *Road level (RDL) to maximum elevation (ZMAX)*: Orifice flow and above-road weir flow. Above RDL, orifice flow occurs through the bridge openings in combination with weir flow through the uncontracted channel cross sectional area above the bridge.

The conveyance for a beam bridge opening is calculated in a similar way, omitting the panel between springer and soffit. The roadway is assumed to be horizontal and flow by-passing the bridge is not represented within the CES-AES.

The AES conveyance rating for a single parabolic arch example is compared with the conveyance curve calculated by HEC-RAS (version 3.1), also using a three panel divided channel method, in Figure 3.40. There are differences as expected between AES and HEC-RAS ratings because AES includes a friction coefficient for the structure and represents the reduction in conveyance when the water surface approaches the soffit, as noted above.

AES water surface profile calculation

The AES calculates the water surface profile through or over the structure in a similar way for both bridges and culverts. Firstly, the conveyance rating curves for the river channel at sections 1 to 4 in Figure 3.35 are estimated using the divided channel method. Then the structure geometry is added to sections 2 and 3, giving additional cross sections named 2D and 3U representing the downstream and upstream openings of the structure. Complete conveyance curves for these sections are then computed from the structure geometry as described above.

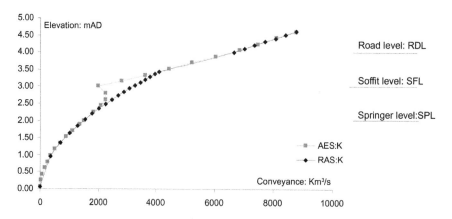

Figure 3.40 Example conveyance curves for a parabolic arch, comparing AES and HEC-RAS (version 3.1).

Upstream (contraction) and downstream (expansion) lengths are estimated based on the required flow conditions, as discussed in more detail below. The standard step backwater method is then applied for the three different flow types:

- Free surface flows below soffit level;
- Free surface flows in the channel combined with pressurized orifice flow through the structure when the upstream water level is between the soffit and road level;
- Overtopping weir flow combined with orifice flow when the upstream water level is above the road level, including modular, non-modular and drowned weir flow.

i   Free surface flow through bridges

For the subcritical, free surface flow modes through the bridge or culvert, the AES uses the standard step backwater method (Sturm, 2001; Montes, 1998). Like the backwater method used in the CES, the AES backwater calculation is based on conservation of energy between cross sections representing discrete channel reaches.

The method assumes that channels can be modelled by section reaches which have relatively small variations in bed slope. Consider the reach illustrated in Figure 3.35. Conservation of energy between a downstream section (subscript 1) and an upstream section (subscript 2) can be expressed as

$$z_2 + \alpha_2 \frac{U_2^2}{2g} = z_1 + \alpha_1 \frac{U_1^2}{2g} + h_e \tag{3.54}$$

where z is the water surface elevation, U is the section average velocity and g is gravitational acceleration. The term $\alpha U^2/2g$ is then the kinetic energy for the section and h_e represents energy losses. The CES backwater calculation is able to make use of the CES quasi-2-D conveyance calculation in its estimation of water surface slope (Eq. 3.48) whereas in the AES it is necessary to specify a kinetic energy correction factor α that accounts empirically for non uniform velocity distribution.

The α coefficient is computed for a section from an abbreviated form of Manning's equation given by

$$Q = AR^{2/3}S^{1/2}/n = KS^{1/2} \tag{3.55}$$

where Q is a steady flow discharge, n is the Manning friction coefficient, A is the cross section area, R is the cross section hydraulic radius (flow area divided by wetted perimeter) and S is the channel energy gradient due to the frictional loss (i.e. the friction slope). The α coefficient can be calculated in terms of K, the conveyance, from Eq (3.53) as

$$\alpha = \sum (K_i^3 / A_i^2) / (K^3 / A^2) \tag{3.56}$$

where now K_i and A_i are the subsection (left or right overbank and channel) conveyance and flow areas respectively. Note that both K and α can be computed for a cross section without knowledge of the friction slope.

The energy loss h_e is assumed to consist of both a channel friction loss (h_f) and a transition loss (h_t) caused by changes in cross section shape and velocity profile, and can be written

$$h_e = h_f + h_t = LS_f + C\left|\alpha_2 \frac{U_2^2}{2g} - \alpha_1 \frac{U_1^2}{2g}\right| \tag{3.57}$$

where S_f represents an average friction slope for the reach of length L, and C is a loss coefficient. Within the AES, the loss coefficient C is important in the transition reaches upstream and downstream of the bridge. Hence there are two such transition loss coefficients, $C = C_C$ the upstream contraction loss coefficient and $C = C_E$, the downstream expansion loss coefficient. It is assumed that Manning's equation for uniform flow can be applied to the total conveyance for between sections, thus

$$Q = K_1 S_1^{1/2} = K_2 S_2^{1/2} \tag{3.58}$$

where $S_1 = Q^2/K_1^2$ is a friction slope for Section 1, $S_2 = Q^2/K_2^2$ is a friction slope for Section 2 and K is the sum of panel conveyances for a given water level. Thus a reach of, say, adverse slope can be approximated to a uniform reach of similar friction and geometry (K) with a bed slope equal to the friction slope ($S = Q^2/K^2$). The average friction gradient for the reach, S_f, can be approximated by the arithmetic ($(S_1 + S_2)/2$), geometric ($(S_1 S_2)^{0.5}$) or harmonic ($2S_1 S_2/(S_1 + S_2)$) mean of the upstream and downstream sections. It has been shown by Montes (1998) that the arithmetic mean is the most accurate approximation for mild slope, subcritical profiles that are expected within the AES.

The solution to the above implicit energy equation is obtained by iteration, as for the CES backwater profile (Section 3.2.3). The solution to the iteration can sometimes produce supercritical flow within the bridge or culvert structure. In these cases, the AES defaults to critical depth to allow the computation to proceed and generates a warning.

ii   Pressure flow

The AES switches to orifice flow when the upstream water level is between the structure soffit and the road level. In this case, the orifice equation in USBPR (1978) is used, which relates the discharge Q to the difference in water surface elevation across the bridge, Δh, as

$$Q = C_d b_N Z (2g\Delta h)^{1/2} \tag{3.59}$$

where b_N is the net width of waterway (excluding piers), Z is the water depth at soffit level and the coefficient of discharge, C_d, is approximately constant at $C_d = 0.8$. This orifice equation is solved explicitly for Δh in the AES. The upstream depth of flow at Section 3, Y_u, is then computed as a function of Δh using the dimensionless relationship given in USBPR (1978) and shown in Figure 3.41, where the depths are scaled by Ybar, the normal flow depth at the bridge determined using Manning's equation.

In summary, for pressure flow the above algorithm is used to solve for the orifice flow depth at Section 3 (upstream face of the structure) and the standard step method for profiles upstream and downstream of the structure.

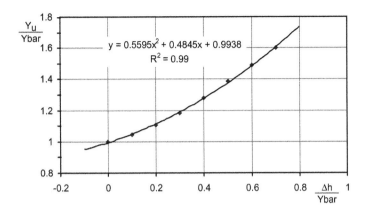

Figure 3.41 Dimensionless relationship between bridge headloss Δh and upstream depth Y_u (after USBPR, 1978).

iii. Overtopping flow when the structure is drowned

The AES assumes weir flow over the structure when the upstream water level is above the supplied road level. The standard weir flow equation is given as

$$Q = CB_{RD}H^{3/2} \tag{3.60}$$

where Q is the total discharge, C the discharge coefficient, B_{RD} the road length across the floodplain, and H the total head upstream measured relative to the road level. Orifice flow through the bridge occurs in addition to the weir flow, and AES iterates both the weir and orifice flow equations until both solutions provide the same upstream energy level.

The major unknown variable in the weir equation is the weir coefficient C. The coefficient varies according to the weir type (broad crested or rectangular) and increases with B_{RD}. Values are summarised in HEC (2004), and vary from about 1.5 to 1.7. In the AES, the roadway length is taken as the distance between the extremities of the floodplain at the road elevation, hence no special allowance is made for approach roads or embankments. The modular weir coefficient is assumed to be 1.6. The modular weir flow coefficient is used until the submergence (the ratio of upstream and downstream water levels above the road) exceeds 85%.

Submerged or non-modular weir flow is taken to begin for a submergence of 85% and ends at a 95% submergence. The same weir flow equation and iterative method are used as for the modular flow, but the weir discharge coefficient is reduced by a discharge reduction factor (DRF) following USBPR (1978), see Figure 3.42.

Transition reaches

The transition energy loss coefficients that appear in the water surface profile calculation relate to energy losses caused by acceleration within the flow as it contracts upstream and expands downstream of the structure.

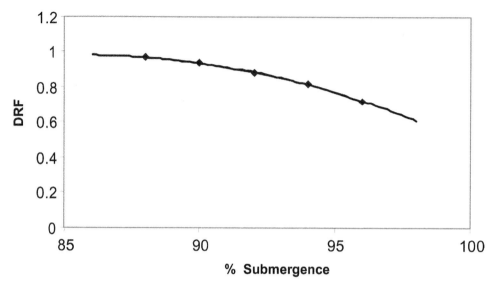

Figure 3.42 Discharge reduction factor for submerged weir flow (after USBPR, 1978).

Section 4 is assumed to be the position of the afflux in Figure 3.35, where the largest difference exists between the water surface profile affected by the structure and normal depth for the unobstructed channel. The flow contracts towards the structure between Sections 4 and 3 for a contraction length L_C. It then exits the structure and expands between Sections 2 and 1 for an expansion length L_E. Section 1 is assumed to be located where the water surface profile influenced by the structure would depart from normal depth for the channel.

For both transitions, the effective lengths vary according to the flow conditions and the geometry and roughness of the structure. The transition lengths affect the appropriate choice of location for cross sections in a backwater model that includes a bridge or culvert. Suitable choices for transition lengths and energy loss coefficients are, therefore, required for an accurate water surface profile calculation through a bridge. For the AES, these parameters have been calibrated using two data sets for flows below soffit level, as follows:

- University of Birmingham laboratory data for bridges in compound channels, described in Chapter 2,
- HEC (1995) two dimensional numerical model data that simulated field data for fixed bed channels.

The data summarized above is considered to be the most suitable, accurate set of records available at the time of the AES development for the analysis of bridge afflux. It comprises detailed measurements of discharge rates and water surface profiles or transition lengths for compound channels and floodplains of varying roughness conditions, for multiple bridge types and skew angles, and at scales ranging across three orders of magnitude (from ~1 m to ~100 m).

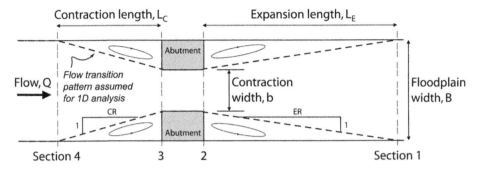

Figure 3.43 Plan view of flow transition reaches. CR is the contraction ratio, ER is the expansion ratio. The structure length is the distance between sections 3 and 2.

i Transition length analysis

Ideally, it is recommended to estimate transition lengths from field observations during high flow conditions (HEC, 1996), but this can rarely be done and so empirical formulae are often used instead. HEC (1995) investigated criteria for estimating transition lengths as a function of the total length of the side constriction caused by the structure abutments, $B - b$, where B is the floodplain width and b the width of the opening across the stream (Figure 3.43). In development of the AES, a dynamic similarity analysis was carried out for dimensionless transition lengths using the Froude Number of the downstream channel flow for scaling. The Froude Number was obtained either from the laboratory data or from numerical model data, in which case the undisturbed stream was replicated using HEC-RAS (version 3.1).

The similarity analysis was based on methods developed by Raju *et al.* (1983) and later used by Brown (1988) in the development of the 'Hydraulics Research Arch Bridges method' in the UK. In simple terms the analysis involves expressing the transition lengths L_C and L_E as follows:

$$L_{(C,E)}/D_N = f\left(Fr, J_D, B\right) \tag{3.61}$$

where D_N is the normal flow depth without the presence of the structure, Fr is the Froude Number of the flow without the presence of the structure and J_D is a blockage ratio to the flow caused by the structure. The blockage ratio J_D is defined as the area of flow blocked by the downstream bridge section divided by the total flow area in the unobstructed channel. The floodplain width, B, is introduced to accommodate information from both the laboratory scale model experiments (Chapter 2) and field scale modeled data from the HEC (1995) study.

From the compound channel laboratory experiments there are measured values of contraction length for 145 tests and expansion length from 130 tests. For some of the experiments, supercritical flow conditions occurred through the structure and thus a flow control was located at the structure as well as downstream. For some cases, a hydraulic jump occurred and the water surface profile was also unsteady. An expansion length cannot be measured for these conditions. For the remaining tests, water levels in the presence of the bridge were compared with the normal depth profile at 0.1 m increments downstream of the bridge in cases where there were large water surface gradients, such as at standing waves, and at 1.0 m increments for smoother profiles. All water level measurements were made along the channel centerline. In each case, a two-point moving average was fitted to the measured water surface to smooth local variations, and to establish an objective intersection with the normal water level profile.

The measured contraction lengths are approximately constant at 0.5 m ($\pm$0.02 m at 95% confidence level). The dimensionless quantity L_C/D_N is weakly related to the undisturbed flow Froude Number and blockage ratio, with a quadratic relationship able to predict L_C/D_N as a function of Froude Number and blockage ratio with a coefficient of determination of approximately 0.6. The relationship appears physically meaningful with L_C approaching zero as the Froude Number approaches zero (i.e. still water) and as the blockage ratio approaches 1.0 (i.e. total blockage). Since the average obstruction length $(B - b)/2$ was approximately 0.4 m for the laboratory data, the contraction ratio $2L_C/(B - b)$ was about 1.25, which is a value similar to that suggested by HEC (1995).

Measurements of expansion length L_E derived from the Birmingham University laboratory data vary more strongly with the unobstructed channel flow Froude Number. Figure 3.44 shows the dimensionless expansion length (scaled by the downstream normal depth) derived from the laboratory measurements and the Froude Number for the unobstructed channel. The data in Figure 3.44 include variation in the blockage ratio, J_D. A quadratic relationship is able to describe the variation in L_E/D_N as a function of Fr and J_D with a coefficient of determination of approximately 0.9.

For larger physical floodplain scales, HEC (1995) estimated transition lengths in data produced using a two dimensional numerical model based on data from three real bridge sites and 76 idealized bridge and channel configurations with a floodplain width of approximately 305 m. The data included various combinations of bridge opening width (b), discharge, overbank and main channel Manning coefficients and bed slopes.

The method for estimating transition lengths reported by HEC (1995) was a visual examination of modeled velocity vector fields, rather than comparison of water surface profiles as for the Birmingham University laboratory data. The limit of the

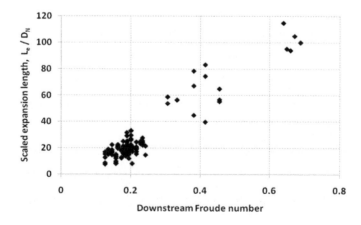

Figure 3.44 Relationship between dimensionless expansion length and Froude Number for compound channel laboratory data for all bridge types tested (see Chapter 2).

expansion reach was defined by HEC (1995) to be just upstream of the location where the velocity vectors became essentially parallel to the channel. This placed the expansion reach limit a small distance upstream of the location where normal depth is re-established (Section 1 in Figures 3.35 and 3.43). However, the two sets of measurements were accepted in development of the AES as alternative approximations of the same physical quantity. For the HEC (1995) data, the relationship found by fitting a model of the form of Eq. 3.61 to the data is similar to that identified from the laboratory data, but with a smaller coefficient of determination.

The AES estimates transition lengths using a composite of the quadratic best fit curves for both the laboratory scale ($B \approx 1$ m) and field scale ($B \approx 300$ m) data sets, described by

$$\frac{L_{(C,E)}}{D_N} = b_0 + b_1 Fr + b_2 J_D + b_3 Fr^2 + b_4 J_D^2 + b_5 Fr J_D \qquad (3.62)$$

where the set of coefficients $b_0, \ldots, b_5$ are defined separately for contraction and expansion, and are of the form $b_i = k_{i,1} B + k_{i,2}$. The fitted estimates are given in Table 3.9.

Figure 3.45 shows data estimated using Eq. 3.62 for expansions and contractions plotted on a common scale against the original experimental values derived from the University of Birmingham laboratory measurements and HEC (1995) two dimensional numerical models. The AES model is shown to approximate the experimental data over the range of flow conditions and scales described in the preceding discussion with little evidence of bias.

ii Transition energy loss coefficients

The transition loss coefficient introduced in Eq. 3.57 is defined as $C = C_C$ for contraction reaches and $C = C_E$ for expansion reaches. The transition coefficients scale the energy losses associated with changes in velocity head over the transition reaches. Their effect is to modify the afflux resulting from the backwater profile calculation through a bridge structure. Figure 3.46 shows comparisons between measured and modelled afflux for the University of Birmingham laboratory experiments that had subcritical flows below soffit level. In Figure 3.46(a) the afflux has been modelled using the standard step backwater method with the transition coefficients C_C and C_E both set to zero. It can be seen that the afflux is systematically underestimated by the model, particularly for the higher flows where the velocity heads in Eq. 3.57 would become

Table 3.9 Polynomial coefficients for AES transition lengths.

Coefficient	Contraction		Expansion	
	$k_1 (m^{-1})$	k_2	$k_1 (m^{-1})$	k_2
b_0	0.0190	−8.9	0.0712	−26.2
b_1	−0.2130	50.8	−0.2940	197.8
b_2	0.1900	26.6	0.0569	15.2
b_3	0.5650	−36.2	−0.8720	−22.2
b_4	−0.1700	−20.4	0.0170	9.9
b_5	0.4180	−39.1	0.3940	−2.8

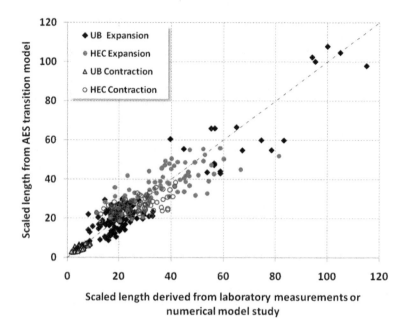

Figure 3.45 Dimensionless transition lengths (scaled by the downstream unobstructed normal depth) showing values estimated from the AES transition length model (Eq. 3.62) plotted against experimental data from University of Birmingham (UB) laboratory flume measurements and HEC (1995) field scale two dimensional numerical modeling.

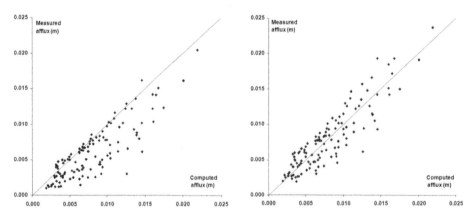

Figure 3.46 Comparison of measured and modeled afflux before (a, left panel) and after (b, right panel) calibration of laboratory scale transition energy loss coefficients.

more important. Correction of this bias therefore provides a convenient method for calibration of the transition energy loss coefficients. An approximate calibration of $C_C = C_E = 0.1$ was adopted, as shown in Figure 3.46(b), giving an ensemble parity of 0.98 for the afflux data with coefficient of determination of 0.80.

When supercritical flow occurs within the AES, the flow depth is set equal to critical depth. Critical flow first occurs at a bridge exit section, if both entry and exit sections have nearly equal conveyance. Upstream levels are solved using the critical condition in effect as a new boundary condition. Defaulting to critical depth at the bridge section increases the water level. The data used for calibration of transition losses did not include cases where critical flow conditions were observed. However, the University of Birmingham bridge data contains 26 critical flow profiles. When the associated afflux is computed using AES with the transition energy losses at the calibrated values for subcritical flow then there is an approximately 10% error in the computed afflux. When the transition energy losses are neglected ($C_C = C_E = 0$) then the results for critical flows agree closely with the experimental measurements, with approximately a 2% bias and a coefficient of determination of 0.9.

The energy loss coefficients were also calibrated using the field scale HEC (1995) data, and the following mean values with their 95% confidence limits were obtained:

$$C_C = 0.12 \pm 0.01 \quad \text{and} \quad C_E = 0.027 \pm 0.03 \tag{3.63}$$

These values were rounded to $C_C = 0.1$ and $C_E = 0.3$ for subcritical flows beneath soffit level. The calibrated field scale value for the contraction energy loss coefficient is consistent with the laboratory scale data, whilst the expansion loss appears to be greater.

The similarity analysis of transition reach lengths would suggest that the transition energy loss coefficients might also be related to undisturbed Froude Number, blockage ratio and the floodplain scale but it is difficult to establish a useful relationship, which is also as reported by HEC (1995). For the available laboratory and field scale data, the blockage ratio varies in the range 0.1 to 0.9, undisturbed Froude Number from about 0.05 to 0.7 and the floodplain width from ~1 m to ~305 m. The mean of the floodplain width values, $B^* = 153$ m, was used as a convenient reference scale and the resulting dimensionless floodplain scale variable B/B^* varies from about 0.01 to 2.0. The variation in floodplain scale is therefore significant in the functional relationship and so C_E is interpolated in the AES as a linear function of floodplain scale between a minimum of $C_E = 0.1$, calibrated using laboratory data, up to the maximum of $C_E = 0.3$, calibrated using field scale data from numerical modeling, such that

$$C_E = \min\left\{0.3, 0.1\frac{B}{B^*} + 0.1\right\}. \tag{3.64}$$

The contraction loss coefficient was set to $C_C = 0.1$, which was found to be a reasonable value for both laboratory and field scale data. For supercritical flow $C_C = C_E = 0$, as determined for the 26 critical flow profiles observed in the laboratory measurements.

Culvert energy losses

The computation of a water surface profile for a culvert structure in the AES is made in a similar way to that for a bridge structure. Firstly, the rating curves for the four surveyed river sections (S1, S2, S3 and S4) are estimated using the divided channel method. The culvert geometry is then added to the upstream and downstream structure sections (S2 and S3) and conveyance rating curves are computed using the same principles as for a bridge, including extension of the conveyance over the culvert road level, and the indeterminate rating between culvert soffit and road level. The standard step water surface profile method is then applied as for a bridge, taking the downstream boundary condition from S1.

For a culvert there are energy losses associated with the inlet and outlet structures, which, in the AES, are applied to the corresponding transition reaches in the backwater calculation. The inlet structure energy losses are calculated using the design formulae and coefficients given in CIRIA (1997) for certain standard culvert inlet configurations. In the CIRIA guide, the inlet control design coefficients and head loss coefficient depend on the shape, type, material and edge form of the inlet structure. Figure 3.47 illustrates the choices available in the AES.

The culvert energy loss variables are defined in Figure 3.48 and the calculation algorithm is summarised in Table 3.10. The AES omits some details of the CIRIA (1997) and FHWA (2001) methods for simplicity, notably energy losses for bends and junctions. Culverts that contain complex features such as junctions and significant changes in cross section should be analysed using more sophisticated modelling software. However, the AES does automatically handle the full range of flow conditions from subcritical flow through the barrel to combined pressure and overtopping flow.

Integration with the CES backwater model

In the combined CES-AES software, the AES appears as a bridge or culvert placed 'within' a channel backwater profile between two CES cross sections, which can

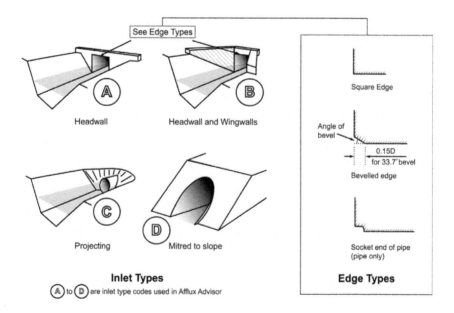

Figure 3.47 Inlet and edge types for culverts (after CIRIA, 1997).

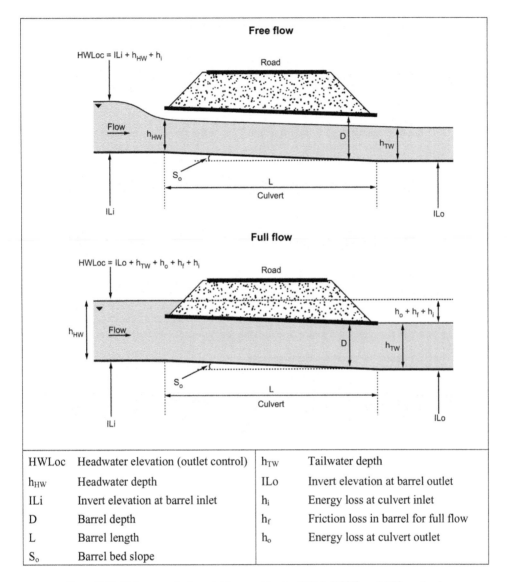

Figure 3.48 Culvert analysis variables used in the CIRIA (1997) and AES methods.

be called CESd for the downstream section and CESu for the upstream section (Figure 3.49). If the CESu and CESd cross sections were located at the appropriate transition reach lengths along the stream, it would be possible simply to use the CES backwater depth at CESd as the downstream boundary condition for the AES model of the bridge or culvert. In the same way, it would then be simple to use the AES water surface results upstream of the bridge as a new downstream boundary for the remaining CES water surface profile calculation.

Table 3.10 Summary of the AES algorithm for culvert hydraulics, based on CIRIA (1997).

Symbol	Free flow	Full flow
Inlet control		
HWLic	An inlet structure design table is used to calculate the headwater elevation under inlet control.	
Outlet control		
h_{TW}	Tailwater depth (h_{TW}) is the maximum of the downstream depth or critical depth for the culvert, h_c.	If the downstream depth (h_{TW}) is less than ($h_c + D$)/2 then use FHWA (2001) approximate backwater assumption, $h_{TW} = (h_c + D)/2$.
h_o	The outlet head loss is calculated from the difference in velocity heads between the culvert barrel and the downstream channel.	
h_i	An inlet structure design Table is used to calculate the inlet head loss.	
h_{HW}	Water depth just inside the culvert barrel entrance is calculated by backwater analysis.	The upstream depth is determined from the inlet, friction and outlet head losses. The friction head loss (h_f) is calculated using Manning's equation.
HWLoc	The headwater elevation under outlet control is calculated as: HWLoc = ILi + h_{TW} + h_i	The headwater elevation under outlet control is calculated as: HWLoc = ILo + h_{TW} + h_o + h_f + h_i

1 If HWLoc > HWLic then use outlet control, else use inlet control. Headwater level is the maximum of HWLic or HWLoc.
2 Inlet control is computed for the entire discharge rating.
3 For inlet control, if the discharge ratio value $1.811 Q/AD^{0.5}$ is between 3.5 and 4, then the HWLic rating is linearly interpolated.
4 Free flow begins at the lowest flow elevation and ends when the flow depth in the barrel is greater than the soffit depth at the culvert inlet.
5 When the head water level is above the road level, the flow is computed sequentially as a modular weir, submerged weir and a drowned weir, each with the appropriate inlet or outlet controlled culvert flow. The weir elevation is computed iteratively since flow continues through the culvert.
6 Energy losses due to flow at culvert bends and at trash screens are omitted for simplicity.

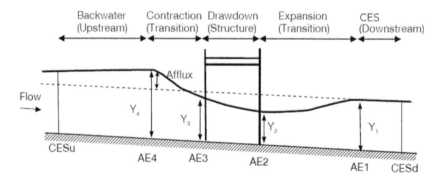

Figure 3.49   Placement of AES calculations within a CES backwater profile.

However, it is unlikely that surveyed channel cross sections will generally be available at the transition lengths (especially because those lengths are themselves a function of the flow conditions, as implied by the presence of the Froude Number in Eq. 3.62, and hence are not fixed). When the surveyed CESu and CESd sections are located beyond the lengths of the AES transition reaches then the CES-AES software introduces interpolated sections in the AES at the transition lengths. The latter are computed by linear interpolation of the cross section geometry between the nearest surveyed cross sections to the structure. If the transition lengths for subcritical sub-soffit flow exceed the distance to the nearest CES cross section or if the structure is drowned then the software currently defaults to use the actual CESu and CESd cross section locations.

3.3 DEALING WITH UNCERTAINTY

3.3.1 Introduction to uncertainty

Uncertainty analysis is required in order to understand the implications for decision makers of limited data, model uncertainties, changes in the flooding system over the long term, incommensurate scales of appraisal and potentially conflicting decision objectives (see Hall *et al.*, 2008; Hall & Solomatine, 2008).

Flood level predictions based on existing hydrodynamic modelling software (e.g. Flood Modeller Pro, HEC-RAS, InfoWorks ICM) provide no indication of the associated uncertainty. In fact, in practice uncertainty analysis and the use of the results are not widespread as Hall *et al.* (2008) points out:

- There is a bewildering proliferation of uncertainty methods published in the academic literature.
- Uncertainty analysis takes time, so adds to the cost of risk analysis, options appraisal and design studies.
- The data necessary for quantified uncertainty analysis are not always available, so new data collection campaigns (perhaps including time-consuming expert elicitation exercises) may need to be commissioned.
- The additional requirements for analysis and computation are rapidly being (more than) compensated for by the availability of enhanced computer processing power. However, computer processing power is only part of the solution, which also requires a step change in the approach to managing data and integrating the software for uncertainty calculations.

The uncertainty in water level prediction for channels arises from many factors including approximations to the fluid mechanics, natural variability in river resistance and the use of judgement or experience in applying models. These all produce differences between the assessment of river channel capacity and its "true" value.

The potential economic benefits of the use of improved methods for conveyance estimation will come from altering design, operation and maintenance practice as the degree of uncertainty is reduced. Strategic decisions made early in the project life cycle can have far reaching consequences and it is at this early stage that uncertainties in information and data are greatest. There is a close relationship between uncertainty and risk in that the greater the uncertainty the greater the probability of the project or maintenance activity not achieving its objective. This is linked to the confidence on the performance of the scheme or process to meet its intended objectives. Thus, optimisation of performance and the confidence with which performance can be delivered are linked inexorably with understanding and controlling uncertainty.

3.3.2   Risk, uncertainty, accuracy and error

Uncertainty has always been inherent in flood defence engineering. Traditionally it was treated implicitly through conservative design equations or through rules of thumb, for example by introduction of freeboard allowances. In recent years the emphasis in the UK has moved from Flood Defence to Flood Risk Management. The introduction of risk-based approaches has enabled more rational treatment of natural variability in loads and responses. It also paved the way for more explicit treatment of uncertainty in the evidence that is used to support risk-based decision making (Hall *et al.*, 2008). A difficulty with the language of risk (Defra/EA, 2002; Gouldby *et al.*, 2005) is that it has been developed across a wide range of disciplines and activities. It is common to describe risk as a combination of the chance of a particular event (i.e. a flood), with the impact that the event or hazard would cause if it occurred. Risk therefore has two components: the chance (or *probability*) of an event occurring and the impact (or *consequence*) associated with the event. The consequence of an event may be either desirable or undesirable. Generally, however, flood management is concerned with protecting society and hence a risk is typically concerned with the likelihood of an undesirable consequence and our ability to manage or prevent it. In some, but not all cases, therefore a convenient measure of the importance of risk is given by the "equation":

Risk = Probability × Consequence

The *Probability* refers to the chance of a particular consequence occurring. The *Consequence* refers to the undesirable outcome of harm that would arise should a risk be realized.

One system model that can be used to identify and assess risks is based upon the causative sequence of sources, pathways and receptors (Sayers *et al.*, 2002; Samuels, 2006). For a risk to arise, there must be a hazard that consists of a source or initiator event (e.g. high rainfall); a receptor (e.g. flood plain properties); and a pathway between the source and receptor (i.e. flood routes including defences, overland flow or landslide). Thus, the source describes the origin of the hazard in terms of the forcing or loading on the system. The source will be attributed quantitative values such as millimeters of rainfall, river discharge or be described in terms of probability of occurrence or exceedance. The receptor is the asset, individual, community, business or habitat etc. that could experience harm if affected by hazard. Whilst contained, the hazard does no damage, for harm and damage to occur a pathway is needed to transmit the source of the hazard to the receptor. However, the identification of a hazard does mean that there is a possibility of harm occurring. Once the receptor has experienced the hazard then adverse consequences occur. For practical analysis it must be recognized that there are likely to be multiple sources, pathways and receptors.

Evaluating risks involves identifying the hazards associated with the risk issue, i.e. what in a particular situation could cause harm or damage, and then assessing the likelihood that harm will actually be experienced by a particular population and what the consequences would be. Thus to evaluate the risk, it is necessary to consider the three generic components:

- the nature and probability of the hazard;
- the degree of exposure of people and assets to the hazard;
- the vulnerability of the people, assets etc to damage should the hazard be realised.

In describing each of these there is likely to be an element of imprecision depending upon the data available and methods employed. For further details on risk and uncertainty in flood management, see Samuels (2006).

Most engineering "failures" (such as those leading to flooding) arise from a complex and often unique combination of events and thus statistical information on their probability and consequence may be scarce or unavailable. Under these circumstances the practitioner has to resort to models and expert judgement. Models will inevitably be an incomplete representation of real life and so will generate results, which are uncertain. Similarly, human expert judgement is subjective and thus uncertain as it is based on mental models and personal experience, understanding and belief about a situation. So in practice every measure of risk has uncertainty associated with it. Uncertainty arises mainly from lack of knowledge or of ability to measure or to calculate and gives rise to potential differences between assessment of some factor and its "true" value. Understanding this uncertainty within our predictions and decisions is at the heart of understanding risk. Within uncertainty it is possible to identify:

- knowledge uncertainty from lack of knowledge of the behaviour of the physical world;
- natural variability from the inherent variability of the real world; and
- decision uncertainty from the complexity of social and organisational values and objectives.

However, this classification is not rigid or unique. For example, uncertainty on weather or climate will be taken as "natural variability" within flood risk management but as "knowledge uncertainty" in the context of climate simulation. It is helpful also to consider the differences between accuracy, error and uncertainty. Accuracy and error differ from uncertainty as defined above but limitations in accuracy or the possibility for human error will contribute to the overall uncertainty.

- Accuracy deals with the precision to which measurement or calculation is carried out; potentially, accuracy can be improved by better technology.
- Errors are mistaken calculations or measurements with quantifiable and predictable differences.

3.3.3 Components of uncertainty in the CES

Principal influences

The uncertainties in conveyance estimation are principally due to natural variability and knowledge uncertainty. Two contributions are normally recognised in knowledge uncertainty; these are:

- Process model uncertainty in any deterministic approach used in the assessment;
- Statistical uncertainty arising from the selection and fitting of statistical distributions and parameters from data within the assessment.

In the CES no use is made of statistical modelling and so issues of statistical uncertainty do not arise. However, in the overall estimation of flood risk there will be an important component of statistical uncertainty arising from any hydrological estimation procedures for flow.

The important components of process model uncertainty in the estimation of conveyance are:

- process model uncertainty arising from the selection and approximation of physical processes and from parameterisation made in the definition of conveyance;
- representation of topography through the density of discrete survey points and interpolation rules between them, i.e. the difference between the profile of the river and of the physical features on the flood plain as represented in the calculations and the river in "real" life;
- uncertainty related to data accuracy from limitations of the survey methods used;
- uncertainties arising from parameter estimation – particularly the experience and expertise of the modellers who set up and calibrate computational models of river flows; and
- uncertainties from the model calculation methods, approximations and rules. These are the domain of traditional numerical analysis.

The contributions to uncertainty from natural variability include

- seasonality of plant growth from variation in temperature, light and nutrients (and hence resistance of vegetation in channels);
- "memory" in the system e.g. from vegetation or river bed conditions being changed by a flood or through episodic input of sediments into the river system washed off the land surface;
- secondary influences of temperature on water viscosity and hence Reynolds' Number and flow and sediment transport capacity of channels.

Importance of contributions

i   Physical processes
In the development of the CES, a specific mathematical process model has been adopted based upon the fundamental equations of fluid mechanics with empirical models for the effects of river meanders and the variation of the water depth across the section. This process model accounts for a greater diversity of the physics of the water movement in rivers than the methods previously used in standard modelling packages. Thus it is to be expected that the contribution of the process model uncertainty due to the formulation of the physics of the flow has been reduced.

ii   Topographic representation
Previous research on the representation of river topography, Defalque *et al.* (1993) indicated that the uncertainties from this source could be controlled by an adequate frequency of survey information. This is achieved through normal survey specifications (e.g. as undertaken by the Environment Agency of England and Wales).

iii   Data accuracy
The development of the CES did not consider specifically the uncertainties from the typical accuracy of the topographic data sources; however, it is expected that these

will be similar to their influence on the uncertainty when using the traditional calculation methods. This has been analysed for the procedures in the US for the prediction of the 100-year flood by Burnham & Davis (1990). With the availability of new data sources (e.g. LiDAR) it would be appropriate to examine the influence of the typical accuracy of these measurements and formulate guidance on the density of the data grid on the consequent flood levels.

iv   Model parameters

The parameters in a model are the "constants", chosen to represent the chosen context and scenario. In general the following types of parameters can be recognised:

- Exact parameters, which are universal constants, such as the mathematical constant π.
- Fixed parameters, which are well determined by experiment and may be considered exact, such as the acceleration of gravity (approximately 9.81 ms^{-2}).
- A-priori chosen parameters, which are parameters that may be difficult to identify by calibration and are assigned certain values. The values of such parameters are associated with uncertainty that must be estimated on the basis of experience, for example from detailed measurements.
- Calibration parameters, which must be established to represent particular circumstances. They are determined by calibrating model results for historical data on both input and output. Parameters are generally chosen to minimise the difference between model results and measurement for the same event. The parameter set for a "satisfactory" calibration is unlikely to be unique.

The CES contains several a-priori chosen parameters in the model of velocity variation across the river channel although there is an option which allows the "expert" user to vary these (Chapter 5). The CES also contains calibration parameters – these are the values of river resistance which are either determined by systematic variation to match observed flow conditions or estimated from the Roughness Advisor (i.e. calibrated against vegetation type or spatial location).

v   Calculation methods

The CES contains several numerical methods to generate (approximate) solutions of the mathematical model for any water depth or flow rate. These include an automatic division of the cross-section into a computational grid which is sufficiently fine to ensure that the uncertainty associated with the differences between the numerical solution and the "true" solution of the mathematical model are much smaller than the uncertainties in the selection of the model parameters.

vi   Natural variability

The Roughness Advisor includes some account of the typical seasonal variation in resistance due to the growth and decay of vegetation. This generic information has been established from a variety of sources and includes an element of professional judgement. There is no guarantee that any particular year will match the average but the user can explore scenarios of advancing or delaying the seasonal growth-decay cycle and include the effects of river "maintenance" activities. No information can be given on the probability of the seasonality.

Other factors classified as natural variability are not incorporated in the standard version of the CES; however, there is an "expert" option for researchers to explore the effects of temperature in experimental facilities. Thus the CES has not reduced the uncertainties in conveyance arising from the erosion, transport or deposition of sediments, system "memory" or temperature.

3.3.4 Representation and assessment of uncertainty

There are many approaches to quantifying and assessing uncertainty. An important distinction is the difference between the gross uncertainties associated with unknowns (e.g. climate, socio-economic growth) and the lesser uncertainties associated with method, model and data. The former are typically handled through scenario analysis to assess the performance of options in terms of robustness, adaptability and sustainability (e.g. Evans *et al.*, 2004; IPCC 2001; Hine & Hall, 2007; de Bruijn *et al.*, 2008) whereas the latter are dealt with through interval probability theory, combined distribution analysis, Monte Carlo or variance-based sensitivity analysis (e.g. see Gouldby & Kingston (2007) for an approach in the context of flood risk). For the case of the CES, only a sub-set of the uncertainties are under consideration rather than the interpretation relying on defined probabilities for all parameters as required for a probabilistic approach. This is appropriate as the outputs are not being used for a detailed economic appraisal but are more likely to inform management practice (e.g. Table 3.11). More detailed approaches such as Monte Carlo are data hungry and computationally expensive.

A key decision in developing advice and estimates of uncertainty was on how the information should be presented. Several approaches are possible as described above but some require considerable additional information and analysis to generate the results. Thus an appropriate starting point was to consider the users of the CES and how they might incorporate information on uncertainty into their decisions. This led to some requirements of the method in practice, it should:

- express uncertainty in water level;
- link to and inform other risk approaches;
- be understandable to the user to facilitate better, more informed decisions in the situations they are responsible for;
- be as simple as possible in presentation;
- not require significant computational resources;
- ideally not require any additional input data; and
- be credible by being based on existing scientific methods for uncertainty where possible.

The potential direct and indirect users of the CES information have different needs for information on uncertainty in the decisions and plans they need to make. Table 3.11 provides an indication of the potential use of the information.

In terms of presenting quantitative information on uncertainty the important choices are whether to:

- recommend the use of Monte-Carlo testing; and/or
- assign probabilities to the information given uncertainty.

It was decided not to recommend either of these because making statements on probability would give an appearance of confidence in the quantification of information which is partly subjective in origin. In addition:

- Monte Carlo testing requires considerable computation, particularly if the testing is embedded in a large flow simulation model;
- the underlying information is not available for the probability distribution of river resistance values which is required by Monte Carlo testing;
- the information on river resistance is not homogeneous in origin;
- the effects of residual process model uncertainties cannot be quantified in any probabilistic manner; and
- some information in the Roughness Advisor came from expert opinion rather than measurement.

However, it is important that the CES includes some quantitative statement on the effects of uncertainty even if probability (or likelihood or frequency) cannot be given. This has been achieved by illustrating the uncertainty in the relationship between water level and discharge through presenting a central estimate and credible upper and lower cases of water level for a given flow rate. The stage discharge curve is usually expected to lie within the upper and lower scenario bands, which may not be symmetrical about the mean. These bands should not be interpreted as minimum/maximum envelopes, but rather 'soft' boundaries within which the 'true' value is likely to occur.

The central estimate represents the "best" assessment without any particular definition of "best". It is important to note that the upper and lower cases

- are not necessarily equally distributed about the central estimate (i.e. not just ± x%);
- are not upper or lower bounds (i.e. it is possible for some real cases to lie outside the range);
- are not for any specific confidence limits (e.g. ±95%); and
- are not simply related to the standard deviation of measured data.

Instead they represent a practical choice of a degree of variability to include in the analysis for flood risk management applications, based upon the range of values found in the compilation of data on river roughness. The concentration of the scenarios on the influence of uncertainty in river roughness is justified in terms of the sensitivity analysis undertaken in the testing phase of the CES research.

The dominant factor to be used in the analysis of the estimation of conveyance is the unit roughness as distributed across the section. A thorough review of the sensitivity of water level, to both the a-priori chosen and calibration parameters, identified roughness as having the most significant contribution to uncertainty in conveyance (Khatibi et al., 1997; Latapie, 2003; Mc Gahey, 2006). The parameters considered were roughness, elevation of the floodplain and main channel, cross-chainage, planform sinuosity, temperature, longitudinal bed-slope and position of top-of-bank markers. The uncertainty of the input survey data depends largely on the method of measurement. The a-priori chosen parameters were the dimensionless eddy viscosity

Table 3.11 Summary of uncertainty users, consequences and potential use of information.

User community	Consequences of uncertainty	Uncertainty information and use	Importance	Suitable presentation of uncertainty
Scheme design	Under capacity of defences leading to potential failure below the design standard or over capacity potentially leading to morphological problems or lower economic return than planned; Over estimation of capacity of defences leading to lack of implementation schemes due to excessive cost.	Change in water level for given flow rate (moving towards probabilistic optimisation for describing uncertainty). Undertake a sensitivity analyses and add a freeboard to allow for under capacity.	Medium to high	Upper and lower estimates Potentially probabilistic information for a full risk-based design appraisal of a major scheme
Strategic planning	Development of conservative strategies which potentially over compensate on data and method uncertainty rather than focussing on the more gross uncertainties associated with the medium to long-term.	Evaluating and comparing options against criteria such as effectiveness, efficiency, robustness, performance.	High	Upper and lower estimates
Maintenance	Inadequate or excessive maintenance activities, possibly unnecessary disruption to aquatic and riparian habitats or insufficient capacity of the watercourse leading to increased flood risk.	Timing, scheduling and prioritisation of vegetation cutting and dredging.	Low to medium	Upper and lower estimates or nothing
Flood forecasting and warning	Under- (over-) estimation of lead times, inexact inundation extent, and incorrect retention times of floods.	Issue a warning? Operational decisions Implement real-time updating procedures.	Low	Possibly an upper estimate of flood level (and early estimate of arrival time)
Hydrometry	Incorrect discharges with potentially large errors, influences flood forecasting and statistical estimation of flood flows for design, impacts upon cost-benefit assessment and decisions to promote flood defence schemes.	Flow rate for a given depth (influences other stakeholders) Undertake a sensitivity analyses and add a freeboard to allow for under capacity.	Medium to high	Upper/lower estimates Probabilistic
Regulation, insurance	Indicative Flood Mapping (IFM) in error – inadequate tool for planning and information on possible flood risk, inadequate (or over-necessary) development control, loss of professional and public confidence in the operating authorities technical abilities.	Water levels, flood outlines, indicative flood risk mapping.	High	Upper/lower estimates

λ, the secondary flow coefficient Γ and the number of lateral divisions for integration. Other possible sources of uncertainty include non-modelled processes such as form losses, vertical accelerations through sudden local changes in depth, energy dissipation from sediment movement over mobile channel beds and lateral variations in surface water level around bends.

As a pragmatic approach it is recommended that the overall uncertainty in a model application may be estimated from an analysis similar to that of Samuels (1995). The uncertainty in water level Δh_u is given by the "root sum of the squares" of the uncertainties in water level due to different parameters (which may be justified for independent Gaussian processes)

$$\Delta h_u = [\Delta h_r^2 + \Delta h_s^2 + \Delta h_c^2 + \Delta h_{\Delta x}^2 + \Delta h_Q^2 ......]^{1/2} \tag{3.65}$$

where the subscripts are:

- r for roughness where little or no calibration data exist
- s for survey data depending upon the method of survey
- c for calibration and extrapolation away from the calibrated state
- Δx for numerical error (grid resolution and physical method)
- Q for flow estimation (outside the use of the CES)
- and ... represents any other commensurate factor.

The assumptions behind this type of formulation are that the sources of uncertainty are mutually independent and that they are roughly normally distributed. The numerical uncertainty $\Delta h_{\Delta x}$ is small (order of magnitude smaller than the other parameters) and well controlled because the grid can be refined to ensure this i.e. cross-section spacing. The flow estimation or uncertainty due to the hydrometric analysis Δh_Q is outside the scope of the CES, but experience shows that this is a key factor in scheme design, contributing about 40% of the overall uncertainty. The remaining factors are thus Δh_r, Δh_s and Δh_c.

Calibration of the flow resistance alters the uncertainty in the estimation of conveyance. Calibration data for flow resistance should supersede the generic, non-site specific information in the Roughness Advisor. Calibration may result in a different value for the central estimate of water level for the calibration flow rate and there is no guarantee that a single roughness value will achieve a perfect calibration for water levels and flow rates observed on different occasions (due to natural variability). However, once the model has been calibrated, Δh_r and Δh_s are likely to decrease, narrowing the range of uncertainty but the exact PDFs for the calibrated and uncalibrated cases will be difficult to determine.

3.4 INTRODUCTION TO THE CES-AES SOFTWARE

Having been introduced to the CES-AES methods, a brief introduction to the software (Figure 3.50) is provided here. The CES-AES Website (*www.river-conveyance.net*) is recommended as an up-to-date source of information, software and tutorials.

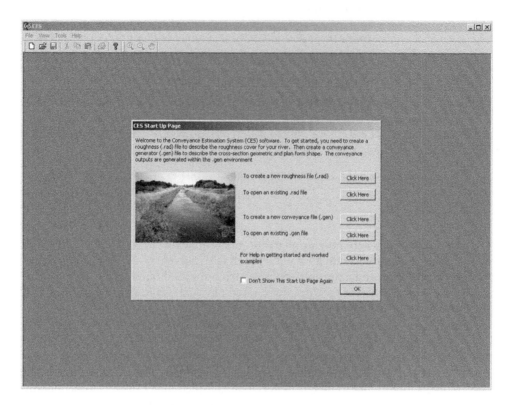

Figure 3.50 CES-AES software introductory page.

The CES-AES software comprises four modules (Figure 3.51):

- *Roughness Advisor* – which is a database of roughness information (Section 3.2.1) including descriptions, photographs and information on seasonal vegetation, cutting and regrowth.
- *Conveyance Generator* – which incorporates the conveyance calculation (Section 3.2.3), giving section-averaged water level (and associated uncertainty, Section 3.4), flow, conveyance, velocity, area, perimeter and Froude and Reynolds Numbers for each flow depth as well as lateral distributions of depth-averaged velocity, boundary shear stress and shear velocity.
- *Backwater Module* – which incorporates the energy-driven backwater calculation (Section 3.2.4) for determining the water surface profile upstream of a control point.
- *Afflux Estimator* – which calculates the afflux upstream of bridges and culverts as well as the energy losses through these structures (Section 3.2.5). It includes arch and beam bridges with up to 20 openings, and pipe, box and arch culverts with up to 10 identical barrels.

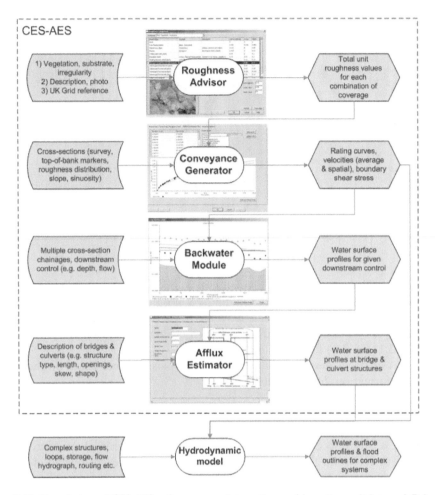

Figure 3.51 Description of CES-AES software user interaction and how it may link to a 1-D hydrodynamic model.

Despite the complexity of the underlying calculations (Section 3.2), a key feature of the CES-AES tool is that the user input is straightforward and not dissimilar to existing 1-D river modelling software such as InfoWorks RS or HEC-RAS. Figure 3.51 describes the inputs and outputs for each calculation module and how these may feed into a more complex 1-D hydrodynamic model. The CES-AES tool is also embedded within Flood Modeller Pro and InfoWorks RS, providing flexibility to switch between this and the more traditional Manning calculation.

3.4.1 Introduction to the Roughness Advisor

The Roughness Advisor (RA) aids the user in building roughness zones which comprise plan form areas of the flood risk system with uniform unit roughness values

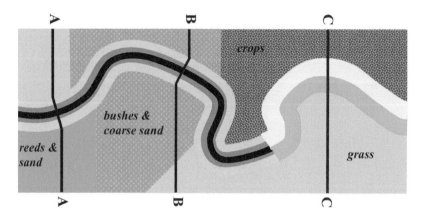

Figure 3.52 Example plan view of roughness zones.

(Figure 3.52). The system may be divided into as many zones as required to describe the channel and floodplain cover, including within a channel cross-section i.e. non-homogenous roughness. Each zone may have up to three component roughness values i.e. n_{sur}, n_{veg}, and/or n_{irr} and the RA then evaluates the total unit roughness n_l (Section 3.2.1, Eq. 3.1) for each zone.

The RA advice is designed for use in a hierarchical manner based on the level of available information, for example:

- *Extensive information* – the user has been to the site and has access to local survey, bed material, information on plant type and cover, detailed measurements for one or more sections (flow, velocity, water surface profile) etc. Here, the RA is used to determine the initial unit roughness values for each zone and thereafter these values are refined based on calibration with measured data. Figures 3.53 and 3.54 provide examples of the substrate and vegetation advice for clay and emergent broad-leaved rooted plants respectively. Figure 3.55 provides an example of the seasonal variation in roughness for submerged fine-leaved plants.
- *Some information* – the user has access to site descriptions, photographs, approximate longitudinal bed slope, roughness types and limited survey data. Here, the RA is used is used together with the descriptions and photographic evidence to estimate the unit roughness value for each zone. Where additional information is present, e.g. historic flood outlines, these may be used to improve confidence in the RA values.
- *No information for sites located in the UK* – the user has no information for sites located in the UK and only knows the location of the channel reach from Ordnance Survey maps. Here, the UK grid reference is provided (e.g. Figure 3.56) and the RA provides advice on which aquatic vegetation is most likely to be present.

The RA roughness zones and descriptions are saved in a .RAD roughness files for use within the conveyance calculation.

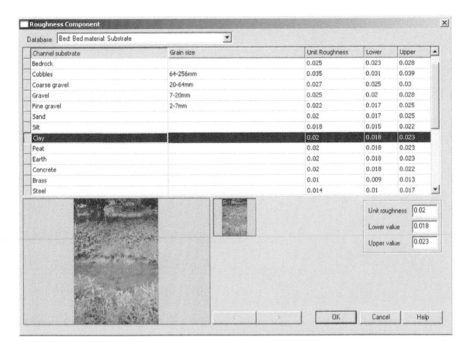

Figure 3.53 Example of RA advice for channel bed substrate (clay).

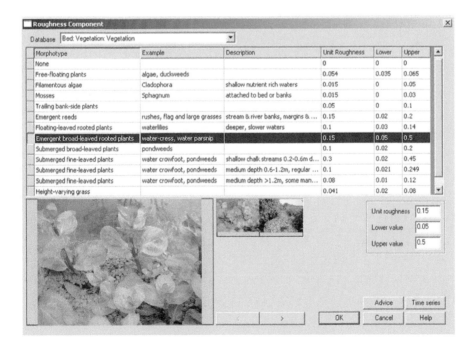

Figure 3.54 Example of RA advice for channel bed aquatic vegetation (emergent broad-leaved plants).

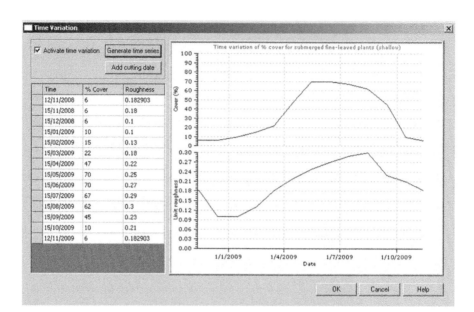

Figure 3.55 Example of RA advice for seasonal variations in vegetation roughness (submerged fine-leaved plants).

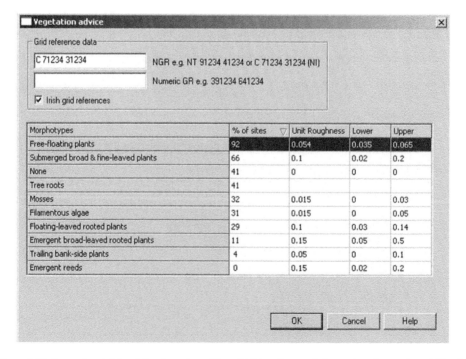

Figure 3.56 Example of RA advice based on River Habitat Survey information for a given UK grid reference.

3.4.2 Introduction to the Conveyance Generator

The Conveyance Generator (CG) aids the user in calculating stage-discharge relationships for individual channel cross-sections. As with traditional 1-D modelling, the user provides information on cross-section survey, top-of-bank positions and longitudinal bed (or water surface) slope (Figure 3.57). The conveyance .GEN file is linked to the relevant roughness .RAD file, enabling the user to match the previously created roughness zones to the cross-section survey (Figure 3.58). For meandering channels, the user indicates the degree of meandering through the plan form sinuosity and assigns inside of bend and outside of bend labels to the top-of-bank markers.

The CG enables the user to explore the various section-averaged (Figure 3.59) and spatially distributed outputs (Figure 3.60) and to compare these to measured data. Outputs include the upper and lower credible scenarios derived from the upper and lower unit roughness values.

The CG allows the user to alter the default calculation parameters through the Advanced Options (Chapter 5), for example, the number of calculation depths, the cross-section discretisation, the value of the main channel lateral eddy viscosity λ_{mc}, etc.

3.4.3 Introduction to the Backwater Module

The Backwater Module (BM) aids the user in calculating the surface water profile upstream of a control i.e. known flow and level. As with 1-D modelling, the user

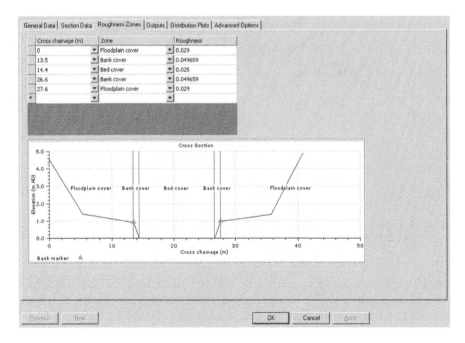

Figure 3.57 Example from Conveyance Generator showing a cross-section with allocated roughness zones for the River Main at Bridge End Bridge in County Antrim.

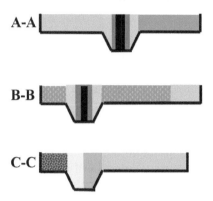

Figure 3.58 Plan form roughness zones from Figure 3.52 allocated to specific cross-sections.

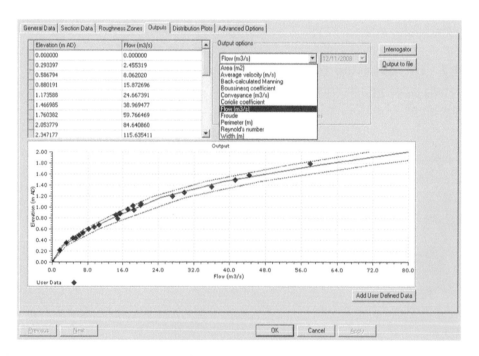

Figure 3.59 Example from Conveyance Generator of predicted flow for the River Main in at Bridge End Bridge in county Antrim and comparison to measured data (red).

creates multiple cross-sections along the channel reach and provides the chainage between these (Figure 3.61). The user provides the downstream depth and the inflow flow to the reach and the backwater is then calculated (Figure 3.62 and 3.63).

The user may impose normal flow conditions at the downstream end. Here, the flow is provided and the corresponding normal flow depth is evaluated at the downstream end and used as the starting point for the backwater calculation.

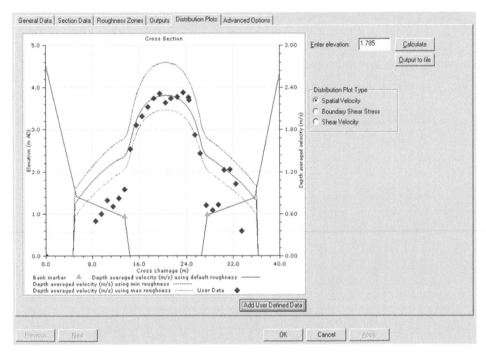

Figure 3.60 Example from Conveyance Generator of predicted velocity for the River Main in at Bridge End Bridge in County Antrim and comparison to measured data (red).

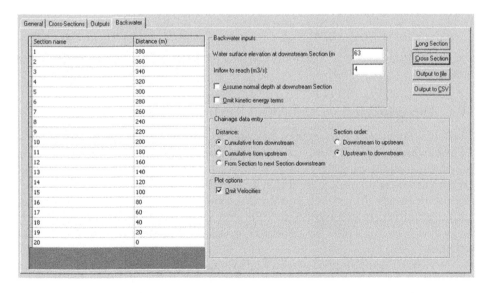

Figure 3.61 Example from Backwater Module showing multiple cross-sections and their associated chainages for the River Blackwater in Hampshire.

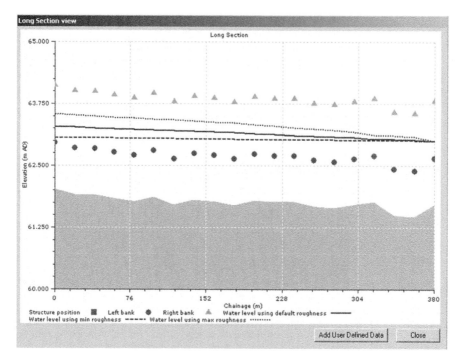

Figure 3.62 Example from Backwater Module showing a long-section with the predicted water surface profiles based on minimum, maximum and expected roughness for the River Blackwater in Hampshire.

The user may elect to omit the velocity term in the energy balance equation (Section 3.2.4). This is a useful alternative for shallow flows, reaches where the velocity changes rapidly along the reach or where the flow in a sub-reach may be supercritical.

3.4.4   Introduction to the Afflux Estimation System

The Afflux Estimator (AE) enables the user to add bridge or culvert units into a CES backwater model. In both cases, the unit appears within the CES software as a single entity containing data entry forms for the channel and structure cross section data at upstream and downstream faces. Figure 3.64 shows opening data for a bridge with identical upstream and downstream faces. Figure 3.65 shows opening data for a culvert. In common with CES river sections, the bridge or culvert data is stored in the CES .GEN file, which is linked to roughness values specified in a .RAD file.

The Afflux Estimator produces section averaged rating curve outputs, in the same way as they are displayed by the Conveyance Generator, for parameters such as conveyance, flow discharge and flow area. Figure 3.66 shows an example conveyance curve for a twin arch bridge. Outputs include upper and lower credible scenarios which are derived from upper and lower roughness ranges for the structure.

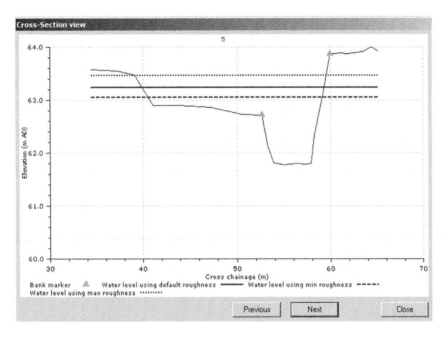

Figure 3.63 Example from Backwater Module showing a long-section with the predicted water surface profiles based on minimum, maximum and expected roughness for the River Blackwater in Hampshire.

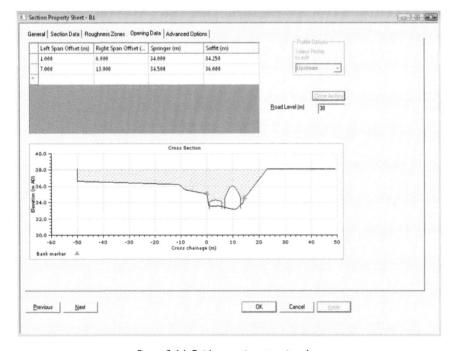

Figure 3.64 Bridge section opening data.

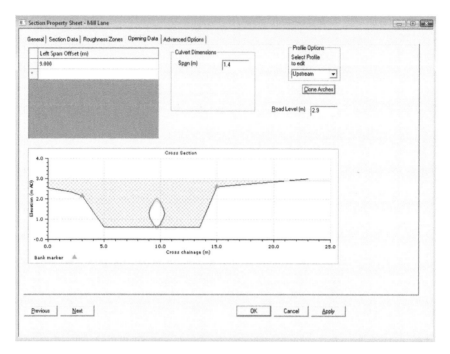

Figure 3.65 Culvert section opening data.

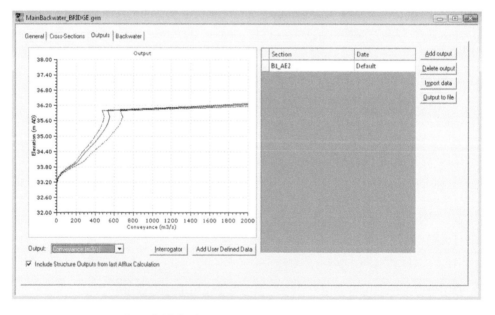

Figure 3.66 Bridge section conveyance curve output.

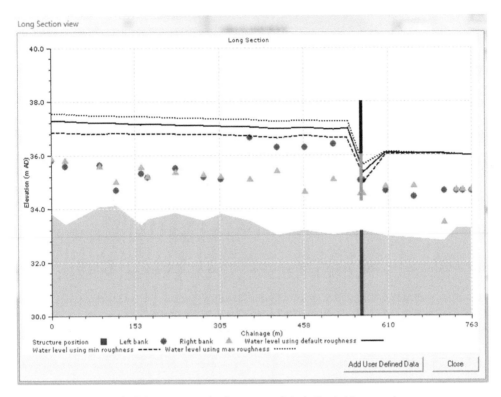

Figure 3.67 Long section backwater reach including bridge or culvert.

When a backwater profile is computed by the CES/AES software, the Afflux Estimator calculates the water surface upstream of the structure and passes this information into the upstream backwater reach, where it is used in effect as a new boundary condition. Figure 3.67 shows an example where a twin arch bridge has been placed into a reach. The water surface profile dips through the arches and then rises upstream of the structure. The plot displays the profile for the defined roughness conditions, plus bounds based on credible limits for the roughness. Velocity data may also be displayed, along with a hypothetical water surface profile for the undisturbed stream (enabling the afflux to be calculated). The vertical grey bar shows the range of bridge soffit levels and the elevation between the highest soffit and the parapet (or 'roadway') level is shown as a black vertical bar.

Chapter 4

Practical issues – roughness, conveyance and afflux

ABSTRACT

This chapter illustrates the use of the CES-AES methods through a series of practical examples involving stage-discharge, velocity, backwater and afflux prediction for a range of channel types (size, shape, cover). Generic guidance on the use of the CES-AES for different flood risk management activities is provided, with specific examples relating to availability of site information, vegetation maintenance and dredging. A series of 'what-if' scenarios is explored to demonstrate the sensitivity of the CES-AES outputs to changes in input data (roughness, slope, sinuosity). An approach for estimating roughness in mountain streams with boulders is presented.

The survey data for the examples are provided in Appendix 3.

4.1 AN OVERVIEW OF THE CES-AES USE IN PRACTICE

The CES-AES software supports practitioners concerned with a range of flood risk management as well as other wider activities through, for example:

- Calculating water levels, flows and velocities for rivers, watercourses and drains;
- Providing upper and lower uncertainty scenarios;
- Assessing flood or extreme water levels, and the sensitivity of these to channel adaptation or management options (particularly dredging and plant management);
- Assessing the impact of timing and nature of vegetation cutting;
- Assessing the impact of blockage due to vegetation or debris;
- Understanding the influence of in-stream structures on water levels;
- Finding holistic solutions which address both environmental and flood risk management or land drainage objectives;
- Implementing guidance and procedures for channel maintenance.

The use of the CES for channel maintenance and flood risk management activities is recognised in Sir Michael Pitt's review following the 2007 widespread flooding (Pitt, 2008):

"7.66 ... To progress its understanding of how seasonal variation in vegetation affects the way in which watercourses behave, the Environment Agency has recently developed a tool called the Conveyance Estimation System (CES), which will help to deliver an improved maintenance programme. ..."

The Association of Drainage Authorities and Natural England's Biodiversity Manual – Integrating Wildlife and Flood Risk Management (Buisson *et al.*, 2008) recognises the potential use of the CES in identifying preferred channel management techniques:

"In choosing the best technique to apply it is essential to examine the effect on flood conveyance in the specific location and circumstances that the technique would be used. This will require judgment informed by experience and one of the available flood risk modelling tools. In many circumstances, the Conveyance Estimation System (www.river-conveyance.net) may provide the information needed. Modelling allows prediction of the effects of management techniques on conveyance and storage and can identify the additional capacity needed to offset any reduction in conveyance caused by additional wildlife habitat created, such as a wider uncut marginal strip of vegetation in the channel."

Table 4.1 provides a summary of some of the core flood risk management activities, the type of hydrodynamic modelling associated with these and advice on the interpretation of outputs for specific activities. Outputs from the CES-AES may be used in modelling types (a) through to (f) in the table, which support a range of activities including strategic planning, scheme design, flood forecasting, flood mapping for regulation and insurance, hydrometry and maintenance.

For all of these activities, the basic building blocks within the CES-AES software include:

- a single cross-section analysis;
- a single structure analysis i.e. bridges and culverts;
- a backwater profile analysis with no structures present;
- a backwater profile analysis with structures present e.g. bridges, culverts; and
- the use of CES-AES within a 1-D modelling environment (see Chapter 5).

4.1.1   Single cross-section analysis

A single cross-section analysis typically involves one or more of the following steps:

1   *Data gathering* – Gather all available data and information about the site, including where possible, survey; water level, discharge and velocity measurements; water surface slope; photographs; channel and floodplain cover; and any historical flood data.

2   *Initial stage-discharge analysis* – In the absence of flow data, apply the CES methodology using default parameters and the Roughness Advisor unit roughness information. Essential information is the surveyed cross-section geometry and the longitudinal water surface slope (ideally) or bed slope. Photographs, site descriptions and UK grid references may be used to assist in selecting the unit roughness values.

Table 4.1 Flood risk management tasks (adapted from EA/Defra, 2002a and Defra/EA, 2004).

Type of modelling	Activity						Comments/Use
	a) 1-D modelling	b) Routing models	c) Backwater models	d) Rating curves	e) Rating surfaces	f) Single levels/ flows	
Strategic planning	x	x	x	x	x		Evaluate and compare options against criteria such as effectiveness, efficiency, robustness and performance. Where small changes in input parameters effect large changes in water level, devote additional resources to under-standing and potentially reducing the cause.
Scheme design	x	x	x	x			New schemes ideally based on a calibrated 1-D model. Alternatively, undertake a cost-benefit analysis for multiple cases with different site roughness values.
Flood forecasting & warning	x	x	x	x	x		Examine the sensitivity of flood wave speed to river resistance using a flood routing model such as that which is available in Flood Modeller Pro/ InfoWorks RS(CES embedded). Additionally, undertake an offline analysis to examine where the flood peaks change significantly with channel roughness.
Flood mapping for regulation/ insurance	x	x	x		x		For broad-scale mapping, inform Agency Development Control that approach is indicative not accurate. Examples of CES use in national flood mapping include the Second Generation Flood Maps for Scotland (Mc Gahey et al., 2005; SEPA, 2006) and Northern Ireland (NIRA, 2005).
Hydrometry	x		x	x			Typically users have high quality calibration data. This may be used for model calibration & extension of ratings.
Maintenance			x	x		x	Detailed local site knowledge (including seasonal vegetation variations). Use CES to predict flow and local velocities.

3   *Calibration using stage-discharge measurements* – If flow data are available, cali-
     brate the stage-discharge curve by varying the n_l values, as the flow rate is par-
     ticularly sensitive to this parameter. The top-of-bank markers may also be varied
     within a realistic range (e.g. ±0.25 m for rivers) to capture the bankfull variations.
4   *Calibration using stage-discharge and velocity measurement* – If velocity data
     are available, refine the lateral distribution of roughness, in particular, the in
     channel, bank and berm n_l values. Here, it is necessary to ensure the changes in
     n_l for a given depth apply to all depths i.e. the calibrated stage-discharge curve is
     preserved. If historic flood data are available, this is a useful means of calibrating
     the top-end of the rating curve, and establishing the floodplain n_l values.
5   *Expert calibration using internal model parameters* – Expert users may wish to
     vary the values of the main channel dimensionless eddy viscosity λ_{mc} (via the CES-
     AES Graphical User Interface, Chapter 5) and the secondary flow term Γ (via
     the source code, Chapter 5) based on local site knowledge, for example, reduc-
     ing λ_{mc} if bankside vegetation reduces the mixing at the main channel floodplain
     interface. Note: this is advocated for research or expert users only as inexperi-
     enced application of this multi-parameter calibration approach may lead to wider
     uncertainty bounds than alternative calculation methods.
6   *Meandering channels* – For meandering channels, Steps 1 to 4 are identical, except
     there are two additional inputs, the plan form sinuosity, σ, and the bend orien-
     tation i.e. left or right hand bend. The σ value should only be varied if there is
     uncertainty associated with the true value, to gauge the sensitivity for the site. The
     bend orientation is a fixed input which is assigned to the top-of-bank markers.

Figure 4.1 provides an example of a simple CES application to a two-stage
cross-section located in a straight reach of the River Penk at Penkridge. The effect
of n_l, λ and Γ on the depth-averaged velocity profile is shown and a small sinuosity is introduced to illustrate what the affect of C_{uv} would be. The calibrated n_l
values are 0.033 and 0.060 for the main channel and floodplains respectively. The
default λ and Γ models are used and an example sinuosity σ of 1.05 ($C_{uv} \approx 1\%$) is
introduced.

4.1.2   Single structure analysis (bridges and culverts)

A single structure analysis typically involves one or more of the following steps:

1   *Data gathering* – Information is required to characterise both the structure and
     the channel surrounding it. Channel cross section information is required as in
     Section 4.1.1. If the bridge structure has abutments or other modifications of the
     surrounding channel then these need to be included in the cross section survey.
     Ideally, the structure should be surveyed (both upstream and downstream faces)
     to provide number of openings and elevation data for springer levels, soffit levels
     and road or parapet levels. The material and condition of the structure should be
     observed, if possible, bearing in mind that the material and condition of the struc-
     ture at the opening might not always be representative of conditions throughout
     its downstream length. For culverts, the entrance structure type and shape should
     be recorded. Consideration should be given to the condition of the culvert barrel

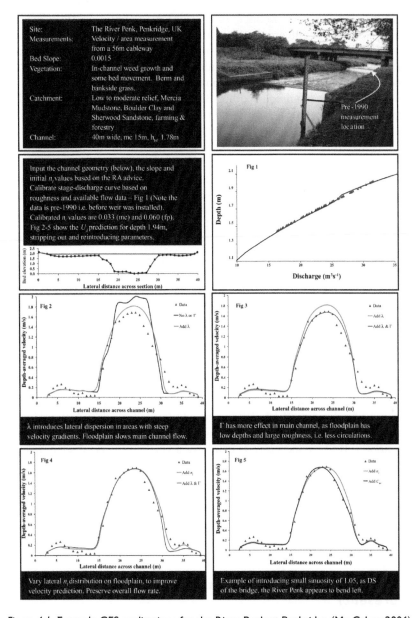

Figure 4.1 Example CES applications for the River Penk at Penkridge (Mc Gahey, 2006).

throughout its length, although this may not be known without a full inspection or CCTV survey.

2   *Initial structure analysis* – The AES includes certain typical bridge and culvert types (e.g. arch and beam bridge, pipe and box culvert). However some structures

may not fit exactly into the defined types. Where this is the case, the closest match should be used. The user defined arch bridge type allows greatest flexibility for opening shapes. For culverts with multiple barrels, the AES standard types allow identically sized openings to be specified. It may be possible to model a culvert with complex openings as a bridge unit if the barrel friction is thought likely to be the hydraulic control in most circumstances.

3   *Data entry* – Care is required in adding bridges or culverts to a CES model. There are a number of geometric and procedural rules enforced by CES-AES. These include:

    a   Openings for bridges and culverts must have valid coordinates within the channel cross section. This means, for example, that culvert openings should not extend below ground level and springer levels for arches must be above ground elevation at a given cross section offset (i.e. arches should not spring directly from ground level at the side of a cross section).

    b   Road elevations must be contained within the cross section.

    c   The CES river cross sections immediately upstream and downstream of a bridge or culvert should have three roughness zones.

    d   In the CES-AES backwater module, the upstream openings of a bridge or culvert are placed at the chainage implied by the reach length downstream of the previous section.

    e   In the CES-AES backwater module, the reach length downstream from the bridge section to the next CES river section is taken to include the bridge length.

4   *Cross section analysis* – The CES-AES software requires a backwater profile to be calculated in order to generate information on conveyance and flow ratings at the bridge or culvert opening sections. The 'Outputs' tab of the software requires the user to check a box 'Include Structure Outputs from last Afflux Calculation' to include data from the structure cross sections.

5   *Calibration* – Bridges and culverts can have a single roughness value (expressed in terms of Manning's *n*) for the structure, which is combined with the channel and overbank roughness values at the structure sections when computing the conveyance. Should there be measurements available to assist with calibration then they may be used as for a normal CES river section. Note, as above, that a backwater profile must be computed to generate conveyance or water level results for bridges or culverts.

4.1.3   Backwater profile analysis (no structures present)

Backwater profiles are evaluated where there is a downstream control (i.e. known head and flow). A backwater analysis typically involves the following steps:

1   *Data gathering* – Information (similar to the single cross-section analysis, Section 4.1.1) for multiple cross-sections is required. An important question is how far upstream of the control the cross-sections are required. A useful rule of thumb (Samuels, 1989b) based on the downstream depth and reach-averaged longitudinal slope may be used to estimate this. Additional inputs include the distance

between each cross-section, the downstream water level and the inflow(s) to the reach. Additional measurements or historic data may include, for example, water levels along the reach for a given inflow.

2. *Initial single section analysis* – Steps 1 to 4 of the single section analysis (Section 4.1.1) are followed for any cross-sections where (i) measurements are available and (ii) the water level measurements are not dominated by the downstreamcontrol. The adopted/calibrated roughness values may then be assigned to the remainder of the reach cross-sections in accordance with the available descriptions.

3. *Backwater analysis* – The downstream control and inflow to the reach are assigned and the backwater calculation is implemented, providing the water surface profile and average velocities along the reach.

4. *Backwater calibration* – The backwater predictions may then be compared to water level and velocity measurements along the reach and these may guide further roughness calibration.

4.1.4   Backwater profile analysis (structures present)

The process for backwater profile analysis when a bridge or culvert is included is the same as above for a reach without structures. As described in Chapter 3, the AES models the water surface profile through a structure using cross sections supplied by the user at the structure entrance and exit, these are labelled AE2 and AE3 in the output files exported by CES-AES following a backwater analysis. Interpolated upstream and downstream sections are also used where the transition lengths surrounding the structure require it. These sections AE1 and AE4 are also reported in the output files.

4.2   ESTIMATING AND USING STAGE-DISCHARGE RELATIONSHIPS AND SPATIAL VELOCITIES

4.2.1   Stage-discharge prediction for the River Trent at Yoxall, Staffordshire, UK

Aim

To predict the stage-discharge and upper and lower uncertainty scenarios for the River Yoxall site using the CES-AES software and available site information. The outputs are compared with measured data to instil confidence in the results.

The Yoxall Site on the River Trent

The Yoxall site (Figure 4.2) is located on a gentle bend of the River Trent, in Yoxall, Staffordshire, immediately downstream of Yoxall road-bridge. The cross-section is 80 m wide, including a 35 m berm, and a cableway for measuring inbank flows is situated across the main channel. The site is bypassed through two flood relief culverts, the flows of which are measured under Yoxall Bridge. There are four overbank

Figure 4.2 Electromagnetic gauging station on the River Trent (a) downstream of Yoxall bridge and (b) an upstream view showing Yoxall Bridge (Courtesy of the EA).

gauged flows, of which two were recorded before and two after the construction of a training bank. The inbank flow measurements are generally considered more reliable than overbank flows (EA, 2005). Measurements from the Environment Agency's gauge station (Station 28012) include water level, discharge and velocity. The reach-averaged longitudinal slope at bankfull is ~0.001, increasing to 0.00275 at a depth of 0.5 m. The channel bed material includes gravels and summer weed growth and the floodplains consist of grasses with occasional bushes.

Roughness and cross-section definition

Based on the description and photographs, the Roughness Advisor values include (Figure 4.3):

- *Channel bed* – The RA coarse gravel is adopted ($n_{sur} = n_l = 0.027$). This may be refined to gravel or fine gravel through calibration. The larger gravel option is incorporated for the initial guess i.e. a conservative approach.
- *Channel banks* – The RA height-varying grass ($n_{veg} = 0.041$) and sand ($n_{sur} = 0.02$) is adopted giving the total unit roughness $n_l = 0.0456$.
- *Floodplain* – The RA turf ($n_{veg} = 0.021$) and sand ($n_{sur} = 0.02$) is adopted giving $n_l = 0.029$. The reasoning is that bank-side vegetation appears thicker and taller (i.e. height-varying grass) than the floodplain grass (i.e. turf).

Results for the Yoxall site

Figure 4.4 provides the CES stage-discharge prediction using these RA roughness values and the measured data. The data are fairly scattered but tend to fall within the upper and lower credible scenarios, defined by the upper and lower roughness values. The data move away from the curve at high overbank flows which may be attributed to the reduced accuracy in flow gauging during times of flood together with uncertainty regarding the floodplain vegetation. A further factor may be the variation in longitudinal water surface slope with depth, which is subject to local topographic

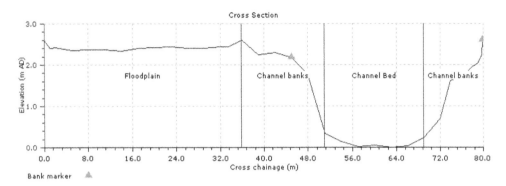

Figure 4.3 River Trent cross-section, roughness zones and top-of-bank marker locations.

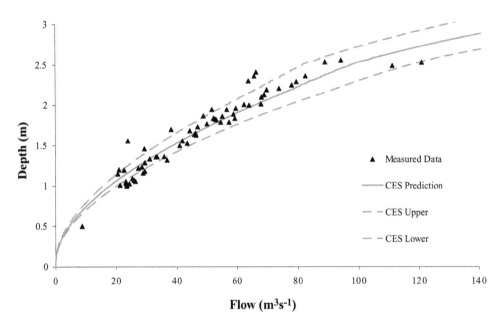

Figure 4.4 Stage-discharge prediction for the River Trent at Yoxall.

variations at low flows, ~0.001 at bankfull (as used in the CES simulation) and more closely linked to the overall valley slope at high flows. This issue may be overcome if the backwater module is used for the reach, as the variations in surface water slope are implicit. Similarly, where the CES is incorporated in a full 1-D model the depth specific surface water slope is incorporated. The stand-alone cross-section analysis is useful in determining and validating the unit roughness values to be adopted for the remainder of the reach.

4.2.2   Stage-discharge and velocity prediction for the River Colorado, La Pampas, Argentina

Aim

To predict a stage-discharge curve and velocity distributions for different flow depths for the River Colorado site using the CES-AES software and available site information. The outputs are compared to measured (flow and velocity) data throughout the depth range to instil confidence in the results.

The River Colorado site

The River Colorado provides a useful example of a wide, rigid bed, simple channel with large flows. It is 885 km long, rising in the Andes and flowing south-east across southern central Argentina to the Atlantic Ocean. It marks the northern limit of Patagonia. It is also a rough boundary between the commercial agriculture to the north and ranching to the south. The main channel is 60 m wide and 3.6 m deep, with a reach-averaged longitudinal bed slope of 0.0013. Typical flow rates are of the order 400–600 m^3s^{-1} and the river often overflows its banks in Spring. Water level, discharge and depth-averaged velocity measurements are available; however, information on channel roughness is limited to the photographic evidence (Figure 4.5).

Roughness and cross-section definition

Based on the photographs, the Roughness Advisor values include (Figure 4.6):

- *Channel bed* – The RA bedrock (n_{sur} = 0.025) and 0–20% boulder coverage (n_{irr} = 0.017) are adopted, giving n_l = 0.030. Bedrock is assumed due to the apparent hard cut-out of the channel side-banks (Figure 4.5) and some boulders are included due to the presence of boulders near the right bank (Figure 4.5c).
- *Channel banks* – The RA bedrock (n_{sur} = 0.025) and height-varying grass (n_{veg} = 0.041) are adopted giving n_l = 0.048. Although the vegetation is not visible on all banks, the added influence of the height varying grass on the banks is likely to be minimal in terms of the overall channel width and hence capacity.
- *Floodplain* – The RA bedrock (n_{sur} = 0.025) and turf (n_{veg} = 0.021) are adopted giving n_l = 0.033.

Results for the Colorado site

Figure 4.7 shows the CES stage-discharge predictions with these roughness values. The flow is described reasonably well throughout the range of depths other than near bankfull (3.6 m), where there is a sharp step or 'kick' in the curve. This is caused by the embedded secondary flow model, which takes the change in orientation and magnitude of the secondary flow circulations from inbank to out-of-bank flow into account, identified in the model by the bank markers (Figure 4.6). The step is sharp as the change in geometry from in-channel vertical banks to floodplain horizontal

Figure 4.5 The River Colorado gauge site showing the instrumentation (a) & (b) and looking (c) downstream and (d) across the full width of the river (Courtesy of Leticia Tarrab).

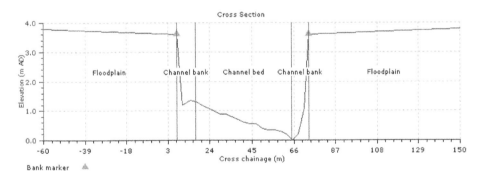

Figure 4.6 River Colorado cross-section, roughness zones and top-of-bank marker locations.

banks is sudden. As survey data were only available for the in-channel section, the floodplain topography is based on the photographic evidence (e.g. Figure 4.5b). The discrepancy between the measured and predicted flows suggests that, in reality, the change from inbank to out-of-bank is perhaps more gradual.

Figure 4.8 provides the depth-averaged velocity predictions for a range of flow depths and the corresponding measured data. The CES tends to under-predict the high

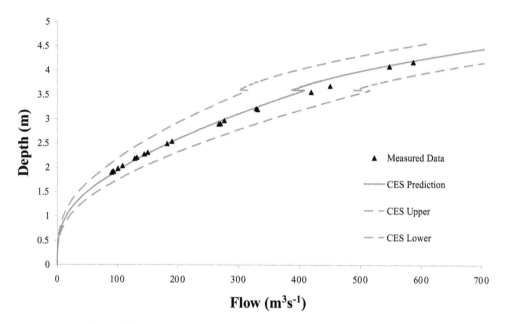

Figure 4.7 Stage-discharge prediction for the River Colorado, Argentina.

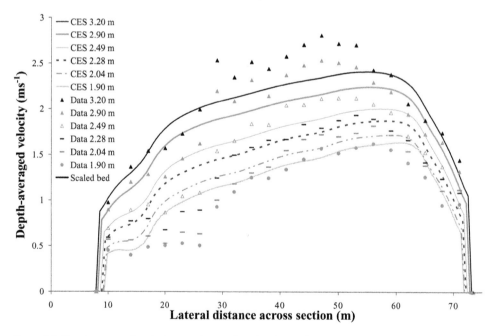

Figure 4.8 Depth-averaged velocity prediction for the River Colorado at various flow depths and comparison to measured data.

velocities (channel centre) at large flow depths and slightly over-predict the velocities at lower depths. This may be a short-coming in the CES approach for adequately resolving the variation of roughness with depth for this wide channel, where bed generated turbulence has a dominant role and the channel banks have less influence. The distribution may be further influenced by the presence of individual boulders (e.g. Figure 4.5c).

4.2.3 Hierarchical approach to estimating roughness (and other flow parameters) for the River Main, County Antrim, UK

Aim

This example is intended to demonstrate the hierarchical approach to estimating roughness within the CES i.e. no information (use of River Habitat Survey); some information (use of Roughness Advisor descriptions); and detailed information (use of photos and measurements to calibrate roughness). The hierarchical approach is based on stage-discharge predictions. Once the detailed information has been used, additional flow parameters are simulated with the CES including velocity, back-calculated Manning *n*, Froude Number and Reynolds Number.

The River Main site

The River Main in County Antrim was introduced in Chapter 2, Section 2.2.2. The reach of interest was reconstructed and realigned in the 1980s to form a double trap-ezoidal channel. The CES is used here to demonstrate its use in predicting various flow parameters at Bridge End Bridge based on available information. Although this includes survey (Figure 4.9), bed slope (0.00297), roughness, photographs (Section 2.2.2) and detailed measurements; this example is set-up to illustrate the RA use with varying degrees of roughness information i.e.:

- *No information* – for the main channel (use of RA RHS advice);
- *Some information* – for the main channel (use of RA based on descriptions);
- *Detailed information* – for the main channel (use of RA including calibrating roughness values to best-fit measurements and use of photographs).

Roughness and cross-section definition and results for the River Main

i. No information
For this case it is assumed there is no information on in-channel cover other than the quarry stone on the channel banks (0.5 tonne weight, 100–200 mm size) – incorpo-rated during the re-design of the channel. Three roughness zones are set-up in the RA:

- *Channel bed* – The UK Grid reference for the River Main is 350100 411000 (NI). Based on this, the RHS suggests the most likely vegetation is filamentous algae

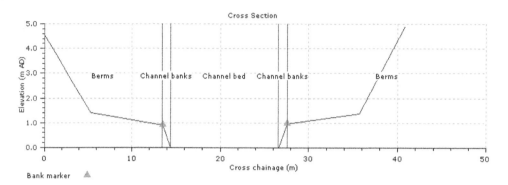

Figure 4.9 River Main cross-section, roughness zones and top-of-bank marker locations.

$(n_{veg} = 0.015)$. This is adopted with sand $(n_{sur} = 0.02)$ as a basic substrate giving $n_l = 0.025$.

- *Channel banks* – The RA rip-rap option is adopted for the quarry stone $(n_{sur} = 0.045)$ augmented with the RHS suggested filamentous algae, giving $n_l = 0.047$.
- *Berms* – Based on the available floodplain information, the RA height-varying grass $(n_{veg} = 0.041)$ option is adopted with sand $(n_{sur} = 0.02)$ as the basic substrate. This gives the total unit roughness $n_l = 0.046$.

The RA roughness zones are combined with the cross-section and slope information to predict the stage discharge (Figure 4.10). These include the upper and lower
credible scenarios which are based on the upper and lower roughness values. The
measured data falls within these scenarios (which are wide); however the curve based
on the expected roughness falls below the data.

ii.   Some information
For this case there is additional information on the roughness cover including: no
vegetation present at the channel bed which consists of coarse gravel and thick grass
protruding through the quarry stone on the banks. As before, three roughness zones
are set-up in the RA composed of:

- *Channel bed* – The RA coarse gravel is adopted $(n_{sur} = n_l = 0.027)$.
- *Channel banks* – The RA rip-rap option is adopted for the quarry stone $(n_{sur} = 0.045)$ together with the RA turf $(n_{veg} = 0.021)$ to simulate the thick grass. This gives the total unit roughness $n_l = 0.050$.
- *Berms* – The values are unaltered i.e. a total unit roughness $n_l = 0.046$.

A revised flow prediction is made based on the updated roughness values
(Figure 4.11). The curve based on the expected roughness value still falls below the measured data albeit closer. A notable difference is the reduced range of uncertainty i.e. the
credible upper and lower scenarios are much narrower. This is largely due to eliminating the filamentous algae and heavy weed growth, which have considerable associated
uncertainties. Some data now fall outside of these scenarios, which is plausible as

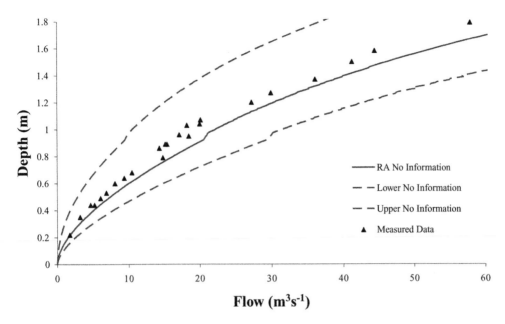

Figure 4.10 Stage-discharge prediction for the River Main with no in-channel information.

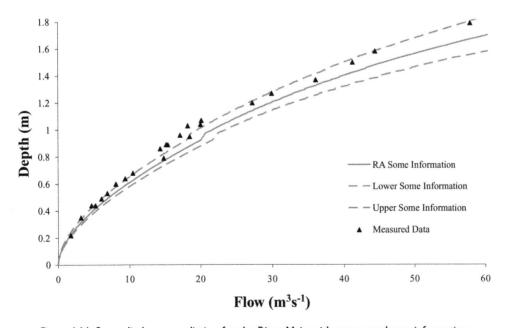

Figure 4.11 Stage-discharge prediction for the River Main with some roughness information.

these are not confidence intervals or envelopes (Chapter 3, Section 3.3) – they simply indicate the area where the data is likely to fall.

iii   Detailed information

For this case there is detailed information available including photographs and measurements (flows and velocities), which can be used to improve the CES simulations. From Figure 4.11 it is apparent that the RA is under-estimating the roughness as the measured data are above the curve, and in some instances, above the upper credible scenario. The photographic evidence (Figure 2.5) suggests that the coarse gravel is very large with cobbles present. The RA unit roughness value for coarse gravels and cobbles are 0.027 and 0.035 respectively, a substantial difference. Here, steered by the measured data and photographic evidence, a value of 0.032 is adopted. Figure 4.12 provides the resulting stage-discharge prediction which follows the measured data well – including the variations at bankfull (~0.95 m). The 'no information' curves are added to illustrate the reduction in uncertainty and improved stage-discharge predictions where more detailed information is available.

These calibrated roughness values are used to predict the depth-averaged velocities for a range of flow depths (Figure 4.13). The predictions follow the data profile reasonably well for greater depths; however, at lower depths the main channel velocity is under-predicted. There appears to be a small degree of skew in the velocity data at low depths, which may arise from the effects of the upstream bend circulations being transported downstream. Since the channel bed material is gravel, with no vegetation present, these higher velocities are plausible but are not captured by the CES

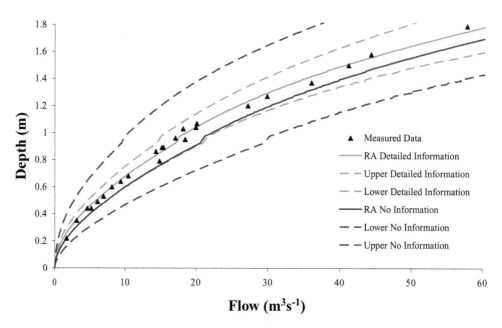

Figure 4.12 Stage-discharge for the River Main with detailed roughness information (and the no information case).

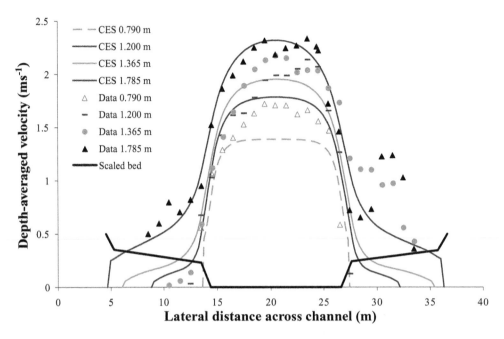

Figure 4.13 Depth-averaged velocity predictions for the River Main at various flow depths and comparison to measured data.

where the effect of the slightly higher calibrated bank side roughness ($n_l = 0.050$) is present even at low flows.

Exploration of additional CES flow parameters

The CES enables users to back-calculate an equivalent Manning n resistance parameter. In modelling practice, a Manning n value is assigned to the channel bed, banks and berms, which is constant for all flow depths. In the field, measurements show how Manning's n varies with flow depth (e.g. Chow, 1959), with a greater resistance at low depths where roughness size is comparable with flow depth. Figure 4.14 provides the CES back-calculated and measured Manning n values with depth. These show a reasonable correlation throughout the depth albeit with the predictions ~15% lower than the measured data.

Figure 4.15 shows the CES predicted Froude Number with depth. The Froude Number is well below 1.0 throughout the depth, indicating subcritical flow conditions, as expected for natural channels. In some instances, supercritical flow may occur, for example, in steep mountain streams, rapidly varied flow downstream of structures or localised supercritical flow in parts of the cross-section. Engineers have also been known to design supercritical channels through cities i.e. steep heavily modified concrete conduits designed to convey floods over shorter distances than the natural plan form channel profile. For the River Main, the Froude Number increases away from the channel bed, reflecting the higher velocities at greater depths. The shape of

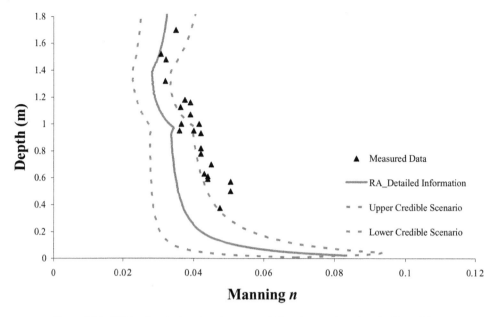

Figure 4.14 CES back-calculated and measured Manning *n* values for the River Main.

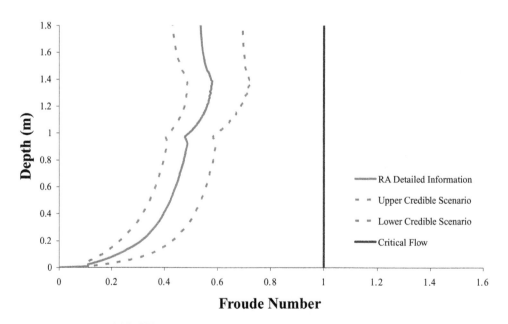

Figure 4.15 CES predicted Froude Number with depth for the River Main.

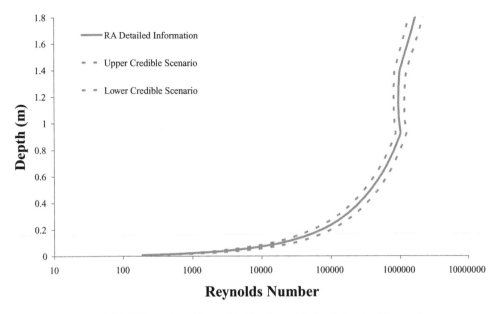

Figure 4.16 CES predicted Reynolds Number with depth for the River Main.

the curve changes as the channel flow moves out of bank onto the berms and again as it comes into contact with the distant berm edges, as these boundaries influence the velocity distribution.

Turbulence occurs at high Reynolds Numbers, where the flow incorporates a three-dimensional eddying or mixing action. Turbulence is dispersive, diffusive, chaotic, and contrary to viscosity, it is a property of the flow and not of the fluid (Reynolds, 1974; Nezu & Nakagawa, 1993). Figure 4.16 shows the CES predicted Reynolds Number with depth which is of the order 10^3 to 10^6 for most of the depth column i.e. characteristic of fully turbulent flow ($Re > 10^3$). Near the channel bed, the Reynolds Number drops substantially indicating the potential for laminar flow close to the boundary (although this is unlikely to occur in natural channels with sizable roughness elements e.g. cobbles).

This single cross-section analysis is a useful means to establish the calibrated rating curve and hence the roughness values to be used for the whole reach. These roughness values are adopted in Section 4.3, where different backwater profiles are explored for the River Main.

4.2.4   Stage-discharge, velocity and roughness predictions for the River Severn at Montford Bridge, UK

Aim

To improve confidence in high flow predictions required for a Flood Risk Assessment climate change scenario. This is achieved through simulating stage-discharge and velocity distributions which capture the measured data at lower and intermediate flows, with the bed slope playing an important role. The back-calculated Manning n values are also compared to measured values to improve confidence. This is achieved using the CES software and available site information.

The Montford Bridge site on the River Severn

The Montford Bridge site along the River Severn is introduced in Chapter 2, Section 2.5.2. To fulfil the requirements of a Flood Risk Assessment climate change scenario, it is necessary to determine the depth at a flow rate well above the available flow measurements. The scenario is to model the 100 year return period, $Q_{100} = 330$ m³s⁻¹, +20% flow to satisfy Defra's Project Appraisal Guidance on indicative sensitivity ranges for peak river flows in 50–100 years time, giving $Q = {\sim}400$ m³s⁻¹. The available site information includes survey, flow and velocity measurements. The description of the floodplains is limited to grass-cover together with some photographic evidence.

Roughness and cross-section definition

Based on the site description, the RA values include (Figure 4.17):

- *Channel bed* – The RA sand ($n_{sur} = n_l = 0.02$) is adopted as the bed material.
- *Channel banks* – The RA sand ($n_{sur} = 0.02$) and medium grass 0.75–1.0 m ($n_{veg} = 0.08$) is adopted giving $n_l = 0.082$.
- *Floodplain* – The RA sand ($n_{sur} = 0.02$) and turf ($n_{veg} = 0.021$) is adopted giving $n_l = 0.029$.

Results for the River Severn at Montford Bridge

For this site, there are measurements available for the water surface slope at each depth of flow, with an average value of 0.0002 at greater depths (>6 m) and the slope decreasing down to as low as 0.00006 for depths of 2–6 m. The initial CES

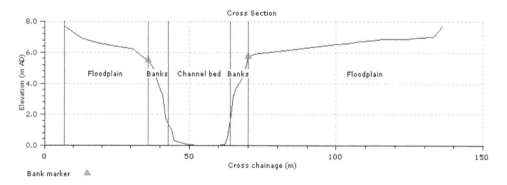

Figure 4.17 River Severn cross-section, roughness zones and top-of-bank marker locations.

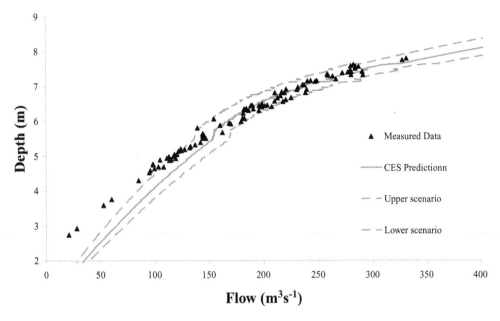

Figure 4.18 Stage-discharge prediction for the River Severn (slope = 0.000195).

stage-discharge predictions (Figure 4.18) are based on the value 0.000195 giving reasonable predictions above 5 m, including the bankfull variations, and reducing the
slope to 0.00015 (Figure 4.19) marginally improves the simulation magnitude at lower
depths. The shape of the curve in the shallower region is poor, which may be related
to the seasonal vegetation. Low flows typically occur in the summer months when the
channel vegetation is dense, resulting in reduced conveyance and large flow depths.

Figure 4.20 shows the depth-averaged velocity predictions and measured data
for a range of flow depths. The data are captured reasonably well, other than the
very high velocities in the centre region of the channel. This may indicate that the
CES model is over-representing the lateral shear (i.e. λ_{mc} is too high), resulting in an
increased retarding effect of the slower floodplain flow on the main channel flow.
This parameter may be altered in the CES-AES software (Chapter 5).

Figure 4.21 shows the CES back-calculated Manning's n values together with the
seasonal data measurements. The CES simulates the average through the scattered
data, capturing the slight increase in roughness where the depth drops from 6.5 m
(out of bank) to 5.5 m (inbank). At lower flow depths, the CES under-predicts the
roughness. This is not unexpected due to the uncertainty about the in-channel vegetation in the summer low-flow months.

As the flow, velocity and roughness information has been reasonably simulated,
the stage-discharge curve (slope = 0.000195) can be extended with reasonable confidence to determine the flow depths for the climate change scenario of $Q_{100} + 20\% =$
400 m³s⁻¹, giving 8.10 m depth with an uncertainty range of 7.9–8.4 m. Note that the
uncertainty on water level is typically smaller than that associated with flow, due to
the shape of the rating curve.

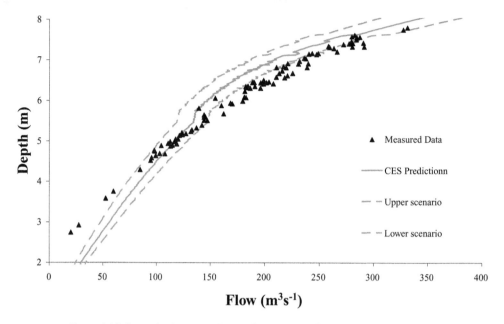

Figure 4.19 Stage-discharge prediction for the River Severn (slope = 0.00015).

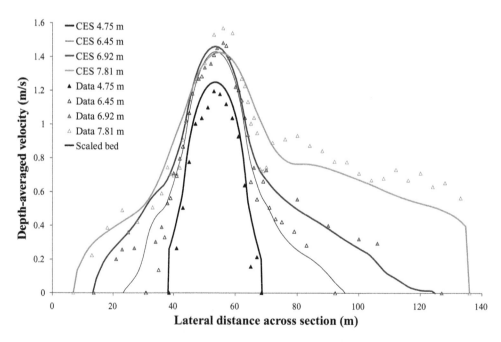

Figure 4.20 Depth-averaged velocity predictions for the River Severn at various flow depths and comparison to measured data (with depth 4.75 m based on slope = 0.00015).

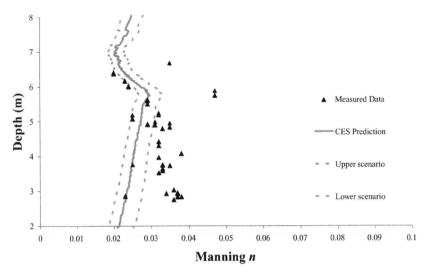

Figure 4.21 CES back-calculated and measured Manning *n* values for the River Severn.

4.2.5   Investigating the influence of roughness, slope and sinuosity on stage-discharge for the River La Suela, Cordoba, Argentina

Aim

To explore a series of 'what-if' scenarios for a simple channel and establish the sensitivity of the CES flow and velocity predictions to roughness, slope and sinuosity. This makes use of the CES-AES software and the available information for the La Suela river site and the outcomes are therefore specific to this site.

The River La Suela site

The River La Suela is a small river situated north-east of Cordoba, Argentina, with approximate dimensions of 25 m wide, 2 m deep and a reach-averaged longitudinal bed slope of 0.001355. Typical flow rates are in the range 50–70 m^{3}s^{-1}. Available measurements include water level, discharge and depth-averaged velocity profiles. Little is known about the roughness other than it consists of gravels, is alluvial in places and there is sparse bank-side vegetation.

Roughness and cross-section definition

Based on the description and a calibration using the unit roughness values, the final assigned roughness values include (Figure 4.22):

- *Channel bed* – the RA coarse gravels ($n_{sur} = 0.027$) and pools ($n_{irr} = 0.020$) giving $n_l = 0.034$.

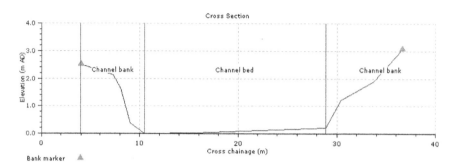

Figure 4.22 River La Suela cross-section, roughness zones and top-of-bank marker locations.

- *Channel banks* – the RA coarse gravels (n_{sur} = 0.027) and height-varying grass (n_{veg} = 0.041) giving n_l = 0.049.

Results for the River La Suela

Figure 4.23 shows the resulting stage-discharge predictions and the measured data – which provide a reasonable fit due to calibration. Figure 4.24 shows the depth-averaged velocity prediction for 1.52 m depth. For both cases the data fall largely within the credible upper and lower uncertainty scenarios.

Exploring what-if scenarios for the River La Suela

As little is known about this river, the La Suela provides a useful example to explore different 'what-if' scenarios for the predicted depths, flows and velocities. The aim is to improve understanding of how the channel behaves through testing the sensitivity of the CES flow predictions to changes in the input parameters. For example:

1. What happens if the channel bed roughness is doubled i.e. n_l = 0.068?
2. What happens if the channel bed roughness is halved i.e. n_l = 0.017?
3. What happens if the channel bank roughness is doubled i.e. n_l = 0.1?
4. What happens if the channel bank roughness is halved i.e. n_l = 0.025?
5. What happens if the channel is lined with concrete i.e. n_l = 0.02?
6. What happens if the concrete-lined channel is made rectangular?
7. What happens if the bed slope is doubled i.e. slope = 0.00271?
8. What happens if the bed slope is halved i.e. slope = 0.000678?
9. What happens if the channel is meandering with a sinuosity of 1.5?
10. What happens if the channel is meandering with a sinuosity of 2.5?

Exploring these scenarios provides some insight into the input parameters which are driving the predictions, for example, the importance of survey versus roughness. This may help direct the effort when obtaining improved local information. For a more detailed CES sensitivity analysis of numerous experimental and natural channels, see Latapie (2003) and Mc Gahey (2006).

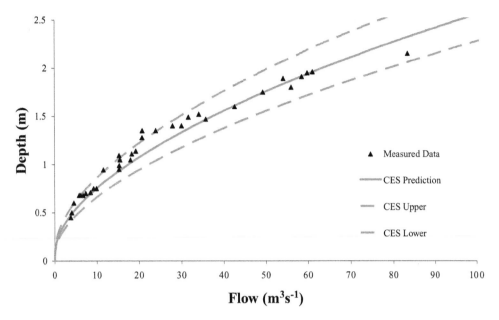

Figure 4.23 Stage-discharge prediction for the River La Suela.

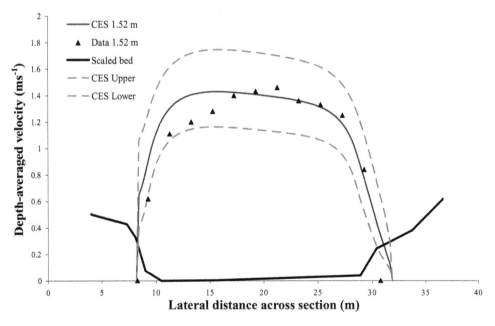

Figure 4.24 Depth-averaged velocity prediction for 1.52 m flow depth showing the uncertainty scenarios (Q_{CES} = 38 m³s⁻¹; Q_{data} = 34 m³s⁻¹).

i   Results for what-if 1 to 6

What-if scenarios 1 to 6 are designed to improve understanding of the influence of channel roughness on flow within the channel. This includes the magnitude of the roughness (e.g. double or half the calibrated roughness value, concrete-lined) and the distribution of the roughness (e.g. the importance of bank cover versus bed cover). The outcomes are peculiar to the La Suela River and should only be considered as indicative for other channels of similar size, shape and roughness.

Figure 4.25 shows the CES predicted stage-discharge, measured data and the revised stage-discharge predictions based on scenarios 1 to 6. Intuitively, any increase in roughness results in a decrease in flow and vice versa for all depths. To this end, the results of scenario 1 and 2 are both shown in solid lines and the results of scenario 3 and 4 are both indicated with dashed lines. Some observations can be made:

- bed roughness has a greater influence on flow than a change in bank roughness;
- a change in bed roughness influences the curve throughout the depth range;
- a change in bank roughness has little influence at low flow depths and increasing influence with increasing depth of flow;
- concrete-lining ($n_l = 0.02$ for the banks and floodplain) increases the flow capacity of the channel to a similar magnitude as scenario 2 (n_l bed = 0.017) as may be expected;
- the rectangular shape appears to improve conveyance; however this is only an artefact of the change in area for a given flow depth (here, the change in area is only zero at 1.52 m);
- there is a large amount of scatter in the measured data, particularly at flow depths of 1.3 to 1.5 m, and none of the predicted curves capture this.

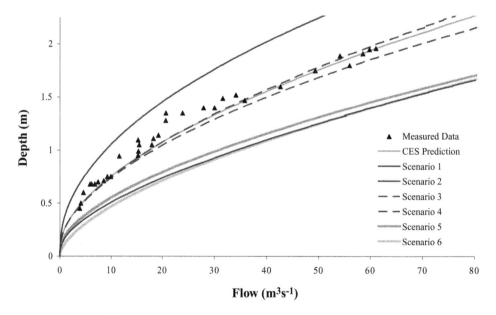

Figure 4.25 Scenario 1 to 6 stage-discharge prediction for the River La Suela.

Figure 4.26 shows the corresponding predicted velocities for scenario 1 to 4 for a 1.52 m depth together with the measured data. The observations include:

- the change in bank roughness influences the velocity profile close to the channel banks;
- the point at which the altered velocity due to the change in bank roughness returns to the original velocity profile mid-stream may differ for different degrees of bank roughness;
- the change in bed roughness has a significant impact on the mid-stream velocity profile which weakens towards the channel banks.

Figure 4.27 shows the predicted velocities for scenario 5 and 6 for 1.52 m depth (i.e. zero change in area) together with the measured data and scaled rectangular geometry. The observations include:

- increased velocities due to the concrete lining for both scenarios;
- change in velocity profile as a result of the change in channel shape.

The main findings for the River La Suela are that altering bed roughness has a more significant impact on the channel conveyance than altering the bank roughness; and that concrete-lining improves the conveyance capacity from that of a natural channel – as expected. These outcomes may well differ for very narrow channels where the channel banks have a more influential role on the mid-stream velocities. Similarly, for very wide channels, it is likely the channel banks will have little influence.

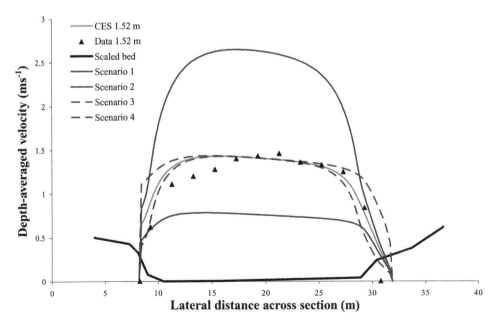

Figure 4.26 Scenario 1 to 4 depth-averaged velocity predictions for 1.52 m flow depth ($Q_{CES} = 38$ m^{3}s^{-1}; $Q_{data} = 34$ m^{3}s^{-1}; Scenarios $Q_1 = 22$ m^{3}s^{-1}; $Q_2 = 69$ m^{3}s^{-1}; $Q_3 = 38$ m^{3}s^{-1}; $Q_4 = 41$ m^{3}s^{-1}).

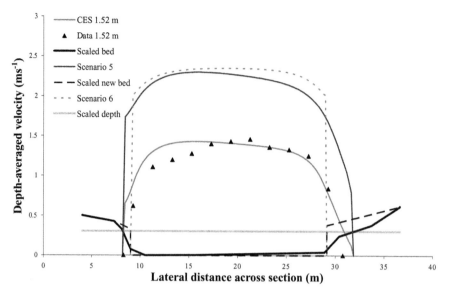

Figure 4.27 Scenario 5 and 6 depth-averaged velocity predictions for 1.52 m flow depth (Q_{CES} = 38 m³s⁻¹; Q_{data} = 34 m³s⁻¹; Scenarios Q_5 = 65 m³s⁻¹; Q_6 = 69 m³s⁻¹).

ii   Results for what-if 7 and 8

What-if scenarios 7 and 8 are designed to improve our understanding of the influence of longitudinal bed slope on channel flow capacity. The scenarios involve doubling (slope = 0.00271) and halving (slope = 0.000678) the bed slope. Figure 4.28 provides the resulting stage-discharge distributions. From this it is apparent that obtaining the correct bed slope is imperative to determining the flow rate (−30% >% change > +45%, Table 4.2).

The calculation of conveyance using the CES approach requires the bed slope as an input as it solves for the unit flow rate (Chapter 3). As conveyance is independent of slope, the conveyance is calculated by dividing the final flow rate by the slope $K = Q / S_o^{1/2}$. Figure 4.29 provides the calculated conveyance for the CES prediction and scenario 7 and 8 as well as the values obtained from the measured data. Here, it is apparent that the conveyance has little dependence on slope and is therefore a robust measure of the channel capacity. For further details on slope testing see Mc Gahey (2006).

iii   Results for what-if 9 and 10

What-if scenarios 9 and 10 are designed to improve our understanding of the influence of sinuosity on in-channel velocity profiles. The River La Suela cross-section is located in a straight portion of the reach and therefore has a sinuosity of 1. Scenarios 9 and 10 involve increasing the sinuosity in a left-bearing bend to 1.5 and 2.5 respectively. Figure 4.30 provides the depth-averaged velocity profiles for these. On the right or outside bend – there is no change in velocity profile. On the left or inside of the bend the velocities are reduced – and the reduction corresponds to the

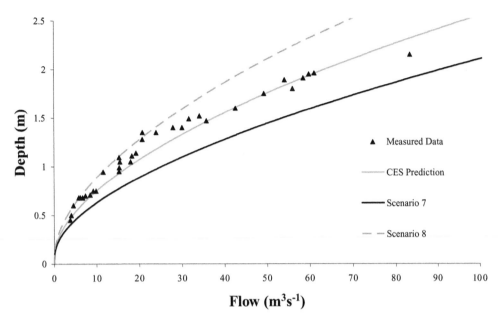

Figure 4.28 Scenario 7 and 8 stage-discharge prediction for the River La Suela.

Table 4.2 Summary of predicted and measured flow rates for 1.52 m and 1.75 m flow depths.

	@ 1.52 m depth		@ 1.75 m depth	
	Flow rate (m³s⁻¹)	Change in flow rate (%)	Flow rate (m³s⁻¹)	Change in flow rate (%)
Measured	34	−11	49	−1
Original CES	38	0	50	0
Scenario 1 – double bed roughness	22	−43	30	−40
Scenario 2 – halve bed roughness	69	80	86	74
Scenario 3 – double bank roughness	38	−2	48	−3
Scenario 4 – halve bank roughness	41	7	54	8
Scenario 5 – concrete-lined channel	65	70	84	69
Scenario 6 – concrete-lined rectangular	69	79	86	73
Scenario 7 – double bed slope	55	44	71	43
Scenario 8 – halve bed slope	28	−28	36	−28
Scenario 9 – sinuosity 1.5	39	2	50	1
Scenario 10 – sinuosity 2.5	37	−4	47	−5

magnitude of the sinuosity. This is typically reversed for overbank flows, with the higher main channel velocities observed in the outside of the bend (as observed in the Flood Channel Facility data). For this case, the impact on the overall flow rate is small (<5% for both cases, Table 4.2).

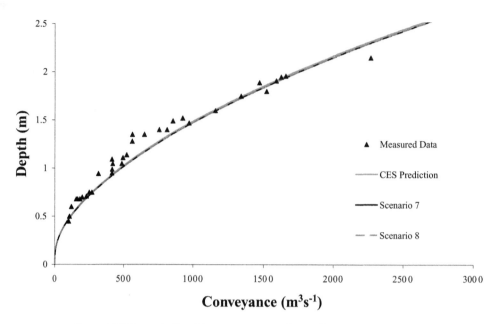

Figure 4.29 Scenario 7 and 8 conveyance prediction for the River La Suela.

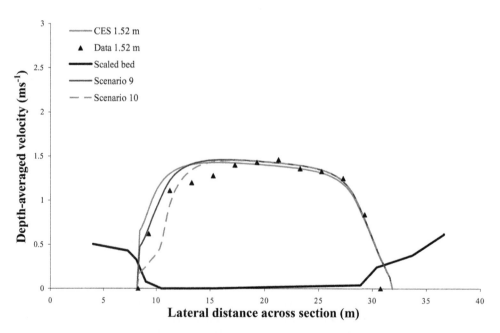

Figure 4.30 Scenario 9 and 10 depth-averaged velocity predictions for 1.52 m flow depth (Q_{CES} = 38 m³s⁻¹; Q_{data} = 34 m³s⁻¹; Scenarios Q_9 = 39 m³s⁻¹; Q_{10} = 37 m³s⁻¹).

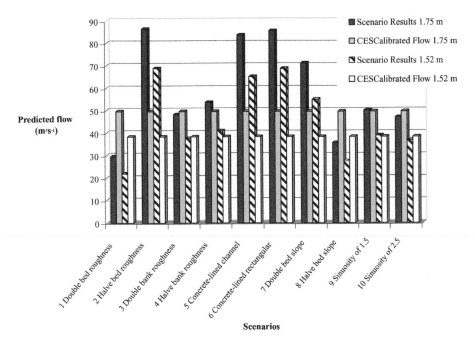

Figure 4.31 Summary of predicted and measured flow rates for 1.52 m and 1.75 m flow depths.

iv   Summary flow results all scenarios

Table 4.2 provides a summary of the predicted and measured flows for Scenarios 1 to 10 at flow depths of 1.52 m and 1.75 m. It also indicates the percentage difference in flow rate relative to the calibrated CES flow rate. Figure 4.31 provides a corresponding visual impression of these results.

From this, it is apparent that Scenarios 1, 2, 5, 6, 7 and 8 have the largest influence on the flow rate for the La Suela i.e. altering the bed roughness, concrete-lining the channel and altering the bed slope. Scenarios 3, 4, 9 and 10 have the least impact i.e. altering the bank roughness and the sinuosity. Although these outcomes are specific to the River La Suela and are likely to be different for other channel types (e.g. irregular, compound, narrow/wide, vegetation, boulders, alluvial etc.), they serve to show the qualitative changes that might be expected.

4.2.6   Application of the CES to a mountain stream with boulders

Aim

CES stage-discharge simulations are carried out for two mountain rivers, the River Waiwakaiho (New Zealand) and the River Tomebamba (Ecuador). These are used to demonstrate the use of the CES for mountain streams with boulders and to propose an alternative roughness law to the current CES prediction of 'f' (Chapter 3,

Colebrook-White law) where large boulders are present. The alternative approach is based on that of Abril and Knight (2004).

Approach for boulders

Abril and Knight (2004) proposed an alternative approach for evaluating boulder roughness in mountain rivers based in part on the work of Ramette (1992), where,

$$f_{mc} = 8 \left[5.75 \log \left(\frac{12H_{mc}}{3d_{90} + \varepsilon_r} \right) \right]^{-2} \tag{4.1}$$

and d_{90} (m) is the sediment dimension, ε_r is a coefficient representing the bed form roughness and f_{mc} is evaluated from,

$$f = f_{mc} \left(0.669 + 0.331 D_r^{-0.719} \right) \tag{4.2}$$

H_{mc} is taken as the maximum local cross-section depth and ε_r is zero for the channels considered.

The CES supports simulations for a wide range of channel types e.g. different section and plan form shapes, sizes, vegetation types, substrate cover, etc. An important component of the calculation is the ability to adapt the model as appropriate to the particular site characteristics. For example, where the roughness varies significantly within a cross-section, the model allows for a description of the local roughness values and these are taken into account in the calculation. However, there are some areas which are more challenging, where the basic science is still emerging, for example:

1   vegetation which changes its behaviour (e.g. form, resistance) as a function of velocity and depth;
2   alluvial channels where the bed roughness is characterised by bed form growth and wash-out;
3   very large channels (e.g. Alto Parana, Argentina, flows upwards of 20,000 m^3s^{-1}) where the roughness laws developed for small channel and pipe flow may not be applicable;
4   steep mountain streams with large boulders – comparable with flow depth.

The following pages now illustrate topic 4 above further, using data from two mountain rivers.

The River Waiwakaiho site

Stage and discharge measurements were taken at station SH3 along the River Waiwakaiho over the nine year period, January 1980 to 1989. The Waiwakaiho serves a catchment area of 58 km^2, has a mean annual flood of 327 m^3s^{-1} and an average flow rate of 0.63 m^3s^{-1}. The observations were made at three cross-sections along a fairly straight 100 m reach, each approximately 40 m wide and 2 m deep. The water surface slope varied from 0.0091 at low depths to 0.0176 at large depths. The bed comprises cobbles and boulders, some as large as 2 m in diameter (Figure 4.32), and the banks consist of boulders with occasional scrub. The observed Manning n values vary with depth in the range 0.047 to 0.180.

River and cross-section definition for River Waiwakaiho site

Based on this description the assigned roughness values include (Figure 4.33):

- *Channel bed* – The RA bedrock ($n_{sur} = 0.025$) and boulders >50% ($n_{irr} = 0.045$) giving $n_l = 0.051$.
- *Channel banks* – The RA cobbles 64–256 mm ($n_{sur} = 0.035$).
- *Channel high banks* – The RA sand ($n_{sur} = 0.02$).

The River Tomebamba site

The River Tomebamba is a natural mountain river located in the Southern Andean region of Ecuador, another tributary to the Paute River. Water level, discharge, velocity and roughness measurements were taken along the Tomebamba River at Monay.

Figure 4.32 River Waiwakaiho at SH3 looking (a) upstream and (b) downstream along the reach (Hicks & Mason, 1998).

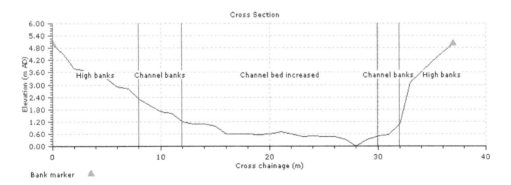

Figure 4.33 River Waiwakaiho cross-section, roughness zones and top-of-bank marker locations.

The channel is 25 m wide and the reach-averaged longitudinal bed slope is 0.0176. Here, the river bed consists of large boulders, approximately 1.3 m in diameter, and the Manning n values, back-calculated from discharge and water level measurements, range from 0.08 to around 0.15 (Abril & Knight, 2004).

River and cross-section definition for River Tomebamba site

Based on this description the assigned roughness value for the whole section is (Figure 4.34):

- *Channel Cover* – The RA bedrock ($n_{sur} = 0.025$) and boulders >50% ($n_{irr} = 0.045$) giving $n_l = 0.051$.

Results for the River Waiwakaiho and the River Tomebamba

Figure 4.35 provides the CES predicted stage-discharge based on the aforementioned RA values. These are labelled as "uncalibrated" and it is clear the CES roughness is too small, resulting in the flow capacity being over-predicted. The roughness values for the two rivers were increased to better simulate the data. The Waiwakaiho calibrated channel bed roughness is $n_l = 0.5$ (2 m boulders) and the Tomebamba calibrated channel roughness is $n_l = 0.25$ (1.3 m diameter boulders). These calibrated curves (dashed lines) fall closer to the data; although the curve shape of the data is not captured. The reason for this is most likely related to the use of the Colebrook-White law in areas where the boulder size is comparable with depth and different flow mechanisms are taking place. Note that a unit roughness value of 0.25 corresponds to a k_s value of 7.4 m, substantially larger than the 1–2 m boulders.

The boulder approach of Abril and Knight (2004) is also used to simulate the data. Here, the initial d_{90} is taken as 2 m and 1.3 m for the Waiwakaiho and Tomebamba. The curves were then calibrated using the sediment dimension and the final d_{90}'s are 2.1 m and 2 m for the Waiwakaiho and Tomebamba. Figure 4.34 shows how the shape of the curve follows the data more closely than the Colebrook-White approach.

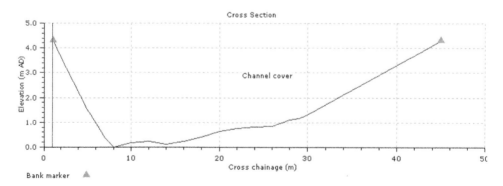

Figure 4.34 River Tomebamba cross-section, roughness zones and top-of-bank marker locations.

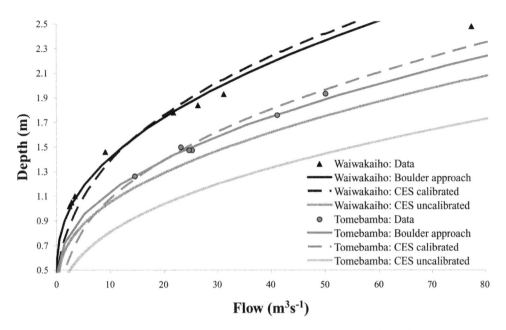

Figure 4.35 Stage-discharge predictions for the River Waiwakaiho and the River Tomebamba using different roughness calculations.

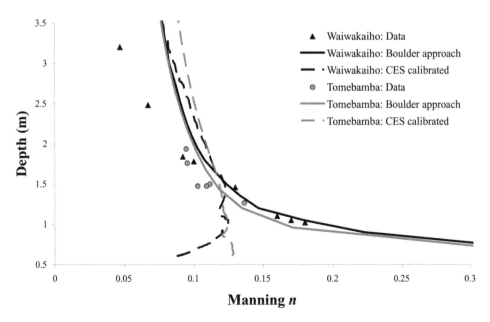

Figure 4.36 Back-calculated Manning *n* values for the River Waiwakaiho and the River Tomebamba using different roughness calculations.

At flow depths greater than 2 m, both approaches provide poor predictions which may be due to the particular boulder layout at the site or, more fundamentally, that the boulder formulae may only be applicable for a limited range e.g. flow up to the level of the sediment diameter, d_{90}. If the boulder approach is adopted in the CES, a transitional roughness rule between the two approaches should be explored.

Figure 4.36 provides the back-calculated Manning n values for both approaches. As before, the boulder approach best captures the shape of the data.

4.3   USE OF BACKWATER MODULE FOR ESTIMATING WATER LEVELS ALONG THE RIVER MAIN

Aim

This example demonstrates the use of the CES-AES backwater calculation to calculate water levels along a reach of the River Main with:

i   an inflow of 60 m³s⁻¹ and a downstream control of 36 mAD; and
ii   an inflow of 10 m³s⁻¹ with normal depth conditions at the downstream end.

For condition (i), the channel section 600 m along the reach needs to convey the 100 year return period storm (Q100 = ~60 m³s⁻¹) below a water level of 38 mAD, despite the new downstream control at 36 mAD.

For condition (ii), the water levels 1400 m along the reach need to be maintained above 35 mAD for flows above 10 m³s⁻¹, to ensure the land drainage pumps can operate. Below this flow, the land drainage pumps are automatically turned off. This is explored for the 12 month period prior to installation of the new 36 mAD downstream control. Thereafter, the water levels will be sufficient to ensure pump operation.

The River Main reach

A 1.6 km reach of the River Main in County Antrim, Northern Ireland, is used for this example (see Chapter 2, Section 2.2.2 for a full description). The inflow to the reach is just downstream of Lisnafillan weir and the reach extends to Gracehill Bridge. The cross-sections are located at approximately 50 meter intervals and the topographical survey data includes channel and floodplain information, measured at right angles to the main flow direction, and measured as far across the floodplain as possible.

Bridge End Bridge is located about 400 m downstream of Lisnafillan weir and the roughness values for the reach are based on the calibrated values for this site (Section 4.2.3).

Note 1: This reach of the River Main has a series of fish groyne structures (approximately every 500–700 m) which control the depths at low flows. These are excluded from the current CES-AES model - but could readily be introduced as additional reach-averaged cross-sections with 'short reaches' or as downstream controls.

Note 2: This reach of the River Main includes 2 bridge structures. These are excluded from the current example but may be introduced using the AES.

Results for the River Main

An initial run is undertaken to assess the water levels along the reach at a more common flow rate of 40 m³s⁻¹. Figure 4.37 provides the model output for this initial run, with inflow 40 m³s⁻¹, and a downstream level of 36 mAD. The surface profile is more-or-less parallel to the ground profile in the upstream reach and in the downstream reach it approaches the controlled level of 36 mAD. The velocity profile indicates the higher velocities in the upstream reaches (in the region 1.25–2.0 ms⁻¹) and lower velocities (<1.25 ms⁻¹) further downstream due to the control. The water levels approximately 600 m and 1400 m downstream of Lisnafillan weir are 37.7 mAD and 36.1 mAD respectively.

Figure 4.38 provides the model outputs for condition (i). The downstream water level is controlled at 36 mAD (assuming no drowning) and the 100 year return period storm, $Q_1 00 = 60$ m³s⁻¹, is set as the inflow to the reach. The water level 600 m downstream of Lisnafillan weir is 37.9 mAD, which is lower than the required 38 mAD for condition (i). The lower and upper credible scenarios (dashed lines) give water levels of 37.7 mAD and 38.0 mAD respectively at this location. As these are close to the required water level of 38 mAD, some additional resource may be invested to better understand any sources of uncertainty. If the water levels are subsequently considered uncertain, possible options may be explored in the CES e.g. improved vegetation maintenance through altering roughness, dredging through altering channel cross-sections etc.

Figure 4.39 provides the model outputs for condition (ii). The inflow to the reach is 10 m³s⁻¹ and normal depth conditions are present at the downstream end. These correspond to a downstream flow depth of 34.3 mAD. The water surface profile follows the ground profile more closely as would be expected at low flows, where the effect

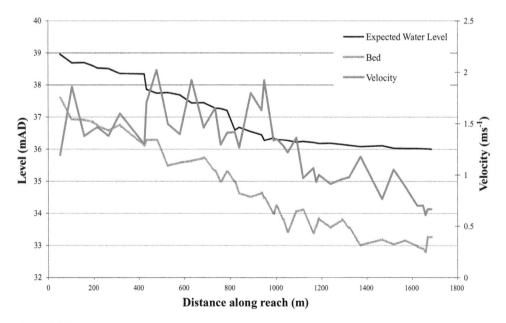

Figure 4.37 Long-section of the River Main reach showing the water surface and velocity profiles for a 40 m³s⁻¹ inflow to the reach and 36 mAD downstream control.

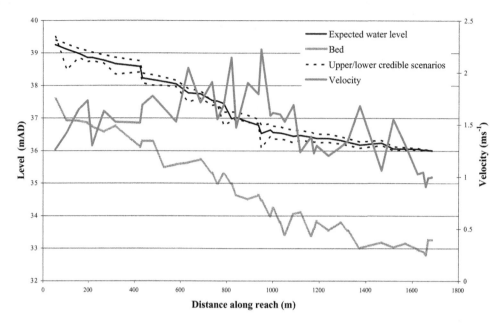

Figure 4.38 Long-section of the River Main reach showing the water surface, upper and lower credible scenarios and velocity profiles for a 60 m³s⁻¹ inflow to the reach and 36 mAD downstream control.

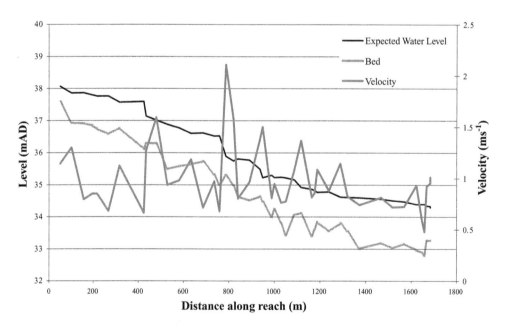

Figure 4.39 Long-section of the River Main reach showing the water surface and velocity profiles for a 10 m³s⁻¹ inflow to the reach with normal depth at downstream end (34.3 mAD).

of the local longitudinal gradient is more dominant. The velocities throughout the reach are in the range 0.5 to 1.5 ms^{-1} as there is no downstream control to reduce these at the downstream end. The velocity peaks (2.1 ms^{-1}) at ~800 m downstream, which is most likely a result of the local topography i.e. bed level and cross-section shape. At 1400 m downstream of Lisnafillan weir, the expected, upper and lower water levels are all 34.6 mAD. This is well below the required 35 mAD for the drainage pumps and it will be necessary to explore different solutions for this e.g. turn the land drainage pump off at a higher flow rate, introduce a temporary downstream control of ≥35 mAD prior to the planned 36 mAD control.

4.4   ESTIMATING AFFLUX AT BRIDGES

4.4.1   Field scale verification of bridge backwater analysis at Pea Creek, Alabama

Aim

To predict the backwater associated with bridge afflux at a field site where comparisons can be made with surveyed data (available at *https://www.usgs.gov/centers/lmg-water*).

Pea Creek

The field site is a heavily vegetated floodplain near Louisville, Alabama. It is one of the field sites that was included in the USGS study of backwater at bridges, begun in 1969, that collected detailed topographic and water level surveys plus flow data for 35 floods at 11 sites in the southern USA. This data set remains one of the primary sources of field observations for the study of bridge afflux. Measurements from the Pea Creek site were made for two flood events in December 1971 and published in the Hydrologic Investigations Atlas HA-608 (Colson *et al.*, 1978). The study area is shown in Figures 4.40 and 4.41.

Roughness definition

The Pea Creek is a tributary to the swamp and creek systems of southern Alabama. Floodplains in the region are densely covered with trees (Figure 4.42 shows typical floodplain cover). Underlying substrates are typically silt, sand and clay deposits. Channels are incised into the broad floodplains and so the Roughness Advisor has been used to set unit roughness zones for channel and floodplain zones.

- *Channel bed* – The RA silt with trailing bank side plants is adopted, resulting in a unit roughness of $n_l = 0.053$.
- *Floodplain* – The RA floodplain heavy stands of trees (with depths below branches) is adopted on sand substrate, with resulting $n_l = 0.102$. The reasoning is that the entire floodplain is densely wooded, but there are unlikely to have been significant trailing leafy branches for the vegetation types observed in the field, especially on the date of the modelled event.

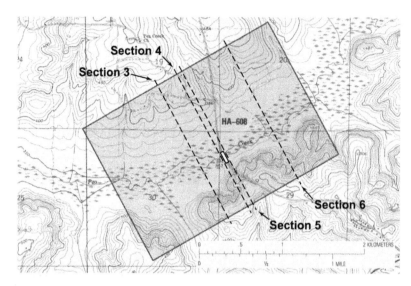

Figure 4.40 Pea Creek site map. Grey box shows extent of study area. Dotted lines show approximate along-stream locations of USGS cross section survey data, section 3 is downstream. Road bridge is situated between sections 4 and 5 (After USGS, 1979).

Figure 4.41 Pea Creek wooded floodplain and road bridge (Photo: USGS).

- *Channel and floodplain roughness at bridge sections* - The AES bridge calculations use Manning's *n* to represent flow resistance, rather than the unit roughness parameter of the CES (see Section 3.2.5). The bridge unit requires a three panel roughness specification (left overbank, main channel and right

Figure 4.42 Choctawhatchee/Pea River channel and vegetation (Photograph: Alabama Clean Water Partnership. Reproduced with the permission of StormCenter Communications, Inc.).

overbank), separated by 'left' and 'right' bank markers. Although the Manning's *n* and unit roughness are conceptually different quantities, for a short, straight, relatively uniform reach the distinction between the CES and the conveyance calculations based on Manning's equation is less important. Hence for the short, straight reach within the bridge the unit roughness values have been adopted.

- *Bridge roughness* – The AES calculates energy losses associated with flow acceleration in the transition reaches upstream and downstream of the bridge, plus energy loss associated with friction within the structure. The bridge at Pea Creek is a deck bridge supported by 16 timber pile bents with centres at 4.57 m spacing. It is assumed that each support is made of 4 wooden piles arranged in a 2 × 2 pattern. The AES treats piers as if they extend continuously from the upstream face to the downstream face of the bridge, which is a common type of structure in the UK. This is unlikely to be the case for the Pea Creek bridge, where the user of more complex hydraulic modelling software would be likely to choose a method based on conservation of momentum to represent form and drag losses around the bridge piers. However, in the simpler AES, only changes in the cross sectional flow area and frictional energy losses are accounted for, and so an 'effective' value of roughness is needed. In this case, the CES floodplain value has been adopted on the basis that, as a crude approximation, dense tree stands on the floodplain may have create similar resistance to flow as closely spaced the timber pile supports within the channel.

Cross section definition

Floodplain and bridge cross section data are as published by Colson *et al.* (1978). The floodplain is in general around 300–400 m wide, but contracts to about 75 m at the bridge.

- *River cross sections* – CES cross sections are set up using USGS channel survey data at the locations shown in Figure 4.40. A simple roughness zonation is adopted, with the RA roughness zones for floodplain and channel used, as shown in Figure 4.43 for Section No. 5, which is located approximately 77 m upstream of the bridge (upstream face).
- *Bridge section* – The AES bridge section is shown in Figure 4.44. The bridge length (upstream to downstream faces) is 8 m and the overall valley gradient is very shallow, hence the AES option to use the same cross section and structure profile for both upstream and downstream faces is selected. The soffit elevation is given as 111.8 m above datum and the road elevation as 112.3 mAD.

Results for Pea Creek bridge

Conveyance rating curves for the modelled reach are shown in Figure 4.45. The bridge section is clearly visible showing reduced conveyance at a given elevation owing to the obstruction to flow within the bridge section combined with the frictional resistance. There is, as expected, a sharp change in the curve between soffit and road level, with little change in conveyance. Above the road level, the increase in conveyance with elevation at the bridge section is roughly parallel with the curves for unobstructed sections.

The USGS Hydrologic Investigations Atlas HA-608 includes detailed measurements for an event on 21 December 1972 (Table 4.3), where the discharge, measured

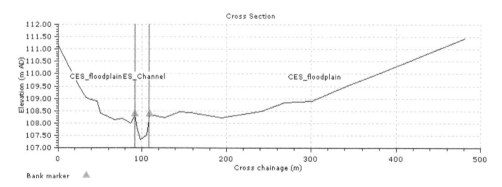

Figure 4.43 Cross-section for Pea Creek site for Section No. 5.

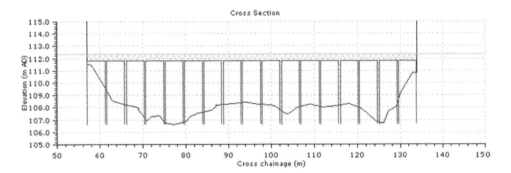

Figure 4.44 Pea Creek AES bridge section.

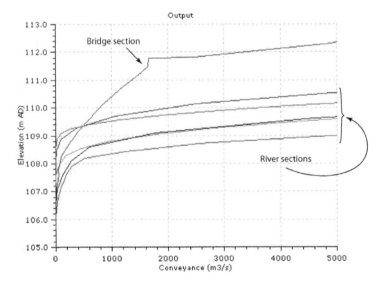

Figure 4.45 CES-AES conveyance curves for Pea Creek.

by velocity area gauging from the bridge, was 50.5 m³s⁻¹. The corresponding level of
the flood outline surveyed by the USGS at Section No. 3 is 108.5 mAD. This provides
a downstream boundary condition for the backwater analysis.

The CES-AES backwater analysis produces water surface profiles as shown in
Figure 4.46. The afflux is clearly visible in the comparison of the profile for the reach
including the bridge with a second profile where the bridge structure is removed from
the model. The modelled water levels can be compared with the surveyed water levels
from the flood outline of 21 December 1972.

The surveyed water levels are derived from ground elevations at the mapped
flood outline. In some cases, these are uncertain because the outline was between
survey points, with no knowledge of the ground elevations in between. In addition,

Table 4.3 Surveyed and modelled flood levels for the 21 December 1972 event at Pea Creek.

Section	Flood level from maximum flood outline	Modelled flood level
3	108.50	108.50 (downstream boundary condition)
4	109.05 (uncertain)	108.68
Bridge	109.23	109.29
5	109.53	109.53
6	109.93	109.60

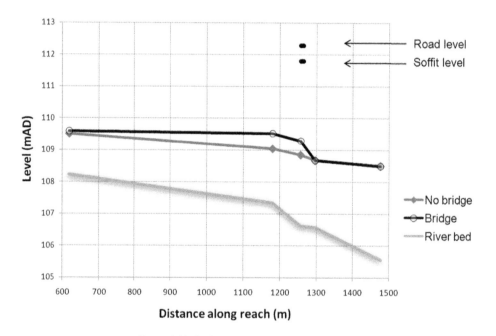

Figure 4.46 Backwater profiles for Pea Creek.

there are numerous surveyed water levels, assumed to be from water marks on trees, which show uncertainty in identifying a cross section average flood water level. Given the uncertainty in the estimate of flood levels from the measurements, it seems that the CES-AES backwater profile offers a good approximation of the water surface profile in the vicinity of the bridge. In particular, the comparison at Section No. 5, close the location of the afflux, shows very close agreement. Figure 4.47 shows the water surface profiles calculated for the minimum and maximum credible roughness values generated by RA.

This example gives details of one field data test of the AES. Further comparisons between afflux data modeled using AES and field measurements by USGS (1978) have been given by Mantz and Benn (2009).

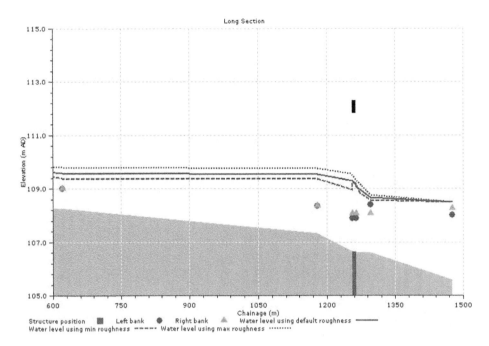

Figure 4.47 Backwater profile for Pea Creek for minimum, default and maximum roughness cases.

4.4.2   Field scale bridge backwater analysis on the River Irwell, UK

Aim

To predict the backwater associated with bridge afflux at a field site on a river in northern England.

Holme Bridge on the River Irwell

The River Irwell flows through Salford close to the city of Manchester in northern England. The Irwell has numerous small tributaries, some of them originating as steep watercourses flowing into small, upland floodplains before joining upstream of Salford. The case study is a twin arched masonry bridge downstream of the town of Rawtenstall. The floodplain and channel gradient is ~0.003 for the reach containing the bridge. The floodplain is about 150 m wide and the main channel approximately 25 m wide. The channel is incised at the bridge. The bridge is one of 40 included in a detailed hydrodynamic river model. There are no direct observations available for flows and water levels at the bridge site, but the river is gauged nearby allowing for confident estimates of design flow rates and corresponding water levels are available from the hydrodynamic model for comparison. The upstream and downstream faces of the bridge are shown in Figure 4.48(a) and (b) respectively.

Figure 4.48(a) Holme Bridge, upstream (Photo: JBA Consulting/Maltby Land Surveys).

Figure 4.48(b) Holme Bridge, downstream (Photo: JBA Consulting/Maltby Land Surveys).

Roughness definition

The reach is an upper branch of the Irwell. The valley has shallow soils underlain by bouldery deposits and bedrock. The floodplain around the bridge is rough grassland with sparse tree cover. The river bank is similar in character to the floodplain. Cross sections upstream and downstream of a bridge in CES-AES require three roughness panels (representing main channel and floodplains on the left and the right). The Roughness Advisor has therefore been used to set roughness zones for the main channel and the floodplain.

- *Channel bed* – The RA coarse gravel with riffles is adopted, resulting in a unit roughness of $n_l = 0.031$.
- *Floodplain* – The RA floodplain light brush with trees is adopted, with resulting $n_l = 0.060$.
- *Channel and floodplain roughness at bridge sections* – The AES bridge calculations use Manning's n to represent flow resistance, rather than the unit roughness parameter of the CES. The bridge unit requires a three panel roughness specification (left overbank, main channel and right overbank), separated by 'left' and 'right' bank markers. Although the Manning's n and unit roughness are conceptually different quantities, for a short, straight, relatively uniform reach the distinction between the CES and the conveyance calculations based on Manning's equation is less important. Hence for the short, straight reach within the bridge the unit roughness values have been adopted.
- *Bridge roughness* – The AES adopts Manning's n values to represent the friction within the arch. In this case a value of $n = 0.020$ is adopted appropriate to an unglazed masonry or brick surface in good condition.

Cross section definition

Floodplain and bridge cross section data were collected by survey.

- *River cross sections* – CES cross sections are set up using the channel survey data at the locations shown in Figure 4.49. A simple roughness zonation is adopted, with the RA roughness zones for floodplain and channel used, as shown in Figure 4.50 for the CES sections located approximately 123 m upstream of the bridge and 89 m downstream.
- *Bridge section* – The AES bridge section is shown in Figure 4.51. The bridge length (upstream to downstream faces) is 3.5 m. The cross section and structure profiles are the same for both upstream and downstream faces. The two arch openings differ slightly in shape but the soffit elevation is approximately 156 m above datum for both. The road elevation is set to equal 157.5 mAD (in fact representing the solid parapet).

Results for Holme Bridge

Conveyance rating curves for the modelled reach are shown in Figure 4.52. The bridge section is clearly visible showing reduced conveyance at a given elevation owing to the

Figure 4.49 Typical bankside and floodplain vegetation around the study reach. (Photo: JBA Consulting/ Maltby Land Surveys).

obstruction to flow within the bridge section combined with the frictional resistance. There is, as expected, a sharp change in the curve between soffit and road level, with changes in conveyance above road level roughly parallel with the curves for unobstructed sections.

Three 'design' flow conditions were analysed using CES-AES, corresponding to flood flows for the return periods of T = 2, 10 and 100 years (or annual excedance probabilities of 50%, 10% and 1%, respectively). The downstream boundary conditions were taken from outputs of the large hydrodynamic river model that included the Holme Bridge. This was a model built using the ISIS Flow software package, with Holme Bridge represented using the arch bridge method of Brown (1988).

The CES-AES backwater analysis produces water surface profiles as shown in Figure 4.53 for each of the three 'design flows'. In each case, the plot shows the water surface profile computed for the model including the bridge (heavy lines) and also a profile computed after removing the bridge structure from the model. The afflux is hence clearly visible by comparing each pair of profiles.

The afflux is clearly seen to increase with increasing flow rate and hence blockage ratio, as the water level rises into the arch openings. At low water levels, the arch bridge has little effect on the flow because only the relatively slender pier causes an obstruction and the bridge structure contributes relatively little to the frictional energy losses through the section. When water rises above the level of the lowest arch

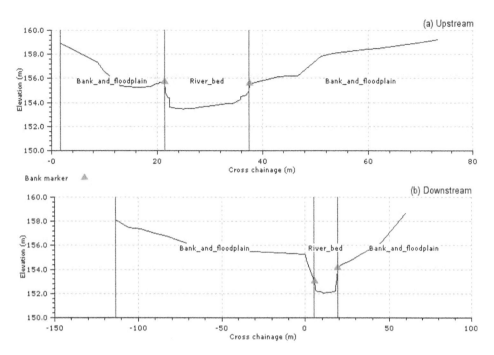

Figure 4.50 TCES channel cross sections (a) 123 m upstream and (b) 89 m downstream of bridge
section.

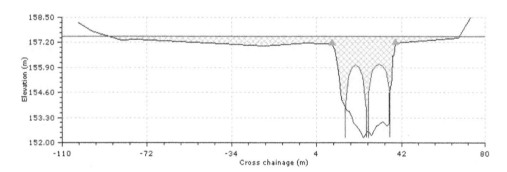

Figure 4.51 Holme Bridge AES bridge section.

springer then the structure begins to have an effect on the water surface profile. The
water surface just rises to soffit level in the modelled 100-year flow condition and so
the friction losses caused by the bridge structure are large.

There are no field measurements available to test these results, but it is of interest to compare them with the outputs of the ISIS hydrodynamic model for the same
design flow conditions. Table 4.4 shows the flow and water level at the downstream
section, the water levels computed at the bridge section in the ISIS simulation and the

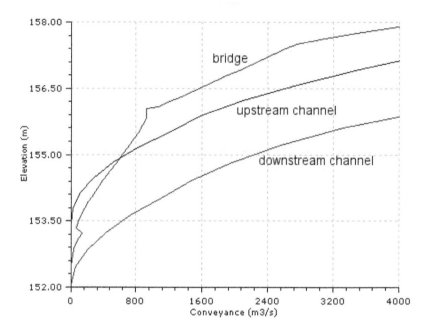

Figure 4.52 CES-AES conveyance curves for Holme Bridge.

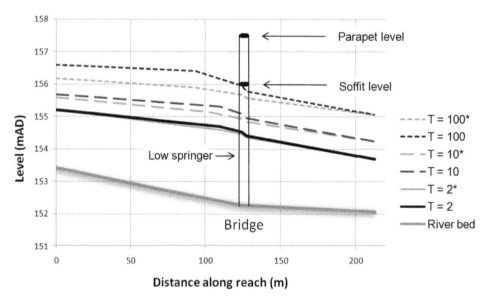

Figure 4.53 Backwater profiles for Holme Bridge for three design flow conditions or return period T = 2, 10 and 100 years. For the profiles labelled with an asterisk (*) the bridge was removed from the model.

Table 4.4 Surveyed and modelled flood levels for T = 2, 10 & 100 years at Holme Bridge.

Return period (years)	Flow rate (m³s⁻¹)	Downstream water level (m AD)	ISIS bridge section water level (m AD)	CES-AES water level, downstream bridge face (m AD)	CES-AES water level, upstream bridge face (m AD)
2	45	153.72	154.20	154.41	154.54
10	71	154.24	154.72	154.95	155.10
100	117	155.07	155.60	155.78	155.96

results from CES-AES for the downstream and upstream bridge faces. The ISIS bridge unit is based on an empirical dimensionless correlation analysis of afflux, calibrated with laboratory data, and of the same general form as the similarity model used in the AES analysis of transition lengths and energy loss coefficients. This model contains no bridge length scale and is therefore unable to represent the backwater profile through the bridge waterway.

The comparison shows that the CES-AES water levels are similar to those computed using ISIS, but slightly higher. A discrepancy of this size can be judged in the light of extensive tests against scaled laboratory data (Atabay, 2008) that have shown the ISIS 'ARCH' bridge unit to have uncertainties in the range of approximately 0.5 to 1.0 meters when estimating an afflux of scale 0 to 3.0 meters.

4.5   ESTIMATING AFFLUX AT CULVERTS

4.5.1   Shallow culvert backwater analysis in a long reach

Aim

To demonstrate backwater analysis for a channel reach including a culvert.

Culvert site

The example is based on the River Main reach described in earlier examples. The reach is 750 m long, and is initially configured for backwater analysis without a structure in place. There are 19 river sections, labeled from 67 (downstream) to 87 (upstream) and a 20 m long culvert is added 560 m downstream. Results are illustrated for a number of example culvert configurations.

Cross section and roughness definition

The channel cross sections immediately upstream and downstream of the culvert are shown in Figure 4.54. The channel upstream and donwnstream has RA unit roughness values $n_i = 0.102$ for the floodplain, $n_i = 0.045$ for the river banks and $n_i = 0.025$ for the main channel. The culvert roughness, expressed in terms of Manning's n, is $n = 0.010$.

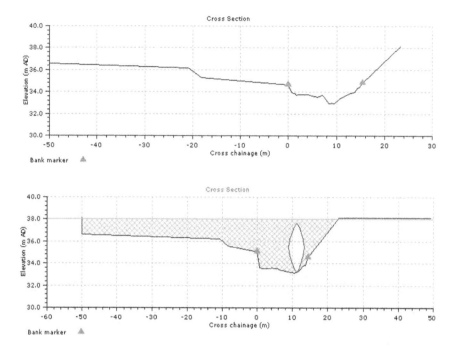

Figure 4.54 Downstream channel and structure cross sections for a pipe culvert in the River Main reach.

Culvert structure

The structure is modelled as a pipe culvert, with a circular concrete inlet set into a headwall (CIRIA type 'A') and square inlet edge. The openings are placed on alignment with the deepest part of the channel. The barrel diameter is 4.5 m.

Results

Conveyance curves are obtained by running a backwater analysis for the reach and plotting the CES-AES output curves, including the results of the afflux calculations (Figure 4.55). It can be seen that, as expected, the conveyance through the culvert is restricted at a given elevation because of the reduction in available flow area. The pipe culvert displays the expected reversal of the conveyance curve as the water level elevation rises into the arch formed in the upper half of the pipe. Above road level, the conveyance increases rapidly with elevation as the channel cross section area expands rapidly above the road.

Water surface profiles are shown in Figure 4.56 for two different downstream boundary conditions, in each case comparing the profiles obtained with and without the culvert. Figure 4.56(a) shows a case where floodplain flow is assumed at the downstream boundary. Here, the culvert is overtops the road and the resulting combined pressure and weir flow results in an afflux upstream of approximately 2 m. Figure 4.56(b) shows a flow just within bank at the downstream section. In this case

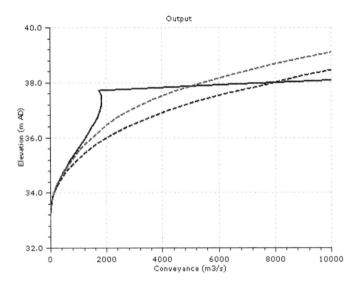

Figure 4.55 Conveyance curves for a pipe culvert in the River Main reach. Solid line shows culvert conveyance at inlet. Dark dashed line shows conveyance in the channel downstream of the culvert, light dashed line upstream.

there is free surface flow through the barrel of the culvert and a smaller afflux of approximately 1 m. Note that there is an increase in velocity through the culvert barrel, as shown by the velocity traces in the CES-AES long section output (Figure 4.57).

4.5.2   Exploratory culvert design and maintenance calculations in CES-AES

Aim

To demonstrate analysis of a trial culvert design. The analysis includes hypothetical scenarios for blockage and deterioration of the culvert barrel condition. The calculations are presented to illustrate an exploratory analysis for a hypothetical culvert design.

Culvert site

The example illustrates simple, exploratory trial calculations for a culvert crossing of a minor road. The required length of the culvert is 12 m and the project requires a box section culvert design. The culvert site is located between two river cross sections in a uniform reach with natural downstream channel control and an average slope of approximately 0.005. The analysis therefore assumes normal depth flow conditions in the reach. The CES-AES model has only three sections. The downstream river section is labelled 1 and the upstream section is labelled 4. Sections 2 and 3 are contained within the culvert unit in CES-AES.

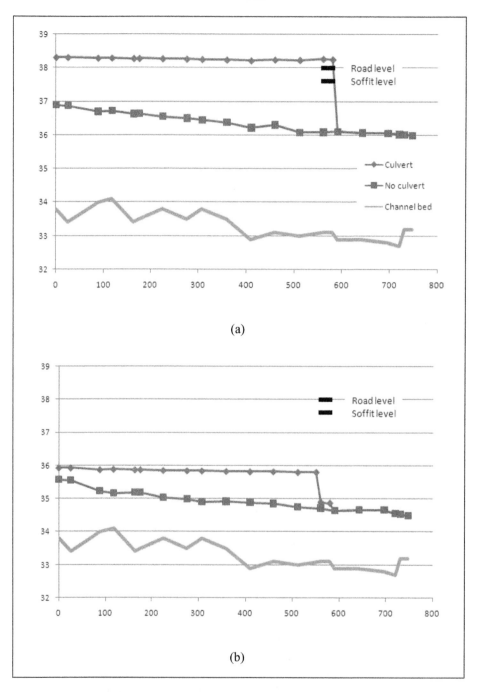

Figure 4.56 Water surface profiles for pipe culvert in the River Main reach (a) out of bank flow = 75 m³s⁻¹, downstream water level = 36 mAD, (b) in bank flow = 19.7 m³s⁻¹, downstream water level = 34.5 mAD. Horizontal axis is distance along stream in meters.

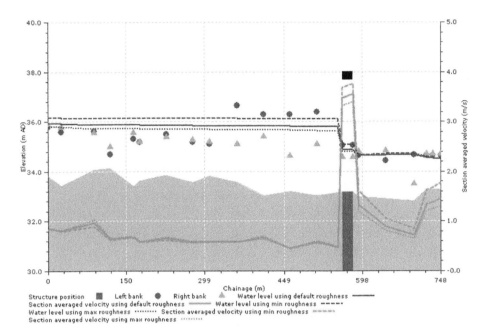

Figure 4.57 Long section backwater profile for reach with 20 m long pipe culvert, showing increased velocity through the culvert barrel.

Table 4.5 Design flow estimates.

Return period (years)	Flow (m³s⁻¹)
10	3.5
25	4.6
50	5.5
100	8.6

Hydrological analysis has provided design flows for four return periods, as shown in Table 4.5. The area upstream is farmland although there is a car park at the edge of the floodplain. The design should keep any increase in flood water levels to no more than 0.10 m for the 10-year return period flow.

Cross section and roughness definition

The channel cross section downstream of the culvert is shown in Figure 4.58. The main channel comprises medium sized cobbles and so the RA unit roughness values $n_l = 0.102$ is adopted. The floodplain is well grazed pasture and so the RA turf vegetation with no irregularities is adopted with unit roughness $n_l = 0.021$. The CES river cross sections upstream and downstream of the culvert unit have to include three roughness zones, and so there is no separate bank roughness. The banks are assumed

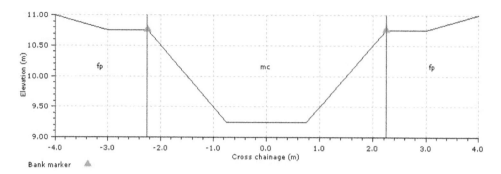

Figure 4.58 Downstream channel cross section. Bed level is 9.25 m.

to have similar roughness to the main channel owing to a combination of substrate and some trailing bank side vegetation.

Culvert structure trial calculations

An estimate of the depth of flow in the channel through the culvert site is obtained from the Conveyance Generator outputs for Section 1. Figure 4.59 shows the flow rating curve for the downstream section. The water level is 10.25 mAD, which corresponds to a flow depth of 1.0 m. The section averaged velocity rating is shown in Figure 4.60. The velocity corresponding to the flow depth of 1.0 m is obtained directly using the outputs interrogator as 1.40 ms^{-1}.

An approximate trial size for the culvert barrel is determined as follows. The section averaged velocity in the culvert should be greater than the channel velocity to discourage excessive siltation. An increase of 20% is desired in this case, giving a design velocity of 1.7 ms^{-1}. The discharge is equal to the cross section averaged downstream velocity multiplied by the flow area. The area required to convey the design flow with a velocity of 1.7 ms^{-1} is therefore 3.5/1.7 = 2.1 m². The depth of flow in the culvert should be similar to the normal depth in the channel, which is 1.0 m. A suitable vertical dimension for the barrel would be 1.3 m, allowing for 0.3 m freeboard. With an assumed flow depth of 1.0 m, the required width of the box section would be 2.1 m. Hence the trial culvert barrel dimensions will be a span of 2.1 m and a rise of 1.3 m.

The culvert barrel will be pre-fabricated concrete. Manning's roughness coefficient n would typically be in the range 0.012 to 0.015 for a new concrete culvert. To allow for some minor degradation in use a design value of 0.018 is adopted. The structure is modelled in AES as a box culvert, with a rectangular concrete inlet including headwall and wingwalls and square inlet edges. A similar structure is shown in Figure 4.61 during high flow conditions. The culvert opening data is shown in Figure 4.62.

Exploratory analysis of the trial design

The trial design is analysed by applying the design flow conditions and generating a backwater profile. Results for the 10 year return period flow are shown in

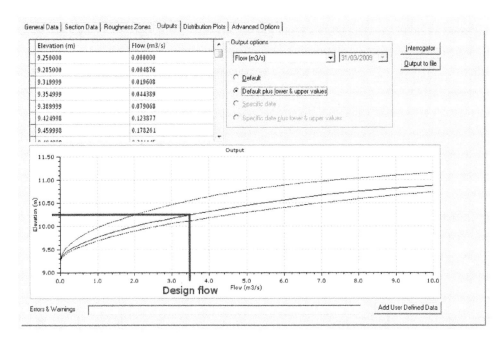

Figure 4.59 Conveyance generator flow rating curve for the downstream section, showing downstream water level for a design flow for the 10-year return period of 3.5 m³s⁻¹.

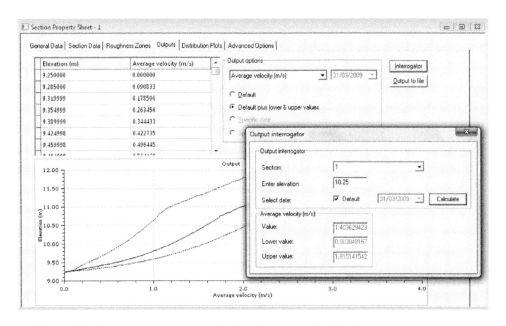

Figure 4.60 Conveyance generator velocity rating curve for the downstream section and 'Interrogator', showing section averaged velocity for a water level of 10.25 mAD.

Figure 4.61 Culvert inlet during high flow conditions (Photo: John Riddell).

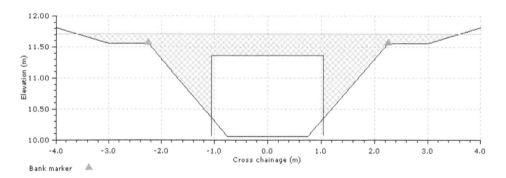

Figure 4.62 Culvert trial opening data in AES.

Figure 4.63. It can be seen that the culvert causes only a very small increase in water levels upstream of approximately 0.03 m, much less than the stipulated value of 0.1 m. Figure 4.63 also shows water surface profiles for the credible uncertainty bounds around the default roughness, as encoded within CES-AES. The range of uncertainty about water levels is much greater than the indicated change in water levels resulting from the introduction of the culvert, confirming the trial design.

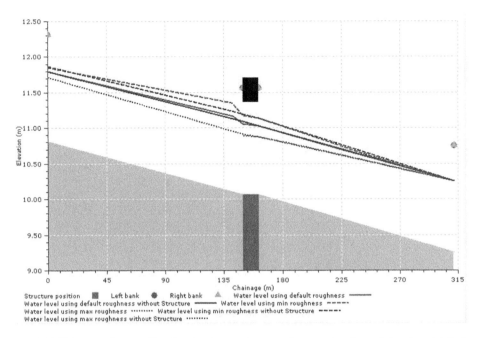

Figure 4.63 Water surface profiles showing best estimate and uncertainty bounds for the trial design, with and without culvert for downstream water level 10.25 m.

Figure 4.64 shows water surface profiles for the 10 year, 50 year and 100 year design flows, both with and without the culvert. At higher flows, the culvert clearly has a greater impact on upstream water levels. For the 50 year design flow the upstream water level just reaches the soffit and a large afflux is produced. For the 100 year flow, the structure is overtopped, flooding the road. The analysis of these larger flow events helps in understanding the robustness of the design. For example, should there be a possibility of development of housing or retail property in the area upstream of the road crossing then the performance of the culvert under higher flow conditions becomes much more important.

Over time there could be deterioration of the condition of the culvert barrel leading to an increase in flow resistance and hence a possible increase in the afflux upstream. An increased roughness value of Manning's $n = 0.035$ has been applied in the AES to test this scenario for future culvert condition. It is also possible that blockage could occur during a high flow event. A simple way to approximate the effects of blockage in AES is to adjust the structure cross section data to reflect a reduction in flow area through the culvert. One such scenario is illustrated in Figure 4.65 where the culvert barrel and opening dimensions have been reduced to a span of 1.6 m and a rise of 1.0 m, which could represent a combination of obstructions within the barrel and floating debris at the inlet. This is a highly simplified representation, but useful for an indicative analysis within the CES-AES.

The impacts of the two culvert condition scenarios are shown in Figure 4.66, which plots rating curves derived from CES-AES backwater profiles for a position

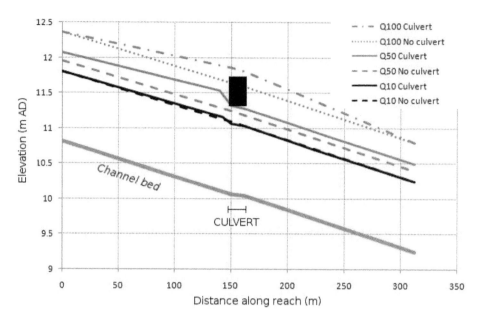

Figure 4.64 Water surface profiles for trial design for design flows with return periods of 10, 50 and 100 years (Q10, Q50 and Q100, respectively).

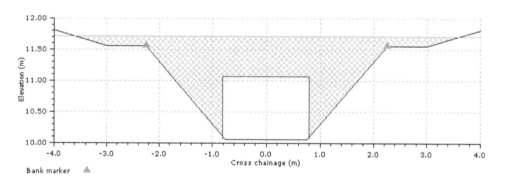

Figure 4.65 Possible blockage scenario represented by a reduction in culvert barrel dimensions.

50 m upstream of the culvert entrance. It can be seen that the modelled scenario for poor barrel condition (increased friction) causes an increased afflux compared to the original trial design, but that this effect diminishes once the structure becomes drowned. This is because the change in barrel friction is less significant once weir flow ensues over the road level. However, the blockage case causes a large increase in water levels for all design flows as a result of the much greater total obstruction to flow in the channel.

Note that the example presents a simplified approach and should not be relied upon as a basis for design calculations. It is intended to demonstrate use of CES-AES

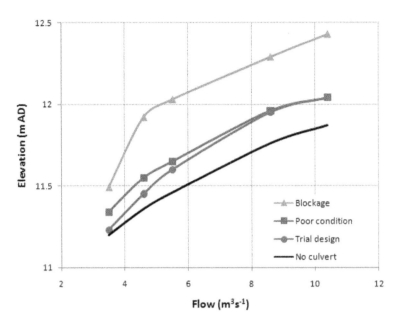

Figure 4.66 Flow rating curves for a position 50 m upstream of the culvert entrance showing two culvert condition scenarios.

as a simple tool to aid understanding of a proposed culvert structure. More detailed calculations are required for a design study.

4.6   DEALING WITH VEGETATION AND MAINTENANCE OF WEEDY RIVERS

Channels provide an important habitat for plants, which in turn provide a range of useful functions such as creating sheltered areas with reduced velocities for fauna, altering the temperature, light penetration and oxygen concentration to promote a variety of species and encouraging siltation. These plants are a vital source of shelter and food for fish, invertebrates and some birds. Where water spills overbank, the drying and wetting of the berms promotes vegetation growth of wetland plants. Vegetation also plays an important role in preventing scour and protecting bed and banks through the binding action of the roots. Channel maintenance is, therefore, multifaceted, incorporating a range of requirements such as providing sufficient capacity to convey flood flows; reducing the seasonal cutting requirements (and hence expenses) where possible; utilising the bank and scour protection function of the vegetation and promoting the natural habitat.

Guidance on channel management provides current 'best practice' for vegetation cutting (e.g. EA, 1998a&b; EA, 2003; CAPM, 1997; Buisson *et al.*, 2008). This typically involves clearing the vegetation in part of the river, for example, cutting along the channel centre or one of the banks. These variations in pattern give rise to spatial variations in channel roughness, and the latter is also linked to the density, type, height and stiffness of the vegetation (Fisher, 2001; Defra/EA, 2003b).

There are a number of existing methods for evaluating the reduced conveyance due to the presence of vegetation (e.g. Cowan, 1956; Whitehead *et al.*, 1992; Gordon *et al.*, 1992; Fisher, 2001). These include approaches such as reducing the cross-section or bankside Manning *n* value according to a simple rule, adapting the Manning equation to account for a reduced flow area due to blockages as well as more complex approaches which aim to relate plant resistance to the biomechanical properties of plants such as form, stiffness, shape and drag (e.g. Petryk & Bosmajian, 1975; Kouwen & Li, 1980; Naden *et al.*, 2006) or to velocity and depth (e.g. Palmer, 1945; Ree & Palmer, 1949; Garton & Green, 1983; Naden *et al.*, 2004; Statzner *et al.*, 2006). In most cases, the main channel and floodplain sections are treated as single units, and the output is an average velocity for each region. As the capacity for aquatic vegetation to support life may be quantified (e.g. invertebrates, Wright *et al.*, 2003), it is possible to estimate the impact to populations of improved conveyance through habitat loss.

The CES conveyance methodology provides scope to describe the local variations in vegetation roughness and it resolves the local depth-averaged velocities and flow depths with due consideration of lateral shearing and boundary layer development. This information can be used to advise on optimum cutting regimes, vegetation patterns, plant types, percentage cut and increased confidence in the prediction of the overall channel flow capacity. This is illustrated through examples:

- River Cole in Birmingham – exploration of different cutting regimes;
- River Avon in Enford – exploration of different cutting regimes to satisfy both ecological and flood risk objectives;
- River Hooke in Maiden Newton – exploration of channel deepening.

4.6.1   Exploration of cutting regimes for the River Cole

Aim

The River Cole site is used to explore the impact on flow and depth-averaged velocity for different vegetation cutting regimes at a flow depth of 1.4 m. These include:

- Scenario 1: no cutting i.e. vegetation across the entire channel;
- Scenario 2: vegetation cut in the channel centre (50% of existing vegetation at cross-section location – in Summer);
- Scenario 3: vegetation cut on the right bank (70% of existing vegetation at cross-section location – in Summer); and
- Scenario 4: vegetation cut on both banks (70% of existing vegetation at cross-section location – in Summer).

A comparison is made with the results produced using one of the existing best practice approaches (EA, 1997), the HR Wallingford (Whitehead *et al.*, 1992) method.

HR Wallingford approach

The empirically derived HR Wallingford (HRW) approach is for inbank flow only. It involves evaluating an increased Manning n value, n_{total}, to allow for the additional retardance due to vegetation in the channel. This is evaluated from:

$$n_{total} = n_{clear} + 0.0239 \left(\frac{K_w}{UR} \right)$$

(4.3)

where n_{clear} is the Manning n value when the channel is clear, K_w is the fraction of the surface area covered by vegetation and U is the average velocity of the clear section. This n_{total} is then used in the Manning equation to evaluate the flow capacity and velocity with vegetation.

The River Cole site

The River Cole in Birmingham is a straight trapezoidal channel, which was over widened for flood control purposes. It consists of a mainly gravel bed with some silt especially amongst the emergent vegetation. In the summer, it has submerged and floating channel vegetation in the main channel and emergent vegetation on both banks. The emergent vegetation traps silt and creates berms, which cause self-meandering within the channel in the summer. Pool riffle sequences create some variation in depth along the channel and urban-debris is common (Figure 4.67). The average water surface slope is 0.00174.

Figure 4.67 The River Cole site in Birmingham showing the channel vegetation, pools and riffles and urban trash (Defra/EA, 2004).

Roughness and cross-section definition

The Roughness Advisor values for the 'as is' case i.e. Scenario 1 are as follows (Figure 4.68):

- *Channel bed* – The RA coarse gravel 20–64 mm (n_{sur} = 0.027), submerged fine-leaved plants (medium) (n_{veg} = 0.1) and pools (n_{irr} = 0.2) are adopted giving n_l = 0.106.
- *Channel banks* – The RA coarse gravel 20–64 mm (n_{sur} = 0.027) and emergent reeds (n_{veg} = 0.15 but taken as 0.09 – average value – as do not appear very thick) are adopted giving n_l = 0.094.
- *Floodplain* – The RA sand (n_{sur} = 0.02) and height-varying grass (n_{veg} = 0.041) are adopted giving n_l = 0.046.

Results for the River Cole site

Scenario 1 is essentially the 'as is' case i.e. if no vegetation cutting takes place. Figure 4.69 provides the CES predicted stage-discharge and the measured data. There are only a few low flow measurements and one high flow (16.7 m³s⁻¹ @ 1.49 m) measurement available. The curve does not capture the local variations of these measurements, but it does pass from the low flows to the high flows. The upper and lower uncertainty scenarios are wide, which is indicative of the uncertainty associated with the submerged and emergent vegetation roughness information. Figure 4.70 provides the back-calculated Manning n values and the corresponding Manning n values from the measured flows. Although the detail is not captured, the trend of the CES curve moves from lower Manning n values at high flows to higher Manning n values at low flows.

The CES flow predictions for Scenario 1 to 4 at a depth of 1.4 m are summarised in Table 4.6 and the corresponding depth-averaged velocity profiles are shown in Figure 4.71.

For the HRW method, K_w is taken as 0.70, 0.36, 0.52 and 0.34 for scenarios 1 to 4 respectively. This is derived from the percentage of the cross-section covered by vegetation for each case at a flow depth of 1.4 m. The n_{clear} is taken as 0.061, giving n_{total} as

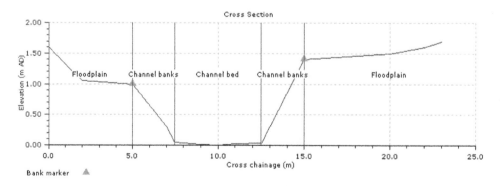

Figure 4.68 River Cole cross-section, roughness zones and top-of-bank marker locations.

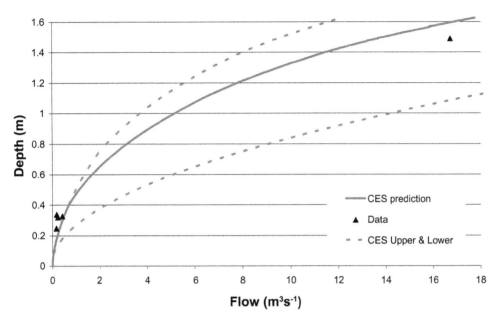

Figure 4.69 Stage-discharge prediction for the River Cole with no channel cutting and comparison to data (data from Defra/EA, 2004).

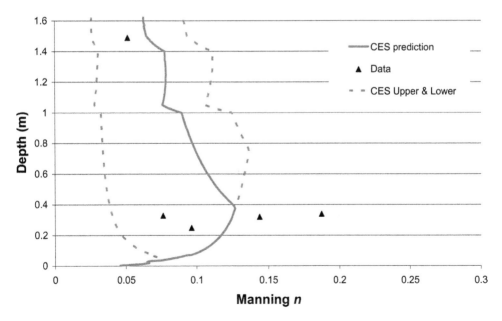

Figure 4.70 Back-calculated Manning *n* values for the River Cole and comparison to data (data from Defra/EA, 2004).

Table 4.6 Summary of flow predictions using the CES and HRW approach at depth 1.4 m.

Vegetation cutting scenario	Flow (m³s⁻¹)		Percentage difference relative to Scenario 1 (%)		Percentage difference in CES HRW value (%)
	CES	HRW	CES	HRW	
Scenario 1: no cutting	11.4	11.4	0	0	0
Scenario 2: main channel cut	12.8	16.5	44	12	23
Scenario 3: right bank cut	12.1	12.6	10	6	4
Scenario 4: right left bank cut	12.8	14.2	24	13	9

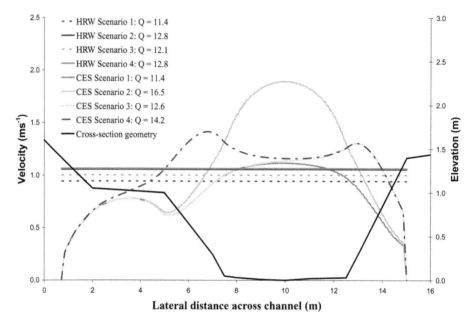

Figure 4.71 Depth-averaged velocity predictions for the River Cole using the CES and the HRW approach.

0.078, selected to ensure the flow rate for the HRW method and the CES method are identical for Scenario 1, the 'as is' case, i.e. $Q_{HRW} = Q_{CES} = 11.4$ m³s⁻¹. The HRW flow rates are provided in Table 4.4 and the average velocities are shown in Figure 4.71.

The following is observed:

- *Vegetation type* – The CES enables a user to select the vegetation roughness associated with the particular plant species (e.g. emergent reeds), including the time of year (e.g. June) and upper and lower scenarios – providing improved confidence in the outputs. Existing methods such as HRW method do not relate the Manning *n* values to vegetation morphotypes.

- *Vegetation location* – The CES enables users to identify the exact location in the cross-section where the vegetation is present e.g. banks, main channel. This information is used to evaluate the local unit flow rate and velocities (Figure 4.71) across the section – providing improved confidence in the final overall flow rate. The HRW approach assumes a single Manning's *n* value for the entire cross-section regardless of different vegetation types and locations within the section and hence average velocities are provided (Figure 4.71). The plan form cover is accounted for through a simple coefficient, K_w.

- *Vegetation cutting, location, nature & timing* – The CES enables users to select the precise location where vegetation is cut, the nature of the cutting (e.g. cut 70% of the vegetation at the location) and the timing of the cut (e.g. June) which influences the resulting unit roughness value. This additional detail provides improved confidence in the results.

- *Flow predictions* – The CES predicted flow capacities are up to 20% larger than those calculated using the HRW method and they provide a clear hierarchy of the cutting regimes with a main channel cut as the most favourable in this case (44% increased flow capacity) followed by cutting on both channel banks (24% increased flow capacity). This information can be considered together with associated costs to ascertain the preferred option.

- *Velocity predictions* – The CES approach provides information on the local velocity variations in the cross-section which is not provided by the more traditional approaches. This information may be used to inform, for example, habitat design, risk to people inclusive during maintenance activities, scour potential.

Further analyses for the River Cole may involve taking the optimum cut, channel bed, and determining the optimum time of year for the cut (current results pertain to a June cut).

To prove the potential added value of the CES outputs outlined above, it is recommended that this information is used in practice to inform a vegetation cutting regime, and the flow capacity should be monitored and compared to that achieved through previous practice.

4.6.2   Exploration of different cutting regimes for the River Avon

Aim

Different cutting regimes are explored for the River Avon to determine the improvement in conveyance with vegetation removal and the potential impact on invertebrate population (Wright *et al.*, 2002). This demonstrates how the trade-off between flood risk management (e.g. greater flow and reduced levels) and ecological (e.g. increased habitat and hence life supported) objectives may potentially be explored.

The River Avon site, roughness and cross-section definition

The Avon comprises a simple channel, 9.2 m wide, with a bankfull depth of ~0.6 m and an average bed slope of 0.0005. The bed consists of gravels and the in-channel section has an abundance of water crowfoot (*Ranunculus pencillatus pseudofluitans*) vegetation (Figure 4.72), a submerged fine-leafed plant. The assigned Roughness Advisor values are (Figure 4.73):

- *Gravel* – The RA gravel 7–20 mm ($n_{sur} = 0.025$) is adopted.
- *Ranunculus* & *gravel* – The RA gravel 7–20 mm ($n_{sur} = 0.025$) and submerged fine-leaved plants ($n_{veg} = 0.10$) are adopted giving $n_l = 0.103$.

Results for the River Avon

Discharge predictions (Figure 4.74) are made for 0, 10, 20, 30, 40, 50, 60, 70, 80, 90 and 100% vegetation coverage, with the gravel coverage at the channel centre for each. The dashed lines show the uncertainty scenarios, which are increasing for increasing vegetation coverage, as this is the primary source of uncertainty in the roughness values. The number of invertebrates supported by the vegetation cover is shown. These are based on densities on *Ranunculus pseudofluitans* (Wright *et al.*, 2002) and are calculated for a 20 m reach. Combining the flow, invertebrate and vegetation information enables different 'what ifs' to be explored. For example, if the vegetation is cut-back from the existing 42% cover to say, 20%, this provides a ~25% increase in bankfull flow capacity but this is at the expense of a 30,000 drop in total invertebrate life supported. These values are, of course, indicative and other influences should also be considered, for example, there may be a further trade-off

Figure 4.72 River Avon site at Enford, Pewsey, Wiltshire (O'Hare *et al.*, 2008; Courtesy Centre for Ecology and Hydrology).

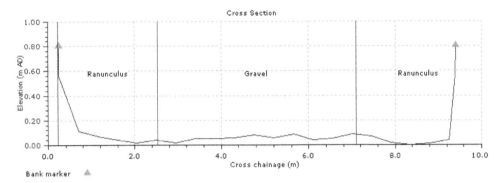

Figure 4.73 River Avon cross-section, roughness zones and top-of-bank marker locations for 50% vegetation present (survey data – O'Hare *et al.*, 2008).

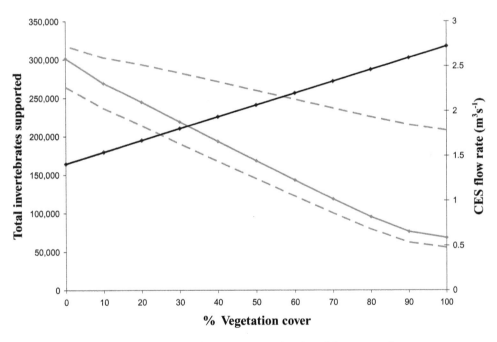

Figure 4.74 Change in number of invertebrates supported and total flow rate with percentage vegetation cover (Mc Gahey *et al.*, 2008).

between vegetation abundance and species diversity. This information can be further enriched through, for example, exploration of different cutting options such as removal of growth on a particular bank and the impact on water levels and flow capacity.

4.6.3 Exploration of channel deepening for the River Hooke

Aim

The River Hooke is used to explore the impact of dredging on local flow depths and depth-averaged velocity profiles, assuming no downstream controls. The existing channel has gravel and water crowfoot scattered throughout the section. The maintenance option will involve removal of the water crowfoot and deepening of the channel section along the central margin.

The River Hooke site, roughness and cross-section definition

The site is found in the village of Maiden Newton, in the Frome catchment, directly upstream of a LOCAR monitoring point, where discharge and other environmental parameters are recorded at high resolution. The channel vegetation is dominated by water crowfoot (*Ranunculus pencillatus pseudofluitans*, Figure 4.75). Directly upstream of the sampling area is a vegetated mid-channel bar, seen on the left of the photograph. The site has been subject to flooding in the past and a flood relief channel

Figure 4.75 River Hooke site at Maiden Newton (O'Hare *et al.*, 2008; Courtesy Centre for Ecology and Hydrology).

has been built just upstream of the mid-channel bar. Water level, velocity, vegetation and roughness measurements are available. The water level slope is 0.0108 at flow depths of 0.32 m.

The 'as is' assigned roughness values are:

- *Substrate only* – The RA gravel 7–20 mm ($n_{sur} = 0.025$)
- *Shallow with vegetation* – The RA gravel 7–20 mm ($n_{sur} = 0.025$) and June value for submerged water crowfoot or fine-leaved plants in shallow water 0.2–0.6 m ($n_{veg} = 0.25$) are adopted giving $n_l = 0.251$.

Results for the River Hooke

Figure 4.76 shows the measured and predicted depth-averaged velocities for the River Hooke. There is much scatter amongst the data, which is not unexpected considering the scattered nature of the vegetation. In the channel centre (chainage 2.5 m–5 m) the measured velocities are low reflecting the influence of the upstream mid-channel bar. The CES velocity profile may be improved if the precise location of the vegetation is defined within the CES cross-section. For this case, the average vegetation values for June in shallow (0.2 m–0.6 m) flow are adopted, and the simulated flow rate is within 2% of the measured data.

The cross-section is then altered to simulate channel deepening (Figure 4.77) with only gravel present in the deepened section. The resulting CES predicted flow capacity

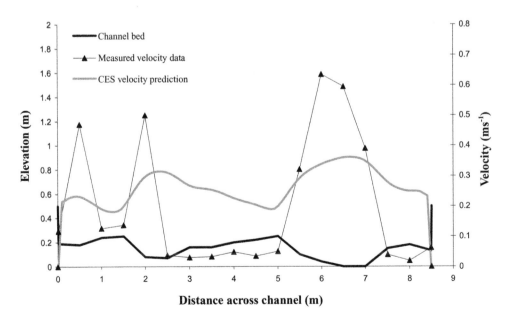

Figure 4.76 Depth-averaged velocity predictions for the River Hooke at flow depth 0.32 m giving a $Q_{CES} = 0.42$ m³s⁻¹ and $Q_{data} = 0.41$ m³s⁻¹ (data – O'Hare et al., 2008).

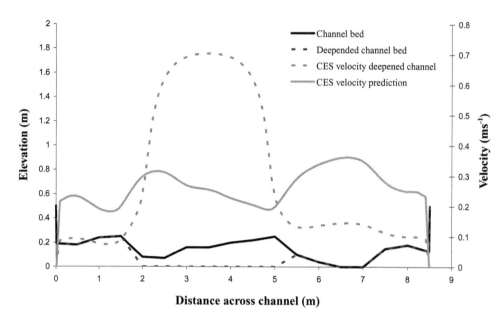

Figure 4.77 Depth-averaged velocity predictions for the River Hooke at flow depth 0.32 m before and after channel deepening and vegetation removal.

at a depth of 0.32 m is 1.8 m³s⁻¹, a 330% increase on the CES flow prediction for the 'as is' channel section. The velocities in the left-hand margin are, as expected, substantially higher reaching 0.7 ms⁻¹. These velocities may be used to inform scour potential and habitat change for the reach.

Chapter 5

Further issues on flows in rivers

ABSTRACT

This chapter considers some further practical issues in river engineering and suggests how the CES-AES software might be used to investigate them. The issues considered relate to river ecology, sediment transport, geomorphology, trash screens and blockages to culverts. A brief overview is given of 1-D, 2-D and 3-D approaches to modelling river flows, and sets the CES-AES software in this context. The chapter concludes with some high level flow charts that indicate how the software might be adapted for other specific research purposes.

5.1 ECOLOGICAL ISSUES

Channel design should be considered in terms of the wider system and its functioning, balancing the ecological biodiversity needs with those of flood risk management. The relationship between channel properties, water levels and habitat provides an opportunity to maximise the biodiversity value of channels (Buisson *et al.*, 2008). Essential to this is the need to incorporate habitats for plants, invertebrates, amphibians, fish, birds, mammals and reptiles where possible, whilst still enabling flood flows to be conveyed with the appropriate standard of protection. For example, aquatic plants in channels provide important environmental benefits (O'Hare *et al.*, 2007), since they increase water depth during summer months, providing wetted habitat for invertebrate and fish species (Hearne *et al.*, 1994). They also boost habitat complexity by providing shelter and varied flow conditions which support large numbers of invertebrates and juvenile salmonid fish (Armitage & Cannan, 2000; Wright *et al.*, 2002). The implications are that all channel design and maintenance should be carried out with sensitivity to wildlife. Thus channel management requires a multi-disciplinary approach, with inputs from a range of experts such as ecologists, water quality specialists, environmentalists and fluvial engineers as well as operating authorities (e.g., in the UK – Local Authorities, Government Agencies, Internal Drainage Boards).

A useful overview of existing legislation, policy and guidance supporting this for drainage channels is provided in Buisson *et al.* (2008). Perhaps most important in the context of European and UK flood risk management are:

- *Water Framework Directive's (2000/60/EEC)* – The primary environmental objectives are to achieve "good ecological and good chemical status" for surface water bodies in general or good ecological potential for the specific case of heavily modified and artificial water bodies. It also seeks that in general "no deterioration" in status of the water body which will require management of the quality, quantity and structure of aquatic environments.
- *Habitats and Species Directive (92/43/EEC)* – This aims to promote the maintenance of biodiversity by requiring Member States to take measures to maintain or restore natural habitats and wild species at a favourable conservation status, introducing robust protection for those habitats and species of European importance. These requirements are now transposed to national laws in the UK, for example, Conservation Regulations 1994, Conservation Regulations for Northern Ireland 1995.
- The *UK Biodiversity Action Plans (1994)* – These are the Governments response to becoming a signatory of the Convention on Biological Diversity. The plan combines new and existing conservation initiatives with an emphasis on a partnership approach. It contains 59 objectives for conserving and enhancing species and habitats as well as promoting public awareness and contributing to international conservation efforts. 391 Species Action Plans and 45 Habitat Action Plans have been published for the UK's most threatened species and habitats, and around 150 Local BAPs have been published (See: *http://jncc.defra.gov.uk/ukbap*).
- *Making Space for Water* (MSfW) – Initiatives such as the UK Government's MSfW advocate flood risk management solutions that satisfy a wider range of objectives such as hydro-morphological, ecological and even social needs.
- *Outcome Measures* – The UK Government has established a framework of flood risk management outcome measures to allocate resources and to guide the activities of flood operating authorities so they reflect MSfW and Government policy in general. These include specific outcome measures for nationally important wildlife sites (Sites of Special Scientific Interest – SSSIs) and for UK Biodiversity Action Plan habitats.

Others include Environmental Impact Assessment; Salmon and Freshwaters Act 1975; Birds Directive 1979; Wildlife and Countryside Act 1981; Habitats Regulation 1994; Land Drainage Act 1994; Natural Environment and Rural Communities Act 2006; UK Priority Species and Habitats 2007.

Software tools such as the CES, the Physical Habitat Simulator (PHABSIM) and River Habitat Simulator (RHABSIM) provide support through enabling exploration of 'what-if' scenarios. Information on local velocities, flow depths and sediment concentrations (Section 5.2) are essential inputs for identifying the likely habitat regime, and for re-engineering channels to support a particular species. Typical measures include (Fisher, 2001):

- creating pool and riffle sequences to alter velocities;
- planting different vegetation morphotypes to create different flow regimes;

Table 5.1 Example habitat information for various UK channel species (adapted from Natural England).

Species	Description	Appearance	Habitat	Example UK habitat
Bullhead (*Cottus gobio*)	The bullhead is the only freshwater cottid found in the UK. A small species, it rarely exceeds 15 cm in length & 28 g in weight. It is easily identified by its large head, with eyes on the top & a dorso-ventrally flattened tapering body.		Sheltered streams with tree roots & vegetation for adults. Young prefer shallow water & stoney riffles. Moderate velocities i.e. 10-40 cms^{-1}, most commonly ~22 cms^{-1}. Water depths not critical, provided > 5 cm.	
Atlantic salmon (*Salmo salar*)	Distinctive & recognizable, well known for agility, strength & persistence. It can leap ~3 m into the air from deep water. Adult males reach up to 1.5 m in length & 36 kg in weight, while females are a maximum of ~120 cm & 20 kg.		Overhanging & submerged vegetation, undercut banks, submerged objects such as logs & rocks, floating debris, deep water, surface turbulence, cobbles & boulders. Water depths 20-40 cm, velocity 60-75 cms^{-1}	
British river lamprey (*Lampetra fluviatilis*)	The average adult length is ~30 cm & weight ~6 0 g but specimens over 40 cm can be found. It is a migratory species, which grows to maturity in estuaries & then moves into fresh water to spawn in clean rivers & streams.		No barriers e.g. weirs. For spawning gravel & interstitial currents. After hatching, sandy silt beds. Average stream gradients 0.2–0.6 mkm^{-1}. Water depth in nursery areas is 'typically 0.1-0.5 m. Spawning depths 20-150 cm.	
Desmoulin's whorl snail (*Vertigo moulinsiana*)	Body is greyish-white with a darker grey to black head & tentacles. The shell is dextrally coiled (mouth on right), ovate to elongate in shape, with a tapering spire & ~5 rounded whorls. Its shell height & breadth is 2.2–2.7 mm & 1.3-1.65 mm.		Climbing species in emergent vegetation e.g. fens & marshes, lakes, ponds, river floodplains. High population: h > 0.25 m (water level above ground); Med population: h > 0 m; Low population h < 0 m (inundation rare).	
Eurasian otter (*Lutra lutra*)	Large, warm-blooded top predators. Insulated against many small-scale environmental factors that impact survival of riparian invertebrates & plants (temperature, rates of flow, water chemistry, etc.).		Otters do not avoid structures e.g. weirs, bank reinforcements unless they impact food availability or pipes with no headroom. Diving has been observed in 2-7 m flow depths. In most UK rivers, otters are capable of foraging throughout the water column.	

- altering timing and nature of vegetation cutting to best suit species;
- cutting bankside vegetation to enhance exposure to natural light;
- channel deepening (dredging);
- creation of wet and dry berms;
- reducing soluble, insoluble and floating pollution; and
- introducing weirs to reduce the flow, including the use of fish (Defra/EA, 2003c) passes to ensure upstream movement of fish.

The CES enables users to describe the local vegetation and roughness cover and to resolve the local depths and velocities whereas more traditional approaches (e.g., PHABSIM) are based on section-averaged calculations of flow and velocity (e.g., Chezy, Manning). This more localised information can be used with habitat information, for example:

- *response curves* – velocity/depth versus species preferences; and
- *LIFE scores* (see Extence & Balbi; 1999) – methods for linking benthic macroinvertebrate data to prevailing flow regimes (example given in Section 4.6.2).

in support of channel design.

Table 5.1 provides some examples from Natural England of species velocity, depth and substrate preferences. This may be used together with CES predictions to identify or promote particular habitat regimes.

5.2   SEDIMENT AND GEOMORPHOLOGICAL ISSUES

Despite being based on a rigid-bed assumption, the conveyance methodology within CES may be used to deal with some loose-boundary topics such as sediment transport and geomorphology. This arises primarily from its capability in being able to estimate the lateral distribution of boundary shear stress around the wetted perimeter of a prismatic channel of any shape, as well as the distribution of depth-averaged velocity. Since τ_b is one of the key parameters that govern sediment behaviour, this is a particularly useful extension of the CES software. Furthermore, many natural channels are characterised by a mobile bed, with bed features such as ripples, dunes, plane bed and antidunes, resulting from sediment entrainment and deposition. Sediment issues also feature in many drainage channels and watercourses. The links between τ_b (or friction factor, f), velocity and bed forms that develop after initiation of motion from a flat bed condition in an alluvial channel are shown schematically in Figure 5.1. This is taken from Raudkivi (1998), and described in an introduction to alluvial resistance by Knight (2006c) and Morvan *et al.* (2008).

Figure 5.2, taken from Shen (1971), shows the impact that such bedforms can have on the stage-discharge relationship, using data from the Padma River in Bangladesh. The lateral distribution of boundary shear stress ($\rho g H S_v$, versus y, where S_v is the valley slope) is shown in the inset diagram at low and high stages. At a relatively low stage, the bedforms vary across the channel beginning with ripples near the left hand bank, then dunes, to transitional plane bed and back to dunes again near the right hand bank, in keeping with the bed shear/local velocity and bed forms indicated in Figure 5.1.

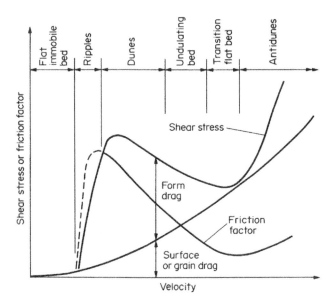

Figure 5.1 Variation of bed shear stress and friction factor with velocity and bedforms for a sand bed river (after Raudkivi, 1998).

Since the local bed shear stress is mainly governed by the local depth of flow, H, the distribution is naturally related somewhat to the geometry of the cross section. At high stage, the bedforms change, in this example by generally a single category, and become dunes, plane bed, antidunes and plane bed again. The effect of this on resistance, and hence the H v Q relationship, is profound. Since dunes have a high resistance, and plane bed (i.e., transitional flow at Froude Numbers approaching unity) have a relatively low resistance, as shown in Figure 5.1, the corresponding H v Q relationship is seen to flatten out at high flows, leading to a more or less constant stage of 27 feet. This is much lower than might have been anticipated had only the low to medium flow data, up to say 75,000 cfs (cubic feet per second) [i.e., 2,124 m³s⁻¹], been used in a regression equation and then subsequently used to extrapolate the H v Q relationship up to flows of say 300,000 cfs [i.e., 8,495 m³s⁻¹]. That estimate would then have been higher than that which actually occurred, arising from the significant proportion of the wetted perimeter experiencing plane bed (i.e., low resistance) conditions at high stages. This illustrates the care that needs to be taken when estimating resistance for sand bed rivers and the difficulties in extrapolating H v Q relationships without taking into account all the relevant physical factors.

There is a large amount of literature on the subject of sediment mechanics, including the flow conditions associated with initiation of motion (e.g., Shields, 1936; Liu, 1957), equilibrium channel conditions (e.g., Ackers, 1992a; Bettess & White, 1987; Blench, 1969; Cao, 1995; Cao & Knight, 1996; Chang, 1988, White *et al.*, 1982), sediment transport rates and concentrations (e.g., Du Boys, 1879; Einstein, 1942; Bagnold, 1966; Engelund & Hansen, 1966; White, 1972; Ackers & White 1973; Chang, 1988; Ackers, 1990), bedforms and resistance (e.g., van Rijn, 1982 &

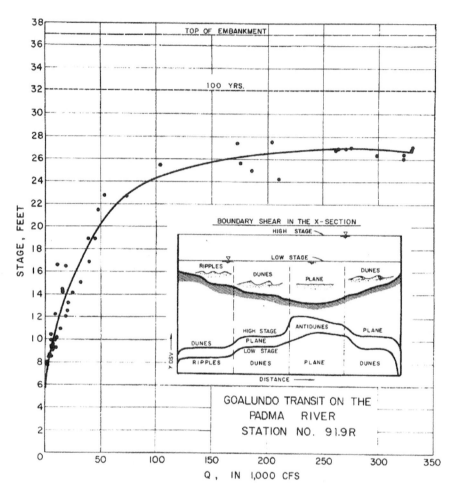

Figure 5.2 Stage-discharge relationship and bedforms in the Padma River, Bangladesh, at low and high discharges (after Shen, 1971).

1984; White *et al.*, 1980) and general behaviour as given in various textbooks (e.g., Chang, 1988; Garde & Ranga-Raju, 1977; Raudkivi, 1998; Shen, 1971; Simons & Senturk, 1992; Thorne *et al.*, 1997; Yalin & da Silva, 2001; Yang, 1987).

With regard to sediment transport rates, most methods typically rely on a resolution of the boundary flow conditions, using bed shear stress, shear velocities, velocity and velocity gradients adjacent to the channel bed as well as the sediment properties such as sediment size, density and fall velocity. This section is now aimed at demonstrating the use of the CES in estimating sediment transport rates, using one of the more recent sediment transport approaches by Ackers & White (1973) and Ackers (1990 & 1993c).

In 1973, Ackers and White first proposed their total material load formulae, which are based on physical considerations and dimensionless analysis. It assumes that for sediment moving as a bed load, transport correlates with grain shear stress, and the grain shear stress becomes less significant while total shear stresses become more significant for finer sediments. Three dimensionless quantities, the sediment transport parameter G_{gr}, the particle mobility number F_{gr} and the particle size number D_{gr}, are defined as

$$G_{gr} = \frac{q_s h}{q d_{50}} \left(\frac{u_*}{U} \right)^{n_{ac}} = C_{ac} \left(\frac{F_{gr}}{A_{ac}} - 1 \right)^{m_{ac}} \tag{5.1}$$

$$F_{gr} = \frac{u_*^{n_{ac}}}{\sqrt{g d_{50} (\rho / \rho_s - 1)}} \left(\frac{U}{\sqrt{32} \log_{10} (10 h / d_{50})} \right)^{(1-n_{ac})} \tag{5.2}$$

$$D_{gr} = d_{50} \left(\frac{g(\rho / \rho_s - 1)}{v^2} \right)^{1/3} \tag{5.3}$$

where q_s (m³s⁻¹m⁻¹) is the volume of sediment transported per second per unit channel width and ρ_s is the sediment density. The coefficients in these formulae, which vary with D_{gr}, were originally based on over 1000 experiments. More recently, these have been updated to include additional data (Ackers, 1990 & 1993), and these coefficients are shown in Table 5.2 according to the value of D_{gr}. A review of the performance of several sediment transport formulae, including that of Ackers & White, is given in Chang (1988).

It should be noted that nearly all sediment transport equations have been derived for inbank flow conditions, typically in a single channel. Their application to overbank flows therefore requires special treatment, as some studies suggest (e.g., Atabay, 2001; Atabay et al., 2004; Atabay et al., 2005; Ayyoubzadeh, 1997; Brown, 1997; Knight et al., 1999; Knight & Brown, 2001; Ikeda & McEwan, 2009; Tang & Knight, 2001). An example is now given to illustrate how the CES can be of use in sediment transport investigations, for both inbank and overbank flows.

The CES shear velocity and unit flow rate outputs may be used to solve these formulae at each point in the cross-section, to evaluate the lateral distribution of q_s. Then by lateral integration of q_s, the total sediment transport rate Q_s (m³s⁻¹) may be obtained, where

$$Q_s = \int_0^B q_s dy \tag{5.4}$$

It is thus possible to determine the unit sediment charge x_s (ppm) i.e. the unit mass flux of sediment per unit mass of flow per unit time from

Table 5.2 Coefficients to be used in the Ackers & White sediment transport equations, (5.1)–(5.3).

Coefficient	$D_{gr} > 60$	$1 < D_{gr} < 60$
n_{ac}	0	$1 - 0.56 \log D_{gr}$
m_{ac}	1.78	$1.67 + 6.83/D_{gr}$
A_{ac}	0.17	$0.14 + 0.23 / \sqrt{D_{gr}}$
C_{ac}	0.025	$\log_{10} C_{ac} = 2.79 \log_{10} D_{gr} - 0.98 (\log_{10} D_{gr})^2 - 3.46$

$$x_s = \frac{q_s}{q}\left(\frac{\rho_s}{\rho}\right)10^6 \text{ (ppm)} \tag{5.5}$$

and the sediment charge Xs (ppm) i.e. the total mass flux of sediment per mass flux of fluid per unit time from

$$X_s = \frac{Q_s}{Q}\left(\frac{\rho_s}{\rho}\right)10^6 \text{ (ppm)} \tag{5.6}$$

Sediment transport is highly non-linear and using an average main channel velocity such as that derived from Divided Channel Methods (DCM) or Ackers' coherence (COHM) approach (Section 2.5), rather than the depth-averaged velocity distribution, may result in substantially different sediment transport rates. In particular, the average main channel velocity does not capture the reduced sediment transport rates at the channel banks, tending to over-predict the average velocity and hence the overall sediment transport. The CES methodology improves the local shear velocity approximation from the more general formula (Eq. (2.31) in Section 2.2.3),

$$U_* = \sqrt{gRS_f} \tag{5.7}$$

as it is based on the locally resolved boundary shear stress, where $U_* = \sqrt{\tau_o/\rho}$.

A hypothetical case is now considered to demonstrate the use of the CES output. This is based on the FCF Phase A Series 2 channel, assuming a sediment d_{50} of 0.8 mm, a sediment density of 2650 kgm^{-3} and a constant flow depth of 0.198 m. Figure 5.3 shows the distribution of X_s with Q for the main channel only, as well as the distribution of X_s assuming no main channel floodplain interaction. This illustrates the importance of capturing the retarding effect of the slower moving floodplain flow on the faster moving main channel flow. The x_s and X_s predictions show similar distributions to those produced by the SKM model of Abril (1997) and Knight & Abril (1996) and the X_s distribution shows a similar tend to that of Ayyoubzadeh (1995), where the COHM method of Ackers was used to account for the floodplain main channel interactions.

As demonstrated, although the CES methodology was primarily intended for predicting the overall flow rate, outputs such as the local boundary shear stress and shear velocities mean it is particularly useful for estimating sediment transport rates for both inbank and overbank flows (Abril & Knight, 2002 & 2004b; Atabay et al., 2004; Brown 1997; Knight & Brown, 2001; Ikeda and McEwan, 2009). Further development and applications of the CES may include: predicting incipient motion based on the critical shear stress or Shields' (1936) function; establishing the transverse sediment gradient for channels with varying sediment sizes; incorporating the bed form roughness as a function of the critical velocity or boundary shear stress; erosion and automatic updating of cross-section geometry based on boundary shear stress distribution. Some of these advances could be incorporated in the current calculation set-up, for example, as a post-processing routine or as a core engine enhancement, e.g., an update to the f value provided by Colebrook-White.

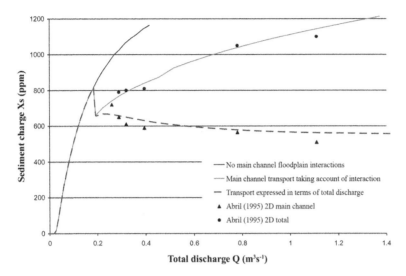

Figure 5.3 The predicted sediment charge X_s for the FCF Phase A Series 2 channel with a d_{50} of 0.8 mm (after Mc Gahey, 2006).

Some work has already been attempted on determining channel shape and bank erosion for inbank flows, as illustrated by Cao & Knight (1998) and Yu & Knight (1998). Geomorphological issues related to overbank flows are dealt with more fully in Ikeda and McEwan (2009), as are sediment processes, computer simulations and design considerations. Work in the FCF on the behaviour of graded sediments in meandering channels is described by Wormleaton *et al.* (2004 & 2005). For information on scour at bridges, see Melville & Coleman (2000).

5.3   TRASH SCREEN AND BLOCKAGE ISSUES

A bridge or culvert structure obstructs flow in the channel, causing an increase in upstream water levels relative to the levels that would be expected for the same channel in the absence of the structure. This afflux can be analyzed using CES-AES and may be considered as part of the hydraulic performance of the structure for specified design conditions. However, additional blockage can increase the afflux, leading to higher than expected upstream water levels. Blockages can occur in several ways, including:

- Collection of floating debris at the abutments/piers or soffit. This is often referred to as 'temporary blockage' because the debris can be removed. Removal more often than not requires human intervention, rather than the flow clearing the debris away;
- Collection of floating debris at a trash screen placed over the inlet of a culvert;
- Collection of floating ice, referred to as 'ice jams';
- Accumulation of bed load at the inlet, outlet, or within a culvert barrel;
- Abrupt blockage by large objects such as shopping trolleys.

There are many causes of blockage, but all are a combination of a source of debris that can cause a blockage and a means of trapping that debris. Table 5.3 provides a list of the most common factors.

The quantitative assessment of blockage potential remains difficult. There is some empirical evidence to support predictive relationships for certain types of blockage. Accumulations of large, floating woody debris (drift) were analyzed by the US Geological Survey (Diehl, 1997) in a Federal Highway Administration study based on 2,577 reported drift accumulations and field investigations of 144 drift accumulations in large, wooded catchments. The maximum width of drift accumulations for any bridge span is about equal to the length of large logs delivered to the bridge. Figure 5.4 shows data used to estimate a 'design log length' that could be used in assessing blockage risk for a given structure. This empirical data is valuable but it is also specific to the particular environment, blockage mechanism and bridge types studied.

Table 5.3 Common factors causing blockage of bridges and culverts.

Factors contributing to debris source potential	Factors contributing to blockage potential
Stream slope	Width/height
Urbanisation	Opening ratio
Vegetation	Shape
Sinuosity of stream	Skew
Upstream structures	Length
Channel width	Existence of piers/multiple culverts
Trash screens	Trash screens
Access to the channel	Trash booms
Time since last flood	Availability of skilled operators
Upstream channel maintenance regime	Auto screen raking
Upstream land use and riparian management	
High rainfall intensity	

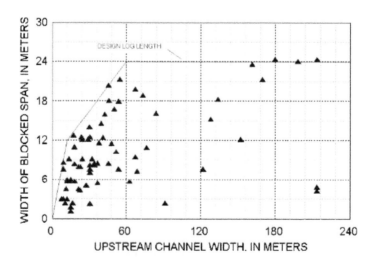

Figure 5.4 Effective width of drift-blocked spans outside the Pacific Northwest (Diehl, 1997).

In the UK, a "Blockage Risk Assessment" method was developed for the Environment Agency South West Region in 1997–8. This method is reviewed by Benn *et al.* (2004). Its main elements are:

- A spreadsheet based "blockage risk model" requiring various basic structural, hydrological and debris type attributes to which individual probability risks can be assigned;
- A decision tree analysis which directs the user to various actions depending on the blockage probability;
- Guidance on the application of hydraulic models to model blockage.

The study identified several major problems with identifying blockage risk. One is the lack of data and the second is that the probability distributions of the variables describing the influence of blockage are not independent, and the associated probabilities are subjective. This problem is not unique to blockage, and requires a consistent risk management framework aligned with the treatment of other risk issues.

Research underway within the Flood Risk Management Research Consortium (FRMRC2, see *gtr.rcuk.ac.uk*) is gathering further field data on blockage and seeking to develop a probabilistic method for predicting the onset and degree of blockage potential at pinch points in the flooding system, such as bridges and culverts.

Trash screens (for example see Figure 5.5) are often placed over the entrance of a culvert either to reduce the amount of debris entering the culvert barrel, where it could cause blockage, or as a security screen to prevent unauthorised access. While a trash screen may prevent blockage within the culvert barrel, where it is difficult to clear debris, it may simply move the location of the blockage from the barrel of the culvert or inlet to the screen.

Figure 5.5 Trash screen on the River Sheaf in Sheffield. The screen can be cleared by a mechanical grab suspended from the gantry visible above the inlet. (Photo: JBA Consulting, Maltby Land Surveys).

Debris can accumulate on a screen, which may be cleared manually or using a mechanical system. Debris accumulation can also be linked to poor inlet conditions (e.g., badly aligned wing-walls, zones of ineffective flow, and piers or other 'obstructions').

Future development of the CES-AES may include a blockage and trash screen module. In the meantime, a pragmatic approach to represent blockage is to adjust a culvert or bridge cross section so as to reduce the available flow area, either by increasing channel bed levels or reducing the opening span and soffit dimensions. Estimates of head loss associated with a screen, h_s, can be calculated based on the change in velocity head associated with the reduction in flow area through the spaces between the bars using

$$h_s = K_{sc} \left(\frac{U_{sc}^2 - U_{mc}^2}{2g} \right) \tag{5.8}$$

where U_{mc} is the section average velocity in the main channel, U_{sc} is the average velocity through the spaces between bars, g is acceleration due to gravity and K_{sc} is an energy loss coefficient. The average velocity through the screen can be computed as $U_{sc} = Q/A_{sc}$ where Q is the discharge and A_{sc} is the flow area between the bars. CIRIA (1997) recommended $K_s \approx 1.5$. Representation of trash screens and blockage in hydraulic models remains an area of on going research. Experimental measurements and numerical modelling indicate that head loss coefficients increase with increasing blockage and that it may be possible to identify dimensionless relationships between flow variables for trash screens (Tsikata *et al.*, 2009) which may assist with the calibration of simplified empirical models for 1-D analysis.

5.4   WIDER MODELLING ISSUES

Having now described the CES-AES and its use, it is appropriate to set it in the context of other 1-D, 2-D & 3-D software systems. This Section deals briefly with some of the wider issues, such as the different types of computational model that are available, their basis and mathematical formulation, their commonly perceived use and purpose, together with a discussion on the different data and calibration requirements for each type of model, especially as the dimensionality increases from 1-D to 3-D. Although an introduction to the use of models in investigating and solving practical river engineering problems has already been given in Sections 2.2.1 & 2.2.2, the reader should refer to specialists textbooks for more information. Those that are particularly relevant to river engineers are, for example, texts by Abbot & Basco (1989), Batchelor *et al.* (2000), Cunge *et al.* (1980), Garcia-Navarro & Playan (2007), Ikeda & McEwan (2009), Nezu & Nakagawa (1993), Versteeg & Malalaskera (1995) and Vreugdenhil (1989).

5.4.1   Types of model

Table 5.4 shows a sample of the codes that are used in practice, or research, for modelling fluid flows. The list is not meant to be comprehensive, but is simply aimed at showing the range of some standard models and the extent of progress in Computational Fluid Dynamics (CFD) over the last few decades. Whereas 1-D codes were just

Table 5.4 Some 1-D, 2-D & 3-D river modelling codes.

1-D models	2-D models	3-D models
Flood Modeller Pro	Flood Modeller Pro	CFX
MIKE11	MIKE21	PHOENICS
HEC-RAS	RMA2	FlUENT
TELEMAC-MASCARET 1D	TELEMAC2D	TELEMAC3D
InfoWorks RS/ICM	InfoWorks RS/ICM	DELFT FlOW3D
NOAH	DIVAST	
SOBEK	SSIM	
	LISFLOOD	
	TUFLOW	
Research codes	Research codes	Research codes

starting in the 1960s, today they are commonplace and 2-D & 3-D codes are now frequently used in solving practical problems. This does not imply that 2-D & 3-D codes have superseded the use of all 1-D models, as each type of model has its own particular use, functionality and purpose. As seen later on, 1-D models are particularly useful in dealing with many river issues since rivers themselves are, by nature, long single-thread systems that convey water in a predominately downstream direction. It is also shown that the results obtained from higher dimensionality 3-D models are not necessarily more accurate and, furthermore, involve considerable effort and cost in obtaining them (Wright *et al.*, 2004). The purpose here is to identify where the CES-AES software fits into this framework and to consider briefly the issues that arise when using software of different spatial dimensionality, i.e., 1-D, 2-D or 3-D. Before discussing these issues an overview of the equations used in each spatial dimension is given, with particular emphasis placed on 1-D unsteady flow models, so that a direct comparison may be made between these and the CES-AES, even though it contains some 2-D and 3-D flow characteristics and is for steady flow only.

i   One-dimensional model equations
As noted in Section 2.2.4, the 1-D equation for flow in an open channel relates the gravitational and frictional forces with the local and convective accelerations. Rather than expressing them in terms of h and U as functions of x and t, as in Eqs. (2.36)–(2.38), in 1-D river modelling codes it is customary to express the continuity and momentum equations in terms of A and Q as functions of x and t. These are known as the St Venant equations, and may be expressed in the form:

$$\frac{\partial A}{\partial t} + \frac{\partial Q}{\partial x} = q^* \tag{5.9}$$

$$\frac{\partial Q}{\partial t} + \frac{\partial}{\partial x}\left(\beta \frac{Q^2}{A}\right) + gA\left(\frac{\partial h}{\partial x} + S_f - S_o\right) = 0 \tag{5.10}$$

where the symbols have their usual meaning, and where q^* = lateral inflow/outflow per unit length and β = momentum correction coefficient. Equations (5.9) & (5.10) thus constitute two equations for two unknowns (A & Q) and may be solved numerically to give A and Q as functions of x and t. The area, A, and discharge, Q, are then known at all points along the river and with time. Examples of unsteady flow where such results might be useful are those involving flood simulations along rivers (flood routing) or tidal motion in long narrow estuaries. Knowing A and Q for all x and t, allows the depth, h, or water level, η, to be determined since the A v h relationship is known at every cross-section. Likewise, the velocity, U, may be calculated at every time step and cross section, since $U = Q/A$. The 1-D nature of the equations thus reflects the mainly one-dimensional flow in a river or estuary. An assessment of the relative importance of the various terms in Eq. (5.10) is given by Knight (1981), based on measurements in the Conwy estuary.

The classification of 1-D flows based on Equations (2.36) and (2.38) in Chapter 2 may now be clarified and linked to the St Venant equations above. Reverting to U and h as variables, then Eq. (2.36) gives the following classification for 1-D flows:

$$S_f = S_o - \frac{\partial h}{\partial x} - \frac{U}{g}\frac{\partial U}{\partial x} - \frac{1}{g}\frac{\partial U}{\partial t} \qquad (5.11)$$

steady uniform flow → |
     steady non-uniform flow → |
          unsteady non-uniform flow → |

Chapter 2 has indicated how uniform flow and non-uniform flow (also known as gradually varied flow) occur in steady flow, and Eq. (5.11) now indicates how unsteady flow adds one extra term into the momentum equation, the so-called local acceleration, $(1/g)(\partial U/\partial t)$. The 3rd and 4th terms on the right hand side of Eq. (5.11) are known as the convective and local accelerations respectively. This method of classifying flows should be distinguished from an alternative method of classification based on Eq. (2.38), which relates particularly to unsteady flows, as follows:

$$Q = Q_n \left[1 - \frac{1}{S_o}\frac{\partial h}{\partial x} - \frac{U}{gS_o}\frac{\partial U}{\partial x} - \frac{1}{gS_o}\frac{\partial U}{\partial t} \right]^{1/2} \qquad (5.12)$$

kinematic wave → |
     diffusion wave → |
          full dynamic wave → |

For the purposes of comparing these two classifications, based on Equations (5.11) & (5.12), with the St Venant equations, (5.9) & (5.10), consider 1-D flow in a frictionless, horizontal prismatic rectangular channel, with no lateral inflow, so that $S_f = 0$, $S_o = 0$, $q^* = 0$ and $A = bh$, where b = constant. Then, by ignoring the convective acceleration term (the second term in Eq. (5.10)), elimination of the terms $\partial A/\partial t$ and $\partial h/\partial x$ by cross differentiation reduces the pair of equations to a single equation, giving the water surface elevation, η, as:

$$\frac{\partial^2 \eta}{\partial t^2} - gh\frac{\partial^2 \eta}{\partial x^2} = 0 \qquad (5.13)$$

This represents a simple progressive wave, since a solution to Eq. (5.13) is of the form:

$$\eta = a\cos(kx - ct) \tag{5.14}$$

where $a =$ amplitude of the wave, $k = 2\pi/\lambda$; $\lambda =$ wavelength; $\omega = 2\pi/T$; $T =$ wave period, and $\eta =$ height of the water surface above a datum level ($= h + z$ as before) and $c =$ wave celerity (speed), given by $\pm\sqrt{(gh)}$ in shallow water. Thus at its simplest, unsteady 1-D flow in open channels may conceived as the movement of a wave-form along a channel. This idea also links up with the concept of flood routing in rivers, based on Eq. (5.12). However the movement of a flood wave is not simply progressive.

Kinematic waves may be regarded as those that translate without distortion of shape, given by combining the first term in Eq. (5.12) with Eq. (5.9). Assuming that A and h are related to Q by single valued functions, then

$$\frac{\partial A}{\partial t} = \frac{\partial A}{\partial Q}\frac{\partial Q}{\partial t} = \frac{\partial Q}{\partial t}\frac{dA}{dQ} \tag{5.15}$$

Substituting Eq. (5.15) into Eq. (5.9) gives

$$\frac{\partial Q}{\partial t} + \frac{dQ}{dA}\frac{\partial Q}{\partial x} = 0 \quad \text{or} \tag{5.16}$$

$$\frac{\partial Q}{\partial t} + c\frac{\partial Q}{\partial x} = 0 \tag{5.17}$$

where dQ/dA is the kinematic wave speed, c, given by

$$c = \frac{dQ}{dA} = \frac{1}{B}\frac{dQ}{dh} \tag{5.18}$$

The solution to Eq. (5.17), the 'kinematic wave' equation, provided c is constant, is

$$Q = f(x - ct) \tag{5.19}$$

where f is any function. Eq. (5.19) is thus seen to represent any shape of wave, travelling in a positive x direction with a speed or celerity, c, given by Eq. (5.18), without change in shape. This involves the inverse of the gradient of the stage-discharge relationship, $1/(dh/dQ)$, and the water surface width, B. Since the CES can provide the stage-discharge relationship for rivers with any prismatic cross-sectional shape, this maybe demonstrates another use to which the software may be put, as described later in Section 5.4.3.

However, in reality the effect of friction tends to attenuate and delay the passage of a flood wave, so that the simple wave form suggested by Eq. (5.19) becomes distorted as the wave translates (moves) along the river. The non-uniformity of the cross-section shape with stage is also particularly important in this respect. To account for this, the first two terms in Eq. (5.12) may be combined with Eq. (5.9) to give the so-called convective-diffusion equation, usually presented in the 'standard' form as

$$\frac{\partial Q}{\partial t} + c\frac{\partial Q}{\partial x} = D\frac{\partial^2 Q}{\partial x^2}$$ (5.20)

in which c = kinematic wave speed, given again by Eq. (5.18) and D = diffusion coefficient, given by

$$D = \frac{Q}{(2BS_o)}$$ (5.21)

It follows from Eq. (5.20) that the discharge in a channel during a flood event has the characteristics of a wave that now translates and attenuates. It should be noted that in the context of river engineering, both c and D in (5.20) are functions of the discharge Q, making the solution of Eq. (5.20) somewhat difficult. However, Cunge (1969) has shown that the Variable Parameter Muskingum-Cunge (VPMC) method is similar to the Preissmann finite-difference scheme used in solving the St Venant equations (Preissmann, 1961). VPMC uses 4 points in x-t, like the Preissmann box scheme but the leading term of the numerical truncation error is tuned to match the diffusion term. This then relates Equations (5.9) & (5.10) with (5.20) and allows for kinematic wave routing methods to be used in practice for channels with bed slopes greater than around 0.002 (See Knight, 2006c, for further details). It also gives a theoretical basis for varying the travel time constant, K, and the distance weighting parameter, ε, in the basic routing equations used in the original Muskingum method (Bedient & Huber, 1988; Shaw, 1994):

$$I - O = \frac{dS}{dt}$$ (5.22)

$$S = K[\varepsilon I + (1 - \varepsilon)O]$$ (5.23)

where I = inflow, O = outflow and S = storage. In the Constant Parameter Muskingum-Cunge method (CPMC), the travel time and distance weighting parameters are regarded as constants, to be determined for a particular river by back analysis of flood level data for at least one typical flood event. In the Variable Parameter Muskingum-Cunge (VPMC) the K and ε values vary for each reach, stage and time step, thus modelling the flood wave more accurately. See Tang et al. (1999a&b, 2001) and Knight (2006c) for further details.

ii   Two-dimensional model equations
The shallow wave water equations are generally used in 2-D models expressed by Morvan et al. (2008) in terms of one mass conservation equation and two momentum conservation equations:

$$\frac{\partial h}{\partial t} + \frac{\partial h U_d}{\partial x} + \frac{\partial h V_d}{\partial y} = Q_l \tag{5.24}$$

$$\frac{\partial h U_d}{\partial t} + \frac{\partial h U_d^2}{\partial x} + \frac{\partial h U_d V_d}{\partial y} + gh\frac{\partial h}{\partial x} = gh(S_{ox} - S_{fx}) \tag{5.25}$$

$$\frac{\partial h V_d}{\partial t} + \frac{\partial h U_d V_d}{\partial x} + \frac{\partial h V_d^2}{\partial y} + gh\frac{\partial h}{\partial y} = gh(S_{oy} - S_{fy}) \tag{5.26}$$

with

$$S_{fx} = \frac{n^2 U_d \sqrt{U_d^2 + V_d^2}}{h^{7/3}} \quad \text{and} \quad S_{fy} = \frac{n^2 V_d \sqrt{U_d^2 + V_d^2}}{h^{7/3}} \tag{5.27}$$

where U_d and V_d are the depth-averaged velocities in the x and y directions respectively. Other terms can be included to represent the effects of turbulence. If the turbulent terms are omitted they are in effect lumped together with the numerical truncation errors and any other calibration coefficient, rather than truly reflecting the physical process. The determination of n is therefore not straightforward and this raises a number of issues, as discussed by Cunge et al. (1980), Samuels (1985) and Morvan et al. (2008).

Equations (5.24)–(5.26) thus constitute three equations for three unknowns $(U_d, V_d$ and $h)$ and may be solved numerically to give these as functions of x, y and t. In this case, rather than use cross-section data, a mesh of $\{x, y\}$ co-ordinates are required to represent the river and any associated floodplains. The individual velocity vectors may be combined to give the magnitude and direction of the depth-averaged velocity at any point with time. This is one advantage of such a model compared with the CES-AES, in which the direction of flow is assumed to be always in the streamwise direction. A 2-D model is therefore more suitable where details of flows over floodplains are required, or where the geometry of a river is such that the flow is complex, as in the case of bends or braided rivers for example. However, the 2-D model, as expressed in Equations (5.24)–(5.26), cannot deal with secondary flows, plan form vorticity or Reynolds stresses, which the CES-AES can do in an approximate, but effective manner.

iii   Three-dimensional model equations
In 3-D models, the three time-averaged RANS equations, together with one equation for mass conservation, are used determine all three velocity components $\{UVW\}$ at every point $\{xyz\}$, together with the depth, h, as a function of $\{xy\}$ in plan form, with time, t. In essence, 4 equations are solved to give the 4 unknowns $\{U, V, W \& h\}$ as functions of $\{x, y, z \& t\}$. A 3-dimensional mesh is therefore now required to represent the model river, with special treatment being taken for the free surface (initial water level throughout the domain, as well as how it is controlled during the computation). See Ikeda & McEwan (2009), Nezu and Nakagawa (1993), Rodi (1980) and Versteeg & Malalaskera (1995) for further details.

Such models are inevitably mathematically complex, and also require a large number of ancillary equations to deal with the turbulent structures, in what is known as the 'turbulence closure' problem. It is referred to as a 'problem' because of the difficulty in describing turbulence in fluid flows, based on our incomplete knowledge of the physical reality, as well as our inability to describe such turbulence in a mathematical set of equations that are universally agreed. Turbulence closure relates the unknown turbulent processes to the larger scale resolved motions. For further information on turbulent energy, cascades and spectra see Tennekes & Lumley (1972) and Versteeg & Malalaskera (1995).

The turbulence closure issue has been addressed by many researchers, usually by formulating an equation (or set of equations), that purports to represent the turbulence. These additional equations may be of many types. The simplest is a zero equation model, based either on the eddy viscosity principle, first proposed by Boussinesq in 1877, or the mixing length hypothesis, first proposed by Prandtl in 1925. The CES-AES in fact uses the same eddy viscosity approach to deal with one aspect of turbulence, as already indicated in Eq. (2.59) or Eq. (3.18). Other types of equations to deal with turbulence closure problem are: a two-equation model, involving differential equations that represent the relationship between energy production and dissipation; a Reynolds stress model that accounts for all three internal turbulent stresses; or, an algebraic stress model that in a somewhat simpler manner accounts for the anisotropy of the Reynolds stresses. Finally, some have even suggested that the time-averaged RANS equations may not even be the right starting place for certain types of problem and that double-averaged equations (i.e., averaging over time and space) should be used instead (Nikora et al., 2008). As an alternative to RANS based models, large eddy simulation (LES) models or Direct Numerical Simulation (DNS) may be used to actually account for the individual turbulent fluctuations. The present position is best summed up by the authors of two books, written some 70 years apart, as cited in Nakato & Ettema (1996, pages 457–458) by one of the present authors:

> "Were water a perfectly non-viscous, inelastic fluid, whose particles, when in motion, always followed sensibly parallel paths, Hydraulics would be one of the most exact of the sciences. But water satisfies none of these conditions, and the result is that in the majority of cases brought before the engineer, motions and forces of such complexity are introduced as baffle all attempts at a rigorous solution. This being so, the best that can be done is to discuss each phenomenon on the assumption that the fluid in motion is perfect, and to modify the results so obtained until they fit the results of experiments, by the introduction of some empirical constant which shall involve the effect of every disregarded factor. It is worthwhile here impressing on the student of the science that, apart from these experimentally-deduced constants, his theoretical results are, at the best, only approximations to the truth, and may, if care be not taken in their interpretation, be actually misleading. On the other hand, it may be well to answer the criticism of those who would cavil at such theoretical treatment, by pointing out that the results so obtained provide the only rational framework on which to erect the more complete structure of hydraulics." (**A.H. Gibson, 1908**)

> "Turbulent motions contribute significantly to the transport of momentum, heat and mass in most flows of practical interest and therefore have a determining influence

on the distribution of velocity, temperature and species concentration over the flow field. It is the basic task of engineers working in the field of fluid mechanics to determine these distributions for a certain problem, and if the task is to be solved by a calculation method, there is no way around making assumptions about the turbulent transport processes. Basically this is what turbulence modelling is about: because the turbulent processes cannot be calculated with an exact method, it must be approximated by a turbulence model which, with the aid of empirical information, allow the turbulent transport quantities to be related to the mean flow field."
(W. Rodi, 1980)

As can be readily appreciated from these comments, as well as from the governing equations, that as one moves from 1-D to 3-D models, so does the level of complexity increase markedly. This is especially so in the ancillary equations required to describe the turbulence in mathematical terms. Another feature of 3-D models is the large number of physical constants that need to be known in order to calibrate, or indeed validate, the model at all. This then suggests that for practical purposes, one should decide on the purpose for using a model, before embarking on any modelling at all. One should then select the most appropriate model carefully, commensurate with the purpose in mind, the model's technical fitness for the task, as well as the ease with which it may be used.

5.4.2 Implications involved in model selection, calibration and use

It is worth stressing again that higher dimensionality of model does not necessarily lead to better accuracy in the results. In certain cases the opposite may be true. The principle of Occam's razor should always be applied – that of starting with the simplest by assuming the least. *Pluralitas non est ponenda sine necessitate*; "Plurality should not be posited without necessity." (William Occam, 1285–1347). The principle gives precedence to simplicity; of two competing theories, the simplest explanation of an entity is to be preferred. The principle may also be expressed as "Entities are not to be multiplied beyond necessity." [Encyclopaedia Britannica, 2003].

In the context of modelling flows in rivers, this suggests caution before embarking on 3-D modelling for solving every type of river engineering problem. As has been seen for flood routing, a 1-D model might be quite adequate for estimating travel times of floods in a river with fairly regular floodplain geometry. However, where a river meanders significantly, and any overbank flow is not aligned with the flow in the main river channel, then a 2-D depth averaged model will be more appropriate and might be sufficient for the purposes of estimating the magnitude and direction of the velocities at various points in the river valley. The trend in Europe and more recently in the UK is a move towards coupled 1-D and 2-D models, where the 1-D model simulates flow in the main river channel and the 2-D model captures the floodplain flow effects. Where detailed flow patterns are required, as for example in pump fore-bays at river intakes, or in heat dissipation and re-circulation modelling studies for a power station, then a 3-D model might be the only possible and sensible option.

Little mention has been made so far about boundary conditions in 1-D, 2-D & 3-D models. These not only differ according to the type of differential equations being solved, but also influence the solution methodology, the results and the data required to run the model at all. For example, in 1-D unsteady flow models, the boundary condition upstream is normally that of an input hydrograph {Q v t} and the boundary condition downstream is frequently that of a fixed depth or a {h v Q} relationship. Initial conditions will also be required throughout the domain, e.g., starting values for depths and discharges (or velocities) at all cross sections. In 2-D models, similar requirements apply, but at open boundaries the velocity field must be specified as well, which may initially be unknown. The same applies in 3-D models where wall functions are used to mimic the boundary conditions at mesh points very close to the wall, and cyclic boundary conditions might be applied at open ends to improve convergence and reduce the run time. For further information on boundary conditions, see Cunge *et al.* (1980), Samuels (1985) and Vreugdenhil (1989).

The numerical approach used in solving any governing differential equation, possibly based on finite-difference, finite-element or finite-volume methods, also needs to be appreciated and understood. The consistency, convergence, stability and accuracy of any adopted numerical method needs to be known, and possibly also the techniques used in investigating them. Such techniques are often based on Fourier series in complex exponential form, but are clearly beyond the scope of this book. See Abbot & Basco (1989), Cunge *et al.* (1980) and Preissmann (1961) for further information.

In 1-D models, it should be noted that even the governing St Venant equations may be discretized in different ways, and that this will inevitably affect the numerical results, as shown by Whitlow and Knight (1992). In 2-D and 3-D models it is imperative to vary the mesh size and to observe at what point of resolution any two solutions agree. In 3-D models, it is not only important to vary the mesh size, but also to check on key parameters at strategic points in the mesh, monitoring them at certain time steps as the computation proceeds. This is because the numerical procedure may take many thousands of iterations to fully converge to a solution, even when solving for a steady flow case. For example, the turbulent intensity or Reynolds stress at a certain point in the mesh is influenced by values at adjacent points in the mesh. Since the solution technique is iterative, the parameters at a single point (e.g., 3 velocity components, {UVW}, mean pressure, 3 turbulent intensities, 3 Reynolds stresses, etc.) may take a long time to converge at that particular point, and at every other point in the mesh (often thousands). The numerical solution for the whole domain is then not complete until similar convergences are obtained at all points in the mesh, to a pre-determined and specified accuracy for each particular variable or parameter.

From the preceding paragraph it is clear that the data required for different types of model varies according to their dimensionality and complexity. A 1-D model will generally only require cross section data at specified intervals, chosen to reflect the terrain, as indicated by Samuels (1990), and more recently by Casellarin *et al.* (2009). Channel roughness (and its representation) is arguably the single most important issue to resolve prior to successful modelling. In 1-D models, the use of channel cross sections makes it relatively easy to compare results with known mean velocities, since $U = Q/A$. In 2-D models, the data required for calibration increases significantly, since two components of velocity are required at every point in the mesh, giving both magnitude and direction locally. Furthermore, in 2-D models the water surface itself may no longer be considered as planar, as for example occurs in bend flow or in the vicinity of bridges. In 3-D models the data requirements multiplies even more significantly, with mean and turbulent parameters required at every location in an {xyz} co-ordinate system.

In most practical case studies there is simply not enough hydraulic or turbulence data at sufficiently detailed spatial and temporal scales to be meaningful. This makes the calibration stage particularly difficult, and reliance has then to be placed on the software user's experience. The dimensionality of the model also affects the run time and data handling. Modern CFD packages usually come with a number of post processing packages, enabling the results to be analysed in several ways. The use of graphs and charts to plot secondary flow cells, strength of vorticity, shear stresses on internal elements, colours to visualize velocity fields within each cross section, etc., make it possible to undertake assessments and any comparisons quickly and efficiently. Comparisons may not always be between measured and numerical results, on account of the scarcity of measured data, but often between different computer runs with differing flow or geometric features. Not to be forgotten are comparisons with any analytical results, if they are available, and with any suitable benchmark laboratory data.

The concept of 'equifinality' is an important one in model calibration, as noted by Beven & Freer (2001), who state that

> "... the concept of equifinality is that the uncertainty associated with the use of models in prediction might be wider, than has hitherto been considered, since there are several (many?) different acceptable model structures or many acceptable parameter sets scattered throughout the parameter space all of which are consistent in some sense with the calibration data. The range of any predicted variables is likely to be greater than might be suggested by a linearised analysis of the area of parameter space around the "optimum". This suggests that the predictions of all the acceptable models should be included in the assessment of prediction uncertainty, weighted by their relative likelihood or level of acceptability."

The use of models outside their calibration range also raises issues in assessing uncertainty. See Beven (2006), Hall & Solamatine (2008), Knight (2013a&b), Samuels (1995) and Sharifi et al. (2009a&b) for further details.

5.4.3   The CES-AES software in context

The CES-AES software may now be reviewed in the light of the two preceding Sections. Firstly it needs to be stressed that the model is for steady flow and is more than a 1-D model, since it has certain 3-D flow features embedded in it. It does therefore not fit neatly into the classification given in Table 5.4. It may be described as a type of lateral distribution model (LDM) for both inbank and overbank flows, as it simulates these in a unified manner by treating the flow as in Nature, as a continuum. The same three calibration parameters f, λ and Γ are used in both types of flow, thus unifying the methodology. It perhaps might be regarded then as a 1.5-D model with 3-D features.

These 3-D features are the ability to model secondary flows for inbank flow via Γ, lateral diffusion via a dimensionless eddy viscosity, λ, planform vorticity in overbank flow via Γ, and boundary shear stresses via a local friction, f. The model was developed on the basis of large scale experiments carried out in the Flood Channel Facility (FCF), and validated against an extensive range of experimental data, including detailed turbulence data on Reynolds stresses, lateral distributions of velocity and boundary shear stress at the same flows in comparable geometries, making the determination of local and zonal friction factors possible, and varying floodplain roughness. The CES has also been tested against a number of full scale rivers in the UK and overseas, for which there is measured data, as well as other laboratory data. See Mc Gahey et al. (2006 & 2008).

The Roughness Advisor (RA) distinguishes the CES-AES from other software, as the resistance term is known to be one of the dominant terms in the St Venant equations, requiring special care. The addition of an Afflux Estimation System (AES) allows the user to include the second most important feature affecting water levels and contributing to head losses, namely bridges and culverts. Together with the Uncertainty Estimator (UE), it allows the CES-AES to be applied to practical river engineering problems with a degree of confidence not matched by comparable systems.

The AES combines many novel features by treating bridges and culverts in a comprehensive way, with multiple arches and backwater effects. One particularly distinguishing feature is that the experimental data for bridge afflux were obtained from experimental flows in compound channels, rather than basing it on flows in simple rectangular channels, as previous authors have done. This is more representative of the actual conditions occurring in practice. The data used herein for both bridge afflux and compound channels are thoroughly documented and available via the two websites *www.flowdata.bham.ac.uk* and *www.river-conveyance.net*.

The examples in Chapter 2, showing how 4 or 6 panels can produce lateral distributions of depth-averaged velocity and boundary shear stress in trapezoidal channels of a comparable standard as produced by any 3-D model, but with much less effort, are remarkable. The number of panels or slices in the depth-averaged model naturally increases when the CES is applied to flows in rivers, especially where the need is to investigate velocity fields due to different vegetation patterns. The ability to estimate lateral distributions of boundary shear stress are particularly useful when dealing with sediment behaviour in rivers and channels. This is perhaps one of the most significant uses to which the CES-AES might be put in the future.

Despite being designed initially for steady flow, some elements of the CES-AES make it applicable to unsteady flows, noticeably the use of the conveyance, K, via Eqs (2.41) and (3.48), and subsequent discharge estimation, provided the correct water surface slope is used. The prediction of an accurate stage-discharge relationship also feeds into the estimation of a reliable wave speed-discharge relationship (c v Q) via Eq. (5.18), as demonstrated by Tang et al. (2001).

5.5 SOFTWARE ARCHITECTURE AND CALCULATION ENGINES

The conveyance and afflux estimation projects both included a requirement to produce a software tool that implemented the new estimation method. This led to the development of the CES/AES software that is available for download at *www.river-conveyance.net*. This software is intended to be a relatively simple application that provides the user with a simple and intuitive means of using the new conveyance and afflux methods on any appropriate set of data.

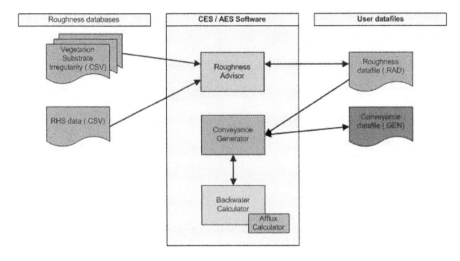

Figure 5.6 Overall structure of the CES-AES software.

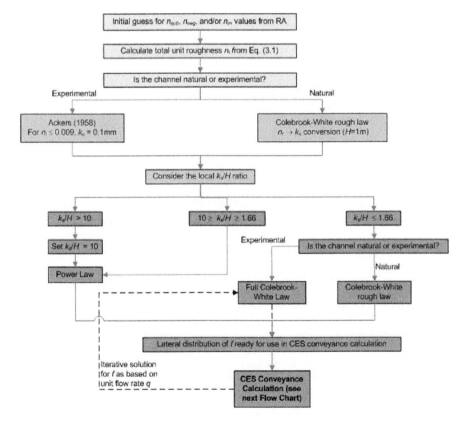

Figure 5.7 Roughness calculation engine.

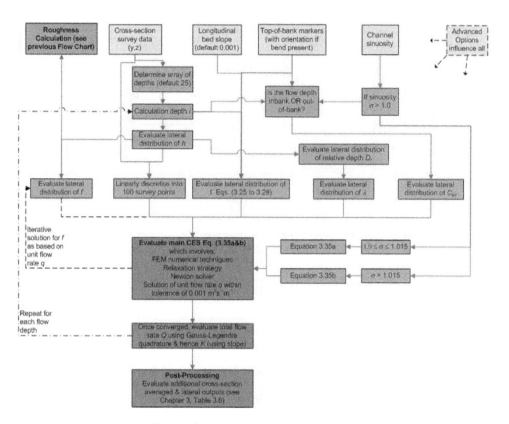

Figure 5.8 Conveyance calculation engine.

The overall structure of the software from a user perspective is shown in Figure 5.6. The software incorporates 3 main elements, the roughness advisor, the conveyance generator and the backwater calculator. Afflux is calculated as part of a backwater as this provides the necessary downstream water level for the afflux method. Data specific to the site of interest is stored in two data files. The .RAD file contains the user specific data relating to vegetation, substrate and irregularity for each of the roughness zones of interest. This file is saved by the user and stores the output of the roughness calculations carried out within the roughness advisor. The .GEN file contains the geometrical data for the channel and any bridge and culvert structures, as well as information on which roughness zones are used for calculating the section conveyance. Together the .GEN and .RAD files contain all of the data relevant to a particular site.

The raw data for the roughness advisor is provided from a number of databases that capture the outputs of the roughness review carried out for the original conveyance estimation project. These databases are in a simple .CSV format and this was selected to allow users the flexibility to edit and update the files if they have access to improved or alternative roughness data, though for most purposes the data should be considered as fixed.

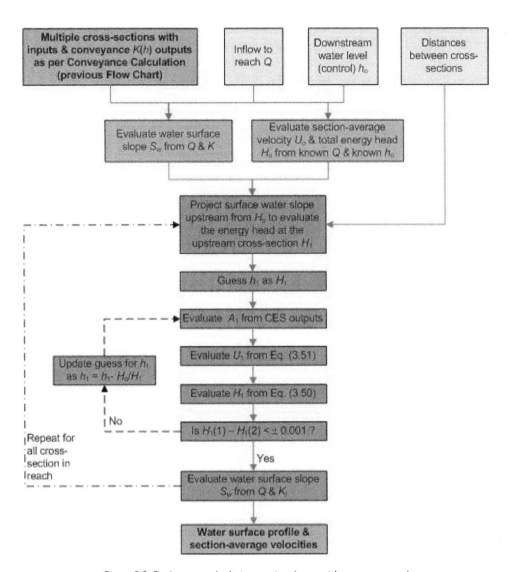

Figure 5.9 Backwater calculation engine (case without structures).

Figures 5.7, 5.8, 5.9 and 5.10 show the flow charts that define the operation of the calculation engines for roughness, conveyance and backwater.

The majority of users of the CES/AES will run the software through the standard user interface. However, it was anticipated in both the conveyance and afflux projects that the source code for the calculation methods would be made available for the purposes of further research. The software design allows for this by separating all the main control and calculation code from the user interface. This is essential, as the user interface contains compiled proprietary code that cannot be distributed. The current structure of the software when compiled is shown in a simplified form in Figure 5.11.

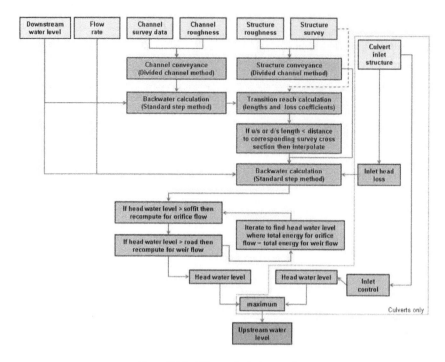

Figure 5.10 Afflux calculation engine.

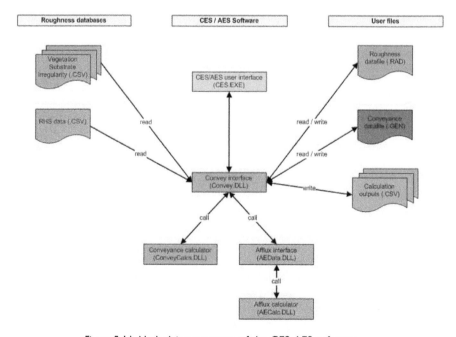

Figure 5.11 Underlying structure of the CES-AES software.

The main software procedures are all handled by the Convey.DLL module, including file access, data structure and manipulation, roughness calculation, backwater calculation and interfacing to the conveyance and afflux calculation modules. This structure effectively allows a programmer with access to the source code to replace the standard user interface with a simple interface of his or her own, and run all of the underlying calculation code.

The Afflux calculation is accessed through an interface (AEData). This structure was adopted as it allowed the afflux calculator to be developed independently from the CES software, with interaction taking place through the agreed software interface.

In practice, the overall structure shown is more complicated at the level of the source code itself, with the code sub-divided into classes in line with good software development procedures. The code has been written in three different languages. The core conveyance calculation engine (ConveyCalcs) is currently written in C for maximum portability, whilst the afflux interface and engine are written in VB. All other parts of the software are written in C++. It is probable that future updates to the software may result in changes to this structure though these will not remove the capability of running the calculation modules from outside the user interface.

Customisation of the software can be carried out at three levels. Firstly, the user can modify the default parameters that are used in the conveyance calculation. The

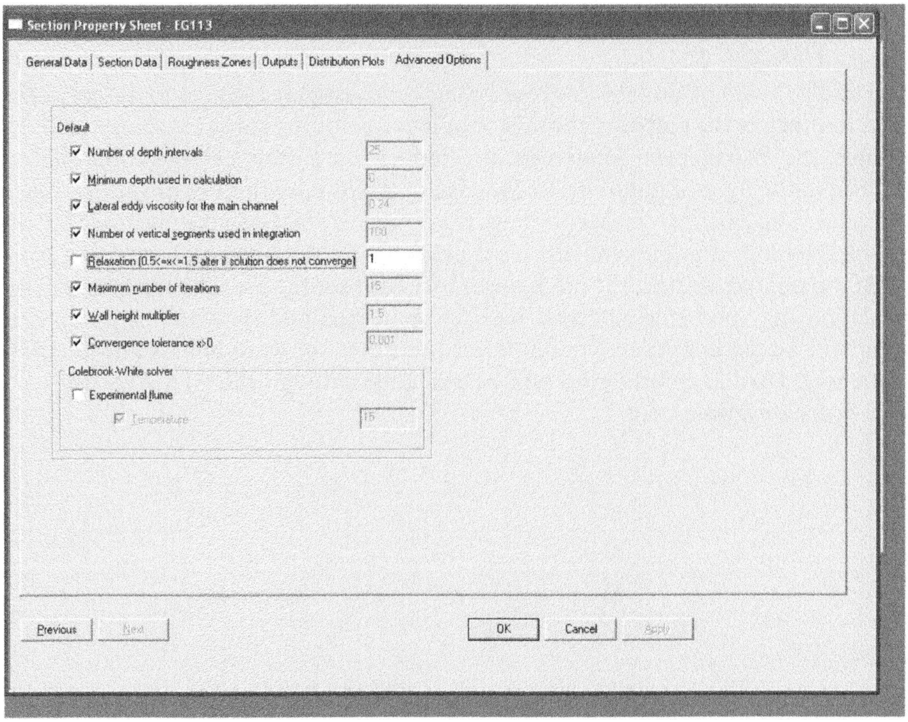

Figure 5.12 Advanced options for conveyance calculation method.

Table 5.5 Advanced options and default values.

Advanced options	Default value	Allowable range
Temperature	15°C	> 0
Number of depth intervals	25	100
Minimum depth used in calculation	Lowest bed elevation	Any depth below maximum entered elevation
Lateral main channel eddy viscosity λ_{mc}	0.24	0.1–0.5
Number of vertical segments used in integration	100	500
Relaxation parameter	1	$0.5 \leq \text{Relaxation} \leq 1.5$
Convergence tolerance	0.001	0.001–1.000
Maximum number of iterations	20	50
Wall height multiplier	1.5	10
Experimental flume	Off	On/off

Advanced Option box shown in Figure 5.12 provides a means for users to alter many of the model parameters via the CES-AES User Interface (Table 5.5).

Secondly, the user may modify the default roughness databases used by the software. These changes might take the form of adjustments to the default roughness values, incorporation of new photographs or addition of new materials and roughnesses. The file formats are fully documented and this documentation can be downloaded from the CES-AES Website.

Thirdly, it will be possible for researchers to investigate changes to the core calculation routines in the software through modification of the software source code. For example, the management models used to implement cutting of vegetation might be modified in the light of improved knowledge of the re-growth rates of various vegetation types. The existing cutting routines are found in the CManagementModel class and the C++ code that implements this is available as ManagementModel.cpp.

At the time of writing, the mechanism by which source code for the CES and AES calculation may be made available had yet to be finalised. However, it is expected that it will be made available under licence to other software houses and bona fide researchers. Further information will be available through the CES-AES website at *www.river-conveyance.net*.

Chapter 6

The Shiono & Knight Method (SKM) for analyzing open channel flows

ABSTRACT

This chapter describes the theoretical basis and physical background to the Shiono & Knight Method (SKM) which was introduced in Chapter 2, Sections 2.3 to 2.5. The derivation of the governing equation, (2.58), is now fully documented, as are how to obtain analytical solutions, with Appendices 4 to 6 providing additional material. The schematization of river cross sections is considered in relation to the physics of the flow, as are the choice of appropriate boundary conditions. The roughness and hydraulic resistance of channels with floodplains are examined, including the drag forces that occur on trees. The distribution of boundary shear stresses around the wetted perimeter of river channels is highlighted, along with the implications for sediment transport and morphology. The SKM is used to illustrate the link between flood wave speed and the channel and floodplain geometry, and its use in extending the wave speed relationship to discharges higher than actually observed. The Chapter concludes with a summary of some further applications of the SKM in river engineering, before considering these in detail in Chapter 7.

6.1 THEORETICAL BACKGROUND TO THE GOVERNING EQUATION USED IN THE SKM

6.1.1 Derivation

As already noted in Sections 2.3 and 3.2.3, the Shiono & Knight Method (SKM) is based on the depth-averaged velocity, U_d, whereas the Conveyance Estimation System (CES) is based on the unit flow rate or discharge per unit width, $q \, (= HU_d)$. Chapter 2 was intended as an introduction to the SKM and Chapter 3 an introduction to the CES, which itself relies heavily on the SKM. Chapter 4 has provided some examples of the use of the CES/AEFS software in estimating stage-discharge relationships, since these are used frequently in the calibration of numerical river basin models, as well as determining water levels at specific locations when dealing with floods in rivers and urban water courses. The purpose of these later chapters, Chapters 6–9, is therefore aimed at providing the more experienced engineer or reader with sufficient knowledge to be able to use and apply these tools and software more effectively. Some additional theoretical concepts and applications related to river engineering are therefore also

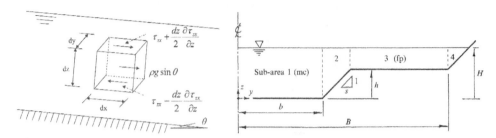

Figure 6.1 Stresses acting on an element.   Figure 6.2 Notation for trapezoidal compound channel.

described. It is hoped that these will enable users to develop both the SKM and the CES further for themselves, possibly more suited to their own particular needs.

The governing equation for the lateral distribution of depth-averaged velocity, U_d, for uniform flow in channels of various prismatic shapes was introduced in Section 2.3 in the form of (2.58). The background to this equation is now described further, with its full derivation given in Appendix 4 and the method for obtaining the analytical solutions shown in Appendix 5. Figure 6.1 (also Figure A4.1 in Appendix) shows the stresses which act on a fluid element in steady uniform flow in a prismatic channel, with no secondary flows, and Figure 6.2 (also Figure A4.2) the notation adopted for overbank flow, since the method was originally focused on rivers with floodplains.

The gravitational body force acting on the mass $\rho(dxdydz)$ in the streamwise direction must be balanced by the surface forces acting on the two sides $(dxdy)$ and the upper and lower surfaces $(dxdz)$. Since any secondary flows are ignored at this stage (i.e. $\Gamma = 0$), this then gives

$$\rho g \sin\theta (dxdydz) + \frac{\partial \tau_{zx}}{\partial z} dz(dxdy) + \frac{\partial \tau_{yx}}{\partial y} dy(dxdz) = 0 \qquad (6.1) \ \& \ (A4.1)$$

Integrating (6.1) over the water depth, H, with $S_0 = \sin\theta$, assuming no wind or other stresses on the water surface $(z = z_s)$, and allowing for bed elevation to vary laterally $(z = z_0(y))$ then the depth-integrated form of (6.1), as shown in Appendix 4, becomes

$$\rho g S_0 H - \tau_{zb} - \frac{\tau_{yb}}{s} + \frac{\partial (H \overline{\tau}_{yx})}{\partial y} = 0 \qquad (6.2) \ \& \ (A4.4)$$

where $\overline{\tau}_{yx}$ is the depth-averaged Reynolds stress and τ_{yb} and τ_{zb} are the shear stresses perpendicular to the y & z directions for an element shown in Figure 6.3. As before, it should be noted that τ_{ij} is the shear stress in the j direction on the plane perpendicular to the i direction. These two shear stresses can be expressed in terms of the shear stress that resists the flow at the bed, τ_b, by considering the shear force balance for a near bed element in region 2 of Figure 6.2, as shown below in Figure 6.3 (also Figure A4.3).

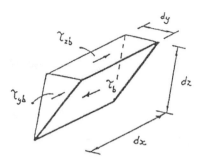

Figure 6.3 Shear stresses on a near-bed element in a side slope region.

Thus $\tau_{yb}(dxdz) + \tau_{zb}(dxdy) = \tau_b dx(dy^2 + dz^2)^{1/2}$ and since $dz/dy = 1/s$, then

$$\frac{\tau_{yb}}{s} + \tau_{zb} = \frac{\sqrt{1+s^2}}{s}\tau_b \quad \text{(for } s > 0\text{)} \tag{6.3) \& (A4.6}$$

Therefore, using the boundary shear stress τ_b at the bed, (6.2) becomes

$$\rho g S_0 H - \frac{\sqrt{1+s^2}}{s}\tau_b + \frac{\partial(H\bar{\tau}_{yx})}{\partial y} = 0 \tag{6.4) \& (A4.7}$$

For a constant depth domain (s = ∞) with H = constant, $\tau_{zb} \Rightarrow \tau_b$ as $s \Rightarrow \infty$, this becomes

$$\rho g S_0 H - \tau_b + \frac{\partial(H\bar{\tau}_{yx})}{\partial y} = 0 \tag{6.5) \& (A4.8}$$

which shows that for a very wide channel in which the lateral shear stresses are ignored,

$$\tau_b = \rho g H S_0 \tag{6.6}$$

Equations (6.4) and (6.7) with $\Gamma = 0$, indicate how the boundary shear stress is generally always affected by lateral shear stresses and secondary flows in open channel flow. This issue is explored further in Section 6.5 as predicting the distribution of boundary shear stresses around the wetted perimeter of a river channel is important for both sediment transport and diffusion processes. For a varying depth domain with $H = f(y)$, the corresponding equations are dealt with in Appendices 4 & 5. The physical background to vorticity and secondary flows is described in Section 6.2.4.

The general governing equation in SKM, including secondary flows ($\Gamma \neq 0$), is repeated here as:

$$\rho g H S_0 - \rho \frac{f}{8} U_d^2 \left(1 + \frac{1}{s^2}\right)^{1/2} + \frac{\partial}{\partial y}\left\{\rho \lambda H^2 \left(\frac{f}{8}\right)^{1/2} U_d \frac{\partial U_d}{\partial y}\right\} = \frac{\partial}{\partial y}\left[H(\rho U V)_d\right]$$

$$(6.7) \ \& \ (2.58)$$

where

$$U_d = \frac{1}{H}\int_0^H U dz; \quad \tau_b = \left(\frac{f}{8}\right)\rho U_d^2; \quad \bar{\tau}_{yx} = \rho \bar{\varepsilon}_{yx}\frac{\partial U_d}{\partial y}; \quad \bar{\varepsilon}_{yx} = \lambda U_* H \quad \text{and}$$

$$\gamma = \left(\frac{2}{\lambda}\right)^{1/2}\left(\frac{f}{8}\right)^{1/4}\frac{1}{H}$$

$$(6.8) \ \& \ (2.59)$$

Letting Γ = RHS of equation (6.7) & (2.58) to represent the 'secondary flow' term, i.e.

$$\frac{\partial}{\partial y}\left[H(\rho U V)_d\right] = \Gamma$$

$$(6.9), (2.60) \ \& \ (3.23)$$

Then (6.7) & (2.58) become

$$\rho g H S_0 - \rho \frac{f}{8} U_d^2 \left(1 + \frac{1}{s^2}\right)^{1/2} + \frac{\partial}{\partial y}\left\{\rho \lambda H^2 \left(\frac{f}{8}\right)^{1/2} U_d \frac{\partial U_d}{\partial y}\right\} = \Gamma$$

$$(6.10)$$

Equation (6.10) is a second order linear partial differential equation (pde). It may be solved analytically, since it is equivalent to a linear ordinary differential equation in the variable U_d^2, as shown by Shiono & Knight (1988 & 1991) and in Appendix 5. Two types of solution were originally developed, one for elements or panels with a non-varying level in the transverse direction, i.e. a horizontal bed with H = constant, and another for those with a varying side slope, i.e. $H = f(y)$. This enables a river cross section to be schematized by a series of linear elements, as already illustrated in Figures 2.30, 2.36, 2.45, 4.3, 4.6 & 4.9. These two types of solution to (6.10) are now summarized below, with the details given in Appendix 5.

6.1.2 Analytical solutions

a. Flow in a constant depth domain with a horizontal bed (H = constant)

Case 1 when there is no secondary flow (i.e. $\Gamma = 0$)
   Let $s = \infty$, so that $1/s = 0$, and H = constant.
   The solution is given by

$$U_d = \left[A_1 e^{\gamma y} + A_2 e^{-\gamma y} + k\right]^{1/2}$$

$$(6.11) \ \& \ (A5.18)$$

where the two constants, γ and k, are given by given by (6.12) and (6.13) respectively.

$$\gamma = \left(\frac{2}{\lambda}\right)^{1/2}\left(\frac{f}{8}\right)^{1/4}\frac{1}{H} \qquad\qquad\qquad (6.12) \ \& \ (A5.11)$$

$$k = \frac{8gS_0 H}{f} \qquad\qquad\qquad\qquad (6.13) \ \& \ (A5.17)$$

These imply that the two physical constants f and λ (the Darcy-Weisbach friction coefficient, defined by (2.24) & (3.10) and the dimensionless eddy viscosity, defined by (2.59) or (A4.8) in Appendix 4) must be known or specified for each panel to which (6.11) or (A5.18) is applied. Table 3.4 has suggested values for λ, explained further in Section 6.2.3, as follows:

$\lambda = 0.07$ for experimental laboratory channels; $\lambda = 0.24$ for rivers.
$\lambda = \lambda_{mc}(-0.2+1.2D_r^{-1.44})$ for floodplains in overbank flow for a relative depth,
   $D_r = (H-h)/H$

The two coefficients, A_1 and A_2, are determined from the boundary conditions for each panel in any specific application, as shown later in Section 6.3. The A_i coefficients are determined for individual panels by solving all panel equations simultaneously, algebraically or numerically, as shown in Section 6.4 and Appendix 6.

Case 2 when there is secondary flow present (i.e. $\Gamma \neq 0$)
   The solution is given by

$$U_d = \left[A_1 e^{\gamma y} + A_2 e^{-\gamma y} + k\right]^{1/2} \qquad\qquad (6.14) \ \& \ (A5.30)$$

where the two constants, γ and k, are given by (6.12) and by (6.15) with (6.16), with suggested values for f, λ and Γ from Table 3.4 as before.

$$k = \frac{8gS_0 H(1-\beta)}{f} \quad \text{(Note know includes the term } \beta\text{)} \qquad (6.15) \ \& \ (A5.29)$$

where

$$\beta = \frac{\Gamma}{\rho g H S_0} \qquad\qquad\qquad\qquad (6.16) \ \& \ (A5.26)$$

The two coefficients, A_1 and A_2, are determined from the boundary conditions for each panel in a specific application, as described later in Section 6.3. The A_i coefficients are determined for individual panels by solving all panel equations simultaneously, algebraically or numerically, as shown in Section 6.4 and Appendix 6.

b. Flow in a variable depth domain with a linear side slope

Case 3 when there is no secondary flow (i.e. $\Gamma = 0$)
   Let s = side slope in transverse direction, as shown in Figure 6.2 (1: s; vertical: horizontal), such that the local depth, ξ, is given by

$$\xi = H - \frac{(y-b)}{s} \tag{6.17) \& (A5.31}$$

and introducing ψ

$$\psi = \left(1 + \frac{1}{s^2}\right)^{1/2} \tag{6.18) \& (A5.32}$$

The solution is given by

$$U_d = \left[A_3 \xi^{\alpha_1} + A_4 \xi^{-(\alpha_1+1)} + \omega\xi\right]^{1/2} \tag{6.19) \& (A5.50}$$

where the two constants, α_1 and ω, are given by (6.20) and (6.21), and λ given again by (6.8), with Table 3.4 providing suggested values.

$$\alpha_1 = -\frac{1}{2} + \frac{1}{2}\left\{1 + \frac{s(1+s^2)^{1/2}}{\lambda}(8f)^{1/2}\right\}^{1/2} \tag{6.20) \& (A5.44}$$

$$\omega = \frac{gS_0}{\frac{(1+s^2)^{1/2}}{s}\frac{f}{8} - \frac{\lambda}{s^2}\left(\frac{f}{8}\right)^{1/2}} \tag{6.21) \& (A5.49}$$

and

$$\gamma = \left(\frac{2}{\lambda}\right)^{1/2}\left(\frac{f}{8}\right)^{1/4}\frac{1}{H} \tag{6.12) \& (A5.11}$$

The two coefficients, A_3 and A_4, are determined from the boundary conditions for each panel in a specific application, as described later in Section 6.3. The A_i coefficients are determined for individual panels by solving all panel equations simultaneously, algebraically or numerically, as shown in Section 6.3 and Appendix 6.

Case 4 when there is secondary flow present (i.e. $\Gamma \neq 0$)
The solution is given by

$$U_d = \left[A_3 \xi^{\alpha_1} + A_4 \xi^{-(\alpha_1+1)} + \omega\xi + \eta\right]^{1/2} \tag{6.22) \& (A5.63}$$

where

$$\alpha_1 = -\frac{1}{2} + \frac{1}{2}\left\{1 + \frac{s(1+s^2)^{1/2}}{\lambda}(8f)^{1/2}\right\}^{1/2} \tag{6.20) \& (A5.44}$$

$$\omega = \frac{gS_0}{\frac{(1+s^2)^{1/2}}{s}\frac{f}{8} - \frac{\lambda}{s^2}\left(\frac{f}{8}\right)^{1/2}}$$

(6.21) & (A5.49)

$$\eta = -\frac{\Gamma}{\frac{\rho f}{8}\left(1+\frac{1}{s^2}\right)^{1/2}}$$

(6.23) & (A5.59)

and

$$\gamma = \left(\frac{2}{\lambda}\right)^{1/2}\left(\frac{f}{8}\right)^{1/4}\frac{1}{H}$$

(6.12) & (A5.11)

The three constants, α_1, ω and η, are given by (6.20) or (A5.44), (6.21) or (A5.49) and (6.23) or (A5.59), with λ and Γ given again by (6.9) with Table 3.4 providing suggested values. This implies that the three physical constants, f, λ and Γ must be known or specified for each panel to which (6.22) or (A5.63) is applied. The two coefficients, A_3 and A_4, are determined from the boundary conditions for each panel in a specific application, as described later in Section 6.3. The A_i coefficients are determined for individual panels by solving all panel equations simultaneously, algebraically or numerically, as shown in Section 6.4 and Appendix 6.

Although Appendices 4–5 explain the derivation of (2.58), based on physical principles, there are a number of further issues that still require comment in order to explain the origins of the SKM. The method was developed initially in conjunction with a series of experiments undertaken in the Flood Channel Facility (FCF) at HR Wallingford, referred to in Section 1.5. Series A was focussed on determining the conveyance capacity of rivers with overbank flow in straight and skewed channels, which required a deeper understanding of the complex interaction between small and large scale turbulent flow structures. In most rivers, these not only affect the velocity field, and hence the discharge within various zones, but also the distribution of boundary shear stresses which govern pollutant mixing and sediment transport or erosion. The scientific study of overbank flow is one that has been undertaken by many researchers on small scale experimental facilities, but a large scale facility, like the FCF, was needed to obtain accurate data on such turbulent structures over a range of floodplain depths that are of practical use to engineers. See Knight (2013a&b).

In addition to the insights gained from the FCF experimental results, work was also focussed on developing both numerical and analytical models, based on sound physics. The reasons for attempting to seek analytical solutions, rather than just numerical ones, was to explore whether it was possible to find some and, secondly, whether it was possible to encapsulate the physics of the flow down into a simple one-dimensional tool. This then might be used independently of any numerical model, thereby possibly saving time and also providing a check on the output of any numerical model. The one-dimensional equation forming the basis of the SKM, (2.58), or its alternative version used in the CES, (3.35), together with the analytical solutions described above in 6.1.2 might be regarded by some as being too 'simple', and that a fully three-dimensional (3D) model should be considered instead as a starting point to capture all the requisite flow physics. However, (2.58) & (3.35) do include certain 3D effects, as explained more fully next in Section 6.2, examining both the 3D equations for fluid flow and some key findings from the FCF experiments on overbank flow. These data are worth exploring a little as they highlight the physical background to these complex flows. See also Knight *et al.* (2010) and Knight (2013a).

6.2   PHYSICAL BACKGROUND TO THE GOVERNING EQUATION USED IN THE SKM

6.2.1   Three-dimensional flow equations

In fluid flow, there are broadly two basic types of flow, one laminar (low Reynolds number, *Re*) and the other turbulent (high *Re*). Since turbulent flows are more common in engineering and in the natural environment than laminar flows, the mathematical treatment of turbulent flow is the only type considered herein as it relates to the SKM. Fortuitously, one can begin with the Navier-Stokes equations for laminar flow, and then adapt them for turbulent flow, to produce the Reynolds-Averaged Navier-Stokes (RANS) equations. The one significant difference is in how the shear stresses and direct stresses are dealt with. In laminar flow the shear stress is proportional to the rate of strain, unlike solids where stress is proportional to strain, and thus related to the velocity gradients by viscosity and Stokes law. However, in turbulent flow, stresses are more complex and need special treatment mathematically.

Lower case characters are used throughout this one section of the book for the velocity components *u v* & *w* in order to distinguish between steady and time varying turbulent velocities later on. As always, it is important to begin with the fundamental equations of fluid flow. Taking the Eulerian form of analysis of flow fields, assuming a fixed frame of reference (as opposed to the Lagrangian analysis that follows the motion of fluid elements), consider an elementary control volume *dxdydz* in a Cartesian system of co-ordinates where normal and shear stresses act on each element face. It follows from Newton's 2nd law, applied in the *x* direction that

$$\rho(dxdydz)\frac{du}{dt} = \rho X(dxdydz) + \frac{\partial \sigma_{xx}}{\partial x}dx(dydz) + \frac{\partial \tau_{yx}}{\partial y}dy(dxdz) + \frac{\partial \tau_{zx}}{\partial z}dz(dxdy) = 0$$

$$(6.24)$$

where the LHS is mass times *du/dt*, the acceleration or the rate of change of momentum, and the RHS contains all the forces, where *X* is a body force per unit mass (e.g. such as gravity), σ_{xx}, is a normal stress and τ_{yx} and τ_{zx} are shear stresses. Corresponding forms of equations apply in the *y* & *z* directions due to the velocities v & *w* in those directions. Dividing these 3 equations by the mass of the element, *ρdxdydz* gives the Euler equations, first derived in 1738, as

$$\frac{du}{dt} = X + \frac{1}{\rho}\left[\frac{\partial \sigma_{xx}}{\partial x} + \frac{\partial \tau_{yx}}{\partial y} + \frac{\partial \tau_{zx}}{\partial z}\right]$$

$$\frac{dv}{dt} = Y + \frac{1}{\rho}\left[\frac{\partial \tau_{xy}}{\partial x} + \frac{\partial \sigma_{yy}}{\partial y} + \frac{\partial \tau_{zy}}{\partial z}\right] \tag{6.25}$$

$$\frac{dw}{dt} = Z + \frac{1}{\rho}\left[\frac{\partial \tau_{xz}}{\partial x} + \frac{\partial \tau_{yz}}{\partial y} + \frac{\partial \sigma_{zz}}{\partial z}\right]$$

By considering the linear and angular strains on the element due to the 3D motion, involving stretching, translation, deformation and rotation, and noting that $\tau_{ij} = \tau_{ji}$, the stresses on the RHS may be related to the velocity components $\{u\ v\ w\}$. For example,

$$\tau_{yx} = \mu\left(\frac{\partial u}{\partial y} + \frac{\partial v}{\partial x}\right) \tag{6.26}$$

It can be shown that for laminar flow the first equation in (6.25) may be developed to give

$$\frac{du}{dt} = X + \frac{1}{\rho}\frac{\partial}{\partial x}\left[-p + 2\mu\frac{\partial u}{\partial x} - \frac{2\mu}{3}\left(\frac{\partial u}{\partial x} + \frac{\partial v}{\partial y} + \frac{\partial w}{\partial z}\right)\right] + \frac{1}{\rho}\frac{\partial}{\partial y}$$

$$\left[\mu\left(\frac{\partial u}{\partial y} + \frac{\partial v}{\partial x}\right)\right] + \frac{1}{\rho}\frac{\partial}{\partial z}\left[\mu\left(\frac{\partial w}{\partial x} + \frac{\partial u}{\partial z}\right)\right] \tag{6.27}$$

where the terms in each [ ] represent σ_{xx}, τ_{yx} and τ_{zx} respectively. Similar equations may be written for the y & z directions. Equation (6.27) may be written alternatively as

$$\frac{du}{dt} = X - \frac{1}{\rho}\frac{\partial p}{\partial x} + \frac{\mu}{\rho}\left[\frac{\partial^2 u}{\partial x^2} + \frac{\partial^2 u}{\partial y^2} + \frac{\partial^2 u}{\partial z^2}\right] + \frac{\mu}{3\rho}\frac{\partial}{\partial x}\left[\frac{\partial u}{\partial x} + \frac{\partial v}{\partial y} + \frac{\partial w}{\partial z}\right] \tag{6.28}$$

Since in general $u = f\{x, y, z, t\}$, the total differential on the LHS results in a series of partial differentials as

$$du = \frac{\partial u}{\partial x}dx + \frac{\partial u}{\partial y}dy + \frac{\partial u}{\partial z}dz + \frac{\partial u}{\partial t}dt$$

$$\frac{du}{dt} = u\frac{\partial u}{\partial x} + v\frac{\partial u}{\partial y} + w\frac{\partial u}{\partial z} + \frac{\partial u}{\partial t} \tag{6.29}$$

The acceleration at any point in the fluid is therefore seen to have two main components: convective and local. Convective acceleration is due to where u varies spatially in any co-ordinate direction, and local acceleration is where u varies only with time, t.

The equation for mass conservation is given by

$$\frac{\partial \rho}{\partial t} + \frac{\partial(\rho u)}{\partial x} + \frac{\partial(\rho v)}{\partial y} + \frac{\partial(\rho w)}{\partial z} = 0 \tag{6.30}$$

Thus for steady flow of an incompressible fluid, with ρ = constant, this simplifies to

$$\frac{\partial u}{\partial x} + \frac{\partial v}{\partial y} + \frac{\partial w}{\partial z} = 0 \tag{6.31}$$

Inserting this into (6.28) and its companion equations for the y and z directions gives

$$\frac{\partial u}{\partial t} + u\frac{\partial u}{\partial x} + v\frac{\partial u}{\partial y} + w\frac{\partial u}{\partial z} = X - \frac{1}{\rho}\frac{\partial p}{\partial x} + \frac{\mu}{\rho}\left[\frac{\partial^2 u}{\partial x^2} + \frac{\partial^2 u}{\partial y^2} + \frac{\partial^2 u}{\partial z^2}\right]$$

$$\frac{\partial v}{\partial t} + u\frac{\partial v}{\partial x} + v\frac{\partial v}{\partial y} + w\frac{\partial v}{\partial z} = Y - \frac{1}{\rho}\frac{\partial p}{\partial y} + \frac{\mu}{\rho}\left[\frac{\partial^2 v}{\partial x^2} + \frac{\partial^2 v}{\partial y^2} + \frac{\partial^2 v}{\partial z^2}\right] \tag{6.32}$$

$$\frac{\partial w}{\partial t} + u\frac{\partial w}{\partial x} + v\frac{\partial w}{\partial y} + w\frac{\partial w}{\partial z} = Z - \frac{1}{\rho}\frac{\partial p}{\partial z} + \frac{\mu}{\rho}\left[\frac{\partial^2 w}{\partial x^2} + \frac{\partial^2 w}{\partial y^2} + \frac{\partial^2 w}{\partial z^2}\right]$$

These momentum equations, (6.32), together with the equation for mass conservation, (6.30), give 4 non-linear partial differential equations for 4 unknowns $\{u\ v\ w\ p\}$ in terms of $\{x\ y\ z\ t\}$, i.e. space and time. They are known as the Navier-Stokes (N-S) equations, first formulated in 1845, and apply to flows of any Newtonian fluid. It is only possible to obtain exact solutions to (6.32) in certain cases, typically for flows in simple geometries, and in cases where certain terms are zero. Approximate analytical solutions may be obtained, notably at low Reynolds numbers (creeping motions with low inertia), and at high Reynolds numbers where boundary layers are thin (high inertia), both of which are particularly useful in dealing with certain practical engineering problems. See Schlichting (1979) and Drazin & Riley (2006) for further details.

In order to keep the analysis concise, consider only the momentum equation in the x-direction for an incompressible fluid. Combining (6.31) with (6.32), by multiplying the continuity equation by u and adding it to (6.32), gives

$$\frac{\partial u}{\partial t} + \frac{\partial(u^2)}{\partial x} + \frac{\partial(uv)}{\partial y} + \frac{\partial(uw)}{\partial z} = X - \frac{1}{\rho}\frac{\partial p}{\partial x} + \frac{\mu}{\rho}\left[\frac{\partial^2 u}{\partial x^2} + \frac{\partial^2 u}{\partial y^2} + \frac{\partial^2 u}{\partial z^2}\right] \tag{6.33}$$

For laminar 2D flow, (6.26) will reduce to

$$\tau_{yx} = \mu\frac{\partial u}{\partial y} \tag{6.34}$$

or

$$\tau_{yx} = \rho v\frac{\partial u}{\partial y} \tag{6.35}$$

since

$$v = \frac{\mu}{\rho} \tag{6.36}$$

The distinction between dynamic viscosity (μ) and kinematic viscosity (v) should be noted.

6.2.2 Three, two and one-dimensional turbulent flow equations

For turbulent flow, the Reynolds-Averaged Navier-Stokes (RANS) equations are mathematical formulations that describe the motion of fluids. They correspond to the Navier-Stokes equations that govern laminar flows, but take into account the additional stresses due to turbulence. Turbulent fluctuations around essentially quasi-steady values are illustrated for the three components of velocity {u v w} in the {x y z} directions in Figure 6.4. These may be formalised by writing time averaged and fluctuating components, including pressure, p, and concentration, c, as

$$u(t) = \bar{u} + u'; v(t) = \bar{v} + v'; w(t) = \bar{w} + w'; p(t) = \bar{p} + p'; c(t) = \bar{c} + c' \tag{6.37}$$

where $\bar{u} = \frac{1}{T}\int_t^{t+T} u\, dt$ and as T becomes very large compared with the timescale of turbulent fluctuations, by definition $\overline{u'} = \frac{1}{T}\int_t^{t+T} u'\, dt = 0$

Inserting $u(t) = \bar{u} + u'$, etc. from (6.37) into (6.33) gives:

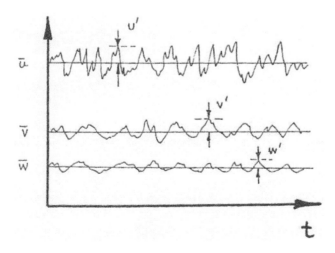

Figure 6.4 Turbulent fluctuations in component velocities.

$$\frac{\partial \bar{u}}{\partial t} + \frac{\partial u'}{\partial t} + \frac{\partial \bar{u}^2}{\partial x} + 2\frac{\partial(\bar{u}u')}{\partial x} + \frac{\partial u'^2}{\partial x} + \frac{\partial \overline{uv}}{\partial y} + \frac{\partial \overline{uw'}}{\partial y} + \frac{\partial u'\bar{v}}{\partial y} + \frac{\partial u'v'}{\partial y} + \frac{\partial \overline{uw}}{\partial z} + \frac{\partial \overline{uw'}}{\partial z} + \frac{\partial u'\bar{w}}{\partial z}$$

$$+ \frac{\partial u'w'}{\partial z} = X - \frac{1}{\rho}\frac{\partial \bar{p}}{\partial x} - \frac{1}{\rho}\frac{\partial p'}{\partial x} + \frac{\mu}{\rho}\left[\frac{\partial^2 \bar{u}}{\partial x^2} + \frac{\partial^2 \bar{u}}{\partial y^2} + \frac{\partial^2 \bar{u}}{\partial z^2}\right] + \frac{\mu}{\rho}\left[\frac{\partial^2 u'}{\partial x^2} + \frac{\partial^2 u'}{\partial y^2} + \frac{\partial^2 u'}{\partial z^2}\right]$$

Averaging each term over a long time interval then gives

$$\frac{\partial \bar{u}}{\partial t} + \frac{\partial \bar{u}^2}{\partial x} + \frac{\partial \overline{u'^2}}{\partial x} + \frac{\partial \overline{uv}}{\partial y} + \frac{\partial \overline{u'v'}}{\partial y} + \frac{\partial \overline{uw}}{\partial z} + \frac{\partial \overline{u'w'}}{\partial z} = X - \frac{1}{\rho}\frac{\partial \bar{p}}{\partial x} + \frac{\mu}{\rho}\left[\frac{\partial^2 \bar{u}}{\partial x^2} + \frac{\partial^2 \bar{u}}{\partial y^2} + \frac{\partial^2 \bar{u}}{\partial z^2}\right]$$

(6.38)

Multiplying the continuity equation by $\bar{u}$

$$\bar{u}\frac{\partial \bar{u}}{\partial x} + \bar{u}\frac{\partial \bar{v}}{\partial y} + \bar{u}\frac{\partial \bar{w}}{\partial z} = 0$$

and subtracting this from the dynamic equation, (6.38) finally gives

$$\frac{\partial \bar{u}}{\partial t} + \bar{u}\frac{\partial \bar{u}}{\partial x} + \bar{v}\frac{\partial \bar{u}}{\partial y} + \bar{w}\frac{\partial \bar{u}}{\partial z} = X - \frac{1}{\rho}\frac{\partial \bar{p}}{\partial x} + \frac{1}{\rho}\frac{\partial}{\partial x}\left[\mu\frac{\partial \bar{u}}{\partial x} - \rho\overline{u'^2}\right] + \frac{1}{\rho}\frac{\partial}{\partial y}\left[\mu\frac{\partial \bar{u}}{\partial y} - \rho\overline{u'v'}\right]$$

$$+ \frac{1}{\rho}\frac{\partial}{\partial z}\left[\mu\frac{\partial \bar{u}}{\partial z} - \rho\overline{u'w'}\right]$$

(6.39)

By comparing (6.39) for turbulent flow with (6.32) for laminar flow, it is apparent that there are additional stress-like terms due to the turbulent motion given by

$$\sigma_{xx} = -\rho\overline{u'^2}; \; \tau_{yx} = -\rho\overline{u'v'}; \; \tau_{zx} = -\rho\overline{u'w'}$$

(6.40)

These are known as Reynolds stresses. The constitutive equations for the normal and tangential turbulent stresses for an incompressible fluid are normally given in the form:

$$\sigma_{xx} = -p + 2\mu\frac{\partial \bar{u}}{\partial x} - \rho\overline{u'^2}; \; \tau_{yx} = \mu\left(\frac{\partial \bar{u}}{\partial y} + \frac{\partial \bar{v}}{\partial x}\right) - \rho\overline{u'v'}; \; \tau_{zx} = \mu\left(\frac{\partial \bar{u}}{\partial z} + \frac{\partial \bar{w}}{\partial x}\right) - \rho\overline{u'w'}$$

(6.41)

which are similar to those shown in (6.26). It may be seen from (6.40) & (6.41) that when the instantaneous fluctuations, u', v' and w' are cross-multiplied at any point in the fluid flow they give instantaneous values of the direct and shear stress at that point. When they are time-averaged they then give values of the Reynolds stresses.

In the FCF experiments referred to in Section 1.5, these fluctuations were simultaneously measured in two coordinate directions, y and z at 20–100 Hz, cross correlated, and then time averaged to give the mean Reynolds stresses τ_{yx} and τ_{zx} throughout the flow field. For steady uniform flow (6.39) becomes

$$-v\frac{\partial\overline{u}}{\partial y}+\overline{w}\frac{\partial\overline{u}}{\partial z} = X - \frac{1}{\rho}\frac{\partial\overline{p}}{\partial x} + \frac{1}{\rho}\frac{\partial\overline{\tau}_{yx}}{\partial y} + \frac{1}{\rho}\frac{\partial\overline{\tau}_{zx}}{\partial z} \qquad (6.42)$$

For 1D uniform flow, since $\partial p / \partial x = 0$, $X = g\sin\theta$ and $\sin\theta = S_o$, (6.42) reduces to

$$\rho\left[\frac{\partial}{\partial y}(\overline{uv})+\frac{\partial}{\partial z}(\overline{uw})\right] = \rho gS_0 + \frac{\partial\overline{\tau}_{yx}}{\partial y} + \frac{\partial\overline{\tau}_{zx}}{\partial z} \qquad (6.43 \;\&\; (2.57)$$

For 1D uniform flow with no secondary flow, then $v = w = 0$, giving

$$\rho gS_0 + \frac{\partial\overline{\tau}_{yx}}{\partial y} + \frac{\partial\overline{\tau}_{zx}}{\partial z} = 0 \qquad (6.44) \;\&\; (A4.1)$$

   In order to proceed further, the way turbulent shear stresses are often considered in engineering now needs to be briefly described. Generally in the literature this topic is called "turbulence closure"; additional assumptions are made which relate the turbulent stresses to characteristics of the primary flow, giving the same number of equations to solve as there are flow variables. By analogy with the kinematic viscosity, v, used in laminar flow, together with the formulation by Reynolds given in (6.41), Boussinesq in 1877 proposed that the Reynolds shear stress could be approximated by

$$\tau_{yx} = \rho v_t\left(\frac{\partial\overline{u}}{\partial y}+\frac{\partial\overline{v}}{\partial x}\right) \qquad (6.45)$$

where v_t is an effective turbulent eddy viscosity, which is a flow property, not a fluid property, unlike v $(= \mu/\rho)$. It thus varies within the fluid flow and is not a constant. For modelling the concentration, c, of some pollutant or sediment in a thin shear layer, the flow is often treated as 2D, leading to gradient type relationships, such as

$$\tau_{yx} = \rho v_m\left(\frac{\partial\overline{u}}{\partial y}\right) \quad \text{and} \quad F_y = v_c\left(\frac{\partial\overline{c}}{\partial y}\right) \qquad (6.46)$$

where the subscripts m and c refer to momentum and concentration respectively, v_m is the eddy viscosity for mass or momentum, v_c is the eddy diffusivity for the pollutant or sediment within the fluid, with perturbations in c of c', and F_c is the flux per unit area $(F_c = \overline{u'c'})$. The ratio between v_m and v_c is known as the turbulent Schmidt number. This number may not necessarily be 1.0, as under certain conditions the mixing of solute, sediment or heat may not occur at the same rate as the mass exchange of fluid between different regions of the flow.

6.2.3   Physical background to a simple turbulence model

Prandtl (1925 & 1926) proposed a very practical and useful distribution law for the eddy viscosity, v_t, based upon a mixing length hypothesis, as illustrated in Figure 6.5.

Prandtl assumed that $u' = \ell_m \dfrac{\partial u}{\partial y}$ and that $v' \approx u'$, thus formulating the Reynolds stress as

$$\tau_{yx} = -\overline{\rho u' v'} = \rho \ell_m^2 \frac{\partial u}{\partial y}\left|\frac{\partial u}{\partial y}\right| \tag{6.47}$$

where ℓ_m is known as the 'mixing length', a distance over which a discrete portion of the fluid moves from one position in the flow to another by virtue of the turbulent motion. Thus for the shear flow example shown in Figure 6.5, if an exchange occurs from level y to a higher level $y + l$, then a positive v' will generally be associated with a negative u', as that parcel of fluid with a lower momentum causes the upper layer to slow down a little. Conversely, if a parcel of fluid in the upper layer moves down towards the boundary, then a negative v' will generally be associated with a positive u', causing the lower layer to speed up a little. This explains why there are negative signs in (6.40) & (6.47), since with $\tau_{yx} = -\overline{\rho u' v'}$, τ_{yx} will be positive. The modulus sign in (6.47) is introduced to ensure that the Reynolds stress is compatible with the sign of the velocity gradient. See Prandtl (1953) for further details.

Comparing (6.46) & (6.47) gives an expression for the Boussinesq eddy viscosity as

$$v_t = \ell_m^2 \frac{\partial u}{\partial y} \tag{6.48}$$

The advantage of (6.48) over (6.46) is that the distribution and size of the mixing length, l_m, within a given flow may be estimated from physical reasoning. For example, near any boundary surface, $l_m \to 0$ as $y \to 0$, and l_m will be larger where large eddies can form. Very near a boundary, it might therefore be appropriate to assume linear distributions

$$l_m = \kappa y \quad \text{or} \quad l_m = \kappa z \tag{6.49}$$

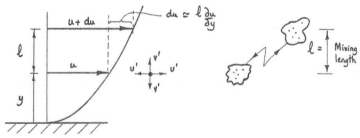

Figure 6.5 Prandtl's mixing length hypothesis.

depending on the distance from the boundary surface, y or z. The scaling parameter, κ, is known as von Karman's constant. Some data presented later in Figure 6.7 indicates that this is a reasonable assumption to make very near the wall or bed of a channel.

Having now established an expression for the turbulent shear stresses in terms of either eddy viscosity or mixing length, it is important to appreciate how these parameters are used in a depth-averaged model. The velocity distributions for smooth and rough surfaces have been given in Section 2.2.5 by (2.51) & (2.47) or (2.48), and are repeated here, with z taken as the distance from the bed surface, i.e.:

Smooth surfaces

$$\frac{u}{u_*} = 5.75 \log_{10}\left(\frac{u_* z}{v}\right) + 5.5 \quad \text{or} \quad \frac{u}{u_*} = A \log_{10}\left(\frac{u_* z}{v}\right) + B \tag{2.51}$$

Rough surfaces

$$\frac{u}{u_*} = 5.75 \log_{10}\left(\frac{z}{k_s}\right) + 8.5 \quad \text{or} \quad \frac{u}{u_*} = A \log_{10}\left(\frac{z}{k_s}\right) + B \tag{2.48}$$

These both stem from Prandtl's law, stated as

$$\frac{u}{u_*} = \frac{1}{\kappa} \ln\left(\frac{z}{z_o}\right) \tag{2.47}$$

Integrating (2.51) & (2.48) these over the depth using (2.52), gives

$$u_d = \frac{1}{h} \int_o^h u \, dz \tag{2.52}$$

$$u_d = \frac{u_*}{h}\left[Ah\left\{\ln\left(\frac{u_* h}{v}\right) - 1\right\} + Bh \right] \tag{6.50}$$

and

$$u_d = \frac{u_*}{h}\left[Ah\left\{\ln\left(\frac{h}{k_s}\right) - 1\right\} + Bh \right] \tag{6.51}$$

where A & B are the two constants adopted in the logarithmic distributions (often taken as 5.75 & 5.5 for smooth surfaces and 5.75 and 8.5 for rough surfaces). Hence the value of z at which $u = u_d$ is given by $z/h = 0.3679$. If the seventh power law, given by (2.55) is used instead, then $z/h = 0.3927$. Thus for practical purposes in river gauging it is often assumed that by measuring the velocity at the position $0.4h$ from the river bed, or measuring at $0.2h$ and $0.8h$ and taking the average, gives the depth mean value for that particular vertical. The distributions of eddy viscosity, v_t, and mixing length, l_m, over the depth may now be determined from the equations for the logarithmic velocity law, as

$$V_t = u_* \kappa z (1 - z/h)$$ (6.52)

and

$$l_m = \kappa z (1 - z/h)^{1/2}$$ (6.53)

Turbulence measurements confirm these, as shown in Figures 6.6 & 6.7, from Nezu & Nakagawa (1993). The wake function, Π, is introduced to deal with the turbulence outside any boundary layer, but in the context of open channel flows this region does not form in fully developed open channel flow where the boundary layer typically extends over the entire depth of flow, from the bed to the free surface. Where it might be significant, in 'unbounded' boundary layers, Π normally lies between 0 and 0.2, but for simplicity is taken as zero in the context of this book.

Depth integrating (6.52) gives

$$\overline{V_t} = \frac{1}{h} \int_0^h u_* \kappa z \left(1 - \frac{z}{h}\right) dz = \frac{\kappa u_* h}{6}$$ (6.54)

From experimental data that give $\kappa \approx 0.41$, then the depth mean eddy viscosity becomes

$$\overline{V_t} = 0.07 u_* h$$ (6.55)

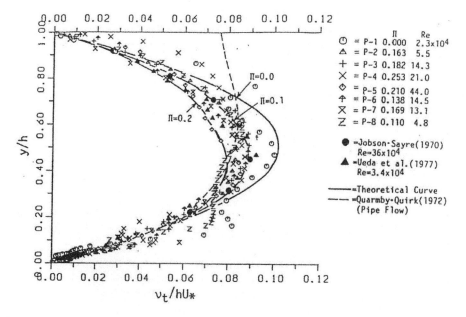

Figure 6.6 Dimensionless eddy viscosity versus relative depth, y/h (from Nezu & Nakagawa, 1993).

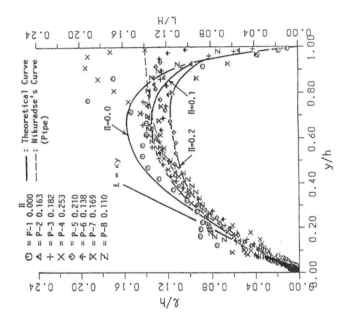

Figure 6.7 Dimensionless mixing length values versus relative depth, y/h (from Nezu & Nakagawa, 1993).

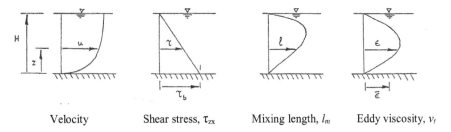

| Velocity | Shear stress, τ_{zx} | Mixing length, l_m | Eddy viscosity, v_t |

Figure 6.8 Distributions of velocity, shear stress, mixing length and eddy viscosity over the depth.

Thus from (6.8), $\lambda = 0.07$, as also indicated by the data shown in Figure 6.6. The data for the corresponding distribution of mixing length, given by (6.53) is shown in Figure 6.7. It should be noted from (6.53) that near the bed, as $z/h \to 0$, $l_m = \kappa z$, as indicated well by the data which show that $l_m = \kappa y$ as $y/h \to 0$. Equation (6.55) forms the basis of selecting one of the key parameters in each panel of a river cross section to which the SKM is applied. It should be noted that $\lambda = 0.07$ is a lower limit and that it will be higher in natural rivers, due to generally greater roughness and increased turbulence, as indicated by (3.20) and Table 3.4.

The physical realities that lie behind the distributions that are frequently used in any depth-averaged model to represent a mean value over a vertical in any open channel flow model, should not therefore be overlooked. The 4 distributions commonly used are shown together in Figure 6.8.

6.2.4 Vorticity and secondary flows

Having introduced the concept of vorticity and secondary flow in open channel flow through Figures 2.14, 2.15 & 2.33 and equation (2.60) in Chapter 2 and Figures 3.3–3.13 and equations (3.23)–(3.27) in Chapter 3, the mathematical and physical backgrounds are now briefly explained beginning with turbulent 3D flows based on the form of (6.39). Using the notation given in Figure 6.9, the RANS equations may be written as

$$U\frac{\partial U}{\partial x}+V\frac{\partial U}{\partial y}+W\frac{\partial U}{\partial z}=-\frac{1}{\rho}\frac{\partial p}{\partial x}-\left(\frac{\partial \overline{u^2}}{\partial x}+\frac{\partial \overline{uv}}{\partial y}+\frac{\partial \overline{uw}}{\partial z}\right)+\nu\left[\frac{\partial^2 U}{\partial x^2}+\frac{\partial^2 U}{\partial y^2}+\frac{\partial^2 U}{\partial z^2}\right]$$

(6.56)

$$U\frac{\partial V}{\partial x}+V\frac{\partial V}{\partial y}+W\frac{\partial V}{\partial z}=-\frac{1}{\rho}\frac{\partial p}{\partial y}-\left(\frac{\partial \overline{uv}}{\partial x}+\frac{\partial \overline{v^2}}{\partial y}+\frac{\partial \overline{vw}}{\partial z}\right)+\nu\left[\frac{\partial^2 V}{\partial x^2}+\frac{\partial^2 V}{\partial y^2}+\frac{\partial^2 V}{\partial z^2}\right]$$

(6.57)

$$U\frac{\partial W}{\partial x}+V\frac{\partial W}{\partial y}+W\frac{\partial W}{\partial z}=-\frac{1}{\rho}\frac{\partial p}{\partial z}-\left(\frac{\partial \overline{uw}}{\partial x}+\frac{\partial \overline{vw}}{\partial y}+\frac{\partial \overline{w^2}}{\partial z}\right)+\nu\left[\frac{\partial^2 W}{\partial x^2}+\frac{\partial^2 W}{\partial y^2}+\frac{\partial^2 W}{\partial z^2}\right]$$

(6.58)

Differentiating (6.57) & (6.58) with respect to z and y respectively, and subtracting one from each other, yields an equation for the component of vorticity in the x co-ordinate direction, ξ, as

$$\underbrace{U\frac{\partial \xi}{\partial x}+V\frac{\partial \xi}{\partial y}+W\frac{\partial \xi}{\partial z}}_{\text{I}}=\underbrace{\nu\left(\frac{\partial^2 \xi}{\partial x^2}+\frac{\partial^2 \xi}{\partial y^2}+\frac{\partial^2 \xi}{\partial z^2}\right)}_{\text{II}}+\underbrace{\xi\frac{\partial U}{\partial x}}_{\text{III}}+\underbrace{\eta\frac{\partial U}{\partial y}+\varsigma\frac{\partial U}{\partial z}}_{\text{IV}}$$
$$+\underbrace{\frac{\partial}{\partial x}\left(\frac{\partial \overline{uv}}{\partial z}-\frac{\partial \overline{uw}}{\partial y}\right)}_{\text{V}}+\underbrace{\frac{\partial^2}{\partial y\partial z}\left(\overline{v^2}-\overline{w^2}\right)}_{\text{VI}}+\underbrace{\left(\frac{\partial^2}{\partial z^2}-\frac{\partial^2}{\partial y^2}\right)\overline{vw}}_{\text{VII}}$$

(6.59)

in which the 3 components of vorticity $\omega=\mathrm{Curl}\{U,\ V,\ W\}=\{\xi,\ \eta,\ \varsigma\}$ [not to be confused with some of the parameter symbols used in SKM] are traditionally given by

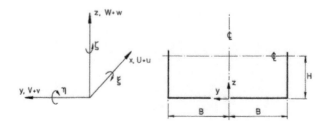

Figure 6.9 Notation for vorticity components $\{\xi\ \eta\ \varsigma\}$ in a rectangular duct.

$$\xi = \partial W / \partial y - \partial V / \partial z; \quad \eta = \partial U / \partial z - \partial W / \partial x \quad \text{and} \quad \varsigma = \partial V / \partial x - \partial U / \partial y$$
$$(6.60)$$

as illustrated in Figure 6.9.

The physical significance of the various terms I-VII in (6.59) is as follows: I = convection, II = dissipation, III = vortex stretching, IV = production through skewing of the mean shear by a transverse pressure gradient, and V to VII = production via turbulence. It therefore follows from (6.59) that for flows in all non-circular channels or ducts, there will inevitability be secondary flow cells formed, particularly in corner regions, due to Terms V to VII. This has already been illustrated in Figure 2.14 using experimental data from flow in a trapezoidal channel. The distortion in the isovels in the central and corner regions indicates mean transverse velocities, referred to as 'secondary flow' since the velocities, V & W, are typically only 2 ~ 4% of the primary velocity, U. These secondary flows tend to be directed into the corners, along angle bisectors, and then away from the boundary again at other places to satisfy continuity. An upward movement may be detected at the centreline of Figure 2.14, by virtue of the increased spacing between isovels there. A pattern of secondary flow cells is thus formed, convecting energy and momentum within the cross-section.

1.   Streamwise vorticity

For inbank flow in rivers, it should be noted that in Figure 2.14 the secondary flows are normal to the predominant streamwise flow, so are frequently referred to as streamwise vorticity. The cells spiral around a streamwise axis in the x-direction. In Figure 3.12, showing flow in a meandering channel, there is one main secondary flow in the main channel that changes direction according to whether the meander is left-handed or right-handed. This is also referred to as streamwise vorticity, since the motion is again normal to the predominant streamwise flow. In classical fluid mechanics Prandtl distinguished between two types of streamwise vorticity. Type 1 is driven by anisotropic turbulence that often occurs in the corner regions of flow in straight non-circular ducts or channels, as shown in Figures 2.14, 6.10 & 6.15. Anisotropic turbulence is where the turbulent fluctuations are not the same in the x, y & z directions, giving rise to stresses as indicated by terms V to VII in (6.59). Type 2 on the other hand is driven by the plan form geometry of the channel, such as occurs in meandering channels shown in Figure 3.12. In both types, the size and number of the secondary flow cells are limited by the depth of flow in which they form, whether in straight or meandering channels, as well as by the channel width.

Figure 6.10 shows a conceptual pattern for secondary flow cells in a straight rectangular duct as the aspect ratio (width/depth) increases. It illustrates that the number of secondary flow cells occur in pairs, that they can increase in number as the aspect ratio increases, and that the size of the cells are largely governed by the depth. The cells tend to be the strongest in the corner regions and weaken towards the centre if the channel is wide. Figure 6.10 also shows the effect of these cells on the lateral distribution of boundary shear stress on the duct floor. Figure 6.11 shows some actual distributions of measured boundary shear stress in a trapezoidal open channel flow for aspect ratios between 1.0 and 10.0, with more or less similar Froude numbers ($Fr = U/\sqrt{(gA/T)}$, where T = top width). These distributions again exhibit a

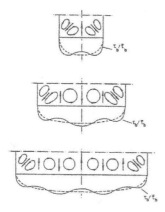

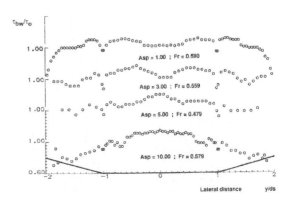

Figure 6.11 Measured boundary shear stresses in a trapezoidal channel for different aspect ratios, Asp = 1, 3, 5 & 10 (after Knight, Yuen & Alhamid, 1994).

Figure 6.10 Conceptual pattern of multiple secondary flow cells forming in a series of rectangular ducts (after Knight & Patel, 1985).

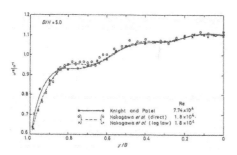

Figure 6.12 Measured boundary shear stresses in rectangular duct, B/H = 5.0 (after Knight & Patel, 1985).

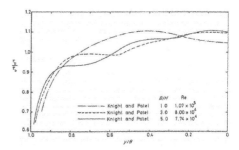

Figure 6.13 Measured boundary shear stresses a rectangular duct, B/H = 1.0, 3.0 & 5.0 (after Knight & Patel, 1985).

number of perturbations commensurate with the presence of secondary flow cells, as already indicated by Figure 6.10. If the bed of the channel is erodible, then longitudinal ridges may develop in the sediment, as shown by Ikeda (1981). Figure 6.12 shows some actual perturbations in the measured data of bed shear stress for a duct with an aspect ratio of 5.0. The sensitivity of boundary shear stress to aspect ratio, and hence the number of cells, is shown in Figure 6.13, with the actual measured data points removed for clarity. As will be shown later, the number and position of these cells is important for accurate simulations using the SKM, as indicated by Figure 2.33 and 2.34. See Omran (2005) and Knight, Omran & Tang (2007) for further details.

Even in overbank flow, in which planform vorticity usually dominates, there may be streamwise vorticity present, as illustrated in Figure 6.14. This Figure shows lateral distributions of boundary shear stress for overbank flow in a small two-stage or compound trapezoidal channel. Despite being overbank flow, there is still evidence

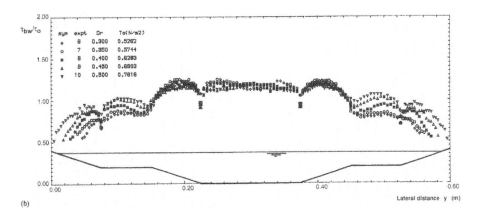

Figure 6.14 Distributions of boundary shear stress in a compound trapezoidal channel for different depths, showing influence of secondary flows (after Knight, Yuen & Alhamid 1994).

of secondary flow cells produced by streamwise vorticity in these data, particularly in the corner regions. The interaction of these two types of flow structure is important for understanding overbank flow, as generally planform vorticity often pre-dominates at low relative depths, and streamwise vorticity at higher depths, as already illustrated by Figure 3.7. At low flow depths on the floodplain (in that Figure for $Dr = 0.180$), it is apparent that planform vorticity is present with large eddies forming in the shear zone between the faster moving flow in the main channel and the slower moving flow on the floodplain. However, these large scale planform vortices appear to break up and eventually disappear at the flow depths on the floodplain increase (in that Figure for $Dr = 0.344$). Figures 3.8 and 3.9 have highlighted the fact that these same flow structures can occur in natural rivers and in curvilinear channels.

2. *Planform vorticity*

In overbank flow in rivers, streamwise vorticity may also occur, but planform vorticity is more likely to be significant, as it is formed by the strong shear layer that may form between the slower and faster moving streams of water on the floodplain and in the main channel respectively. In this case, the cells rotate around the z-axis, a nearly vertical axis, as already shown in Figures 2.15, 3.5, 3.6 & 3.10 at the edges of floodplains. The typical size of these vortices is large, often extending over the entire width of the floodplain, as shown in Figures 3.5 to 3.7 for experimental channels and in Figure 3.8 for the River Tone. In the latter example, the two floodplains are over 200 m wide and the overall width of the river is approximately ~ 600 m. The rationale behind the definition of Γ defined by (6.7) or (6.9), together with the values given in Table 3.4 for both inbank and overbank flows are now explained more fully, including why Γ might apply to both planform and streamwise vorticity.

Figures 6.15 & 6.16 show distributions of boundary shear stress and isovels in the lower half of a wind tunnel with a cross-shaped duct. This may be regarded as simulating 'overbank' flow in a river with two symmetric floodplains, without any free surface effects (such as surface tension), due to the plane of symmetry. Wind tunnels were especially built to this configuration, as shown by Knight & Lai (1985) and Rhodes & Knight (1994 & 1995), in order to study these types of flow in detail. See Patel (1984), Lai (1987), and Rhodes (1991) for further information. Figure 6.15 illustrates how at a low relative depth (Dr ~0.15), streamwise vorticity is apparent in each corner of the main channel, giving rise to a predictable dip in the boundary shear stress at the centreline. Figure 6.16 shows flow at a much larger relative depth (Dr ~0.50), illustrating how a pair of cells are now present much higher up, at each re-entrant corner between the main channel and the floodplain.

In order to illustrate planform vorticity and its effect on overbank flow, some detailed data from some large scale experiments undertaken in the UK Flood Channel

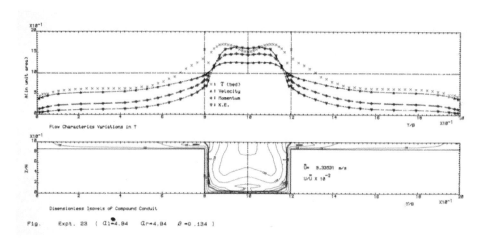

Figure 6.15 Isovels in a compound rectangular duct (low depth, $Dr = 0.134$) (after Lai, 1987, Experiment 23).

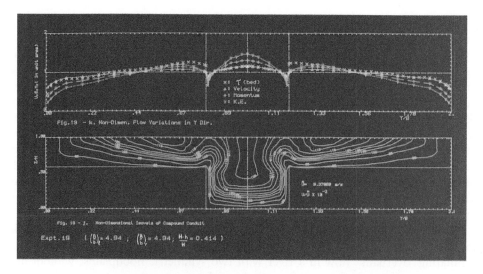

Figure 6.16 Isovels in a compound rectangular duct (large depth, $Dr = 0.414$) (after Lai, 1987, Experiment 19).

Facility (FCF) are now presented. Figure 2.67 has already shown the FCF, set up with a straight, skewed or meandering channel within it, one with a roughened flood-plain and one with a smooth floodplain. Figure 6.17 shows an example of overbank open channel flow with $Dr = 0.250$ (experiment 080501), comparable to the duct flow examples in Figures 6.15 & 6.16, with a rectangular shaped main channel with vertical walls ($s = 0$). It illustrates how the boundary shear stresses vary laterally, decreasing from a value of $\tau_b \approx 1.6$ Nm^{-2} in the main channel to $\tau_b = 0.634$ Nm^{-2} on the floodplain. This reduction is due to the strong planform vorticity at the edge of the floodplain, as illustrated in Figures 2.15, 3.5 & 3.6. Figure 6.18 illustrates some numerical simulations of both U_d and τ_b for an experiment in Series 2 (020501) with a trapezoidal main channel and s = 1.0, taken from Knight & Abril (1996). From such data it is evident that on the flood plains

$$\frac{\partial \tau_b}{\partial y} = 0 : \frac{\partial U_d}{\partial y} = 0 \text{ and that } \tau_b \neq \rho g (H - h), \text{ where (H-h) is the depth on the}$$

floodplain.

By re-arranging (6.7) & (2.58) this might be expected, since

$$\rho g H S_0 - \tau_b \left(1 + \frac{1}{s^2} \right)^{1/2} = \frac{\partial}{\partial y} \left\{ H \left((\rho U V)_d - \overline{\tau}_{yx} \right) \right\} \tag{6.61}$$

This shows that τ_b will differ from the standard two-dimensional value ($\rho g H S_0$) due to the additional stresses arising from the effects of secondary flow and lateral shear. The RHS of (6.61) is clearly non-zero whenever either or both of these effects are present. Figure 6.19 illustrates one complete set of non-dimensionalised data boundary shear stress (Series 2). In every case the non-dimensionalised τ_b values decrease systematically, but are all non zero. See Shiono & Knight (1991) for further details.

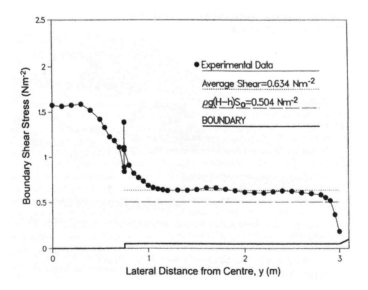

Figure 6.17 Lateral distribution of boundary shear stress for overbank flow (experiment 080501).

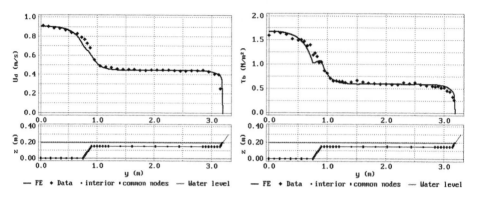

Figure 6.18 Overbank flow simulations of FCF experiment 020501 (Knight & Abril, 1996).

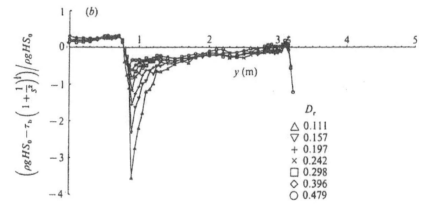

Figure 6.19 Lateral distribution of non-dimensional boundary shear stresses for overbank flow, FCF
Series 2 (from Shiono & Knight, 1991).

The depth-mean apparent shear stress acting on a vertical interface, $\overline{\tau}_a$, at any
distance y from the centre ($y = 0$), was calculated in each experiment by integrating
(6.61) in the form of (6.62):

$$\overline{\tau}_a = -\frac{1}{H}\int_0^y \left[\rho gHS_0 - \tau_b\left(1+\frac{1}{s^2}\right)^{1/2}\right]dy \qquad (6.62)$$

One set of data for $\overline{\tau}_a$ are shown in Figure 6.20, using the boundary shear data in
Figure 6.19. This particular apparent shear stress has two quite distinct components,
one arising from secondary flows and the other from turbulence. In order to separate
these out, measurements were made of the Reynolds stresses, τ_{yx} and τ_{zx} throughout
the flow field. The lateral distributions of individual τ_{yx} are shown in Figures 6.21
& 6.22 for various z values for relative depths of 0.152 and 0.250 respectively. Turbulence measurements were only made in the FCF for just the four lowest overbank

depths tested in each series, as the shear layer is the strongest at these low floodplain depths. These data from Series 1–15 of Phase A also illustrate the value of laboratory experiments focused on a specific set of objectives and how they can be used in developing numerical models.

Values of τ_{yx} were then depth-averaged over each vertical to give values of $\bar{\tau}_{yx}$, as shown in Figure 6.23. The values are approximately zero for $0 < y < 0.6$ m (in the main channel) and $y > 0.6$ m (on the floodplain), but increase rapidly to reach a maximum at the edge of the beginning of the floodplain at $y = 0.9$ m, increasing as the depth,

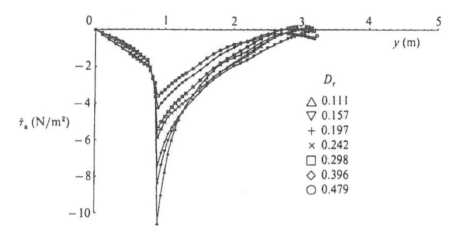

Figure 6.20 Lateral distribution of depth-mean apparent shear stresses for overbank flow, FCF Series 2 (from Shiono & Knight, 1991).

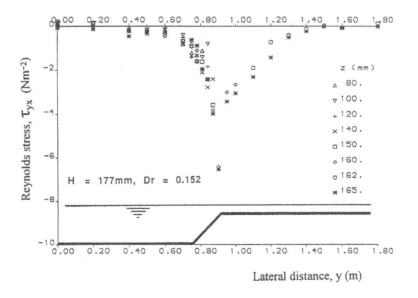

Figure 6.21 Lateral variation of τ_{yx} near the main channel/floodplain for FCF Experiment 0203010 (after Knight & Shiono, 1996).

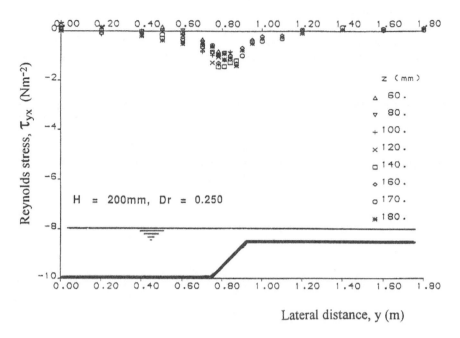

Figure 6.22 Lateral variation of τ_{yx} near the main channel/floodplain for FCF Experiment 0205010 (after Knight & Shiono, 1996).

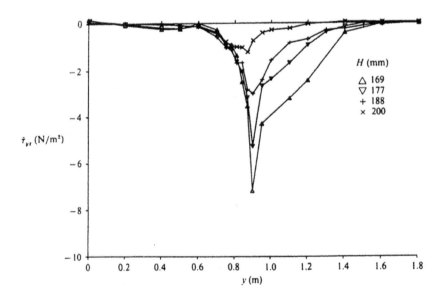

Figure 6.23 Lateral variation of depth-averaged Reynolds stresses, $\overline{\tau}_{yx}$, for H = 169–200 mm in the FCF experiments, Series 02 (Knight *et al.*, 2009).

H, decreases. These depth-averaged values were then inserted into (6.63) to establish values for the stress term $(\rho UV)_d$ and the shear force term per unit length, $H(\rho UV)_d$ at each lateral position. These are shown in Figures 6.24 & 6.25 respectively.

$$-(\rho UV)_d = -\frac{1}{H}\int_0^y\left[\rho gHS_0 - \tau_b\left(1+\frac{1}{s^2}\right)^{1/2}\right]dy - \overline{\tau}_{yx} \tag{6.63}$$

Figure 6.24 shows that the maximum value of the apparent shear stress $(\rho UV)_d$ occurs again at the edge of the floodplain ($y = 0.9$ m) and that it decreases almost linearly with y for all relative depths, Dr, shown. This may indicate the presence of just one large planform secondary flow on the flood plain, almost extending to its outer limit, as indicated by the sketches in Figures 3.5 & 3.10. In the main channel the decrease is again almost linear. However in the corner region ($0.75 < y < 0.9$ m) it is more complex, no doubt due to streamwise vorticity in the side sloping region, as observed earlier in wind tunnel experiments shown in Figures 6.15 & 6.16.

The lateral variation of $(\rho UV)_d$ with y is important as the secondary flow term, Γ, is defined by its differential, as proposed in (2.60), (3.23) & (6.9). Figure 6.24 shows that the variation of $(\rho UV)_d$ with y is approximately linear in the main channel and on the floodplain, with the sign of $\partial(\rho UV)_d / \partial y$ being negative in one region and positive in the other. It was the perceived linearity in these distributions that formed the basis of the assumption behind (2.60), (3.23) & (6.9).

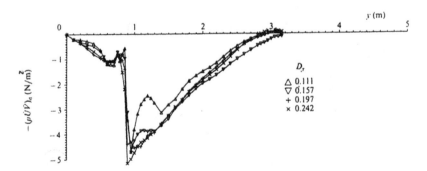

Figure 6.24 Lateral variation of apparent shear stress $(\rho UV)_d$ for different relative depths, Dr [= $(H-h)/H$], in the FCF experiments, Series 02 (after Shiono & Knight, 1991).

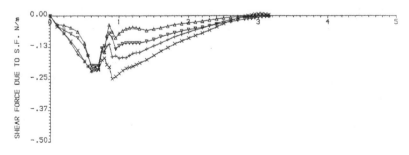

Figure 6.25 Lateral variation of apparent shear force per unit length, $H(\rho UV)_d$ for different relative depths, Dr [= $(H-h)/H$], in in the FCF experiments, Series 02 (after Shiono & Knight, 1991).

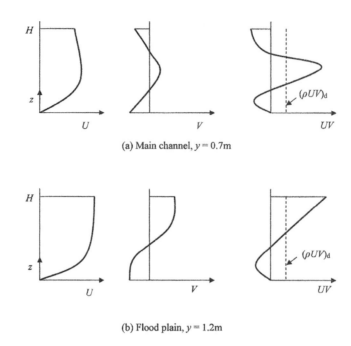

(a) Main channel, $y = 0.7$m

(b) Flood plain, $y = 1.2$m

Figure 6.26 Vertical distributions of primary and transverse velocities, U & V with $(\rho UV)_d$.

The effect of secondary flow cells on the vertical distributions of horizontal veloc-ity, U, and transverse velocity, V, in the vicinity of the main channel side slope is shown in Figure 6.26 for FCF experiment 020501. In this case, $(\rho UV)_d$ is positive at both locations ($y = 0.7$ & 1.2 m), which is consistent with the data already shown in Figure 6.24. The streamwise velocity, U, is generally always positive, with values larger near the free surface than near the bed, and the transverse velocity, V, is generally only a fraction of U, and may be either positive or negative at different z values. The sign of the depth-averaged term $(\rho UV)_d$ therefore depends not only on the distribution of V over the depth, but is also be linked the rotational sense of the secondary flow cell.

In the case of inbank flow in a simple trapezoidal channel, as already shown in Figure 2.33, the secondary flow cells are much stronger and the sign of $(\rho UV)_d$ var-ies alternately in the transverse direction, as adjoining cells rotate either clockwise or anti-clockwise, as illustrated in Figure 6.27. See Omran (2005) and Knight, Omran & Tang (2007) for further details.

The effect of these secondary flow cells on the depth-mean velocity is clearly important and the experimental data from both overbank and inbank flow studies suggests that the Γ term might adequately represent both planform and the stream-wise vorticity in a one-dimensional model, as commented upon earlier, and already used in Chapter 2, in the section dealing with flow in simple engineered channels (Section 2.3 and Figures 2.28 & 2.33).

It follows from detailed numerical simulations of all the FCF data by Knight & Abril (1996) and Abril & Knight (2004), and the evidence presented in Figures 6.21 &

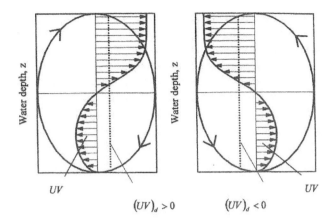

Figure 6.27 Sign of the depth-averaged term, $(UV)_d$ (after Knight, Omran & Tang, 2007).

6.22, that the influence of $\overline{\tau}_{yx}$ within the shear layer is mainly significant at the main channel/floodplain interface. This also agrees with the three-dimensional large eddy simulations (LES) by Thomas & Williams (1995). The value of the term $\partial\overline{\tau}_{yx}/\partial y$ in (6.61) may be considered to be negligible outside $0.6 < y < 1.6$ m. Consequently on the horizontal floodplain, with $1/s = 0$, and assuming average lateral values of boundary shear, τ_{avg}, and of the secondary flow contribution, Γ^*_{avg}, then (6.61) may be written as

$$\rho gHS_0 - \tau_{avg} = \frac{\partial}{\partial y}\left\{H\left(\rho UV\right)_d\right\} = H\Gamma^*_{avg} \qquad (6.64)$$

where

$$\Gamma^*_{avg} = \frac{\Gamma}{H}\ (\text{i.e. } \Gamma\ /\ \text{local depth}, H, H-h \text{ or } \xi \text{ in SKM}) \qquad (6.65)$$

Substituting $\tau_{avg} = k\tau_{2d} = k\rho gHS_0$ in (6.64) gives
$H\Gamma^*_{avg} = \rho gHS_0 - k\tau_{2d} = \rho gHS_0(1-k)$ hence

$$\Gamma^*_{avg} = \rho gS_0(1-k) \qquad (6.66)\ \&\ (3.24)$$

Figure 6.28 shows that the behaviour of k depends on both the relative depth and the geometry of the floodplain, with the main channel values, k_{mc}, remaining fairly constant with depth for all the FCF Series shown. For relative depths greater than 0.2, the average values of k_{mc} & k_{fp} are 0.85 & 1.25, respectively. The secondary flow term may therefore be roughly estimated from

$$\Gamma^*_{mc} = 0.15\rho gS_0 \quad \text{and} \quad \Gamma^*_{fp} = -0.25\rho gS_0 \qquad (6.67)$$

or

$$\Gamma_{mc} = 0.15H\rho gS_0 \quad \text{and} \quad \Gamma_{fp} = -0.25(H-h)\rho gS_0 \qquad (6.68)\ \&\ (3.24)\text{--}(3.26)$$

It should be noted that these coefficients are only applicable for channels with homogeneous roughness and for relative depths greater than 0.2. Figure 6.28 indicates that Γ^*_{fp} values should be increased if this method is used for lower relative depths for constant depth floodplain. Experiments also suggest that the main channel Γ^*_{mc} values need to be increased for those cases in which the floodplain is rougher than the main channel, as shown by Abril & Knight (2004) in their Figure 13. Figure 6.29 shows one comparison between experimental and numerical simulations of these k factors, and how they can assist in estimating the boundary shear stresses on floodplains reasonably well over the whole relative depth range. Details of the numerical simulations are given in Abril & Knight (2004. Further laboratory and field work is obviously needed to understand how Γ^*_{fp} values vary in natural rivers with floodplains that have a crossfall, i.e. where the depth varies laterally, and also where there is heterogeneous roughness between main channel and floodplain.

Having dealt with the lateral variations of velocity, boundary shear stress, Reynolds stresses, τ_{yx}, and depth-averaged Reynolds stresses, $\overline{\tau}_{yx}$, it is now appropriate to consider the vertical variation of the Reynolds stresses, τ_{zx}, in order to ascertain the relative effects of both secondary flows and vertical shear. One set of data are shown in Figures 6.30 & 6.31 for experiment 020301 in which the relative depth was 0.157. It can be seen from Figure 6.31, that the vertical distribution of τ_{zx} is approximately linear at the centerline of the channel ($y = 0$), as would be expected from either (6.6) or (6.69), taken from Knight & Shiono (1990), when derivatives with respect to y are zero.

$$\tau_{zx} = \rho g(H - z)\sin\theta + \int_z^H \frac{\partial \tau_{yx}}{\partial y}dz + \int_z^H \frac{\partial(\rho UV)}{\partial y}dz \qquad (6.69)$$

In the vicinity of the main channel river bank ($y = 0.81$ m), the distributions are highly non-linear, in agreement with the terms of the various terms in (6.69). At large values of y ($y > 1.5$ m), i.e. outside the region in which τ_{yx} is non zero, but still within the region where vorticity and secondary flow are influential, the vertical distributions revert to a linear form. Thus in the shear layer region it is clear that there

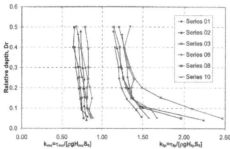

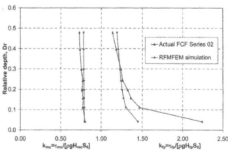

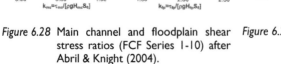

Figure 6.28 Main channel and floodplain shear stress ratios (FCF Series 1-10) after Abril & Knight (2004).

Figure 6.29 Simulated average main channel and floodplain shear stress ratios (FCF Series 2) after Abril & Knight (2004).

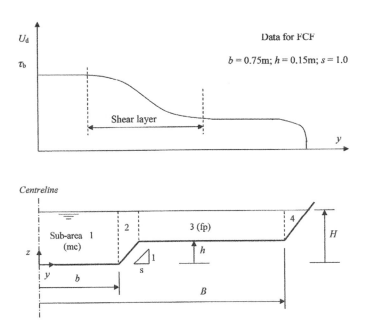

Figure 6.30 Notation used for overbank flow in Series 2 of the FCF experiments.

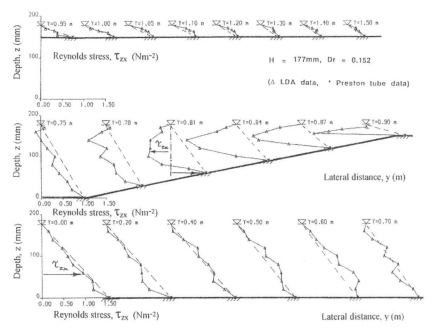

Figure 6.31 Reynolds stresses, τ_{zx}, in the FCF experiment 020301 (H = 0.1778 m, Dr = 0.157) in sub areas 1 & 2 and initial portion of 3, as shown in Fig. 6.30 (after Knight & Shiono, 1996). N.B. * = Preston tube measurement of τ_b.

are strong three-dimensional effects and that it is not possible to predict boundary shear stress from simply assuming logarithmic velocity profiles over the depth in that region.

For example, it should be noted that the τ_{zx} values mid way along the sloping main channel side wall at $y = 0.81$ m are both negative and positive within the same vertical. This is very different from the linear distributions associated with the standard logarithmic velocity profile, when extrapolating the vertical distributions of τ_{zx} down to the bed at $z = 0$ gives the boundary shear stress. It was therefore both surprising and encouraging in the experimental work that when the Reynolds stress distributions of τ_{zx} were extrapolated down to the bed at $z = 0$, as shown in Figure 6.31, the local boundary shear stress values obtained by extrapolation agreed with the values measured independently by the Preston tube method, marked in the Figure with the symbol * at the bed. The two sets of measurements agree well, giving confidence in both sets of data. Experimental errors in the measurement of any parameter in the FCF work were strictly controlled and monitored daily throughout the project and are described elsewhere. Generally errors in the measurement of water surface slopes were kept within 1–2% limits to ensure uniform flow, discharges were checked immediately after each experiment by comparing the values from installed orifice plates with the value obtained by numerical integration of all the local point velocity data (normally kept below 2–3%) and likewise the overall mean boundary shear stress ($\tau_o = gRS_0$) was compared with the mean value obtained by laterally integrating all the local boundary shear stresses, τ_b, around the wetted perimeter measured by a Preston tube (normally kept within 3–6%). In this way, by careful planning, control and daily analysis only a few experiments had to be repeated when these limits were not achieved at the first attempt.

Strict control of the errors in any of the turbulence measurements was also of paramount importance, particularly the separation of Reynolds stresses, τ_{zx} and τ_{yx}, from the large scale planform vortices forming in the shear layer. This can be achieved by using low pass and high pass filters to separate out and distinguish between high frequency fluctuations as intensive turbulence, and low frequency fluctuations as large scale planform eddies. The Reynolds stresses, τ_{yx} and τ_{zx}, shown in Figures 6.21 and 6.31 were however not calculated using such separation filters. Although low frequency fluctuations can be observed in time series at the measurable depth 12 mm below the water surface by LDA (see Knight and Shiono, 1990), more larger scale of horizontal eddies would be expected nearer the water surface according to visualizations during floods in Tone River and Ikeda experiments that large planform eddies are extended to the whole floodplain width on water surface. To complete the analysis of such combined fluctuations accurately, including small and large scales of horizontal eddies on the surface, it is necessary to capture all fluctuations over the whole of the water depth, crucially including near or on the water surface, and then to separate out low and high frequencies for Reynolds stresses. In recent years, measurement techniques have been enhanced/improved to capture micro and macro movement of flow, such as Particle Image Velocimetry (PIV) which can be used to measure all fluctuations. It is interesting to see the contribution of each one to the Reynolds stresses, hence the Reynolds stresses can be separated at least into two components for use in SKM.

Having now explained both the theoretical and physical background to the SKM, as well as the CES, it is hoped that the reader will be able to understand the basis of these two tools. It appears that the SKM is able to deal with some complex 3D flow physics, having taken particular care over the depth-averaging of key physical parameters used within the model. It is a fairly simple lateral distribution model, based on depth-averaged velocities, but despite its shortcomings, can give good estimates for the lateral distributions of depth-averaged velocity, U_d, and boundary shear stress τ_b, for overbank flow in moderately straight reaches of river channels with any cross section, as indicated by Figure 6.18. This is of considerable practical use in estimating the stage-discharge relationship, H v Q, for rivers in flood and in the analysis of other hydraulic phenomena, as explored further in the following sections.

6.3   BOUNDARY CONDITIONS

The application of either the SKM or the CES methodology requires the river cross section to be discretised into a number of fluid elements, typically rectangular, trapezoidal or triangular in shape, arising from vertical slicing, as shown in Figures 2.45, 3.30, 3.33 & 3.38. This mimics the commonly adopted approach when river gauging by velocity measurements over a number of fixed verticals, and is appropriate when applying a depth-averaged model. The division of a prismatic channel into a number of discrete panels or domains should ideally take into account any significant geometrical characteristics (e.g. main channel or floodplain), the roughness properties in different regions (e.g. one type or variable) and the presence of different types of flow structure (as described in Section 2.3). As well as selecting the appropriate number of panels and coefficients for resistance, lateral shear and secondary flow within each element or panel, care also needs to be taken over the boundary conditions between each panel and at the channel edges.

These are outlined for the CES methodology in Section 3.2.3 and (3.44) as being

$$q = 0 \text{ (at boundary nodes);} \quad q_i = q_{i+1} \text{ (at element boundaries);} \tag{6.70}$$

In the SKM, they are based on the depth-averaged velocity, U_d, and not q ($= U_d H$), and take into account the lateral shear forces between panels at each interface ($= \overline{\tau_{yx}} H$), and are:

$$U_d = 0 \text{ (at boundary nodes);} \qquad U_{di} = U_{di+1} \text{ (at element boundaries);}$$
$$H^{(1)} \overline{\tau_{yx}^{(1)}} = H^{(2)} \overline{\tau_{yx}^{(2)}} \text{ (forces at element boundaries)} \tag{6.71}$$

These are compatible when q, U_d & H are smooth continuous functions across the channel if the boundary is continuous. Where the boundary is discontinuous, as for example for a vertically stepped boundary, then an appropriate force balance is required as shown by (6.76) and discussed later

The solution technique adopted in the CES requires an iterative strategy when determining q for each panel, whereas in the SKM, U_d is determined for each panel directly, providing that the interfacial shear stresses that are based on the Reynolds stresses and the three calibration parameters are known for all panels. In the absence of secondary flows, the first requirement is always satisfied, and where secondary flows are present, their contribution may be allowed for by compensatory adjustment of individual calibration parameters. The second requirement, that of knowing the 3 calibration parameters, f, λ & Γ, in each panel before beginning a simulation, which is also required in the CES, is one that is strictly impossible to satisfy, since the only reliable data on boundary shear stress come from experimental studies, such as those undertaken in the FCF, as described in Section 1.5. It is therefore inevitable that both methods will require some iterative procedure at the calibration stage, in which the values of the 3 calibration parameters, f, λ & Γ, in each panel are adjusted to simulate satisfactorily given lateral distributions of both U_d & τ_b simultaneously. Furthermore, it is recognised that the choice of values may compensate for model structural weaknesses in both the CES and the SKM. Indeed, the input of poorly known parameters, such as the dimensionless eddy viscosity, makes it difficult to establish whether either method is more physically realistic than any other model. The SKM was developed primarily to give accurate representations of the stage-discharge relationship, secondly to give good estimates for the lateral distribution of depth-averaged velocity, and then thirdly to give good representations of the lateral distribution of boundary shear stress. This might form a convenient hierarchy for ranking the outputs from different models or methods, since simulating the lateral distributions of τ_b well are more difficult to achieve than distributions of U_d, due to the square law that inevitably shows up poor simulations more starkly. The theoretical basis behind (2.58), as outlined in Section 6.2, was therefore formulated with these particular objectives in mind.

Accordingly, with lateral distributions of boundary shear stress in mind, the shear force term $(H\bar{\tau}_{yx})$ was developed from (2.59) or (A4.9) as follows:

$$H\bar{\tau}_{yx} = H\rho\nu_t\frac{\partial U_d}{\partial y} = H\rho\{\lambda U_*H\}\frac{\partial U_d}{\partial y} = \rho\lambda H^2 U_*\frac{\partial U_d}{\partial y} = \rho\lambda H^2\left\{\left(\frac{f}{8}\right)^{1/2}U_d\right\}\frac{\partial U_d}{\partial y}$$

Hence

$$H\bar{\tau}_{yx} = \rho\lambda H^2\left(\frac{f}{8}\right)^{1/2}U_d\frac{\partial U_d}{\partial y} \quad \text{or} \quad H\bar{\tau}_{yx} = \frac{1}{2}\rho\lambda H^2\left(\frac{f}{8}\right)^{1/2}\frac{\partial U_d^2}{\partial y} \tag{6.72}$$

Thus, where H & U_d are continuous at an interface, this boundary condition may be stated as:

$$\left[\lambda\left(\frac{f}{8}\right)^{1/2}\frac{\partial U_d}{\partial y}\right]^{(1)} = \left[\lambda\left(\frac{f}{8}\right)^{1/2}\frac{\partial U_d}{\partial y}\right]^{(2)} \tag{6.73}$$

or more succinctly by

$$\left(\mu\frac{\partial U_d}{\partial y}\right)^{(1)} = \left(\mu\frac{\partial U_d}{\partial y}\right)^{(2)} \tag{6.74}$$

where

$$\mu = \lambda \sqrt{f/8} \tag{6.75}$$

It therefore follows that if $\mu_1 = \mu_2$, the velocity gradients in adjoining panels at an interface will be the same. Where f & λ are different in adjoining panels, then $(\partial U_d/\partial y)^{(1)} \neq (\partial U_d/\partial y)^{(2)}$, and there is not a smooth transition at the interface boundary.

Where H is not continuous, then the force balance in (6.71) should be modified to become

$$H^{(1)} \overline{\tau_{yx}^{(1)}} = H^{(2)} \overline{\tau_{yx}^{(2)}} \pm h\tau_w$$
$$H^{(1)} \overline{\tau_{yx}^{(1)}} = \left(H^{(1)} \pm h\right) \overline{\tau_{yx}^{(2)}} \pm h\tau_w \tag{6.76}$$

where τ_w is the mean shear stress on the vertical wall (step either up or down). Since τ_w is not usually known *a priori* before a simulation, regardless of whether SKM or CES is used, a method to overcome this issue needs to be supplied, as shown later. Furthermore, there is an additional difficulty with vertical walls at the extreme boundaries of a cross section, where a possible conflict arises in the fact that $U_d = 0$, but the shear stress and shear force are non-zero. This boundary shear force on a vertical wall arises from the resistance to the flow provided by the wall, in which case it must be determined from the results of the simulation or, alternatively, it might be directly imposed at the boundary location, in which case it must be specified as a boundary condition before beginning the simulation. Before dealing with this issue in Section 6.3.3, consider the application of the boundary conditions, (6.71) & (6.75), to different types of flow in channels of various shapes where q, U_d & H are smooth continuous functions across the channel.

6.3.1 Symmetric flow

For symmetric flow (either imposed or assumed) in a trapezoidal compound channel, as shown in Figure 6.32, then with 4 panels, 8 boundary conditions are required.

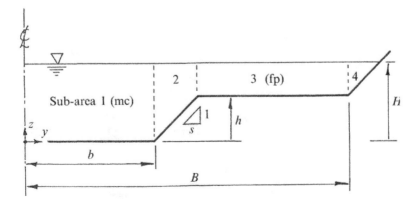

Figure 6.32 Notation and panel numbering for overbank flow in a compound trapezoidal channel.

The boundary conditions are:

$U_d = 0$ (at edge of free surface on floodplain side slope, sub-area 4);
$\partial U_d / \partial y = 0$ (at centerline, due to imposed symmetry of flow, sub-area 1);
$U_{di} = U_{di+1}$ and $\mu_i \partial U_{di} / \partial y = \mu_{i+1} \partial U_{di+1} / \partial y$ (at the 3 internal joints between panels)

These will produce the same results as (6.70) in the CES, provided U_d and H are continuous with y.

6.3.2 Asymmetric flow

For asymmetric flow in a simple trapezoidal channel with differing side slopes, as shown in Figure 6.33, then with 3 panels, 6 boundary conditions are required.
The boundary conditions are:

$$U_d^{(2)}|_{\xi_2=0} = 0 \quad \text{and} \quad U_d^{(3)}|_{\xi_3=0} = 0$$

$$U_d^{(1)}|_{y=b} = U_d^{(2)}|_{\xi_2=H} \quad \text{and} \quad \mu_1 \frac{\partial U_d^{(1)}}{\partial y}\bigg|_{y=b} = \mu_2 \frac{\partial U_d^{(2)}}{\partial y}\bigg|_{\xi_2=H}$$

$$U_d^{(1)}|_{y=-b} = U_d^{(3)}|_{\xi_3=H} \quad \text{and} \quad \mu_1 \frac{\partial U_d^{(1)}}{\partial y}\bigg|_{y=-b} = \mu_3 \frac{\partial U_d^{(3)}}{\partial y}\bigg|_{\xi_3=H}$$

in which the local depths on the sloping side walls are $\xi_2 = H - \dfrac{y-b}{s_2}$, and $\xi_3 = H + \dfrac{y+b}{s_3}$

It should be noted that $A_4 \xi^{-(\alpha_1+1)}$ of the analytical solution in the sloping bed case does not exist at the edges, since $\xi = 0$ and $\eta = 0$, there being no secondary flow there also. It should also be noted that in panel 1 it is not assumed that $\partial U_d / \partial y = 0$ at $y = 0$ (i.e. at the centerline, as drawn) since the channel shape is not symmetric and therefore it is unlikely that the flow conditions will be similar in domains

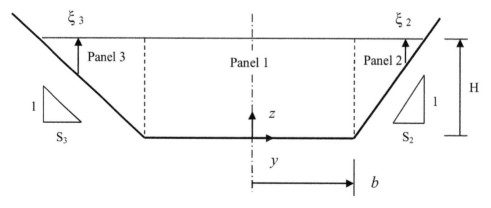

Figure 6.33 Trapezoidal channel with different side slopes.

2 & 3, and definitely not so if one wall is rougher than the other, leading to $f_2 \neq f_3$, even before considering possible differences that there are likely to be in the λ & Γ values between panels 2 & 3. Despite this, it can be seen that two boundary conditions are required for each panel. Although the boundary conditions concerning the depth-averaged velocity and its lateral gradient at each panel junction, as originally proposed by Shiono and Knight (1988, 1990), are still sometimes applied in order to obtain numerical solutions, as for example by Ervine *et al.* (2000), their potential limitations need to be appreciated.

Where there are discontinuities in the roughness distribution across a river, but no step changes in depth, there will be a difference in the values of f & λ in adjoining panels, which (6.75) indicates will affect the μ values in each panel. This then implies that there will be a difference in the lateral gradients of U_d at panel interfaces, since in these cases $\mu_i \neq \mu_{i+1}$. Furthermore, if the roughness varies laterally with y, as it usually does in rivers, then the f values will vary between panels and lead to a step change in boundary shear stress, as a consequence of defining the bed shear in terms of U_d^2, as indicated by (2.59) or (2.61). Since U_d is always the same at the interface between adjoining panels, except where internal vertical walls exist, τ_b will vary in step changes in direct response to any lateral change in f between adjoining panels. The physics of flow at interfaces between rough and smooth surfaces, side by side on a horizontal bed, is known to be complex. Strong secondary flow cells are known to form at the discontinuity in roughness, leading to anisotropy in the turbulence and distortion of the isovels. Some experimental results on flows with heterogeneous roughness are shown in Figures 6.34–6.36, taken from Tominaga & Nezu (1991). Figure 6.34 shows a 400 mm wide flume with 5 mm square strips roughening one half of the width. Two series of tests were undertaken with the lateral spacing of the strip roughness, L/k, varied from 4 to 8, and measurements made of all three components of velocity, turbulence intensities and

Reynolds stresses. It was estimated that the equivalent sand roughness size, k_s, for the strip roughness pattern was approximately 45 mm, making $k_s/k = 9$. This is roughly comparable with the value shown in Figure 10 of Knight & Macdonald (1979) for $L/k = 8$ and the listed H/k values of between 8 & 16. Figure 6.34 shows isovels for the primary mean velocity flow in experiment A3, normalized by U_{max}, which indicate significant lateral gradients for all values of y/H, as might be expected. Figure 6.35 shows the pattern of strong secondary flow cells in the shear layer, with streamwise vorticity that direct flows downwards at the boundary between smooth

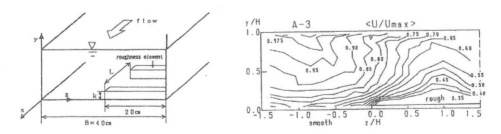

Figure 6.34 Heterogeneous roughness pattern and corresponding isovels (from Tominaga & Nezu, 1991).

and rough beds and upwards at $z/H = -1.5$, regardless of roughness spacing. As a result of the downward flow, the bed shear stresses increase greatly at the smooth/ rough boundary, peaking just inside the roughened bed portion.

Figure 6.36 shows the spanwise distribution of normalised Reynolds stresses depth averaged over the middle depth zone ($0.4 < y/H < 0.7$ in their notation). These correspond to $\overline{\tau}_{yx}$ in the notation used elsewhere in this book and are somewhat comparable with the distributions shown in Figure 6.20. The data from Tominaga & Nezu (1991) underscore the need to appreciate and understand the physical aspects of

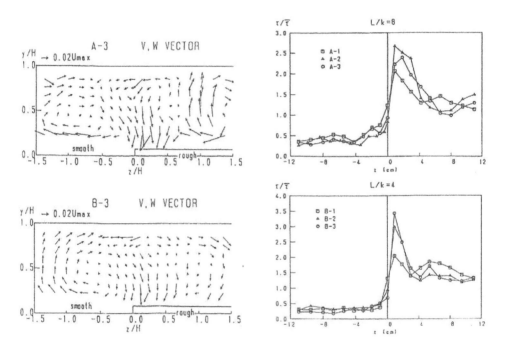

Figure 6.35 Secondary flow cells and boundary shear stresses (from Tominaga & Nezu, 1991).

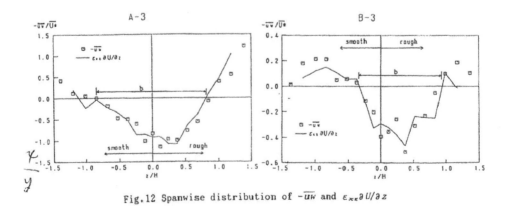

Fig. 12 Spanwise distribution of $-\overline{uw}$ and $\varepsilon_{xx}\partial U/\partial z$

Figure 6.36 Lateral shear stresses normalised by shear velocity (from Tominaga & Nezu, 1991).

flow structures in open channel flow before making any of the assumptions that are inevitably required in developing a simple depth-averaged flow model. Depth averaging will always gloss over certain flow details, but it is incumbent on any model developer to appreciate the approximations that must inevitably be made and to base their model as closely as possible on the flow physics. For further details on strip roughness see Knight & Macdonald (1979) and Morvan et al. (2008). It also reinforces the lesson that extra care needs to be taken whenever accurate estimates for the lateral distributions of U_d and τ_b are required in such cases.

One approach which does not require full 3D turbulence modelling might be to insert more panels in the region of interest, in order to produce smoother lateral profiles of τ_b in an attempt to represent the finer details of the flow. Another approach, and one that is adopted by Knight et al. (2007), is to vary the value of f linearly across individual panels, maintaining their mean values, which although valid in a finite difference method, but not strictly valid in the analytical solution where f should be constant. This will smooth out any saw-tooth pattern in the lateral distributions of τ_b, and can be very effective, as already shown in Figures 2.25–2.29 and 2.31–2.34. However, this procedure should be used cautiously.

6.3.3 Vertical walls

The issue of how to deal with cross sections which have step changes in the bed level, or vertical walls, is now considered, as these may occur in some engineered river channels and frequently in laboratory flumes. Lai (1987) and Rhodes (1991) performed many series of experiments in compound ducts, to study the variation of depth-averaged velocity, U_d, and boundary shear stress, τ_b across different floodplain/main channel interfaces, using small and large wind tunnels respectively. Measurements were focused on determining as accurately as possible the lateral distributions of τ_b and U_d, from which the gradients, dU_d/dy and dq/dy were obtained, as well as the distribution of the apparent shear force (ASF). The experimental data were obtained with a very fine lateral spacing interval, Δy, in the vicinity of the main channel/floodplain interface, in some cases as small as 1 mm. Furthermore, the vertical interval, Δz, was such that the U_d values were obtained from between 6 and 17 local streamwise values averaged over the depth, depending on the geometry of the cross section. These very high quality sets can therefore be used to illustrate some of the physical characteristics of boundary conditions between adjoining panels which might be adopted in depth-averaged flow models.

Figure 6.37 shows results from Experiment 14 (Rhodes, 1991), in which a large compound duct (1231.5 mm × 300 mm × 25 m) with an internal wall element with a vertical side, approximately 21 mm high, was used, giving a step in the bed elevation at approximately $y = 400$ mm from one side. These show that only the unit discharge q is continuous at the main channel/floodplain, interface, as would be expected from Samuels (1989), whereas U_d, dU_d/dy & H are not. Although q is continuous, there is a relatively large change at the interface position. See Knight, Omran & Abril (2004) for more details.

Figure 6.38 shows results from Experiment 20 in which the same duct had one sloping side wall installed at the same position as in Experiment 14. These show that U_d, H, q and dU_d/dy are continuous at the main channel/floodplain interface.

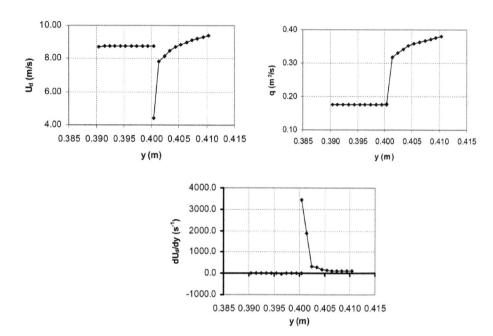

Figure 6.37 Variation of U_d, q & $\partial U_d/\partial y$ from flow in a rectangular compound duct from experiment 14 by Rhodes (1991), after Knight et al. (2004).

However, there is a change in the gradient at $y = 0.412$ m, but if the data over just 5 mm between 0.410 m and 0.415 m were missing, the change in dU_d/dy might not be noticed. The original boundary conditions in the SKM regarding the continuity of U_d & dU_d/dy were therefore not unreasonable to specify, although it is better to adopt those in (6.75) with μ factors, since slight changes in f & λ will produce slight changes in dU_d/dy, as shown in Figure 6.38.

Figure 6.39 shows more clearly that dU_d/dy is not equal at an interface where there is a vertical step change in depth, shown here between the main channel and floodplain panels joined by an internal vertical wall. However, as in the previous example, there are some circumstances in which it might appear to be valid, even near vertical walls, either where insufficient care is taken over acquiring data or where the data is 'smoothed' too much in any subsequent processing, as illustrated in Figure 6.39. The data are taken from Experiment 17 of Rhodes (1991), where velocity data in the main channel ($y > 0.400$ m) is depth-averaged in the vicinity of the step change of 20 mm in bed level, creating an internal vertical wall. Data were collected at increments of 1 mm either side from the step change. The measured velocity data, given by the 'Not smoothed' curve, shows a very marked decline very close to the step. However, if only three data points over the first 2 mm of the main channel were ignored, then the 'smoothed' trend line would result. This shows the level of accuracy required in experimental work in order to see physical details and why the continuity of q was chosen for the CES, as indicated by (6.70). Despite this, surprisingly, Figure 6.39 shows that smoothing U_d near the step, might not be too unreasonable a decision to make, in the absence of actual data. It is better however, to either take the

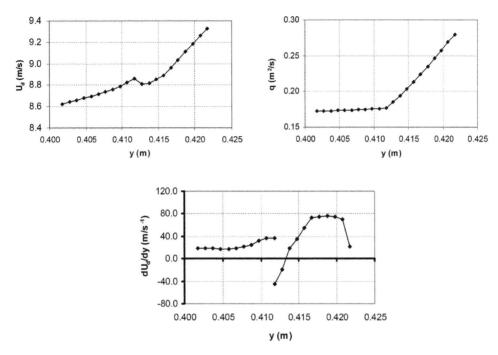

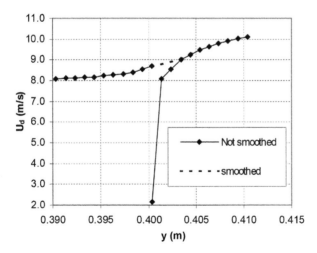

Figure 6.38 Variation of U_d, q & $\partial U_d/\partial y$ from flow in a compound duct with sloping sidewalls from experiment 20 by Rhodes (1991), after Knight *et al.* (2004).

Figure 6.39 Variation of U_d very close to a vertical step in a compound rectangular duct from experiment 17 by Rhodes (1991) after Knight *et al.* (2004).

continuity of q at the step, as in CES, or to use μ factors to vary the gradients of U_d. It is essential however, to use the force balance in (6.76) in the SKM to get τ_b accurately on the bed and τ_w on any vertical wall. Since vertical walls are a common feature in many experimental laboratory flumes, in which experiments on sediment transport, erosion and deposition processes, and other associated sediment phenomena, are frequently carried out, there is a need to estimate bed shear stresses accurately on the bed of the flume, discounting the effects of the two sidewalls. The SKM is shown to be a good alternative to using standard sidewall correction procedures and is therefore dealt with in detail later in Sections 6.5 and 7.1.2.

Further information concerning boundary conditions in relation to the SKM may be found in Abril & Knight (2004) and Omran *et al.* (2008). Knight *et al.* (2004) and Omran (2005) investigated the variation of the apparent shear stress (ASS) across channels and showed that where a discontinuity exists in the bed levels, the discontinuity in the apparent shear force (ASF) at the floodplain/main channel interface is equal to the mean shear force on the vertical wall. This is illustrated in Fig. 3 of Knight *et al.* (2004) using data from Lai (1987). The apparent shear stress (ASS) at any point may be determined by applying (6.62) and the ASF is easily obtained by multiplying by the appropriate depth (ASF $= \overline{\tau}_a H$). Some depth-averaged apparent shear stresses have already been shown for overbank flow in Figure 6.20. As noted in the comments below (6.62) this ASS, and the corresponding ASF has two quite distinct components, one arising from secondary flows and the other from turbulence. These two components are often lumped together, as done by Knight & Demetriou (1983), Myers (1978) and Wormleaton (1988).

For the rectangular compound channel case with a vertical internal wall, the ASF affects the bed shear stresses, both in the main channel and on the floodplain. There is a reduction in bed shear stress compared with the 2-D flow case in the main channel, and an increase on the floodplain. This can be seen in the bed shear stress distributions of Series 8 of the FCF data, as illustrated by Figure 6.17. The ASF at the interface demonstrates clear momentum transfer from the main channel to the floodplain.

6.3.4 Symmetric flow in a trapezoidal compound channel with a very steep internal wall

In order to illustrate how to model vertical walls in the SKM, Tang & Knight (2008) used a hypothetical symmetric compound channel with a very steep internal wall, as shown in Figure 6.40.

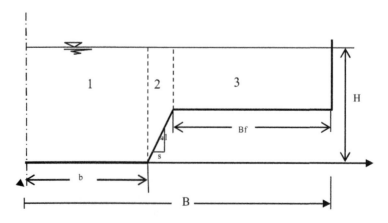

Figure 6.40 Symmetric compound channel with a very steep internal wall.

The channel cross section was modelled in two ways, one as shown, with 3 panels, and secondly as a rectangular compound with s = 0, hence just 2 panels, and applying (6.76). For simplicity, in both cases the flow was assumed to be symmetric. Thus with 3 panels and a sloping side wall, the boundary conditions are:

$\partial U_d / \partial y = 0$ (at centerline, $y = 0$, due to imposed symmetry of flow);
$U_d = 0$ (at edge of free surface on floodplain edge, $y = B$);
$U_{di} = U_{di+1}$ and $\mu_i \partial U_{di} / \partial y = \mu_{i+1} \partial U_{di+1} / y$ (at the 2 internal joints)

Alternatively, if this case was modelled as a rectangular compound channel, with 2 panels and a vertical wall, then the boundary conditions are:

$\partial U_d / \partial y = 0$ (at centerline, $y = 0$, due to imposed symmetry of flow);
$U_d = 0$ (at edge of free surface on floodplain edge, $y = B$);

$$\left(H\overline{\tau_{yx}}\right)^i_{y=b} + b\tau_w = \left(H\overline{\tau_{yx}}\right)^{i+1}_{y=b} \quad \text{and} \quad \mu_i \partial U_{di} / \partial y = \mu_{i+1} \partial U_{di+1} / \partial y \quad (6.77)$$

where τ_w is the mean shear stress on the vertical wall. In these simulations, Tang & Knight (2008) used the second form of (6.72) as

$$H\overline{\tau_{yx}} = \rho \lambda H^2 \left(\frac{f}{8}\right)^{1/2} U_d \frac{\partial U_d}{\partial y} \quad \text{or} \quad H\overline{\tau_{yx}} = \frac{1}{2}\rho \lambda H^2 \left(\frac{f}{8}\right)^{1/2} \frac{\partial U_d^2}{\partial y} \quad (6.72)$$

Thus the force balance (6.77) becomes

$$\left(\phi \frac{\partial U_d^2}{\partial y}\right)^{(1)}_{y=b} = \left(\varphi \frac{\partial U_d^2}{\partial y}\right)^{(2)}_{y=b} - b\tau_w \quad (6.78)$$

where

$$\phi = \frac{1}{2}\rho \lambda H^2 \sqrt{f/8} \quad \text{and} \quad \tau_w = \rho f_w U_{d(at\,y=b)}^2 / 8 \quad (6.79)$$

Table 6.1 lists 5 possible boundary conditions used to produce the various results in Figure 6.41. They are labelled [A]–[E], as in the original paper by Knight (2013b), with some good choices and others deliberately poor ones, in order to illustrate how different boundary conditions will affect the results. The results of applying them to the hypothetical case are presented in Fig. 6.40. It is clear that boundary condition [A] with the continuity of unit force is technically the most suitable, as would be expected from (6.72). A possible alternative to [A] is one in which an adjustment factor, φ, is selected to retain the usual μ factor approach. Then bc [A] and bc [B] are shown to give identical results, provided φ is chosen appropriately. The bc [C] is a composite of q and simple velocity gradient, $\partial U_d / \partial y$, equivalence, which is obviously incorrect. The bc [D] & [E] over-estimate or underestimate U_d in the main channel.

Because bc [A] suffers from the drawback that the wall shear stress needs to be known, the approximation is made that the friction factor in the main channel and the

Table 6.1 Possible boundary conditions for internal walls in rectangular compound channels.

Form	U_d or q continuity	U_d gradient or unit force continuity	Notes
[A]	$U_d^{(1)} = U_d^{(2)}$	$\left(\phi\dfrac{\partial U_d^2}{\partial y}\right)_{y=b}^{(1)} = \left(\phi\dfrac{\partial U_d^2}{\partial y}\right)_{y=b}^{(2)} - h\tau_w$	$\varphi = \dfrac{1}{2}\rho\lambda H^2 \sqrt{f/8}$ $\tau_w = \rho f_w U_d^2(y=b)/8$
[B]	$U_d^{(1)} = U_d^{(2)}$	$\left(\mu\dfrac{\partial U_d}{\partial y}\right)_{y=b}^{(1)} = \left(\varphi\mu\dfrac{\partial U_d}{\partial y}\right)_{y=b}^{(2)}$	$\mu = H^2\lambda\sqrt{f}$ with an adjustment factor φ
[C]	$[HU_d]^{(1)} = [HU_d]^{(2)}$	$\left(\dfrac{\partial U_d}{\partial y}\right)_{y=b}^{(1)} = \left(\dfrac{\partial U_d}{\partial y}\right)_{y=b}^{(2)}$	
[D]	$U_d^{(1)} = U_d^{(2)}$	$\left(\mu\dfrac{\partial U_d}{\partial y}\right)_{y=b}^{(1)} = \left(\mu\dfrac{\partial U_d}{\partial y}\right)_{y=b}^{(2)}$	$\mu = H^2\lambda\sqrt{f}$
[E]	$[U_d]^{(1)} = [U_d]^{(2)}$	$\left(\dfrac{\partial U_d}{\partial y}\right)_{y=b}^{(1)} = \left(\dfrac{\partial U_d}{\partial y}\right)_{y=b}^{(2)}$	

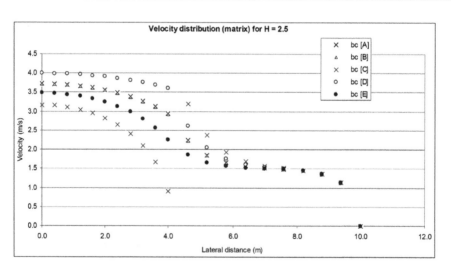

Figure 6.41 Effect of different boundary conditions on U_d for a symmetric rectangular compound channel for $H = 2.5$ m ($S_o = 0.001$, $b = 4$ m, $B = 10$ m, $h = 2$ m; $f_1 = f_w = 0.01$ & $f_2 = 0.02$; $\lambda_1 = 0.01$ & $\lambda_2 = 0.2$; $\Gamma_1 = 1.0$ & $\Gamma_2 = -0.75$), from Knight (2013b).

depth-averaged velocity at the interface can be used to estimate the wall shear force. In the subsequent model simulations, this was achieved by assuming the main channel friction factor could be used in (6.76) to determine the wall shear force.

Thus bc [A] was subsequently used in a 2 panel simulation of flow in the same hypothetical case as before and checked for several depths using a 3 panel trapezoidal

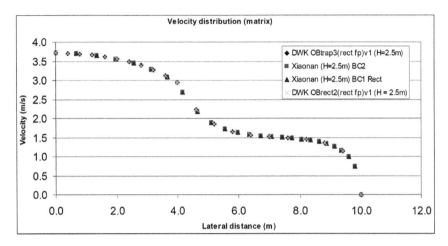

Figure 6.42 Comparison between U_d distributions for a trapezoidal compound channel with a nearly vertical internal wall with that for a rectangular compound channel with bc [A] for $H = 2.5$ m ($S_o = 0.001$, $b = 4$ m, $B = 10$ m, $h = 2$ m; $f_1 = f_2 = 0.01$ & $f_3 = 0.02$; $\lambda_1 = \lambda_2 = 0.1$ & $\lambda_3 = 0.2$; $\Gamma_1 = \Gamma_2 = 1.0$ & $\Gamma_3 = -0.75$), from Knight (2013b).

simulation with a very steep internal wall (s = 0.001), as shown in Figure 6.40. Figure 6.42 shows that the simulations in a 2 panel rectangular channel using bc [A] agree well with those with a steep internal wall.

The balance between satisfying both $U_d = 0$ and the shear wall shear force criterion at $y = B$, was investigated for simple channels by Chlebek & Knight (2006). For inbank flows this is more straightforward problem to solve since the bed shear force may be determined by integration of the bed shear stress distribution and then subtracted from the total shear force (= ρgAS_o) to obtain the wall shear force directly. For compound channels this is not possible, and alternative methods have to be found, as illustrated for an internal vertical wall.

Finally, there is the issue of establishing an automatic testing procedure for optimum parameter values and hence the solutions. This was undertaken, several years later, using multi-objective evolutionary algorithms, as shown by Sharifi *et al.* (2008 & 2009). The results of these optimisation techniques were then compared with the corresponding results undertaken using visual inspection of the various output graphs, and the efficacy of both methods assessed. Despite some advantages in numerical assessments, the focus on a single determinant for a multi-objective function may cause one to overlook some important physical feature in either model or data.

6.4   ANALYTICAL SOLUTIONS USING THE SKM

This section illustrates how the SKM may be used to give the lateral distributions of depth-averaged velocity and boundary shear stress, U_d & τ_b, for flows in various shaped channels. Some simple rectangular and trapezoidal shapes are considered first, in order to illustrate the solution procedure, followed by more complex shapes. Cross sections of natural rivers are schematised by a series of linear elements with the flow area changing markedly with stage, depending on whether the flows are inbank or overbank. Analytical solutions to the governing SKM equation, (2.58), are now presented for 12 cases, selected to illustrate the solution procedure to obtain the various coefficients, A_i, required in the velocity distribution equations for each panel. To make it easier to follow an individual case, the equations are numbered (1)-(n) in this particular section, and not sequentially as in the rest of the chapter. The accompanying Figures are likewise numbered according to the particular case. Some of the cases considered, together with some additional cases, are re-visited in Chapter 7, in a summarised form ready for use. They are given there without μ factors, in their original form, so as to satisfy the strict condition that λ & f should not vary within a panel, as required by the third differential term of (2.58). This implies that at every interface the velocity gradients are the same in adjoining panels, even where λ & f may differ between adjacent panels. This has been shown in Section 6.3 to be a reasonable approximation to physical reality, but where refined distributions of U_d & τ_b are required the SKM can be adjusted to suit particular conditions. For example, Figure 2.34 has already shown how schematising the cross section with the number and location of the secondary flows cells in mind, as well as linearly varying f within each panel, can give very smooth profiles of τ_b that match known boundary shear stress data. This forcing of a depth-averaged model to mimic 3D flow effects is admittedly an approximation, but justified on the grounds of simplicity and being practically useful. As illustrated in Sections 6.2, 6.3 & 6.5, the physics of the flow should always be understood first, followed by use of the most appropriate theoretical equations, in order to simulate any fluid flow.

Section 6.5 considers boundary shear stress distributions in open channel flow from a physical perspective, as these are often neglected by modellers in preference for just simulating the velocity distributions. However, achieving good simulations of the lateral distributions of bed shear stress, τ_b, is considerably more difficult to achieve than the depth-averaged velocity, U_d. Moreover, they are needed in the simulation of U_d, by virtue of using depth-averaged bed friction factors, f_b, as indicated by (2.61). Furthermore, a wall shear stress or force might be required as a boundary condition, and secondly the shear velocity (related to skin friction) is important in many sediment transport equations.

Section 6.6 shows how the analytical solutions presented may be extended to deal with roughness other than skin friction, by considering the additional drag forces that arise when trees are present on a floodplain. It also deals with other boundary roughness issues from a more general perspective, as roughness is usually of critical importance in the calibration stage of any model simulating flows in rivers. Chapter 7 builds on these introductory 15 examples, or case studies, and illustrates how the SKM may be applied to natural rivers and urban drainage systems, to estimate not only the stage-discharge relationship and velocity distributions, but to deal with other phenomena such as sediment motion and flood wave speed.

6.4.1 Flows in rectangular channels

Although the rectangular shape might be considered to be the simplest one to apply the SKM to, it is used here to illustrate boundary conditions and flow asymmetry.

Four cases are now considered, beginning with Case R1, which utilizes just one panel to represent half the cross section, assuming flow symmetry around the centre-line. Next, Case R2 considers a slightly more complex flow in which, for example, the roughness may vary laterally, and therefore two panels of unequal width are now used to represent half the cross section. Thirdly, Case R3, utilizes two panels of equal width, and shows how this may be derived directly or by simply making the panels in Case R2 of equal width. Finally, Case R4 using two unequal sized panels offset from the centre-line to simulate those cases where there is no flow symmetry. These 4 cases are labelled Cases R1 to R4 respectively, and are followed by similar cases for flows in trapezoidal and triangular channels (T1–T7), one for flow in an urban drainage channel (U1) and 3 for overbank flows in compound channels (C1–C3).

Case R1 Symmetric flow in a simple rectangular channel, using one panel only

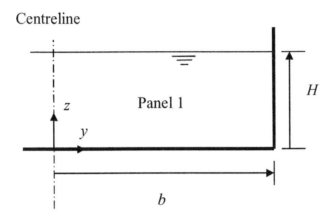

Figure R1 Rectangular channel with one panel representing half the cross section.

With a single panel, as shown in Figure R1, the lateral distribution of U_d will be given by:

$$U_d^{(1)} = \left[A_1 e^{\gamma_1 y} + A_2 e^{-\gamma_1 y} + k_1 \right]^{1/2} \tag{1}$$

where

$$k_1 = \frac{8gHS_0}{f_1}(1 - \beta_1); \quad \beta_1 = \frac{\Gamma_1}{\rho g H S_0}; \quad \text{and} \quad \gamma_1 = \sqrt{\frac{2}{\lambda}\left(\frac{f_1}{8}\right)^{1/4} \frac{1}{H}} \tag{2}$$

thus

$$U_d = \left[A_1 e^{\gamma_1 y} + A_2 e^{-\gamma_1 y} + \frac{8gHS}{f_1}(1 - \beta_1) \right]^{1/2} \tag{3}$$

bc (1)

$$U_d^{(1)} = 0 \quad \text{at } y = b \, (\text{wall}) \tag{4}$$

bc (2)

$$\frac{\partial U_d^{(1)}}{\partial y} = 0 \quad \text{at } y = 0 \text{ (centerline), if flow is symmetric} \tag{5}$$

In general for a flat bed, differentiating (3) gives

$$\frac{\partial U_d}{\partial y} = \frac{\gamma_1 \left[A_1 e^{\gamma_1 y} - A_2 e^{-\gamma_1 y} \right]}{2 \left[A_1 e^{\gamma_1 y} + A_2 e^{-\gamma_1 y} + k_1 \right]^{1/2}} \tag{6}$$

Applying bc (2)

$$\frac{\partial U_d^{(1)}}{\partial y} = \frac{\gamma_1 \left[A_1 - A_2 \right]}{\left[A_1 + A_2 + k_1 \right]} = 0 \tag{7}$$

$$A_1 - A_2 = 0 \quad \text{hence} \quad A_1 = A_2$$

Applying bc (1)

$$\left[A_1 e^{\gamma_1 b} + A_2 e^{-\gamma_1 b} + k_1 \right]^{1/2} = 0 \tag{8}$$

$$A_1 \left(e^{\gamma_1 b} + e^{-\gamma_1 b} \right) + k_1 = 0$$

$$A_1 = - \frac{k_1}{2 \cosh(\gamma_1 b)} \quad (= A_2) \tag{9}$$

$$U_d^{(1)} = \left[\frac{-k_1}{2 \cosh(\gamma_1 b)} \left\{ e^{\gamma_1 y} + e^{-\gamma_1 y} \right\} + k_1 \right]^{1/2}$$

$$U_d^{(1)} = \left[\frac{-k_1 \cosh(\gamma_1 y)}{\cosh(\gamma_1 b)} + k_1 \right]^{1/2} \tag{10}$$

$$U_d^{(1)} = \left[C_1 \cosh(\gamma_1 y) + C_2 \right]^{1/2} \tag{11}$$

where

$$C_1 = \frac{-k_1}{\cosh(\gamma_1 b)} ; C_2 = k_1 = \frac{8gHS_0 \left(1 - \beta_1 \right)}{f_1} \tag{12}$$

$$U_d^{(1)} = k_1^{1/2} \left[1 - \frac{\cosh(\gamma_1 y)}{\cosh(\gamma_1 b)} \right]^{1/2} \tag{13}$$

$$U_d^{(1)} = \left\{ \frac{8gHS_0 \left(1 - \beta_1 \right)}{f_1} \right\}^{1/2} \left[1 - \frac{\cosh(\gamma_1 y)}{\cosh(\gamma_1 b)} \right]^{1/2} \tag{14}$$

Since the bed shear stress distribution, τ_b, is given by

$$\tau_b = \rho \frac{f}{8} U_d^2 \tag{15}$$

then

$$\tau_b = \rho \frac{f_1}{8}\left[C_1 \cosh(\gamma_1 y) + C_2\right] \tag{16}$$

or

$$\tau_b = \rho g H S_0 (1 - \beta_1)\left[1 - \frac{\cosh(\gamma_1 y)}{\cosh(\gamma_1 b)}\right] \tag{17}$$

The bed shear stress at the centreline $(y = 0)$ is then given by

$$\tau_b = \rho \frac{f_1}{8}\left[C_1 + C_2\right] = \rho g H S_0 (1 - \beta_1)\left[1 - \frac{1}{\cosh(\gamma_1 b)}\right] \tag{18}$$

The bed corner shear stress $(y = b)$ is given by

$$\tau_b = \rho \frac{f_1}{8}\left[C_1 \cosh(\gamma_1 y) + C_2\right]\Big|_{y=b} = 0 \tag{19}$$

By integration of (16) or (17), the mean bed shear stress, $\overline{\tau}_b$, is given by

$$\overline{\tau}_b = \frac{\rho f_1}{8}\left[\frac{C_1}{\gamma_1 b}\sinh(\gamma_1 b) + C_2\right] \tag{20}$$

or

$$\overline{\tau}_b = \rho g H S_0 (1 - \beta_1)\left[1 - \frac{\sinh(\gamma_1 b)}{\gamma_1 b \cosh(\gamma_1 b)}\right] \tag{21}$$

The mean shear force on the bed, SF, is then $\overline{\tau}_b \times 2b$, giving

$$SF_b = 2b \rho g H S_0 (1 - \beta_1)\left[1 - \frac{\sinh(\gamma b)}{\gamma_1 b \cosh(\gamma_1 b)}\right] \tag{22}$$

The mean wall shear force, SF_w, (both walls) is then given by the overall force balance:

$$\tau_o P = SF_b + SF_w \tag{23}$$

where P is the overall wetted perimeter $(= 2b + 2H)$ and τ_o is the overall shear stress $(\tau_o = \rho g R S_f)$.

Thus for both walls,

$$SF_w = \rho g A S_0 - SF_b \tag{24}$$

The percentage of total shear force on the walls is then given by

$$\%SF_w = 100 \times SF_w / \rho g A S_0 \tag{25}$$

Comments on Case R1:
Eqs (22) to (25) may be used to produce a theoretical $\%SF_w$ v $2b/H$ (aspect ratio) relationship for flows in a rectangular channel. This is considered further in Sections 6.5 and 6.6. However, it should be noticed at this point that although the boundary condition at the vertical wall is that $U_d = 0$, this strictly only applies at the wall as the velocity will be zero through the depth of the boundary layer at the wall, giving the depth-average U_d as identically zero. However, this does not mean that the wall shear stress, $\tau_{w,}$ is zero, since even for small transverse distances from the vertical wall the velocity is considerable, leading to high wall shear stresses, especially near the free surface. This incompatibility is resolved by observing that the wall shear stress is calculable, and typically significant, obviously with $\%SF_w$ equalling 50% for flow in a closed square duct. Using a depth-averaged model to deal with vertical walls and the forces has been introduced in Section 6.3, but because of its importance in sidewall correction procedures, it is dealt with more fully in Section 6.5. Some examples of typical distributions of U_d & τ_b are shown in Figures 2.25–2.29 in Section 2.3.

Case R2 Symmetric flow in a rectangular channel with non-uniform roughness, lateral shear & secondary flow, using two panels of unequal width

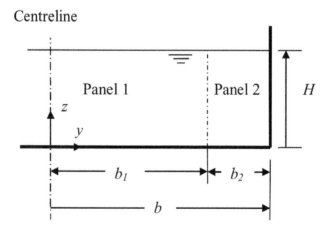

Figure R2 Rectangular channel with two panels of unequal width representing half the cross section.

With two panels, as shown in Figure R2, the following boundary conditions apply:

$$U_d^{(2)}\big|_{y=b} = 0 \quad \text{at } y = b(\text{wall}) \tag{1}$$

$$U_d^{(1)}\big|_{y=b_1} = U_d^{(2)}\big|_{y=b_1} \quad \text{at } y = b_1, \text{ the interface between panels 1 \& 2} \tag{2}$$

$$\mu_1 \frac{\partial U_d^{(1)}}{\partial y}\bigg|_{y=b_1} = \mu_2 \frac{\partial U_d^{(2)}}{\partial y}\bigg|_{y=b_1} \quad \text{at } y = b_1, \text{ the interface, based on (6.74)} \tag{3}$$

$$\frac{\partial U_d^{(1)}}{\partial y}\bigg|_{y=0} = 0 \quad \text{at } y = 0 \text{ (centerline), if flow is symmetric} \tag{4}$$

where

$$\mu = \lambda \sqrt{\frac{f}{8}} \quad \text{from (6.75) in Section 6.3} \tag{5}$$

The lateral distribution of U_d will be given by

$$U_d^{(1)} = \left[A_1 e^{\gamma_1 y} + A_2 e^{-\gamma_1 y} + k_1 \right]^{1/2} \tag{6}$$

where

$$k_1 = \frac{8gHS_0}{f_1}(1 - \beta_1) \tag{7}$$

$$\gamma_1 = \sqrt{\frac{2}{\lambda_1}} \left(\frac{f_1}{8} \right)^{1/4} \frac{1}{H} \tag{8}$$

$$\beta_1 = \frac{\Gamma_1}{\rho g H S_0} \tag{9}$$

$$U_d^{(2)} = \left[A_3 e^{\gamma_2 y} + A_4 e^{-\gamma_2 y} + k_2 \right]^{1/2} \tag{10}$$

$$k_2 = \frac{8gHS_0}{f_2}(1 - \beta_2) \tag{11}$$

$$\gamma_2 = \sqrt{\frac{2}{\lambda_2}} \left(\frac{f_2}{8} \right)^{1/4} \frac{1}{H} \tag{12}$$

$$\beta_2 = \frac{\Gamma_2}{\rho g H S_0} \tag{13}$$

$$\frac{\mu_1}{2} \frac{\left[A_1 \gamma_1 e^{\gamma_1 y} - A_2 \gamma_1 e^{-\gamma_1 y} \right]}{\left[A_1 e^{\gamma_1 y} + A_2 e^{-\gamma_1 y} + k_1 \right]^{1/2}}\bigg|_{y=b_1} = \frac{\mu_2}{2} \frac{\left[A_3 \gamma_2 e^{\gamma_2 y} - A_4 \gamma_2 e^{-\gamma_2 y} \right]}{\left[A_3 e^{\gamma_2 y} + A_4 e^{-\gamma_2 y} + k_2 \right]^{1/2}}\bigg|_{y=b_1} \tag{14}$$

where $\mu_1 = \lambda_1 \sqrt{\dfrac{f_1}{8}}$ and $\mu_2 = \lambda_2 \sqrt{\dfrac{f_2}{8}}$

But from (2), (6) & (10) the denominators in (14) are the same, hence

$$\left(\frac{\mu_1}{\mu_2}\right)\left[A_1\gamma_1 e^{\gamma_1 b_1} - A_2\gamma_1 e^{-\gamma_1 b_1}\right] = \left[A_3\gamma_2 e^{\gamma_2 b_1} - A_4\gamma_2 e^{-\gamma_2 b_1}\right] \tag{15}$$

bc (1)

$$A_3 e^{\gamma_2 b} + A_4 e^{-\gamma_2 b} + k_2 = 0 \tag{16}$$

bc (2)

$$A_1 e^{\gamma_1 b_1} + A_2 e^{-\gamma_1 b_1} + k_1 = A_3 e^{\gamma_2 b_1} + A_4 e^{-\gamma_2 b_1} + k_2 \tag{17}$$

bc (4)

$$A_1\gamma_1 - A_2\gamma_1 = 0 \tag{18}$$

Elimination process

a.   From (18)

$$A_1 = A_2 \tag{19}$$

b.   Eliminate A_4 from (15) & (17) by $\gamma_2 \times (17) + (15)$

$$\gamma_2\left[A_1 e^{\gamma_1 b_1} + A_2 e^{-\gamma_1 b_1} + k_1\right] + \left(\frac{\mu_1}{\mu_2}\right)\left[A_1\gamma_1 e^{\gamma_1 b_1} - A_2\gamma_1 e^{-\gamma_1 b_1}\right] = 2A_3\gamma_2 e^{\gamma_2 b_1} + \gamma_2 k_2$$

hence

$$\left[\gamma_2 + \gamma_1\left(\frac{\mu_1}{\mu_2}\right)\right]e^{\gamma_1 b_1} A_1 + \left[\gamma_2 - \gamma_1\left(\frac{\mu_1}{\mu_2}\right)\right]e^{-\gamma_1 b_1} A_2 = 2A_3\gamma_2 e^{\gamma_2 b_1} + \gamma_2\left(k_2 - k_1\right) \tag{20}$$

c.   Eliminate A_4 from (15) & (16) by $\gamma_2 e^{-\gamma_2 b_1} \times (16) - e^{-\gamma_2 b} \times (15)$

$$\gamma_2 e^{-\gamma_2 b_1}\left[A_3 e^{\gamma_2 b} + A_4 e^{-\gamma_2 b} + k_2\right] - e^{-\gamma_2 b}\left(\frac{\mu_1}{\mu_2}\right)\left[A_1\gamma_1 e^{\gamma_1 b_1} - A_2\gamma_1 e^{-\gamma_1 b_1}\right]$$

$$= -e^{-\gamma_2 b}\left[A_3\gamma_2 e^{\gamma_2 b_1} - A_4\gamma_2 e^{-\gamma_2 b_1}\right] - \gamma_1\left(\frac{\mu_1}{\mu_2}\right)e^{(\gamma_1 b_1 - \gamma_2 b)} A_1$$

$$+ \gamma_1\left(\frac{\mu_1}{\mu_2}\right)e^{(-\gamma_1 b_1 - \gamma_2 b)} A_2 + k_2\gamma_2 e^{-\gamma_2 b_1} = -\gamma_2 A_3\left[e^{\gamma_2(b - b_1)} + e^{-\gamma_2(b - b_1)}\right] \tag{21}$$

Note that $b - b_1 = b_2$

d.   Eliminate A_3 from (20) & (21)   by $(20) \times \left[e^{\gamma_2(b - b_1)} + e^{-\gamma_2(b - b_1)}\right] + (21) \times 2e^{\gamma_2 b_1}$

$$\left[e^{\gamma_2(b - b_1)} + e^{-\gamma_2(b - b_1)}\right]\left\{\left[\gamma_2 + \gamma_1\left(\frac{\mu_1}{\mu_2}\right)\right]e^{\gamma_1 b_1} A_1 + \left[\gamma_2 - \gamma_1\left(\frac{\mu_1}{\mu_2}\right)\right]e^{-\gamma_1 b_1} A_2\right\} + 2e^{\gamma_2 b_1}$$

$$....X....\left\{-\gamma_1\left(\frac{\mu_1}{\mu_2}\right)e^{(\gamma_1 b_1 - \gamma_2 b)} A_1 + \gamma_1\left(\frac{\mu_1}{\mu_2}\right)e^{(-\gamma_1 b_1 - \gamma_2 b)} A_2 + k_2\gamma_2 e^{-\gamma_2 b_1}\right\}$$

$$= \left[e^{\gamma_2(b - b_1)} + e^{-\gamma_2(b - b_1)}\right]\gamma_2\left(k_2 - k_1\right)$$

hence

$$\left\{\left[e^{\gamma_2(b-b_1)}+e^{-\gamma_2(b-b_1)}\right]\left[\gamma_2+\gamma_1\left(\frac{\mu_1}{\mu_2}\right)\right]e^{\gamma_1 b_1}-2\gamma_1\left(\frac{\mu_1}{\mu_2}\right)e^{(\gamma_1 b_1-\gamma_2 b+\gamma_2 b_1)}\right\}A_1+....$$

$$....\left\{\left[e^{\gamma_2(b-b_1)}+e^{-\gamma_2(b-b_1)}\right]\left[\gamma_2-\gamma_1\left(\frac{\mu_1}{\mu_2}\right)\right]e^{-\gamma_1 b_1}+2\gamma_1\left(\frac{\mu_1}{\mu_2}\right)e^{-(\gamma_1 b_1-\gamma_2 b+\gamma_2 b_1)}\right\}A_2 \qquad (22)$$

$$= \gamma_2(k_2-k_1)\left[e^{\gamma_2(b-b_1)}+e^{-\gamma_2(b-b_1)}\right]-2\gamma_2 k_2$$

then, since $A_1 = A_2$, this simplifies to

$$\left[e^{\gamma_2(b-b_1)}+e^{-\gamma_2(b-b_1)}\right]\left\{\left[\gamma_2+\gamma_1\left(\frac{\mu_1}{\mu_2}\right)\right]e^{\gamma_1 b_1}+\left[\gamma_2-\gamma_1\left(\frac{\mu_1}{\mu_2}\right)\right]e^{-\gamma_1 b_1}\right\}A_1$$

$$-2\gamma_1\left(\frac{\mu_1}{\mu_2}\right)\left[e^{(\gamma_1 b_1-\gamma_2(b-b_1))}-e^{-(\gamma_1 b_1-\gamma_2(b-b_1))}\right]A_1 = \gamma_2(k_2-k_1)\left[e^{\gamma_2(b-b_1)}+e^{-\gamma_2(b-b_1)}\right]-2\gamma_2 k_2$$

Hence

$$A_1 = \frac{\gamma_2(k_2-k_1)\left[e^{\gamma_2(b-b_1)}+e^{-\gamma_2(b-b_1)}\right]-2\gamma_2 k_2}{\left[e^{\gamma_2(b-b_1)}+e^{-\gamma_2(b-b_1)}\right]\left\{\left[\gamma_2+\gamma_1\left(\frac{\mu_1}{\mu_2}\right)\right]e^{\gamma_1 b_1}+\left[\gamma_2-\gamma_1\left(\frac{\mu_1}{\mu_2}\right)\right]e^{-\gamma_1 b_1}\right\}} \qquad (23)$$

$$-2\gamma_1\left(\frac{\mu_1}{\mu_2}\right)\left[e^{\gamma_1 b_1-\gamma_2(b-b_1)}-e^{-\gamma_1 b_1-\gamma_2(b-b_1)}\right]$$

From (19)

$$A_2 = A_1 \qquad\qquad\qquad (24) \ \& \ (19)$$

From (20)

$$A_3 = \frac{1}{2\gamma_2 e^{\gamma_2 b_1}}\left\{\left[\gamma_2+\gamma_1\left(\frac{\mu_1}{\mu_2}\right)\right]e^{\gamma_1 b_1}A_1+\left[\gamma_2-\gamma_1\left(\frac{\mu_1}{\mu_2}\right)\right]e^{-\gamma_1 b_1}A_2-\gamma_2(k_2-k_1)\right\} \qquad (25)$$

From (16)

$$A_4 = -e^{\gamma_2 b}\left[A_3 e^{\gamma_2 b}+k_2\right] \qquad\qquad\qquad (26)$$

The coefficients (A_1 to A_4) may be inserted into (6) & (10) to give the velocity distribution.

Comments on Case R2:
Figure 2.29 has shown some examples of flow in a 10 m wide rectangular channel using a variety of panel widths, b_1 & b_2. Details of these simulations are given in the accompanying text in Section 2.3. In order to illustrate how the coefficients for some other simple cases may be derived from (23) to (26), Case R3 now considers a rectangular channel with uniform roughness, lateral shear & secondary flow, using two equal sized panels.

Case R3 Symmetric flow in a rectangular channel with uniform roughness, lateral shear & secondary flow, using two equal sized panels

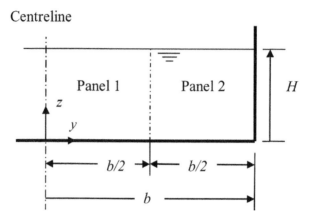

Figure R3 Rectangular channel with two panels of equal width representing half the cross section.

Let $b_1 = b_2 = b/2$; $f_1 = f_2 = f$; $\lambda_1 = \lambda_2$ & $\Gamma_1 = \Gamma_2$, so $k_1 = k_2 = k$; $\gamma_1 = \gamma_2 = \gamma$ & $\beta_1 = \beta_2 = \beta$, (i.e. uniform roughness, lateral shear & secondary flow). Then in Case R2, (23) – (26) become:

$$A_1 = -\frac{k}{2\cosh(\gamma b)} = A_2 = A_3 = A_4 \quad (\text{cp } (9) \text{ of } Case\ R1) \tag{1}$$

Comments on Case R3:
Case R3 shows how it is relatively straightforward to obtain a simpler flow type from a more complex one by adapting another solution, and also how the symmetry of the flow greatly reduces the algebra.

Case R4 Asymmetric flow in a rectangular channel, using two unequal sized panels

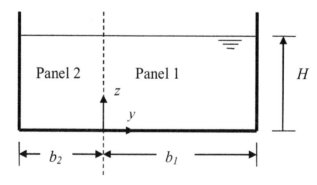

Figure R4 Rectangular channel with two panels of unequal width offset from the centreline.

With two panels, as shown in Figure R4, the following boundary conditions apply:

$$U_d^{(1)}\big|_{y=b_1} = 0 \quad \text{at } y = b_1 (\text{wall}) \tag{1}$$

$$U_d^{(2)}\big|_{y=-b_2} = 0 \quad \text{at } y = -b_2 (\text{wall}) \tag{2}$$

$$U_d^{(1)}\big|_{y=0} = U_d^{(2)}\big|_{y=0} \quad \text{at } y = 0, \text{ the interface between panels 1 \& 2} \tag{3}$$

$$\mu_1 \frac{\partial U_d^{(1)}}{\partial y}\bigg|_{y=0} = \mu_2 \frac{\partial U_d^{(2)}}{\partial y}\bigg|_{y=0} \quad \text{at } y = 0, \text{ the interface, based on (6.74)} \tag{4}$$

where

$$\mu = \lambda \sqrt{\frac{f}{8}} \quad \text{from (6.75) in Section 6.3} \tag{5}$$

The lateral distribution of U_d will be given by

$$U_d^{(1)} = \left[A_1 e^{\gamma_1 y} + A_2 e^{-\gamma_1 y} + k_1 \right]^{1/2} \tag{6}$$

where

$$k_1 = \frac{8gHS_0}{f_1}(1 - \beta_1) \tag{7}$$

$$\gamma_1 = \sqrt{\frac{2}{\lambda_1}} \left(\frac{f_1}{8} \right)^{1/4} \frac{1}{H} \tag{8}$$

$$\beta_1 = \frac{\Gamma_1}{\rho g H S_0} \tag{9}$$

$$U_d^{(2)} = \left[A_3 e^{\gamma_2 y} + A_4 e^{-\gamma_2 y} + k_2 \right]^{1/2} \tag{10}$$

$$k_2 = \frac{8gHS_0}{f_2}(1 - \beta_2) \tag{11}$$

$$\gamma_2 = \sqrt{\frac{2}{\lambda_2}} \left(\frac{f_2}{8} \right)^{1/4} \frac{1}{H} \tag{12}$$

$$\beta_2 = \frac{\Gamma_2}{\rho g H S_0} \tag{13}$$

$$\frac{\mu_1}{2} \frac{\left[A_1 \gamma_1 e^{\gamma_1 y} - A_2 \gamma_1 e^{-\gamma_1 y} \right]}{\left[A_1 e^{\gamma_1 y} + A_2 e^{-\gamma_1 y} + k_1 \right]^{1/2}}\bigg|_{y=0} = \frac{\mu_2}{2} \frac{\left[A_3 \gamma_2 e^{\gamma_2 y} - A_4 \gamma_2 e^{-\gamma_2 y} \right]}{\left[A_3 e^{\gamma_2 y} + A_4 e^{-\gamma_2 y} + k_2 \right]^{1/2}}\bigg|_{y=0} \tag{14}$$

where $\mu_1 = \lambda_1 \sqrt{\dfrac{f_1}{8}}$ and $\mu_2 = \lambda_2 \sqrt{\dfrac{f_2}{8}}$

But from (2), (6) & (10) the denominators in (14) are the same, hence
From (4)

$$\mu_1\gamma_1[A_1 - A_2] = \mu_2\gamma_2[A_3 - A_4] \tag{15}$$

From (1)

$$A_1 e^{\gamma_1 b_1} + A_2 e^{-\gamma_1 b_1} + k_1 = 0 \tag{16}$$

From (2)

$$A_3 e^{-\gamma_2 b_2} + A_4 e^{\gamma_2 b_2} + k_2 = 0 \tag{17}$$

From (3)

$$A_1 + A_2 + k_1 = A_3 + A_4 + k_2 \tag{18}$$

Elimination process

a. Eliminate A_4 from (15) & (18) by $\mu_2\gamma_2 \times (18) + (15)$

$$\mu_2\gamma_2\left(A_1 + A_2 + k_1\right) - \mu_2\gamma_2\left(A_3 + A_4 + k_2\right) + \mu_1\gamma_1\left(A_1 - A_2\right) - \mu_2\gamma_2\left(A_3 - A_4\right) = 0$$
$$\left(\mu_1\gamma_1 + \mu_2\gamma_2\right)A_1 - \left(\mu_1\gamma_1 - \mu_2\gamma_2\right)A_2 + \mu_2\gamma_2\left(k_1 - k_2\right) - 2\mu_2\gamma_2 A_3 = 0 \tag{19}$$

b. Eliminate A_4 from (17) & (18) by $e^{\gamma_2 b_2} \times (18) + (17)$

$$e^{\gamma_2 b_2}\left(A_1 + A_2 + k_1\right) - e^{\gamma_2 b_2}\left(A_3 + A_4 + k_2\right) + A_3 e^{-\gamma_2 b_2} + A_4 e^{\gamma_2 b_2} + k_2 = 0$$
$$A_3\left(e^{\gamma_2 b_2} - e^{-\gamma_2 b_2}\right) = \left[A_1 + A_2 + k_1 - k_2\right]e^{\gamma_2 b_2} + k_2$$
$$A_3 = \frac{\left[A_1 + A_2 + k_1 - k_2\right]e^{\gamma_2 b_2} + k_2}{2\sinh(\gamma_2 b_2)} \tag{20}$$

c. Eliminate A_3 from (19) & (20)

$$\frac{\left(\mu_1\gamma_1 + \mu_2\gamma_2\right)A_1}{2\mu_2\gamma_2} - \frac{\left(\mu_1\gamma_1 - \mu_2\gamma_2\right)A_2}{2\mu_2\gamma_2} + \frac{\left(k_1 - k_2\right)}{2} = \frac{\left[A_1 + A_2 + k_1 - k_2\right]e^{\gamma_2 b_2} + k_2}{2\sinh(\gamma_2 b_2)}$$
$$\left[\frac{\left(\mu_1\gamma_1 + \mu_2\gamma_2\right)\sinh(\gamma_2 b_2)}{\mu_2\gamma_2} - e^{\gamma_2 b_2}\right]A_1 - \left[\frac{\left(\mu_1\gamma_1 - \mu_2\gamma_2\right)\sinh(\gamma_2 b_2)}{\mu_2\gamma_2} + e^{\gamma_2 b_2}\right]A_2 + \cdots$$
$$\cdots\left(k_1 - k_2\right)\sinh(\gamma_2 b_2) - \left(k_1 - k_2\right)e^{\gamma_2 b_2} - k_2 = 0 \tag{21}$$

d. Eliminate A_2 from (16) & (21) using

$$A_2 = -\left[\frac{A_1 e^{\gamma_1 b_1} + k_1}{e^{-\gamma_1 b_1}}\right] \tag{22}$$

$$\left[\frac{(\mu_1\gamma_1 + \mu_2\gamma_2)\sinh(\gamma_2 b_2)}{\mu_2\gamma_2} - e^{\gamma_2 b_2}\right]A_1 + \left[\frac{(\mu_1\gamma_1 - \mu_2\gamma_2)\sinh(\gamma_2 b_2)}{\mu_2\gamma_2} + e^{\gamma_2 b_2}\right] \times$$

$$\left[\frac{A_1 e^{\gamma_1 b_1} + k_1}{e^{-\gamma_1 b_1}}\right] + (k_1 - k_2)\sinh(\gamma_2 b_2) - (k_1 - k_2)e^{\gamma_2 b_2} - k_2 = 0$$

Hence

$$A_1 =$$
$$\frac{2\mu_2\gamma_2 e^{-\gamma_1 b_1}\left[(k_2 - k_1)\cosh(\gamma_2 b_2) - k_2\right] + k_1\left[(\mu_1\gamma_1 + \mu_2\gamma_2)e^{\gamma_2 b_2} + (\mu_2\gamma_2 - \mu_1\gamma_1)e^{-\gamma_2 b_2}\right]}{(\mu_1\gamma_1 - \mu_2\gamma_2)\sinh(\gamma_1 b_1 - \gamma_2 b_2) - (\mu_1\gamma_1 + \mu_2\gamma_2)\sinh(\gamma_1 b_1 + \gamma_2 b_2)} \tag{23}$$

From (22)

$$A_2 = -\left[A_1 e^{\gamma_1 b_1} + k_1\right]e^{\gamma_1 b_1} \tag{24}$$

From (20)

$$A_3 = \frac{\left[A_1 + A_2 + k_1 - k_2\right]e^{\gamma_2 b_2} + k_2}{2\sinh(\gamma_2 b_2)} \tag{25}$$

From (17)

$$A_4 = -\left[A_3 e^{-\gamma_2 b_2} + k_2\right]e^{-\gamma_2 b_2} \tag{26}$$

Comments on Case R4:
Case R4 shows how asymmetric flows may be simulated using the SKM and how the boundary conditions are adjusted accordingly.

6.4.2 Flows in trapezoidal channels

The trapezoidal shape requires both types of solution available within the SKM to be applied. One is for flow on a horizontal bed and the other for flow on a side slop-ing bed, as shown in Figure T1 and by equations (8) & (12) below. Four trapezoidal cases are considered initially, followed by 3 triangular ones, these being special cases of a trapezoid with $b = 0$. Case T1, which is for symmetric flow, utilizes two panels to represent half the cross section, as shown in Figure T1 below. This is the simplest case, with the flows symmetric around the centre-line. Case T2 considers asymmetric flow for a trapezoidal channel with different side slopes, using 3 panels. Case T3 is similar to Case T2, but dispenses with the μ factors. Case T4 shows how Case T2 may be simplified down to Case T1, by imposing equal side slopes and symmetry of flow.

Three further cases are then considered for the special case when $b = 0$, making the channels triangular in shape, as these shaped channels are commonly used in many urban drainage systems, such as those alongside motorways. Case T5 is for asymmetric flow in a triangular channel with different side slopes, using two panels only. Case T6 considers asymmetric flow in a triangular channel with the same side slopes, and Case T7.

Case T1 Symmetric flow in a simple trapezoidal channel, using two panels

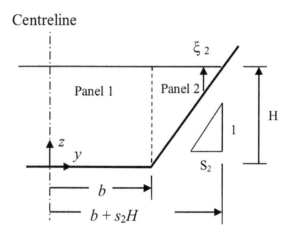

Figure T1 Trapezoidal channel with two panels representing half the cross section.

With two panels, as shown in Figure T1, the following boundary conditions apply:

$$U_d^{(2)} \big|_{\xi_2=0} = 0 \quad \text{at edges where depths are zero} \tag{1}$$

$$U_d^{(1)} \big|_{y=b} = U_d^{(2)} \big|_{y=b, \xi_2=H} \quad \text{at interface between panels 1 \& 2} \tag{2}$$

$$\mu_1 \frac{\partial U_d^{(1)}}{\partial y} \bigg|_{y=b} = \mu_2 \frac{\partial U_d^{(2)}}{\partial y} \bigg|_{\xi_2=H} \quad \text{at interface between panels 1 \& 2} \tag{3}$$

$$\frac{\partial U_d^{(1)}}{\partial y} \bigg|_{y=0} = 0 \quad \text{at centerline (y=0), due to symmetry} \tag{4}$$

where

$$\mu = \lambda \sqrt{\frac{f}{8}} \tag{5}$$

and the local depths, ξ_2 & ξ_3, on the sloping side walls are

$$\xi_2 = H - \frac{(y-b)}{s_2} \tag{6}$$

and for a slope increasing with y,

$$\frac{\partial \xi}{\partial y} = -\frac{1}{s_2} \tag{7}$$

The lateral distribution of U_d in panels 1 & 2 will be given by

$$U_d^{(1)} = \left[A_1 e^{\gamma_1 y} + A_2 e^{-\gamma_1 y} + k_1 \right]^{1/2} \tag{8}$$

where

$$k_1 = \frac{8gHS_0}{f_1}(1 - \beta_1) \tag{9}$$

$$\gamma_1 = \sqrt{\frac{2}{\lambda_1}} \left(\frac{f_1}{8} \right)^{1/4} \frac{1}{H} \tag{10}$$

$$\beta_1 = \frac{\Gamma_1}{\rho g H S_0} \tag{11}$$

$$U_d^{(2)} = \left[A_3 \xi_2^{\alpha_2} + A_4 \xi_2^{-(\alpha_2+1)} + \omega_2 \xi_2 + \eta_2 \right]^{1/2} \tag{12}$$

$$\alpha_2 = -\frac{1}{2} + \frac{1}{2} \left[1 + \frac{s_2 (1 + s_2^2)^{1/2}}{\lambda_2} (8 f_2)^{1/2} \right]^{1/2} \tag{13}$$

$$\omega_2 = \frac{g S_0}{\dfrac{(1 + s_2^2)^{1/2}}{s_2} \left(\dfrac{f_2}{8} \right) - \dfrac{\lambda_2}{s_2^2} \left(\dfrac{f_2}{8} \right)^{1/2}} \tag{14}$$

$$\eta_2 = -\frac{\Gamma_2}{\dfrac{(1 + s_2^2)^{1/2}}{s_2} \rho \left(\dfrac{f_2}{8} \right)} \tag{15}$$

From (7) & (12)

$$\frac{\partial U_d^{(2)}}{\partial y} = \frac{\partial U_d^{(2)}}{\partial \xi} \frac{\partial \xi}{\partial y} \tag{16}$$

$$\frac{\partial U_d^{(2)}}{\partial y} = -\frac{\left[A_3 \alpha_2 \xi_2^{\alpha_2-1} - A_4 (\alpha_2 + 1) \xi_2^{-\alpha_2-2} + \omega_2 \right]}{2 s_2 \left[A_3 \xi_2^{\alpha_2} + A_4 \xi_2^{-\alpha_2-1} + \omega_2 \xi_2 + \eta_2 \right]^{1/2}} \tag{17}$$

From (12) and bc (1) as $\xi \to 0$ at the edge in panel 2, so U_d must $\to 0$. This must also be so because of the decrease in U in the boundary layer. Thus from (12)

bc (1)

$$A_4 + \eta_2 \xi_2^{\alpha_2+1} \to 0 \quad \text{as } \xi_2 \to 0, \therefore A_4 = 0 \tag{18}$$

bc (2)

$$A_1 e^{\eta b} + A_2 e^{-\eta b} + k_1 = A_3 H^{\alpha_2} + \omega_2 H + \eta_2 \tag{19}$$

bc (3)

$$\frac{\mu_1 \gamma_1 (A_1 e^{\eta b} - A_2 e^{-\eta b})}{2 \left[A_1 e^{\eta b} + A_2 e^{-\eta b} + k_1 \right]^{1/2}} = -\frac{\mu_2 \left[A_3 \alpha_2 H^{\alpha_2-1} + \omega_2 \right]}{2 s_2 \left[A_3 H^{\alpha_2} + \omega_2 H + \eta_2 \right]^{1/2}} \tag{20a}$$

From (19), this becomes

$$\left(\frac{\mu_1}{\mu_2} \right) s_2 \gamma_1 \left[A_1 e^{\eta b} - A_2 e^{-\eta b} \right] = -\left[A_3 \alpha_2 H^{\alpha_2-1} + \omega_2 \right] \tag{20b}$$

Elimination process

a. From bc (4) and (8)
$$A_1 = A_2 \tag{21}$$

b. Eliminate A_3 from (19) & (20b) by $\alpha_2 \times$ (19) $+ H \times$ (20b)

$$\alpha_2 \left[A_1 e^{\eta b} + A_2 e^{-\eta b} + k_1 \right] - \alpha_2 \left[A_3 H^{\alpha_2} + \omega_2 H + \eta_2 \right]$$
$$+ \left(\frac{\mu_1}{\mu_2} \right) s_2 \gamma_1 H \left[A_1 e^{\eta b} - A_2 e^{-\eta b} \right] + H \left[A_3 \alpha_2 H^{\alpha_2-1} + \omega_2 \right] = 0$$

$$\left[\left\{ \alpha_2 + \left(\frac{\mu_1}{\mu_2} \right) s_2 \gamma_1 H \right\} e^{\eta b} \right] A_1 + \left[\left\{ \alpha_2 - \left(\frac{\mu_1}{\mu_2} \right) s_2 \gamma_1 H \right\} e^{-\eta b} \right] A_2$$
$$= (\alpha_2 - 1) \omega_2 H - k_1 \alpha_2 + \alpha_2 \eta_2 \tag{22}$$

c. Eliminate A_2 from (21) & (22)

$$A_1 = \frac{(\alpha_2 - 1) \omega_2 H - k_1 \alpha_2 + \alpha_2 \eta_2}{\left[\left\{ \alpha_2 + \left(\frac{\mu_1}{\mu_2} \right) s_2 \gamma_1 H \right\} e^{\eta b} \right] + \left[\left\{ \alpha_2 - \left(\frac{\mu_1}{\mu_2} \right) s_2 \gamma_1 H \right\} e^{-\eta b} \right]} \tag{23}$$

$$A_2 = A_1 \tag{24}$$

$$A_3 = \frac{1}{H^{\alpha_2}} \left[A_1 e^{\eta b} + A_2 e^{-\eta b} + k_1 - \omega_2 H - \eta_2 \right] \quad \text{and} \quad A_4 = 0 \tag{25}$$

Equations (8) and (12) give the lateral distributions of U_d with y.

Comments on Case T1:
Comparisons could be made also between flow in rectangular channels and those in trapezoidal ones with a very steep side slope. This issue is considered later in compound channel cases where an internal vertical wall may be present.

Case T2 Asymmetric flow in a trapezoidal channel with different side slopes, using three panels

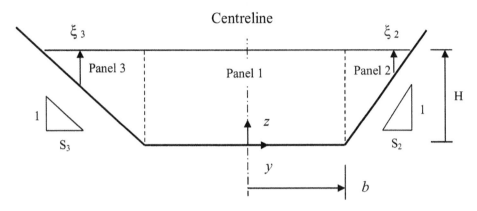

Figure T2 Trapezoidal channel with different side slopes, using 3 panels.

With three panels, as shown described in Section 6.3, the following boundary conditions apply:

$$U_d^{(2)}\,|_{\xi_2=0}=0 \quad \text{and} \quad U_d^{(3)}\,|_{\xi_3=0}=0 \quad \text{at edges where depths are zero} \tag{1}$$

$$U_d^{(1)}\,|_{y=b}=U_d^{(2)}\,|_{\xi_2=H} \quad \text{and} \quad \mu_1\frac{\partial U_d^{(1)}}{\partial y}\bigg|_{y=b}=\mu_2\frac{\partial U_d^{(2)}}{\partial y}\bigg|_{\xi_2=H} \tag{2}$$

at interfaces between panels 1 & 2

$$U_d^{(1)}\,|_{y=-b}=U_d^{(3)}\,|_{\xi_3=H} \quad \text{and} \quad \mu_1\frac{\partial U_d^{(1)}}{\partial y}\bigg|_{y=-b}=\mu_3\frac{\partial U_d^{(3)}}{\partial y}\bigg|_{\xi_3=H} \tag{3}$$

at interfaces between panels 2 & 3

$$\text{where} \quad \mu=\lambda\sqrt{\frac{f}{8}} \tag{4}$$

and the local depths, ξ_2 & ξ_3, on the sloping side walls are

$$\xi_2=H-\frac{y-b}{s_2} \quad \text{and} \quad \xi_3=H+\frac{y+b}{s_3} \tag{5}$$

The lateral distribution of U_d in panels 1 & 2 will be given by

$$U_d^{(1)}=\left[A_1e^{\gamma_1 y}+A_2e^{-\gamma_1 y}+k_1\right]^{1/2} \tag{6}$$

where

$$k_1=\frac{8gHS_0}{f_1}(1-\beta_1) \tag{7}$$

$$\gamma_1 = \sqrt{\frac{2}{\lambda_1}} \left(\frac{f_1}{8}\right)^{1/4} \frac{1}{H} \tag{8}$$

$$\beta_1 = \frac{\Gamma_1}{\rho g H S_0} \tag{9}$$

$$U_d^{(2)} = \left[A_3 \xi_2^{\alpha_2} + A_4 \xi_2^{-(\alpha_2+1)} + \omega_2 \xi_2 + \eta_2\right]^{1/2} \tag{10}$$

$$\alpha_2 = -\frac{1}{2} + \frac{1}{2}\left[1 + \frac{s_2\left(1+s_2^2\right)^{1/2}}{\lambda_2}(8f_2)^{1/2}\right]^{1/2} \tag{11}$$

$$\omega_2 = \frac{g S_0}{\frac{(1+s_2^2)^{1/2}}{s_2}\left(\frac{f_2}{8}\right) - \frac{\lambda_2}{s_2^2}\left(\frac{f_2}{8}\right)^{1/2}} \tag{12}$$

$$\eta_2 = -\frac{\Gamma_2}{\frac{(1+s_2^2)^{1/2}}{s_2}\rho\left(\frac{f_2}{8}\right)} \tag{13}$$

From (10) and its companion equation for panel 3, noting from (5) that $\dfrac{\partial \xi_2}{\partial y} = -\dfrac{1}{s_2}$ and $\dfrac{\partial \xi_3}{\partial y} = \dfrac{1}{s_3}$

$$\frac{\partial U_d^{(1)}}{\partial y} = \frac{\gamma_1\left[A_1 e^{\gamma_1 b} - A_2 e^{-\gamma_1 b}\right]}{2\left[A_1 e^{\gamma_1 b} + A_2 e^{-\gamma_1 b} + k_1\right]^{1/2}} \tag{14}$$

$$\frac{\partial U_d^{(2)}}{\partial y} = -\frac{\left[A_3 \alpha_2 \xi_2^{\alpha_2-1} - A_4(\alpha_2+1)\xi_2^{-\alpha_2-2} + \omega_2\right]}{2 s_2 \left[A_3 \xi_2^{\alpha_2} + A_4 \xi_2^{-\alpha_2-1} + \omega_2 \xi_2 + \eta_2\right]^{1/2}} \tag{15}$$

$$\frac{\partial U_d^{(3)}}{\partial y} = \frac{\left[A_5 \alpha_3 \xi_3^{\alpha_3-1} - A_6(\alpha_3+1)\xi_3^{-\alpha_3-2} + \omega_3\right]}{2 s_3 \left[A_5 \xi_3^{\alpha_3} + A_6 \xi_3^{-\alpha_3-1} + \omega_3 \xi_3 + \eta_3\right]^{1/2}} \tag{16}$$

From (1)

$$A_4 + \eta_2 \xi_2^{\alpha_2+1} \to 0 \text{ as } \xi_2 \to 0, \therefore A_4 = 0 \text{ and } A_6 + \eta_3 \xi_3^{\alpha_3+1} \to 0 \text{ as } \xi_3 \to 0, \therefore A_6 = 0$$

$$A_4 = A_6 = 0 \tag{17}$$

From (2)

$$A_1 e^{\gamma_1 b} + A_2 e^{-\gamma_1 b} + k_1 = A_3 H^{\alpha_2} + \omega_2 H + \eta_2 \tag{18}$$

$$\frac{\mu_1 \gamma_1\left[A_1 e^{\gamma_1 b} - A_2 e^{-\gamma_1 b}\right]}{2\left[A_1 e^{\gamma_1 b} + A_2 e^{-\gamma_1 b} + k_1\right]^{1/2}} = -\frac{\mu_2\left[A_3 \alpha_2 H^{\alpha_2-1} + \omega_2\right]}{2 s_2\left[A_3 H^{\alpha_2} + \omega_2 H + \eta_2\right]^{1/2}} \tag{19}$$

From (18), this becomes

$$\mu_1 \gamma_1 s_2 \left[A_1 e^{\gamma_1 b} - A_2 e^{-\gamma_1 b} \right] = -\mu_2 \left[A_3 \alpha_2 H^{\alpha_2 - 1} + \omega_2 \right]$$

$$\left(\frac{\mu_1}{\mu_2} \right) \gamma_1 s_2 \left[A_1 e^{\gamma_1 b} - A_2 e^{-\gamma_1 b} \right] = -\left[A_3 \alpha_2 H^{\alpha_2 - 1} + \omega_2 \right] \tag{20}$$

From (3)

$$A_1 e^{-\gamma_1 b} + A_2 e^{\gamma_1 b} + k_1 = A_5 H^{\alpha_3} + \omega_3 H + \eta_3 \tag{21}$$

$$\frac{\mu_1 \gamma_1 \left[A_1 e^{-\gamma_1 b} - A_2 e^{\gamma_1 b} \right]}{2 \left[A_1 e^{-\gamma_1 b} + A_2 e^{\gamma_1 b} + k_1 \right]^{1/2}} = \frac{\mu_3 \left[A_5 \alpha_3 H^{\alpha_3 - 1} + \omega_3 \right]}{2 s_3 \left[A_5 H^{\alpha_3} + \omega_3 H + \eta_3 \right]^{1/2}} \tag{22}$$

From (21), this becomes

$$\left(\frac{\mu_1}{\mu_3} \right) s_3 \gamma_1 \left[A_1 e^{-\gamma_1 b} - A_2 e^{\gamma_1 b} \right] = \left[A_5 \alpha_3 H^{\alpha_3 - 1} + \omega_3 \right] \tag{23}$$

Elimination process

a. Eliminate A_3 by $\alpha_2 \times (18) + H \times (20)$

$$\alpha_2 \left\{ A_1 e^{\gamma_1 b} + A_2 e^{-\gamma_1 b} + k_1 - A_3 H^{\alpha_2} - \omega_2 H - \eta_2 \right\}$$
$$+ H \left\{ \left(\frac{\mu_1}{\mu_2} \right) \gamma_1 s_2 \left[A_1 e^{\gamma_1 b} - A_2 e^{-\gamma_1 b} \right] + \left[A_3 \alpha_2 H^{\alpha_2 - 1} + \omega_2 \right] \right\} = 0$$

Hence

$$\left[\left(\alpha_2 + \left(\frac{\mu_1}{\mu_2} \right) \gamma_1 s_2 H \right) e^{\gamma_1 b} \right] A_1 + \left[\left(\alpha_2 - \left(\frac{\mu_1}{\mu_2} \right) \gamma_1 s_2 H \right) e^{-\gamma_1 b} \right] A_2 + $$
$$\alpha_2 \left(k_1 - \omega_2 H - \eta_2 \right) + \omega_2 H \tag{24}$$

b. Eliminate A_5 by $\alpha_3 \times (21) - H \times (23)$

$$\alpha_3 \left\{ A_1 e^{-\gamma_1 b} + A_2 e^{\gamma_1 b} + k_1 - A_5 H^{\alpha_3} - \omega_3 H - \eta_3 \right\}$$
$$- H \left\{ \left(\frac{\mu_1}{\mu_3} \right) s_3 \gamma_1 \left[A_1 e^{-\gamma_1 b} - A_2 e^{\gamma_1 b} \right] - \left[A_5 \alpha_3 H^{\alpha_3 - 1} + \omega_3 \right] \right\} = 0$$

Hence

$$\left[\left(\alpha_3 - \left(\frac{\mu_1}{\mu_3} \right) s_3 \gamma_1 H \right) e^{-\gamma_1 b} \right] A_1 + \left[\left(\alpha_3 + \left(\frac{\mu_1}{\mu_3} \right) s_3 \gamma_1 H \right) e^{\gamma_1 b} \right] A_2 + $$
$$\alpha_3 \left(k_1 - \omega_3 H - \eta_3 \right) + \omega_3 H = 0 \tag{25}$$

c. Eliminate A_2 by $(24) \times \left[\left(\alpha_3 + \left(\dfrac{\mu_1}{\mu_3} \right) s_3 \gamma_1 H \right) e^{\eta b} \right] - (25) \times \left[\left(\alpha_2 - \left(\dfrac{\mu_1}{\mu_2} \right) \gamma_1 s_2 H \right) e^{-\eta b} \right]$

$$\left[\left(\alpha_3 + \left(\frac{\mu_1}{\mu_3} \right) s_3 \gamma_1 H \right) e^{\eta b} \right] \left\{ \left[\left(\alpha_2 + \left(\frac{\mu_1}{\mu_2} \right) s_2 \gamma_1 H \right) e^{\eta b} \right] A_1 + \left[\left(\alpha_2 - \left(\frac{\mu_1}{\mu_2} \right) s_2 \gamma_1 H \right) e^{-\eta b} \right] A_2 + \right.$$

$$\alpha_2 \left(k_1 - \omega_2 H - \eta_2 \right) + \omega_2 H \Big\} - \left[\left(\alpha_2 - \left(\frac{\mu_1}{\mu_2} \right) s_2 \gamma_1 H \right) e^{-\eta b} \right] \left\{ \left[\left(\alpha_3 - \left(\frac{\mu_1}{\mu_3} \right) s_3 \gamma_1 H \right) e^{-\eta b} \right] A_1 + \right.$$

$$\left[\left(\alpha_3 + \left(\frac{\mu_1}{\mu_3} \right) s_3 \gamma_1 H \right) e^{\eta b} \right] A_2 + \alpha_3 \left(k_1 - \omega_3 H - \eta_3 \right) + \omega_3 H \right\} = 0$$

(26)

Hence

$$A_1 \left\{ \left(\alpha_3 + \left(\frac{\mu_1}{\mu_3} \right) s_3 \gamma_1 H \right) \left(\alpha_2 + \left(\frac{\mu_1}{\mu_2} \right) s_2 \gamma_1 H \right) e^{2\eta b} \right\} - \left(\alpha_2 - \left(\frac{\mu_1}{\mu_2} \right) s_2 \gamma_1 H \right)$$

$$\left(\alpha_3 - \left(\frac{\mu_1}{\mu_3} \right) s_3 \gamma_1 H \right) e^{-2\eta b} + \cdots \left(\alpha_3 + \left(\frac{\mu_1}{\mu_3} \right) s_3 \gamma_1 H \right) e^{\eta b} \left[\alpha_2 \left(k_1 - \omega_2 H - \eta_2 \right) + \omega_2 H \right]$$

$$- \cdots \left(\alpha_2 - \left(\frac{\mu_1}{\mu_2} \right) s_2 \gamma_1 H \right) e^{-\eta b} \left[\alpha_3 \left(k_1 - \omega_3 H - \eta_3 \right) + \omega_3 H \right] = 0 \qquad (27)$$

$$A_1 = \frac{\begin{aligned} &\left(\alpha_2 - \left(\frac{\mu_1}{\mu_2} \right) s_2 \gamma_1 H \right) e^{-\eta b} \left[\alpha_3 \left(k_1 - \eta_3 \right) - \omega_3 H \left(\alpha_3 - 1 \right) \right] \\ &- \left(\alpha_3 + \left(\frac{\mu_1}{\mu_3} \right) s_3 \gamma_1 H \right) e^{\eta b} \left[\alpha_2 \left(k_1 - \eta_2 \right) - \omega_2 H \left(\alpha_2 - 1 \right) \right] \end{aligned}}{\begin{aligned} &\left(\alpha_3 + \left(\frac{\mu_1}{\mu_3} \right) s_3 \gamma_1 H \right) \left(\alpha_2 + \left(\frac{\mu_1}{\mu_2} \right) s_2 \gamma_1 H \right) e^{2\eta b} \\ &- \left(\alpha_2 - \left(\frac{\mu_1}{\mu_2} \right) s_2 \gamma_1 H \right) \left(\alpha_3 - \left(\frac{\mu_1}{\mu_3} \right) s_3 \gamma_1 H \right) e^{-2\eta b} \end{aligned}}$$

(28)

From (24)

$$A_2 = \frac{e^{\eta b}}{\left(\alpha_2 - \left(\frac{\mu_1}{\mu_2} \right) \gamma_1 s_2 H \right)} \left[\omega_2 H \left(\alpha_2 - 1 \right) - \alpha_2 \left(k_1 - \eta_2 \right) - \left(\alpha_2 + \left(\frac{\mu_1}{\mu_2} \right) s_2 \gamma_1 H \right) e^{\eta b} A_1 \right]$$

(29)

From (18)

$$A_3 = \frac{1}{H^{\alpha_2}} \left[A_1 e^{\eta b} + A_2 e^{-\eta b} + k_1 - \omega_2 H - \eta_2 \right] \qquad (30)$$

From (21)

$$A_5 = \frac{1}{H^{\alpha_3}}\left[A_1 e^{-\gamma_1 b} + A_2 e^{\gamma_1 b} + k_1 - \omega_3 H - \eta_3\right] \tag{31}$$

Comments on Case T2:
This illustrates how the volume of algebraic manipulation increases as the number of panels increase. This and some other solutions in this Chapter are utilized in Chapter 7, in order to appreciate the numerical output for particular cases.

Case T3 Asymmetric flow in a trapezoidal channel, as T2 but with no μ factors

From (28), check original solution without μ factors by putting $\mu_1 / \mu_2 = 1$ and $\mu_1 / \mu_3 = 1$. Hence

$$A_1 = \frac{(\alpha_2 - s_2\gamma_1 H)e^{-\gamma_1 b}\left[\alpha_3(k_1 - \eta_3) - \omega_3 H(\alpha_3 - 1)\right] - (\alpha_3 + s_3\gamma_1 H)e^{\gamma_1 b}\left[\alpha_2(k_1 - \eta_2) - \omega_2 H(\alpha_2 - 1)\right]}{(\alpha_3 + s_3\gamma_1 H)(\alpha_2 + s_2\gamma_1 H)e^{2\gamma_1 b} - (\alpha_2 - s_2\gamma_1 H)(\alpha_3 - s_3\gamma_1 H)e^{-2\gamma_1 b}}$$

$$\tag{1}$$

From (29) of T2

$$A_2 = \frac{e^{\gamma_1 b}}{(\alpha_2 - s_2\gamma_1 H)}\left[\omega_2 H(\alpha_2 - 1) - \alpha_2(k_1 - \eta_2) - (\alpha_2 + s_2\gamma_1 H)e^{\gamma_1 b})A_1\right] \tag{2}$$

And A_3 & A_4 are as before

$$A_3 = \frac{1}{H^{\alpha_2}}\left[A_1 e^{\gamma_1 b} + A_2 e^{-\gamma_1 b} + k_1 - \omega_2 H - \eta_2\right] \tag{3}$$

$$A_5 = \frac{1}{H^{\alpha_3}}\left[A_1 e^{-\gamma_1 b} + A_2 e^{\gamma_1 b} + k_1 - \omega_3 H - \eta_3\right] \tag{4}$$

Comments on Case T3:
This again shows how it is relatively easy to adapt one solution to another when a component part is preferred over another.

CaseT4 Symmetric flow in a trapezoidal channel with the same side slopes, without μ factors

Check on originals for symmetric case. Put $\mu_1 / \mu_2 = 1$, $\mu_1 / \mu_3 = 1$, $s_3 = s_2$, $\eta_3 = \eta_2$, $\omega_3 = \omega_2$, $\alpha_3 = \alpha_2$

From (28)

$$A_1 = \frac{(\alpha_2 - s_2\gamma_1 H)e^{-\gamma_1 b}\left[\alpha_2(k_1 - \eta_2) - \omega_2 H(\alpha_2 - 1)\right] - (\alpha_2 + s_2\gamma_1 H)e^{\gamma_1 b}\left[\alpha_2(k_1 - \eta_2) - \omega_2 H(\alpha_2 - 1)\right]}{(\alpha_2 + s_2\gamma_1 H)(\alpha_2 + s_2\gamma_1 H)e^{2\gamma_1 b} - (\alpha_2 - s_2\gamma_1 H)(\alpha_2 - s_2\gamma_1 H)e^{-2\gamma_1 b}}$$

$$A_1 = \frac{\left[\alpha_2(k_1 - \eta_2) - \omega_2 H(\alpha_2 - 1)\right]\{(\alpha_2 - s_2\gamma_1 H)e^{-\eta b} - (\alpha_2 + s_2\gamma_1 H)e^{\eta b}\}}{\left[(\alpha_2 + s_2\gamma_1 H)e^{\eta b}\right]^2 - \left[(\alpha_2 - s_2\gamma_1 H)e^{-\eta b}\right]^2}$$

$$A_1 = \frac{\left[\alpha_2(k_1 - \eta_2) - \omega_2 H(\alpha_2 - 1)\right]\{(\alpha_2 - s_2\gamma_1 H)e^{-\eta b} - (\alpha_2 + s_2\gamma_1 H)e^{\eta b}\}}{\left[(\alpha_2 + s_2\gamma_1 H)e^{\eta b} - (\alpha_2 - s_2\gamma_1 H)e^{-\eta b}\right]\left[(\alpha_2 + s_2\gamma_1 H)e^{\eta b} + (\alpha_2 - s_2\gamma_1 H)e^{-\eta b}\right]}$$

$$A_1 = -\frac{\left[\alpha_2(k_1 - \eta_2) - \omega_2 H(\alpha_2 - 1)\right]}{\left[(\alpha_2 + s_2\gamma_1 H)e^{\eta b} + (\alpha_2 - s_2\gamma_1 H)e^{-\eta b}\right]}$$

$$A_1 = \frac{\left[\omega_2 H(\alpha_2 - 1) - \alpha_2(k_1 - \eta_2)\right]}{\left[(\alpha_2 + s_2\gamma_1 H)e^{\eta b} + (\alpha_2 - s_2\gamma_1 H)e^{-\eta b}\right]} \tag{1}$$

$$A_2 = A_1 \tag{2}$$

And A_3 & A_4 are as before

$$A_3 = \frac{1}{H\alpha_2}\left[A_1 e^{\eta b} + A_2 e^{-\eta b} + k_1 - \omega_2 H - \eta_2\right] \tag{3}$$

$$A_5 = \frac{1}{H\alpha_3}\left[A_1 e^{-\eta b} + A_2 e^{\eta b} + k_1 - \omega_3 H - \eta_3\right] \tag{4}$$

Comments on Case T4:
This demonstrates again that it is relatively straightforward to adapt the solution for one case to another case, provided it is done with due care. Since a triangular channel may be created by allowing $b \to 0$ in the trapezoidal shaped channel cases just shown, 3 further examples are now given by retaining panels 2 & 3 in Figure 6, omitting panel 1, as indicated in Figure 7 below.

Case T5 Asymmetric flow in a triangular channel with unequal side slopes, using two panels

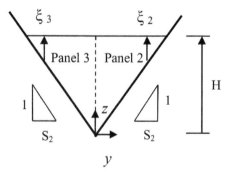

FigureT5 Triangular channel.

Direct solution from first principles:

bc (1)   At edges where depths are zero ($\xi_2 = 0$ or $y = s_2 H$ and $\xi_3 = 0$ or $y = -s_3 H$)

$$U_d^{(2)}\big|_{\xi_2=0}=0 \quad \text{and} \quad U_d^{(3)}\big|_{\xi_3=0}=0 \tag{1}$$

bc (2)   At interface between panels,

$$U_d^{(2)}\big|_{y=0}=U_d^{(3)}\big|_{\xi=H} \quad \text{and} \quad \mu_2\frac{\partial U_d^{(2)}}{\partial y}\bigg|_{y=0,\xi_2=H}=\mu_3\frac{\partial U_d^{(3)}}{\partial y}\bigg|_{y=0,\xi_3=H} \tag{2}$$

where

$$\mu=\lambda\sqrt{\frac{f}{8}} \tag{3}$$

and the local depths, ξ_2 & ξ_3, on the sloping side walls are

$$\xi_2=H-\frac{y}{s_2} \quad \text{and} \quad \xi_3=H+\frac{y}{s_3} \tag{4}$$

The lateral distribution of U_d in panels 2 & 3 are given by

$$U_d^{(2)}=\left[A_3\xi_2^{\alpha_2}+A_4\xi_2^{-(\alpha_2+1)}+\omega_2\xi_2+\eta_2\right]^{1/2} \tag{5}$$

and

$$U_d^{(3)}=\left[A_5\xi_3^{\alpha_3}+A_6\xi_3^{-(\alpha_3+1)}+\omega_3\xi_3+\eta_3\right]^{1/2} \tag{6}$$

where

$$\alpha_2=-\frac{1}{2}+\frac{1}{2}\left[1+\frac{s_2\left(1+s_2^2\right)^{1/2}}{\lambda_2}(8f_2)^{1/2}\right]^{1/2} \tag{7}$$

and

$$\alpha_3=-\frac{1}{2}+\frac{1}{2}\left[1+\frac{s_3\left(1+s_3^2\right)^{1/2}}{\lambda_3}(8f_3)^{1/2}\right]^{1/2} \tag{8}$$

$$\omega_2=\frac{gS_0}{\dfrac{(1+s_2^2)^{1/2}}{s_2}\left(\dfrac{f_2}{8}\right)-\dfrac{\lambda_2}{s_2^2}\left(\dfrac{f_2}{8}\right)^{1/2}} \tag{9}$$

and

$$\omega_3=\frac{gS_0}{\dfrac{(1+s_3^2)^{1/2}}{s_3}\left(\dfrac{f_3}{8}\right)-\dfrac{\lambda_3}{s_3^2}\left(\dfrac{f_3}{8}\right)^{1/2}} \tag{10}$$

$$\eta_2 = -\frac{\Gamma_2}{\dfrac{(1+s_2^2)^{1/2}}{s_2}\rho\left(\dfrac{f_2}{8}\right)} \tag{11}$$

and

$$\eta_3 = -\frac{\Gamma_3}{\dfrac{(1+s_3^2)^{1/2}}{s_3}\rho\left(\dfrac{f_3}{8}\right)} \tag{12}$$

From (3) & (4) and noting that $\dfrac{\partial \xi_2}{\partial y} = -\dfrac{1}{s_2}$ and $\dfrac{\partial \xi_3}{\partial y} = \dfrac{1}{s_3}$

$$\frac{\partial U_d^{(2)}}{\partial y} = -\frac{\left[A_3\alpha_2\,\xi_2^{\alpha_2-1} - A_4(\alpha_2+1)\xi_2^{-\alpha_2-2} + \omega_2\right]}{2s_2\left[A_3\xi_2^{\alpha_2} + A_4\xi_2^{-(\alpha_2+1)} + \omega_2\xi_2 + \eta_2\right]^{1/2}} \tag{13}$$

$$\frac{\partial U_d^{(3)}}{\partial y} = \frac{\left[A_5\alpha_3\,\xi_3^{\alpha_3-1} - A_6(\alpha_3+1)\xi_3^{-\alpha_3-2} + \omega_3\right]}{2s_3\left[A_5\xi_3^{\alpha_3} + A_6\xi_3^{-(\alpha_3+1)} + \omega_3\xi_3 + \eta_3\right]^{1/2}} \tag{14}$$

From (5) & (6), as $\xi \to 0$, $\left(A_4\xi_2^{-(\alpha_2+1)} + \eta_2\right)$ must also $\to 0$, so

$$A_4 + \eta_2\xi_2^{\alpha_2+1} \to 0 \text{ as } \xi_2 \to 0, \therefore A_4 = 0 \text{ and } A_6 + \eta_3\xi_3^{\alpha_3+1} \to 0 \text{ as } \xi_3 \to 0, \therefore A_6 = 0$$
$$A_4 = A_6 = 0 \tag{15}$$

bc (1) gives

$$A_3 H^{\alpha_2} + \omega_2 H + \eta_2 = A_5 H^{\alpha_3} + \omega_3 H + \eta_3 \tag{16}$$

bc (2) gives

$$-\frac{\mu_2\left[A_3\alpha_2\,\xi_2^{\alpha_2-1} + \omega_2\right]}{2s_2\left[A_3\xi_2^{\alpha_2} + \omega_2\xi_2 + \eta_2\right]^{1/2}} = \frac{\mu_3\left[A_5\alpha_3\,\xi_3^{\alpha_3-1} + \omega_3\right]}{2s_3\left[A_5\xi_3^{\alpha_3} + \omega_3\xi_3 + \eta_3\right]^{1/2}}$$

$$s_3\left(\frac{\mu_2}{\mu_3}\right)\left[A_3\alpha_2\,H^{\alpha_2-1} + \omega_2\right] = -s_2\left[A_5\alpha_3\,H^{\alpha_3-1} + \omega_3\right] \tag{17}$$

Elimination process

a.   Eliminate A_5 by $\alpha_3 s_2 \times (16) + H \times (17)$

$$\alpha_3 s_2\left\{A_3 H^{\alpha_2} + \omega_2 H + \eta_2 - A_5 H^{\alpha_3} - \omega_3 H - \eta_3\right\} + H\left\{s_3\left(\frac{\mu_2}{\mu_3}\right)\left[A_3\alpha_2\,H^{\alpha_2-1} + \omega_2\right]\right.$$

$$\left. + s_2\left[A_5\alpha_3\,H^{\alpha_3-1} + \omega_3\right]\right\} = 0$$

$$\left(\alpha_3 s_2 H^{\alpha_2} + \left(\frac{\mu_2}{\mu_3}\right)\alpha_2 s_3 H^{\alpha_2}\right)A_3 + \alpha_3 s_2\left(\omega_2 H + \eta_2 - \omega_3 H - \eta_3\right) + \left(\frac{\mu_2}{\mu_3}\right)\omega_2 s_3 H + \omega_3 s_2 H = 0$$

$$\left(\alpha_3 s_2 + \left(\frac{\mu_2}{\mu_3}\right)\alpha_2 s_3\right)H^{\alpha_2}A_3 + \alpha_3 s_2 H\left(\omega_2 - \omega_3\right) + \alpha_3 s_2\left(\eta_2 - \eta_3\right) + \omega_3 s_2 H + \omega_2 s_3 H\left(\frac{\mu_2}{\mu_3}\right) = 0$$

$$A_3 = \frac{\left(\omega_3 - \omega_2\right)\alpha_3 s_2 H - \left\{\omega_2 s_3\left(\frac{\mu_2}{\mu_3}\right) + \omega_3 s_2\right\}H + \left(\eta_3 - \eta_2\right)\alpha_3 s_2}{\left[\alpha_3 s_2 + \left(\frac{\mu_2}{\mu_3}\right)\alpha_2 s_3\right]H^{\alpha_2}} \tag{18}$$

From (16)

$$A_5 = \frac{1}{H^{\alpha_3}}\left[A_3 H^{\alpha_2} + \omega_2 H + \eta_2 - \omega_3 H - \eta_3\right] = \frac{1}{H^{\alpha_3}}\left[A_3 H^{\alpha_2} + \left(\omega_2 - \omega_3\right)H + \left(\eta_2 - \eta_3\right)\right]$$

$$A_5 = \frac{1}{H^{\alpha_3}}\left\{\left[\frac{\left(\omega_3 - \omega_2\right)\alpha_3 s_2 H - \left\{\omega_2 s_3\left(\frac{\mu_2}{\mu_3}\right) + \omega_3 s_2\right\}H + \left(\eta_3 - \eta_2\right)\alpha_3 s_2}{\left[\alpha_3 s_2 + \left(\frac{\mu_2}{\mu_3}\right)\alpha_2 s_3\right]}\right]\right.$$

$$\left.+ \left(\omega_2 - \omega_3\right)H + \left(\eta_2 - \eta_3\right)\right]$$

The numerator is

$$\left(\omega_3 - \omega_2\right)\alpha_3 s_2 H - \left\{\omega_2 s_3\left(\frac{\mu_2}{\mu_3}\right) + \omega_3 s_2\right\}H + \left(\eta_3 - \eta_2\right)\alpha_3 s_2 + \left(\omega_2 - \omega_3\right)H\left[\alpha_3 s_2 + \left(\frac{\mu_2}{\mu_3}\right)\alpha_2 s_3\right] +$$

$$\left(\eta_2 - \eta_3\right)\left[\alpha_3 s_2 + \left(\frac{\mu_2}{\mu_3}\right)\alpha_2 s_3\right]$$

$$= -\left\{\omega_2 s_3\left(\frac{\mu_2}{\mu_3}\right) + \omega_3 s_2\right\}H + \left(\omega_2 - \omega_3\right)H\alpha_2 s_3\left(\frac{\mu_2}{\mu_3}\right) + \left(\eta_2 - \eta_3\right)\alpha_2 s_3\left(\frac{\mu_2}{\mu_3}\right)$$

$$= \left(\omega_2 - \omega_3\right)H\alpha_2 s_3\left(\frac{\mu_2}{\mu_3}\right) - \left\{\omega_2 s_3\left(\frac{\mu_2}{\mu_3}\right) + \omega_3 s_2\right\}H + + \left(\eta_2 - \eta_3\right)\alpha_2 s_3\left(\frac{\mu_2}{\mu_3}\right)$$

$$A_5 = \frac{\left(\omega_2 - \omega_3\right)H\alpha_2 s_3\left(\frac{\mu_2}{\mu_3}\right) - \left\{\omega_2 s_3\left(\frac{\mu_2}{\mu_3}\right) + \omega_3 s_2\right\}H + \left(\eta_2 - \eta_3\right)\alpha_2 s_3\left(\frac{\mu_2}{\mu_3}\right)}{\left[\alpha_3 s_2 + \left(\frac{\mu_2}{\mu_3}\right)\alpha_2 s_3\right]H^{\alpha_3}} \tag{19}$$

Comments on Case T5:
This demonstrates again that it is relatively straightforward to develop algebraic expressions for the coefficients in the SKM for simple cases.

CaseT6 Asymmetric flow in a triangular channel with equal side slopes, using two panels

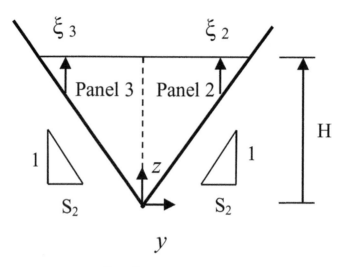

Figure T6 Triangular channel.

Ignore μ factors, but retain previous panel notation, even though $s_2 = s_3$ and let $\mu_2 / \mu_3 = 1$

bc (1) At edges where depths are zero ($\xi_2 = 0$ or $y = s_2 H$ and $\xi_3 = 0$ or $y = -s_3 H$)

$$U_d^{(2)} \big|_{\xi_2=0} = 0 \quad \text{and} \quad U_d^{(3)} \big|_{\xi_3=0} = 0 \tag{1}$$

bc (2) At interface between panels,

$$U_d^{(2)} \big|_{y=0} = U_d^{(3)} \big|_{\xi=H} \quad \text{and} \quad \frac{\partial U_d^{(2)}}{\partial y}\bigg|_{y=0,\xi_2=H} = \frac{\partial U_d^{(3)}}{\partial y}\bigg|_{y=0,\xi_3=H} \tag{2}$$

where

$$\mu = \lambda \sqrt{\frac{f}{8}} \tag{3}$$

and the local depths, ξ_2 & ξ_3, on the sloping side walls are

$$\xi_2 = H - \frac{y}{s_2} \quad \text{and} \quad \xi_3 = H + \frac{y}{s_3} \tag{4}$$

The lateral distribution of U_d in panels 2 & 3 are given by

$$U_d^{(2)} = \left[A_3 \xi_2^{\alpha_2} + A_4 \xi_2^{-(\alpha_2+1)} + \omega_2 \xi_2 + \eta_2 \right]^{1/2} \quad (5) \quad \text{and} \quad U_d^{(3)} = \left[A_5 \xi_3^{\alpha_3} + A_6 \xi_3^{-(\alpha_3+1)} + \omega_3 \xi_3 + \eta_3 \right]^{1/2} \quad (6)$$

where

$$\alpha_2 = -\frac{1}{2} + \frac{1}{2}\left[1 + \frac{s_2(1+s_2^2)^{1/2}}{\lambda_2}(8f_2)^{1/2} \right]^{1/2} \quad (7) \quad \text{and} \quad \alpha_3 = -\frac{1}{2} + \frac{1}{2}\left[1 + \frac{s_3(1+s_3^2)^{1/2}}{\lambda_3}(8f_3)^{1/2} \right]^{1/2} \quad (8)$$

$$\omega_2 = \frac{gS_0}{\dfrac{(1+s_2^2)^{1/2}}{s_2}\left(\dfrac{f_2}{8}\right) - \dfrac{\lambda_2}{s_2^2}\left(\dfrac{f_2}{8}\right)^{1/2}} \quad (9) \quad \text{and} \quad \omega_3 = \frac{gS_0}{\dfrac{(1+s_3^2)^{1/2}}{s_3}\left(\dfrac{f_3}{8}\right) - \dfrac{\lambda_3}{s_3^2}\left(\dfrac{f_3}{8}\right)^{1/2}} \quad (10)$$

$$\eta_2 = -\frac{\Gamma_2}{\dfrac{(1+s_2^2)^{1/2}}{s_2}\rho\left(\dfrac{f_2}{8}\right)} \quad (11) \quad \text{and} \quad \eta_3 = -\frac{\Gamma_3}{\dfrac{(1+s_3^2)^{1/2}}{s_3}\rho\left(\dfrac{f_3}{8}\right)} \quad (12)$$

From (3) & (4) and noting that $\dfrac{\partial \xi_2}{\partial y} = -\dfrac{1}{s_2}$ and $\dfrac{\partial \xi_3}{\partial y} = \dfrac{1}{s_3}$

$$\frac{\partial U_d^{(2)}}{\partial y} = -\frac{\left[A_3\alpha_2 \xi_2^{\alpha_2-1} - A_4(\alpha_2+1)\xi_2^{-\alpha_2-2} + \omega_2 \right]}{2s_2\left[A_3\xi_2^{\alpha_2} + A_4\xi_2^{-(\alpha_2+1)} + \omega_2\xi_2 + \eta_2 \right]^{1/2}} \quad (13)$$

$$\frac{\partial U_d^{(3)}}{\partial y} = \frac{\left[A_5\alpha_3 \xi_3^{\alpha_3-1} - A_6(\alpha_3+1)\xi_3^{-\alpha_3-2} + \omega_3 \right]}{2s_3\left[A_5\xi_3^{\alpha_3} + A_6\xi_3^{-(\alpha_3+1)} + \omega_3\xi_3 + \eta_3 \right]^{1/2}} \quad (14)$$

From (5) & (6), as $\xi \to 0$, $\left(A_4\xi_2^{-(\alpha_2+1)} + \eta_2 \right)$ must also $\to 0$, so

$A_4 + \eta_2\xi_2^{\alpha_2+1} \to 0$ as $\xi_2 \to 0$, $\therefore A_4 = 0$ and $A_6 + \eta_3\xi_3^{\alpha_3+1} \to 0$ as $\xi_3 \to 0$, $\therefore A_6 = 0$
$A_4 = A_6 = 0$ (15)

bc (1) gives

$$A_3 H^{\alpha_2} + \omega_2 H + \eta_2 = A_5 H^{\alpha_3} + \omega_3 H + \eta_3 \quad (16)$$

bc (2) gives

$$-\frac{\left[A_3\alpha_2\,\xi_2^{\alpha_2-1}+\omega_2\right]}{2s_2\left[A_3\xi_2^{\alpha_2}+\omega_2\xi_2+\eta_2\right]^{1/2}}=\frac{\left[A_5\alpha_3\,\xi_3^{\alpha_3-1}+\omega_3\right]}{2s_3\left[A_5\xi_3^{\alpha_3}+\omega_3\xi_3+\eta_3\right]^{1/2}}$$

$$s_3\left[A_3\alpha_2\,H^{\alpha_2-1}+\omega_2\right]=-s_2\left[A_5\alpha_3\,H^{\alpha_3-1}+\omega_3\right] \tag{17}$$

a.　Eliminate A_5 by $\alpha_3 s_2$ x (16) + H x (17)

$$\alpha_3 s_2\left\{A_3 H^{\alpha_2}+\omega_2 H+\eta_2-A_5 H^{\alpha_3}-\omega_3 H-\eta_3\right\}+H\left\{s_3\left[A_3\alpha_2\,H^{\alpha_2-1}+\omega_2\right]\right.$$
$$\left.+s_2\left[A_5\alpha_3\,H^{\alpha_3-1}+\omega_3\right]\right\}=0$$

$$\left(\alpha_3 s_2 H^{\alpha_2}+\alpha_2 s_3 H^{\alpha_2}\right)A_3+\alpha_3 s_2\left(\omega_2 H+\eta_2-\omega_3 H-\eta_3\right)+\omega_2 s_3 H+\omega_3 s_2 H=0$$

$$\left(\alpha_3 s_2+\alpha_2 s_3\right)H^{\alpha_2}A_3+\alpha_3 s_2 H\left(\omega_2-\omega_3\right)+\alpha_3 s_2\left(\eta_2-\eta_3\right)+\omega_2 s_3 H+\omega_3 s_2 H=0$$

$$A_3=\frac{\left(\omega_3-\omega_2\right)\alpha_3 s_2 H-\left\{\omega_2 s_3+\omega_3 s_2\right\}H+\left(\eta_3-\eta_2\right)\alpha_3 s_2}{\left[\alpha_3 s_2+\alpha_2 s_3\right]H^{\alpha_2}} \tag{18}$$

From (16)

$$A_5=\frac{1}{H^{\alpha_3}}\left[A_3 H^{\alpha_2}+\left(\omega_2-\omega_3\right)H+\left(\eta_2-\eta_3\right)\right]$$

$$A_5=\frac{1}{H^{\alpha_3}}\left[\left\{\frac{\left(\omega_3-\omega_2\right)\alpha_3 s_2 H-\left\{\omega_2 s_3+\omega_3 s_2\right\}H+\left(\eta_3-\eta_2\right)\alpha_3 s_2}{\left[\alpha_3 s_2+\alpha_2 s_3\right]}\right\}+\left(\omega_2-\omega_3\right)H+\left(\eta_2-\eta_3\right)\right]$$

$$A_5=\left\{\frac{\begin{array}{l}\left(\omega_3-\omega_2\right)\alpha_3 s_2 H-\left\{\omega_2 s_3+\omega_3 s_2\right\}H+\left(\eta_3-\eta_2\right)\alpha_3 s_2+\\\left(\alpha_3 s_2+\alpha_2 s_3\right)\left(\omega_2-\omega_3\right)H+\left(\alpha_3 s_2+\alpha_2 s_3\right)\left(\eta_2-\eta_3\right)\end{array}}{\left[\alpha_3 s_2+\alpha_2 s_3\right]H^{\alpha_3}}\right\}$$

The numerator is

$$\left(\omega_3-\omega_2\right)\alpha_3 s_2 H-\left\{\omega_2 s_3+\omega_3 s_2\right\}H+\left(\eta_3-\eta_2\right)\alpha_3 s_2+\left(\omega_2-\omega_3\right)H\left[\alpha_3 s_2+\alpha_2 s_3\right]$$
$$+\left(\eta_2-\eta_3\right)\left[\alpha_3 s_2+\alpha_2 s_3\right]=-\left\{\omega_2 s_3+\omega_3 s_2\right\}H+\left(\omega_2-\omega_3\right)H\alpha_2 s_3+\left(\eta_2-\eta_3\right)\alpha_2 s_3$$

$$A_5=\frac{\left(\omega_2-\omega_3\right)H\alpha_2 s_3-\left\{\omega_2 s_3+\omega_3 s_2\right\}H+\left(\eta_2-\eta_3\right)\alpha_2 s_3}{\left[\alpha_3 s_2+\alpha_2 s_3\right]H^{\alpha_3}} \tag{19}$$

Comments on Case T6:
This allows for asymmetry in the flow even when the side slopes are the same, due to possibly frictional resistance parameters being different in each panel.

Case T7 Symmetric flow in a triangular channel, using one panel only

For this case $\omega_2 = \omega_3 = \omega$; $\eta_2 = \eta_3 = \eta$; and $\alpha_2 = \alpha_3 = \alpha$, thus from (18) & (19) in Case T6

$$A_3 = -\frac{(s_3 + s_2)\omega H}{(s_2 + s_3)\alpha H^\alpha}; \quad A_5 = -\frac{(s_3 + s_2)\omega H}{(s_2 + s_3)\alpha H^\alpha} \quad \text{and since } s_2 = s_3, \text{ then}$$

$$A_3 = -\frac{\omega}{\alpha}H^{1-\alpha} : A_5 = -\frac{\omega}{\alpha}H^{1-\alpha} \tag{1}$$

Note that the same result for flow in a triangular channel may be obtained from the trapezoidal channel case, Case T4, by letting $b = 0$, and retaining only two panels, with $s_3 = s_2$. Thus from (3) in T4,

$$A_3 = \frac{1}{H^\alpha}[A_1 + A_2 + k_1 - \omega H - \eta] = \frac{1}{H^\alpha}[2A_1 + k_1 - \omega H - \eta]$$

From (1) in T4,

$$A_1 = \frac{[(\alpha - 1)\omega H - \alpha(k - \eta)]}{2\alpha}$$

Hence

$$A_3 = \frac{1}{H^\alpha}\frac{-\omega H}{\alpha} = -\frac{\omega}{\alpha}H^{1-\alpha} \tag{2}$$

Comments on Case T7:
This result may of course be noted directly from (13) of T5 with $A_4 = 0$. This is a simple case used frequently in urban drainage. A more common urban drainage channel where both sediment transport and flow capacity are required is a simple extension of a triangular channel with vertical side walls, as shown in Figure U1, described next.

Case U1 Urban drainage type channel as a triangular shape with vertical walls, using one panel

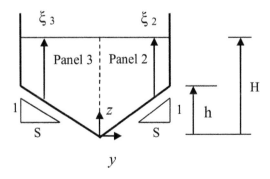

Figure U1 Urban drainage section or bridge deck drainage channel.

For symmetric flow, then panel 2 the velocity distribution is given (5) in T6 as

$$U_d^{(2)} = \left[A_3 \xi_2^{\alpha_2} + A_4 \xi_2^{-(\alpha_2+1)} + \omega_2 \xi_2 + \eta_2 \right]^{1/2}$$

With boundary conditions as $U_d^{(2)} \big|_{\xi_2 = H - h} = 0$ (edge) and $\dfrac{\partial U_d^{(2)}}{\partial y} \bigg|_{y=0,\, \xi_2 = H}$ (centerline)
Hence

$$\frac{\partial U_d^{(2)}}{\partial y} = -\frac{\left[A_3 \alpha_2 \xi_2^{\alpha_2-1} - A_4(\alpha_2+1)\xi_2^{-\alpha_2-2} + \omega_2 \right]}{2 s_2 \left[A_3 \xi_2^{\alpha_2} + A_4 \xi_2^{-(\alpha_2+1)} + \omega_2 \xi_2 + \eta_2 \right]^{1/2}}$$

$$A_3 \alpha H^{\alpha-1} - A_4(\alpha+1)H^{-\alpha-2} + \omega = 0 \tag{1}$$

and

$$A_3(H-h)^{\alpha} + A_4(H-h)^{-(\alpha+1)} + \omega(H-h) + \eta = 0 \tag{2}$$

Eliminate A_4 by $(1) \times (H-h)^{-(\alpha+1)} + (2) \times (\alpha+1)H^{-(\alpha+2)}$

$$(H-h)^{-(\alpha+1)} \left\{ A_3 \alpha H^{\alpha-1} + \omega \right\} + (\alpha+1)H^{-(\alpha+2)} \left\{ A_3(H-h)^{\alpha} + \omega(H-h) + \eta \right\} = 0$$

Hence

$$A_3 = -\frac{\omega(H-h)^{-(\alpha+1)} + \omega(\alpha+1)(H-h)H^{-(\alpha+2)} + \eta(\alpha+1)H^{-(\alpha+2)}}{\alpha(H-h)^{-(\alpha+1)} H^{(\alpha-1)} + (\alpha+1)(H-h)^{\alpha} H^{-(\alpha+2)}} \tag{3}$$

$$A_4 = \frac{A_3 \alpha H^{(\alpha-1)} + \omega}{(\alpha+1)H^{-(\alpha+2)}} \tag{4}$$

or

$$A_4 = -\frac{A_3(H-h)^{\alpha} + \omega(H-h) + \eta}{(H-h)^{-(\alpha+1)}} \tag{4b}$$

Note that multiplying the numerator and denominator of (3) by $(H-h)^{(\alpha+1)}$ gives

$$A_3 = -\frac{\omega + \omega(\alpha+1)(H-h)^{\alpha+2} H^{-(\alpha+2)} + \eta(\alpha+1)(H-h)^{(\alpha+1)} H^{-(\alpha+2)}}{\alpha H^{(\alpha-1)} + (\alpha-1)(H-h)^{2\alpha+1} H^{-(\alpha+2)}} \tag{3b}$$

so that when $(H-h) = 0$, which is a triangular channel, then (3b) becomes

$$A_3 = -\frac{\omega}{\alpha} H^{(1-\alpha)} \tag{5}$$

which agrees with (1) in Case T7

Comments on Case U1:
This illustrates that care needs to be taken in simplifying the algebra down from a more complex case in order to produce another result, and that it is often easier to obtain the simpler case directly from first principles. Other shapes of urban drainage channels are typically circular pipes running part full, with or without sediment beds, which makes the application of the SKM more challenging unless sufficient linear elements are provided to make up the non-linear boundary. Attempts at doing this may be found in Knight & Sterling (2000) and in Sterling & Knight (2000, 2001 & 2002).

6.4.3   Flows in compound channels

Three cases are considered, one for flow in a symmetric trapezoidal compound channel and two for flows in a compound channel with either a very steep internal wall or a vertical wall, making it a rectangular compound channel.

Case C1 Symmetric flow in a trapezoidal compound channel, using four panels

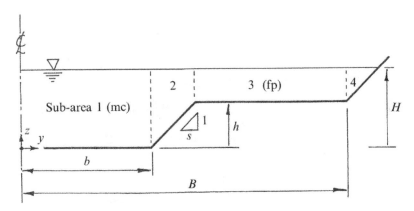

Figure C1 Symmetric compound trapezoidal channel, using 4 panels.

Let s_2 = main channel side slope; s_4 = floodplain side slope; Let $b_1 = b + s_2 h$
The velocities in panels 1 – 4 are as follows:

$$U_d^{(1)} = \left[A_1 e^{\gamma_1 y} + A_2 e^{-\gamma_1 y} + k_1 \right]^{1/2} \quad \text{for} \quad 0 \le y \le b \quad \text{depth} = H \tag{1}$$

$$U_d^{(2)} = \left[A_3 \xi_2^{\alpha_2} + A_4 \xi_2^{-(\alpha_2+1)} + \omega_2 \xi_2 + \eta_2 \right]^{1/2} \quad \text{for} \quad b \le y \le (b + s_2 h)$$
$$\text{or} \quad H \ge \xi_2 \ge (H - h) \tag{2}$$

$$U_d^{(3)} = \left[A_5 e^{\gamma_3 y} + A_6 e^{-\gamma_3 y} + k_3 \right]^{1/2} \quad \text{for} \quad (b + s_2 h) \le y \le B \quad \text{or} \quad H_3 = (H - h) \tag{3}$$

$$U_d^{(4)} = \left[A_7 \xi_4^{\alpha_4} + A_8 \xi_4^{-(\alpha_4+1)} + \omega_4 \xi_4 + \eta_4 \right]^{1/2} \quad \text{for} \quad B \le y \le B + s_4 (H - h)$$
$$\text{or} \quad (H - h) \xi_4 \ge 0 \tag{4}$$

The boundary conditions are:

$\partial U_d / \partial y = 0$ (at centerline, due to imposed symmetry of flow, sub-area 1);   (5)

$U_{di} = U_{di+1}$ and $\mu_i \partial U_{di} / \partial y = \mu_{i+1} \partial U_{di+1} / \partial y$ (at the 3 internal joints between panels)   (6)

$U_d = 0$ (at edge of free surface on floodplain side slope, sub-area 4);   (7)

For a flat bed $\dfrac{\partial U_d^{(1)}}{\partial y} = \dfrac{\gamma_1 \left[A_1 e^{\gamma_1 y} - A_2 e^{-\gamma_1 y} \right]}{2 \left[A_1 e^{\gamma_1 y} + A_2 e^{-\gamma_1 y} + k_1 \right]^{1/2}}$ (8)

For a sloping bed $\dfrac{\partial U_d^{(2)}}{\partial y} = -\dfrac{\left[A_3 \alpha_2 \xi_2^{\alpha_2 - 1} - A_4 (\alpha_2 + 1) \xi_2^{-\alpha_2 - 2} + \omega_2 \right]}{2 s_2 \left[A_3 \xi_2^{\alpha_2} + A_4 \xi_2^{-\alpha_2 - 1} + \omega_2 \xi_2 + \eta_2 \right]^{1/2}}$ (9)

Hence there are 8 equations for 8 unknowns

From (1), (5) & (8)

$A_1 = A_2$ (10)

From (6)

$A_1 e^{\gamma_1 b} + A_2 e^{-\gamma_1 b} + k_1 = A_3 H^{\alpha_2} + A_4 H^{-(\alpha_2 + 1)} + \omega_2 H + \eta_2$ (11)

and

$\dfrac{\mu_1 \gamma_1 \left[A_1 e^{\gamma_1 b} - A_2 e^{-\gamma_1 b} \right]}{2 \left[A_1 e^{\gamma_1 b} + A_2 e^{-\gamma_1 b} + k_1 \right]^{1/2}} = -\dfrac{\mu_2 \left[A_3 \alpha_2 H^{\alpha_2 - 1} - A_4 (\alpha_2 + 1) H^{-(\alpha_2 + 2)} + \omega_2 \right]}{2 s_2 \left[A_3 H^{\alpha_2} + A_4 H^{-(\alpha_2 + 1)} + \omega_2 H + \eta_2 \right]^{1/2}}$

So, using (11)

$\mu_1 \gamma_1 s_2 \left[A_1 e^{\gamma_1 b} - A_2 e^{-\gamma_1 b} \right] + \mu_2 \alpha_2 A_3 H^{\alpha_2 - 1} - \mu_2 (\alpha_2 + 1) A_4 H^{-(\alpha_2 + 2)} + \mu_2 \omega_2 = 0$ (12)

From (6)

$A_3 H_3^{\alpha_2} + A_4 H_3^{-(\alpha_2 + 1)} + \omega_2 H_3 + \eta_2 - A_5 e^{\gamma_3 b_1} - A_6 e^{-\gamma_3 b_1} - k_3 = 0$ (13)

$\mu_2 \alpha_2 A_3 H_3^{\alpha_2 - 1} - \mu_2 (\alpha_2 + 1) A_4 H_3^{-(\alpha_2 + 2)} + \mu_2 \omega_2 + \mu_3 \gamma_3 s_2 A_5 e^{\gamma_3 b_1}$
$\quad - \mu_3 \gamma_3 s_2 A_6 e^{-\gamma_3 b_1} = 0$ (14)

From (6)

$A_5 e^{\gamma_3 B} + A_6 e^{-\gamma_3 B} + k_3 - A_7 H_3^{\alpha_4} - \omega_4 H_3 - \eta_4 = 0$ (15)

$\mu_3 \gamma_3 s_4 A_5 e^{\gamma_3 B} - \mu_3 \gamma_3 s_4 A_6 e^{-\gamma_3 B} + \mu_4 \alpha_4 A_7 H_3^{\alpha_4 - 1} + \mu_4 \omega_4 = 0$ (16)

From (7)

$A_8 = 0$ (17)

In matrix form these are

$$
\begin{bmatrix}
1 & -1 & 0 & 0 & 0 \\
e^{\gamma_1 b} & e^{-\gamma_1 b} & -H^{\alpha_2} & -H^{-(\alpha_2+1)} & 0 \\
\mu_1\gamma_1 s_2 e^{\gamma_1 b} & -\mu_1\gamma_1 s_2 e^{-\gamma_1 b} & \mu_2\alpha_2 H^{\alpha_2-1} & -\mu_2(\alpha_2+1)H^{-(\alpha_2+2)} & 0 \\
0 & 0 & H_3^{\alpha_2} & H_3^{-(\alpha_2+1)} & -e^{\gamma_3 b_1} \\
0 & 0 & \mu_2\alpha_2 H_3^{\alpha_2-1} & -\mu_2(\alpha_2+1)H_3^{-(\alpha_2+2)} & \mu_3\gamma_3 s_2 e^{\gamma_3 b_1} \\
0 & 0 & 0 & 0 & e^{\gamma_3 B} \\
0 & 0 & 0 & 0 & \mu_3\gamma_3 s_4 e^{\gamma_3 B} \\
0 & 0 & 0 & 0 & 0
\end{bmatrix}
$$

$$
\begin{bmatrix}
0 & 0 & 0 \\
0 & 0 & 0 \\
0 & 0 & 0 \\
-e^{-\gamma_3 b_1} & 0 & 0 \\
-\mu_3\gamma_3 s_2 e^{-\gamma_3 b_1} & 0 & 0 \\
e^{-\gamma_3 B} & -H_3^{\alpha_4} & 0 \\
-\mu_3\gamma_3 s_4 e^{-\gamma_3 B} & \mu_4\alpha_4 H_3^{\alpha_4-1} & 0 \\
0 & 0 & 1
\end{bmatrix}
\times
\begin{bmatrix}
A_1 \\ A_2 \\ A_3 \\ A_4 \\ A_5 \\ A_6 \\ A_7 \\ A_8
\end{bmatrix}
=
\begin{bmatrix}
0 \\
\omega_2 H + \eta_2 - k_1 \\
-\mu_2\omega_2 \\
k_3 - \omega_2 H_3 - \eta_2 \\
-\mu_2\omega_2 \\
\omega_4 H_3 + \eta_4 - k_3 \\
-\mu_4\omega_4 \\
0
\end{bmatrix}
$$

These may be solved by using Excel, or some other means. If they are solved algebraically, then the 8 coefficients are:

$$
A_1 = \frac{\left[(\alpha_2-1)\mu_2\omega_2 H + (\eta_2-k_1)\mu_2\alpha_2\right]m_2 + \mu_2(2\alpha_2+1)m_3}{\left[(\mu_2\alpha_2+\mu_1\gamma_1 s_2 H)m_2 + \mu_2(2\alpha_2+1)m_1\right]e^{\gamma_1 b} + \left[(\mu_2\alpha_2-\mu_1\gamma_1 s_2 H)m_2 + \mu_2(2\alpha_2+1)m_1\right]e^{-\gamma_1 b}}
$$

$$
A_2 = A_1
$$

$$
A_3 = \frac{\left[\mu_2(\alpha_2+1)-\mu_1\gamma_1 s_2 H\right]e^{\gamma_1 b}A_1 + \left[\mu_2(\alpha_2+1)+\mu_1\gamma_1 s_2 H\right]e^{-\gamma_1 b}A_2 - \mu_2(\alpha_2+2)\omega_2 H - \mu_2(\alpha_2+1)(\eta_2-k_1)}{\mu_2(2\alpha_2+1)H^{\alpha_2}}
$$

$$
A_4 = \frac{\left[(\mu_2\alpha_2+\mu_1\gamma_1 s_2 H)e^{\gamma_1 b} + (\mu_2\alpha_2-\mu_1\gamma_1 s_2 H)e^{-\gamma_1 b}\right]A_1 - (\alpha_2-1)\mu_2\omega_2 H + (\eta_2-k_1)\mu_2\alpha_2}{\mu_2(2\alpha_2+1)H^{-(\alpha_2+1)}}
$$

$$
A_5 = \frac{(\alpha_4-1)\mu_4\omega_4 H_3 - (k_3-\eta_4)\mu_4\alpha_4 - (\mu_4\alpha_4-\mu_3\gamma_3 s_4 H_3)e^{-\gamma_3 B}A_6}{(\mu_4\alpha_4+\mu_3\gamma_3 s_4 H_3)e^{\gamma_3 B}}
$$

$$
A_6 = \frac{\left[(\mu_2\alpha_2+\mu_3\gamma_3 s_2 H_3)H_3^{\alpha_2-1}\right]A_3 - \left[(\mu_2(\alpha_2+1)-\mu_3\gamma_3 s_2 H_3)\right]H_3^{-(\alpha_2+2)}A_4 - \mu_3\gamma_3 s_2\left[k_3-(\omega_2 H_3+\eta_2)\right] + \mu_2\omega_2}{2\mu_3\gamma_3 s_2 e^{-\gamma_3 b_1}}
$$

$$
A_7 = H_3^{-\alpha_4}\left[A_5 e^{\gamma_3 B} + A_6 e^{-\gamma_3 B} + k_3 - \omega_4 H_3 - \eta_4\right]
$$

$$A_8 = 0$$

where

$$m_1 = t_1 H_3^{\alpha_2-1} H^{-\alpha_2}$$

$$m_2 = t_2 H_3^{-(\alpha_2+2)} H^{\alpha_2+1} - t_1 H_3^{\alpha_2-1} H^{-\alpha_2}$$

$$m_3 = t_1 H_3^{\alpha_2-1} H^{-\alpha_2} \left[\omega_2 H + \eta_2 - k_1\right] + t_3$$

$$t_1 = \left\{ \begin{array}{l} 2\mu_3\gamma_3 s_2 H_3 \left(\mu_4\alpha_4 + \mu_3\gamma_3 s_4 H_3\right) e^{\gamma_3(B-b_1)} + \cdots \\ \cdots \left(\mu_2\alpha_2 + \mu_3\gamma_3 s_2 H_3\right)\left[\left(\mu_4\alpha_4 - \mu_3\gamma_3 s_4 H_3\right) \\ -\left(\mu_4\alpha_4 + \mu_3\gamma_3 s_4 H_3\right)e^{2\gamma_3(B-b_1)}\right]e^{\gamma_3(b_1-B)} \end{array} \right\}$$

$$t_2 = \left\{ \begin{array}{l} 2\mu_3\gamma_3 s_2 H_3 \left(\mu_4\alpha_4 + \mu_3\gamma_3 s_4 H_3\right) e^{\gamma_3(B-b_1)} - \cdots \\ \cdots \left(\mu_2\left(\alpha_2+1\right) - \mu_3\gamma_3 s_2 H_3\right)\left[\left(\mu_4\alpha_4 - \mu_3\gamma_3 s_4 H_3\right) \\ -\left(\mu_4\alpha_4 + \mu_3\gamma_3 s_4 H_3\right)e^{2\gamma_3(B-b_1)}\right]e^{\gamma_3(b_1-B)} \end{array} \right\}$$

$$t_3 = \left\{ \begin{array}{l} -\left[\mu_2\omega_2 + \mu_3\gamma_3 s_2 \left(\omega_2 H_3 + \eta_2 - k_3\right)\right]\left[\left(\mu_4\gamma_4 - \mu_3\gamma_3 s_4 H_3\right) \\ -\left(\mu_4\alpha_4 + \mu_3\gamma_3 s_4 H_3\right)e^{2\gamma_3(B-b_1)}\right]e^{\gamma_3(b_1-B)} + \cdots \\ \cdots 2\mu_3\gamma_3 s_2 \left(\mu_4\alpha_4 + \mu_3\gamma_3 s_4 H_3\right)\left[k_3 - \left(\omega_2 H_3 + \eta_2\right)\right]e^{\gamma_3(B-b_1)} \\ + 2\mu_3\gamma_3 s_2 \left(\alpha_4-1\right)\mu_4\omega_4 H_3 - \cdots 2\mu_3\gamma_3 s_2 \left(k_3 - \eta_4\right)\mu_4\alpha_4 \end{array} \right\}$$

Comments on Case C1:
This illustrates that the 8 governing equations are easily derived, but their solution algebraically becomes somewhat tedious if the coefficients A_1 to A_8 are required for other purposes. The most panels that have been attempted by the authors is 8, using 9 nodes, as this can create a reasonably good fit to many natural rivers with two flood-plains, as shown in Figure 7.1.

Case C2 Symmetric flow in a rectangular compound channel, using two panels

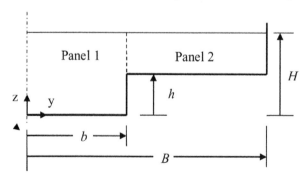

Figure C2 Symmetric rectangular compound channel, using 2 panels.

With two panels, as shown in Figure C2, and from (6.77) the following boundary conditions apply:

$\partial U_d / \partial y = 0$ (at centerline, $y = 0$, due to imposed symmetry of flow);

$U_d = 0$ (at edge of free surface on floodplain edge, $y = B$);

$$\left(H\overline{\tau_{yx}}\right)^i_{y=b} + b\tau_w = \left(H\overline{\tau_{yx}}\right)^{i+1}_{y=b} \quad \text{and} \quad \mu_i \partial U_{di} / \partial y = \mu_{i+1} \partial U_{di+1} / \partial y \quad \text{from} \tag{1}$$

where τ_w is the mean shear stress on the vertical wall. In these simulations, Tang & Knight (2008) used the second form of (6.72) as

$$H\overline{\tau_{yx}} = \rho\lambda H^2 \left(\frac{f}{8}\right)^{1/2} U_d \frac{\partial U_d}{\partial y} \quad \text{or} \quad H\overline{\tau_{yx}} = \frac{1}{2}\rho\lambda H^2 \left(\frac{f}{8}\right)^{1/2} \frac{\partial U_d^2}{\partial y} \tag{2}$$

Thus the force balance (6.77) becomes

$$\left(\phi \frac{\partial U_d^2}{\partial y}\right)^{(1)}_{y=b} = \left(\phi \frac{\partial U_d^2}{\partial y}\right)^{(2)}_{y=b} - b\tau_w \tag{3}$$

where

$$\phi = \frac{1}{2}\rho\lambda H^2 \sqrt{f/8} \quad \text{and} \quad \tau_w = \rho f_w U_{d(at\ y=b)}^2 / 8 \tag{4}$$

where

$$\mu = \lambda \sqrt{\frac{f}{8}} \quad \text{from (6.74) in Section 6.3} \tag{5}$$

The lateral distribution of U_d will be given by

$$U_d^{(1)} = \left[A_1 e^{\gamma_1 y} + A_2 e^{-\gamma_1 y} + k_1\right]^{1/2} \tag{6}$$

where

$$k_1 = \frac{8gHS_0}{f_1}(1 - \beta_1) \tag{7}$$

$$\gamma_1 = \sqrt{\frac{2}{\lambda_1}\left(\frac{f_1}{8}\right)^{1/4} \frac{1}{H}} \tag{8}$$

$$\beta_1 = \frac{\Gamma_1}{\rho g H S_0} \tag{9}$$

$$U_d^{(2)} = \left[A_3 e^{\gamma_2 y} + A_4 e^{-\gamma_2 y} + k_2\right]^{1/2} \tag{10}$$

$$k_2 = \frac{8gHS_0}{f_2}(1-\beta_2)$$

(11)

$$\gamma_2 = \sqrt{\frac{2}{\lambda_2}}\left(\frac{f_2}{8}\right)^{1/4}\frac{1}{H}$$

(12)

$$\beta_2 = \frac{\Gamma_2}{\rho gHS_0}$$

(13)

$$\frac{\mu_1}{2}\frac{\left[A_1\gamma_1e^{\gamma_1 y}-A_2\gamma_1e^{-\gamma_1 y}\right]}{\left[A_1e^{\gamma_1 y}+A_2e^{-\gamma_1 y}+k_1\right]^{1/2}}\bigg|_{y=b_1} = \frac{\mu_2}{2}\frac{\left[A_3\gamma_2e^{\gamma_2 y}-A_4\gamma_2e^{-\gamma_2 y}\right]}{\left[A_3e^{\gamma_2 y}+A_4e^{-\gamma_2 y}+k_2\right]^{1/2}}\bigg|_{y=b_1}$$

(14)

$$A_1 - A_2 = 0$$

(10)

$$A_3e^{\gamma_2 B}+A_4e^{-\gamma_2 B}+k_2 = 0$$

(11)

$$A_1e^{\gamma_1 b}+A_2e^{-\gamma_1 b}+k_1-A_3e^{\gamma_2 b}-A_4e^{-\gamma_2 b}-k_2 = 0$$

(12)

$$\left(\phi\frac{\partial U_d^2}{\partial y}\right)^{(1)}_{y=b} = \left(\phi\frac{\partial U_d^2}{\partial y}\right)^{(2)}_{y=b} - h\overline{\tau}_w$$

(13)

$$\left\{\frac{1}{2}\rho\lambda_1 H^2\sqrt{\frac{f_1}{8}}\right\}\gamma_1\left[A_1e^{\gamma_1 b}-A_2e^{-\gamma_1 b}\right]-\left\{\frac{1}{2}\rho\lambda_2(H-h)^2\sqrt{\frac{f_2}{8}}\right\}\gamma_2\left[A_3e^{\gamma_2 b}-A_4e^{-\gamma_2 b}\right]$$

$$+h\rho\frac{f_w}{8}\left[A_1e^{\gamma_1 b}+A_2e^{-\gamma_1 b}+k_1\right] = 0$$

(14)

$$\left[\left(\phi_1\gamma_1+\rho\frac{f_w}{8}h\right)e^{\gamma_1 b}\right]A_1 - \left[\left(\phi_1\gamma_1-\rho\frac{f_w}{8}h\right)e^{-\gamma_1 b}\right]A_2 - \phi_2\gamma_2e^{\gamma_2 b}A_3 + \phi_2\gamma_2e^{-\gamma_2 b}A_4$$

$$+\rho\frac{f_w}{8}hk_1 = 0$$

(15)

From (11)

$$A_4 = -A_3e^{2\gamma_2 B}-k_2e^{\gamma_2 B}$$

(16)

Substituting (16) & (10) into (12) gives

$$A_1\left[e^{\gamma_1 b}+e^{-\gamma_1 b}\right]+k_1-A_3e^{\gamma_2 b}-e^{-\gamma_2 b}\left[-A_3e^{2\gamma_2 B}-k_2e^{\gamma_2 B}\right]-k_2 = 0$$

$$A_3 = \frac{A_1\left[e^{\gamma_1 b}+e^{-\gamma_1 b}\right]+k_1-k_2\left(1-e^{\gamma_2(B-b)}\right)}{e^{\gamma_2 b}-e^{\gamma_2(2B-b)}}$$

(17)

Substituting (16) & (10) and re-arranging (15) gives

$$\left[\phi_1\gamma_1\left(e^{\gamma_1 b}-e^{-\gamma_1 b}\right)+\rho\frac{f_w}{8}h\left(e^{\gamma_1 b}+e^{-\gamma_1 b}\right)\right]A_1 - \phi_2\gamma_2e^{\gamma_2 b}A_3$$

$$+\phi_2\gamma_2e^{-\gamma_2 b}A_4+\rho\frac{f_w}{8}hk_1 = 0$$

(15a)

$$\left[\phi_1\gamma_1\left(e^{\gamma_1 b}-e^{-\gamma_1 b}\right)+\rho\frac{f_w}{8}h\left(e^{\gamma_1 b}+e^{-\gamma_1 b}\right)\right]A_1-\left[\phi_2\gamma_2 e^{\gamma_2 b}+\phi_2\gamma_2 e^{\gamma_2(2B-b)}\right]A_3$$

$$-\phi_2\gamma_2 k_2 e^{\gamma_2(B-b)}+\rho\frac{f_w}{8}hk_1=0$$

$$A_3=\frac{\left[\phi_1\gamma_1\left(e^{\gamma_1 b}-e^{-\gamma_1 b}\right)+\rho\frac{f_w}{8}h\left(e^{\gamma_1 b}+e^{-\gamma_1 b}\right)\right]A_1-\phi_2\gamma_2 k_2 e^{\gamma_2(B-b)}+\rho\frac{f_w}{8}hk_1}{\phi_2\gamma_2\left(e^{\gamma_2 b}+e^{\gamma_2(2B-b)}\right)}\tag{18}$$

Eliminating A_3 from (17) & (18)

$$A_1\left[e^{\gamma_1 b}-e^{-\gamma_1 b}\right]\phi_2\gamma_2\left(e^{\gamma_e b}+e^{\gamma_2(2B-b)}\right)+\phi_2\gamma_2\left(e^{\gamma_2 b}+e^{\gamma_2(2B-b)}\right)\left[k_1-k_2\left(1-e^{\gamma_2(B-b)}\right)\right]-$$
$$\left(e^{\gamma_2 b}-e^{\gamma_2(2b-b)}\right)\left[\phi_1\gamma_1\left(e^{\gamma_1 b}-e^{-\gamma_1 b}\right)+\rho\frac{f_w}{8}h\left(e^{\gamma_1 b}+e^{-\gamma_1 b}\right)\right]A_1+$$
$$\left(e^{\gamma_2 b}-e^{\gamma_2(2B-b)}\right)\left[\phi_2\gamma_2 k_2 e^{\gamma_2(B-b)}-\rho\frac{f_w}{8}hk_1\right]=0$$

$$A_1=\frac{\phi_2\gamma_2\left(e^{\gamma_2 b}+e^{\gamma_2(2B-b)}\right)\left[k_2\left(1-e^{\gamma_2(B-b)}\right)-k_1\right]-\left(e^{\gamma_2 b}-e^{\gamma_2(2B-b)}\right)\left[\phi_2\gamma_2 k_2 e^{\gamma_2(B-b)}-\rho\frac{f_w}{8}hk_1\right]}{\phi_2\gamma_2\left(e^{\gamma_1 b}+e^{-\gamma_1 b}\right)\left(e^{\gamma_2 b}+e^{\gamma_2(2B-b)}\right)-\phi_1\gamma_1\left(e^{\gamma_1 b}-e^{-\gamma_1 b}\right)\left(e^{\gamma_2 b}-e^{\gamma_2(2B-b)}\right)}\tag{19}$$
$$-\rho\frac{f_w}{8}h\left(e^{\gamma_1 b}+e^{-\gamma_1 b}\right)\left(e^{\gamma_2 b}-e^{\gamma_2(2B-b)}\right)$$

Back substitution into (10), (17) & (16) gives A_2, A_3 & A_4.

Comments on Case C2:
Care is needed over the force balance at the vertical wall. It is possible to use U_d to get very close to the true solution, for the reasons given in Section 6.3, but using q is better.

Case C3 Symmetric flow in a compound channel with a very steep internal step, using three panels

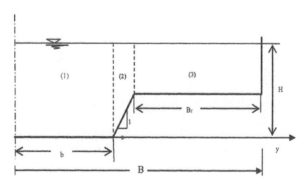

Figure C3 Symmetric compound channel with a very steep internal wall.

Let $b_1 = b + s_2 h$

$$U_d^{(1)} = \left[A_1 e^{\gamma_1 y} + A_2 e^{-\gamma_1 y} + k_1\right]^{1/2} \quad \text{for } 0 \le y \le b \tag{1}$$

$$U_d^{(2)} = \left[A_3 \xi_2^{\alpha_2} + A_4 \xi_2^{-(\alpha_2+1)} + \omega_2 \xi_2 + \eta_2\right]^{1/2} \quad \text{for } 0 \le y \le b + s_2 h \text{ or } H \ge \xi_2 \ge (H - h) \tag{2}$$

$$U_d^{(3)} = \left[A_5 e^{\gamma_3 y} + A_6 e^{-\gamma_3 y} + k_3\right]^{1/2} \quad \text{for } (b + s2h) \le y \le B \text{ or } \xi_3 = (H - h) \tag{3}$$

The boundary conditions are:

$\partial U_d / \partial y = 0$ (at centerline, due to imposed symmetry of flow, sub-area 1); (4)

$U_{di} = U_{di+1}$ and $\mu_i \partial U_{di} / \partial y = \mu_{i+1} \partial U_{di+1} / \partial y$ (at the 2 internal joints between panels) (5)

$U_d = 0$ (at edge of free surface on floodplain, $y = B$, sub-area 3); (6)

For a flat bed

$$\frac{\partial U_d^{(1)}}{\partial y} = \frac{\gamma_1 \left[A_1 e^{\gamma_1 b} - A_2 e^{-\gamma_1 b}\right]}{2\left[A_1 e^{\gamma_1 b} + A_2 e^{-\gamma_1 b} + k_1\right]^{1/2}} \tag{7}$$

For a sloping bed

$$\frac{\partial U_d^{(2)}}{\partial y} = -\frac{\left[A_3 \alpha_2 \xi_2^{\alpha_2-1} - A_4(\alpha_2+1)\xi_2^{-\alpha_2-2} + \omega_2\right]}{2 s_2 \left[A_3 \xi_2^{\alpha_2} + A_4 \xi_2^{-\alpha_2-1} + \omega_2 \xi_2 + \eta_2\right]^{1/2}} \tag{8}$$

Hence there are 6 equations for 6 unknowns
From (1) & (7)

$$A_1 = A_2 \tag{9}$$

From (5)

$$A_1 e^{\gamma_1 b} + A_2 e^{-\gamma_1 b} + k_1 - A_3 H^{\alpha_2} - A_4 H^{-(\alpha_2+1)} - \omega_2 H - \eta_2 = 0 \tag{10}$$

and

$$\frac{\mu_1 \gamma_1 \left[A_1 e^{\gamma_1 b} - A_2 e^{-\gamma_1 b}\right]}{2\left[A_1 e^{\gamma_1 b} + A_2 e^{-\gamma_1 b} + k_1\right]^{1/2}} = -\frac{\mu_2 \left[A_3 \alpha_2 H^{\alpha_2-1} - A_4(\alpha_2+1)H^{-(\alpha_2+2)} + \omega_2\right]}{2 s_2 \left[A_3 H^{\alpha_2} + A_4 H^{-(\alpha_2+1)} + \omega_2 H + \eta_2\right]^{1/2}}$$

So, using (10)

$$\mu_1 \gamma_1 s_2 \left[A_1 e^{\gamma_1 b} - A_2 e^{-\gamma_1 b}\right] + \mu_2 \alpha_2 A_3 H^{\alpha_2-1} - \mu_2 (\alpha_2+1) A_4 H^{-(\alpha_2+2)} + \mu_2 \omega_2 = 0 \tag{11}$$

From (5)

$$A_3 H_3^{\alpha_2} + A_4 H_3^{-(\alpha_2+1)} + \omega_2 H_3 + \eta_2 - A_5 e^{\gamma_3 b_1} - A_6 e^{-\gamma_3 b_1} - k_3 = 0 \tag{12}$$

$$\mu_2 \alpha_2 A_3 H_3^{\alpha_2-1} - \mu_2(\alpha_2+1)A_4 H_3^{-(\alpha_2+2)} + \mu_2 \omega_2 + \mu_3 \gamma_3 s_2 A_5 e^{\gamma_3 b_1} - \mu_3 \gamma_3 s_2 A_6 e^{-\gamma_3 b_1} = 0 \tag{13}$$

From (6)

$$A_5 e^{\gamma_3 B} + A_6 e^{-\gamma_3 B} + k_3 = 0 \tag{14}$$

In matrix form these are

$$
\begin{bmatrix}
1 & -1 & 0 & 0 & 0 & 0 \\
e^{\gamma_1 b} & e^{-\gamma_1 b} & -H^{\alpha_2} & -H^{-(\alpha_2+1)} & 0 & 0 \\
\mu_1 \gamma_1 s_2 e^{\gamma_1 b} & -\mu_1 \gamma_1 s_2 e^{-\gamma_1 b} & \mu_2 \alpha_2 H^{(\alpha_2-1)} & -\mu_2(\alpha_2+1)H^{-(\alpha_2+2)} & 0 & 0 \\
0 & 0 & H_3^{\alpha_2} & H_3^{-(\alpha_2+1)} & -e^{\gamma_3 b_1} & -e^{-\gamma_3 b_1} \\
0 & 0 & \mu_2 \alpha_2 H_3^{(\alpha_2-1)} & -\mu_2(\alpha_2+1)H_3^{-(\alpha_2+2)} & \mu_3 \gamma_3 s_2 e^{\gamma_3 b_1} & -\mu_3 \gamma_3 s_2 e^{-\gamma_3 b_1} \\
0 & 0 & 0 & 0 & e^{\gamma_3 B} & e^{-\gamma_3 B}
\end{bmatrix}
$$

$$
\times
\begin{bmatrix}
A_1 \\ A_2 \\ A_3 \\ A_4 \\ A_5 \\ A_6
\end{bmatrix}
=
\begin{bmatrix}
0 \\
\omega_2 H + \eta_2 - k_1 \\
-\mu_2 \omega_2 \\
k_3 - \omega_2 H_3 - \eta_2 \\
-\mu_2 \omega_2 \\
-k_3
\end{bmatrix}
$$

These may be solved directly or algebraically to give;

$$
A_1 = \frac{\left(t_2 H^{\alpha_2} - t_1 H^{-(\alpha_2+1)}\right)\left[(\alpha_2-1)\mu_2 \omega_2 H + (\eta_2-k_1)\mu_2 \alpha_2\right] + \mu_2(2\alpha_2+1)}{\left(t_2 H^{\alpha_2} - t_1 H^{-(\alpha_2+1)}\right)\left[(\mu_2 \alpha_2 + \mu_1 \gamma_1 s_2 H)e^{\gamma_1 b} + (\mu_2 \alpha_2 - \mu_1 \gamma_1 s_2 H)e^{-\gamma_1 b}\right]} \\
\frac{H^{-(\alpha_2+1)}\left[t_1(\omega_2 H + \eta_2 - k_1)+t_3 H^{\alpha_2}\right]}{+\mu_2(2\alpha_2+1)H^{-(\alpha_2+1)}\left[t_1 e^{\gamma_1 b} + t_1 e^{-\gamma_1 b}\right]}
$$

$$A_2 = A_1$$

$$A_3 = \frac{e^{\gamma_1 b} A_1 + e^{-\gamma_1 b} A_2 - H^{-(\alpha_2+1)} A_4 - (\omega_2 H + \eta_2 - k_1)}{H^{\alpha_2}}$$

$$A_4 = \frac{(\mu_2 \alpha_2 + \mu_1 \gamma_1 s_2 H)e^{\gamma_1 b} A_1 + (\mu_2 \alpha_2 - \mu_1 \gamma_1 s_2 H)e^{-\gamma_1 b} - (\alpha_2-1)\mu_2 \omega_2 H}{-(\eta_2-k_1)\mu_2 \alpha_2}$$

$$A_5 = \frac{H_3^{\alpha_2} A_3 + H_3^{-(\alpha_2+1)} A_4 - e^{-\gamma_3 b_1} A_6 - (k_3 - \omega_2 H_3 - \eta_2)}{e^{\gamma_3 b_1}}$$

$$A_6 = \frac{(\mu_2 \alpha_2 + \mu_3 \gamma_3 s_2 H_3)H_3^{\alpha_2-1} A_3 - \left[\mu_2(\alpha_2+1)-\mu_3 \gamma_3 s_2 H_3\right]H_3^{-(\alpha_2+2)} A_4}{2\mu_3 \gamma_3 s_2 e^{-\gamma_3 b_1}}$$

where

$$t_1 = \left\{\left(e^{-\gamma_3(B-b_1)} - e^{\gamma_3(B-b_1)}\right)\left(\mu_2\alpha_2 + \mu_3\gamma_3 s_2 H_3\right)H_3^{\alpha_2-1} + 2\mu_3\gamma_3 s_2 e^{\gamma_3(B-b_1)}H_3^{\alpha_2}\right\}$$

$$t_2 = \left\{-\left(e^{-\gamma_3(B-b_1)} - e^{\gamma_3(B-b_1)}\right)\left(\mu_2(\alpha_2+1) - \mu_3\gamma_3 s_2 H_3\right)H_3^{-(\alpha_2+2)} + 2\mu_3\gamma_3 s_2 e^{\gamma_3(B-b_1)}\right.$$
$$\left. H_3^{-(\alpha_2+1)}\right\}$$

$$t_3 = \left\{\left(e^{-\gamma_3(B-b_1)} - e^{\gamma_3(B-b_1)}\right)\left(\mu_3\gamma_3 s_2\left[k_3 - (\omega_2 H_3 + \eta_2)\right] - \mu_2\omega_2\right) + 2\mu_3\gamma_3 s_2 e^{-\gamma_3 b_1}\right.$$
$$\left. \left(e^{\gamma_3 B}\left[k_3 - (\omega_2 H_3 + \eta_2)\right] - e^{\gamma_3 b_1}k_3\right)\right\}$$

Comments on Case C3:
Figure 6.44 has already shown that the simulations in a 2 panel rectangular channel using bc [A] agree well with those with a steep internal wall.

6.5   BOUNDARY SHEAR STRESS DISTRIBUTIONS IN CHANNEL FLOW AND SHEAR FORCES ON BOUNDARY ELEMENTS

6.5.1   Experimental data on boundary shear stress distributions with uniform roughness

Some knowledge concerning the distribution of boundary shear stresses is important in understanding several hydraulic issues, from bank erosion, sediment transport, the mixing of pollutants, resistance to flow and in adjusting for the influence of sidewalls

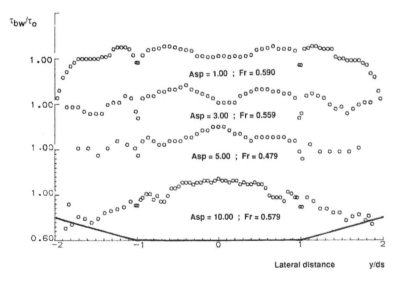

Figure 6.43 Boundary shear stress distributions for subcritical flows (Froude numbers, Fr = 0.48 – 0.59) in a trapezoidal channel at different aspect ratios. From Knight, Yuen & Alhamid (1994).

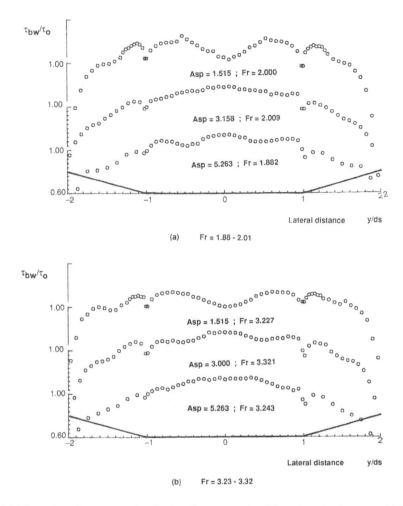

Figure 6.44 Boundary shear stress distributions for super-critical flows in a simple trapezoidal channel (Froude numbers, Fr = 1.88 − 3.32) at different aspect ratios, Knight et al. (1994).

in studies involving laboratory flumes. It is also difficult to measure accurately and difficult to predict unless a 3D CFD model is used, so one of the aims in developing the SKM was to provide practitioners with a simple tool that can estimate boundary shear stresses in rivers and channels easily. In order to appreciate the task in hand, some experimentally determined boundary shear stress distributions are shown in Figures 6.43–6.45 for straight prismatic simple and compound trapezoidal channels, taken from Knight, Yuen & Alhamid (1994). The local boundary shear stresses on the bed or walls have been non-dimensionalized by the section mean value, τ_0 $(= gRS_f)$ and the length scale has been non-dimensionalized by base width. The most obvious features of Figure 6.43 are how the distributions change with aspect ratio and the number and magnitude of the perturbations due to streamwise vorticity and the number of

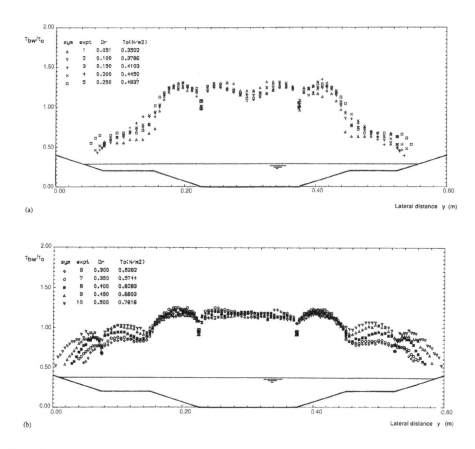

(a)

(b)

Figure 6.45 Boundary shear stress distributions for subcritical flows in a compound trapezoidal channel at different relative depths (0.05–0.50). From Knight, Yuen & Alhamid (1994).

secondary flow cells. In subcritical flow the τ_{bw}/τ_0 values at the centerline change from 1.02 to 1.20 as the aspect ratio changes from 1.0 to 5.0. Figure 6.44 shows the corresponding distributions in super-critical flow at comparable aspect ratios. Figure 6.45 shows one set of data for subcritical flows in a small compound trapezoidal channel at different relative depths, ranging from 0.05–0.50. Many similar sets exist for the large scale FCF work in compound channels of width up to 10 m, as already shown in Figures 6.17 to 6.19.

For simple trapezoidal channels, the mean wall and bed shear stresses, $\overline{\tau_w}$ and $\overline{\tau_b}$, were obtained by integrating the individual values along the walls or bed, and then multiplying them by the appropriate length, P_w and P_b to give the shear forces on the bed and the two walls, SF_w and SF_b, respectively. The total shear force per unit length along the channel, SF_T, is the sum of the two and was generally within about ±3% of their mean normal depth values τ_0 ($= gRS_f$). The water surface slopes were set precisely to within ±2% of the bed slope and involved a preliminary series of experiments for each discharge, measuring a number of M1 and M2 profiles prior to each series of boundary shear stress or velocity measurements. The same procedure was adopted for all the FCF experiments too, using the Preston tube technique (1954) and Patel's (1965) calibration. For a trapezoidal channel of base width B, depth H, and side slopes s (1:s, vertical:horizontal, or angle θ, where $s = \cot \theta$), then the percentage of the total shear force carried by the walls or bed may be determined from

$$SF_T = SF_w + SF_b; \quad SF_w = P_w \overline{\tau_w}; \quad SF_b = P_b \overline{\tau_b}; \quad P_w = 2H(1+s^2)^{1/2}; \quad P_b = B;$$

$$\%SF_w = 100 SF_w / SF_T \quad \text{and} \quad \%SF_b = 100 SF_b / SF_T \tag{6.80}$$

The $\%SF_w$ values will obviously decrease with increasing B/H, due to the walls having less influence in wider channels or ducts. Analysis of rectangular and trapezoidal data shows that it is better to use the wetted perimeter ratio, P_b/P_w, rather than B/H, as shown in Figures 6.46 to 6.48. The variation of $\%SF_w$ with P_b/P_w, is expressed in Figure 6.47 by the equation

$$\%SF_w = C_{sf} e^\alpha \tag{6.81}$$

where the shape factor, $C_{sf} = 1.0$, for $P_b/P_w < 6.546$, and
for $P_b/P_w > 6.546$, $C_{sf} = 0.5857 (P_b/P_w)^{0.28471}$ and

$$\alpha = -3.23 \log_{10} \left(\frac{P_b}{P_w C_2} + 1.0 \right) + 4.6052 \text{ with } C_2 = 1.50 \tag{6.82}$$

Although this appears to fit the sub-critical flow data quite well, it makes no distinction between open channel flow data and duct flow data. For $P_b/P_w = 1.0$, i.e. $B/H = 2.0$, $\%SF_w = 48.8\%$ somewhat less than the 50% for duct flow, due to free surface effects. For super-critical flow, the data are best simulated by simply adjusting the shape factor if such flows are required to be analysed, together with the coefficient C_2, using instead

$$C_{sf} = 1.0 \text{ for } P_b/P_w < 4.374, \text{ else } C_{sf} = 0.6603(P_b/P_w)^{0.28125} \quad \text{and} \quad C_2 = 1.38 \tag{6.83}$$

The weak dependence of $\%SF_w$ on Froude number, Fr, and Reynolds number, Re, are shown by the data for 3 aspect ratios in Figure 6.49.

The mean wall and bed shear stress variations may be determined from (6.80), since

$$\%SF_w = 100 SF_w / SF_T = 100(P_w \overline{\tau_w}) / \rho g A S_f \tag{6.84}$$

or

$$\%SF_w = 100 / \left[1 + \frac{P_b}{P_w} \frac{\overline{\tau_b}}{\overline{\tau_w}} \right] \tag{6.85}$$

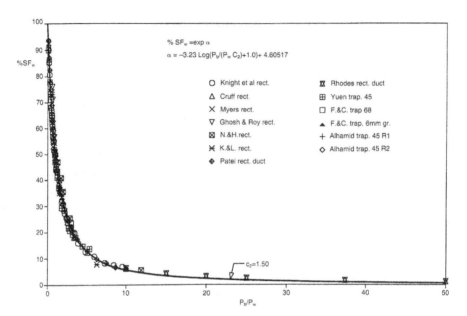

Figure 6.46 Variation of %SF$_w$ values with wetted perimeter ratio, P_b/P_w in sub-critical flow for P_b/P_w < 50 or aspect ratios, B/H, up to 100 (N.B. mixed open channel and duct data). From Knight, Yuen & Alhamid (1994).

Equation (6.85) indicates that %SF$_w$ only varies in proportion to the perimeter length ratio P_b/P_w if $\tau_b = \tau_w$, which is generally not true for open channel flow in trapezoidal channels. However, for closed pipe flow in rectangular ducts one might expect from (6.85) that %SF$_w$ will vary according to [1+b/H], as Rhodes & Knight (1994) show and therefore seek an alternative to the exponential variation represented by (6.81) & (6.82) for the special case of rectangular duct flow.

For open channel flow in simple trapezoidal channels, (6.85) may be adapted to give equations for practical design purposes, in the fom

$$\frac{\bar{\tau}_w}{\rho g H S_f} = 0.01\%SF_w \left[\frac{s}{2\left(1+s^2\right)^{1/2}} + \frac{P_b}{P_w} \right] \quad or \quad \frac{\bar{\tau}_w}{\rho g R S_f} = 0.01\%SF_w \left[1 + \frac{P_b}{P_w} \right]$$

(6.86)

$$\frac{\bar{\tau}_b}{\rho g H S_f} = \left(1 - 0.01\%SF_w\right)\left[1 + \frac{s}{2\left(1+s^2\right)^{1/2}} \frac{P_w}{P_b} \right] \quad or$$

$$\frac{\bar{\tau}_b}{\rho g R S_f} = \left(1 - 0.01\%SF_w\right)\left[1 + \frac{1}{\left(P_b / P_w\right)} \right]$$

(6.87)

Then, if the %SF$_w$ versus P_b/P_w relationships from (6.82) & (6.83) for sub-critical and super-critical flow are inserted into (6.86)–(6.87), then useful information results concerning the design of canals or culverts. The resulting equations are shown in

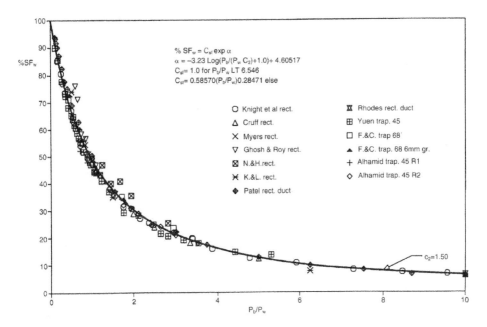

Figure 6.47 Variation of %SF_w values with wetted perimeter ratio, P_b/P_w in sub-critical flow. From Knight, Yuen & Alhamid (1994).

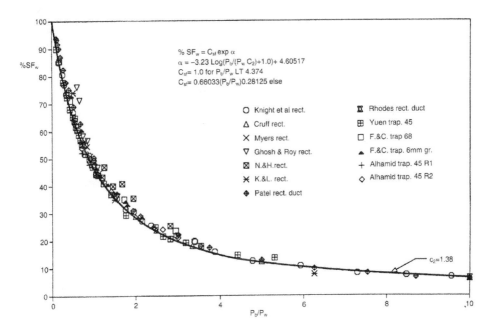

Figure 6.48 Variation of %SF_w values with wetted perimeter ratio, P_b/P_w in super-critical flow. From Knight, Yuen & Alhamid (1994).

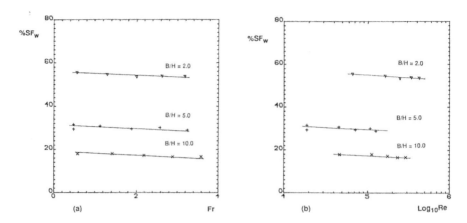

Figure 6.49 Evidence of weak dependence of %SF$_w$ on Fr and Re. From Knight, Yuen & Alhamid (1994).

Figure 6.50, where it may be seen that they predict the mean bed and wall stresses for homogeneously roughened trapezoidal channels quite well over a wide range of aspect ratios and for different side slopes. Similar equations are shown in Figure 6.51 for predicting the maximum wall and bed shear stresses, as fitted by Alhamid (1991), in the form

$$\frac{\overline{\tau}_{w\,max}}{\rho g HS_f} = 0.01\%SF_w\left[2.37\left(\frac{P_b}{P_w}\right)^{0.85}\right] \quad \text{or}$$

$$\frac{\overline{\tau}_{w\,max}}{\rho g RS_f} = 0.01\%SF_w\frac{2.37}{\left(\dfrac{P_b}{P_w}+\dfrac{s}{2\left(1+s^2\right)^{1/2}}\right)}\left[\left(\frac{P_b}{P_w}\right)^{1.85}+\left(\frac{P_b}{P_w}\right)^{0.85}\right] \quad (6.88)$$

$$\frac{\overline{\tau}_{b\,max}}{\rho g HS_f} = \left(1-0.01\%SF_w\right)\left[0.8+\left(\frac{P_b}{P_w}\right)^{-0.35}\right] \quad \text{or}$$

$$\frac{\overline{\tau}_{b\,max}}{\rho g RS_f} = \left(1-0.01\%SF_w\right)\left\{\frac{0.8+\left(\dfrac{P_b}{P_w}\right)^{-0.35}}{\left(\dfrac{P_b}{P_w}+\dfrac{s}{2\left(1+s^2\right)^{1/2}}\right)}\right\}\left[\frac{P_b}{P_w}+1\right] \quad (6.89)$$

These might be compared with the USBR tractive force approach (see Chang 1988), that suggests that the maximum allowable boundary shear stresses on the walls and bed are given by

$$\tau_{bmax} = 1.37\rho g RS_o \quad \text{and} \quad \tau_{wmax} = 1.08\rho g RS_o \quad (6.90)$$

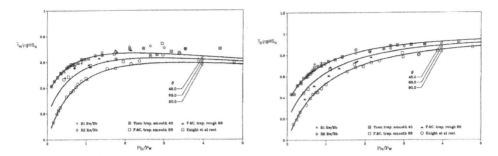

Figure 6.50 Mean wall and bed shear stresses for smooth and rough trapezoidal channels with different side slopes, with equations (6.86) & (6.87), after Knight *et al.* (1994).

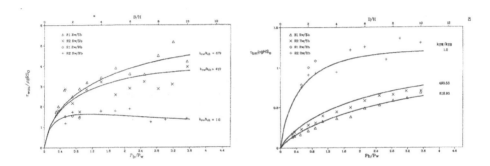

Figure 6.51 Maximum wall and bed shear stresses for uniformly and differentially roughened channels (after Alhamid, 1991).

which are demonstrably sensible for average stresses, but not so for maximum stresses, as shown by Figure 6.51. See Knight *et al.* (1994), Flintham & Carling (1988), Thorne *et al.* (1998), USBR (1952) and Alhamid (1991) for further details. Boundary shear stress data for flow in circular conduits may be found in Knight & Sterling (2000) and Sterling & Knight (2000, 2001 & 2002), as mentioned in 6.4.1, Case *U1*.

6.5.2   Experimental data on boundary shear stress distributions with non-uniform roughness

For trapezoidal channels in which the bed or the walls are differentially roughened with respect to each other, then an additional parameter, such as the ratio of Nikuradse roughness length, k_s, for the walls and bed, k_{sw}/k_{sb}, needs to be added to the previous equations. Figures 6.51–6.53 show some further %SF_w v P_b/P_w experimental data from Knight *et al.* (1992), for trapezoidal channels with smooth beds and the roughened walls. These showed that the coefficient C_2 in (6.82) varied from 1.5 for homogeneous roughness (k_{sw}/k_{sb} = 1.0) to 4.87 & 6.84 for the heterogeneous roughness cases (k_{sw}/k_{sb}, = 419 & 679). The k_{sw} values were determined from a separate series of experiments in uniformly roughened channels using the same two gravel sizes used in the differential roughened channels ($d_{84} = 9.3$ & 18 mm). Although the precise relationship between C_2 and k_{sw}/k_{sb} could not be determined accurately on the basis of just 3 series of experiments, an attempt was made by expressing C_2 in terms of k_{sw}/k_{sb} by either a linear or a power law relationship, as indicated by (6.91) & (6.92).

$$C_2 = 1.492 + 0.008(k_{sw}/k_{sb}) \tag{6.91}$$

or

$$C_2 = 1.50 \, (k_{sw}/k_{sb})^{0.2115} \tag{6.92}$$

It should be noted that in both cases $C_2 = 1.50$ when $k_{sw}/k_{sb} = 1.0$. Additional data from 4 series of experiments by Flintham & Carling (1988) were also used in this analysis. In their experiments the beds of the trapezoidal channels were roughened relative to the walls ($k_{sb}/k_{sw} = 2.1–91.1$). Neither formulation (6.91) or (6.92) was found to be entirely satisfactory for all cases, with (6.91) being the best for ($k_{sw}/k_{sb} \geq 1$) and (6.92) for ($k_{sw}/k_{sb} < 1$). Figure 6.53 shows that the mean wall and bed shear stresses are well represented by (6.86) & (6.87) for the two differentially roughened

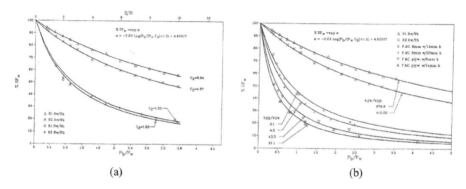

(a)   (b)

Figure 6.52 Percentage shear force %SF_w v P_b/P_w for differentially roughened channels (a) fitted with different C_2 values (from Knight, Alhamid & Yuen 1992), or with (b) equations (4), (5) & (6) for $k_{sw}/k_{sb} \geq 1$) and (4), (5) & (8) for $k_{sb}/k_{sw} >1$).

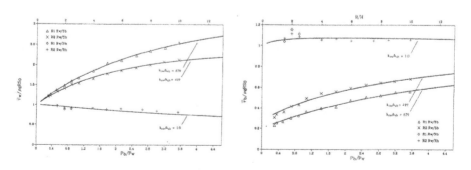

Figure 6.53 Mean wall and bed shear stresses for uniformly and differentially roughened channels (after Alhamid, 1991).

and the two uniformly roughened channels, using C_2 given by (6.91). Further research and data are clearly needed on this important topic.

Knowing the boundary shear stress distributions around the wetted perimeter allows one to determine the depth-mean apparent shear force on any vertical by applying (6.62). This balances the resolved gravitational force with the opposing shear force over the section of the channel up to the particular vertical selected, lumping together the net effects of turbulence, secondary flows and viscous shear. The depth-averaged eddy viscosity may then be obtained by applying (6.93), in an analogous way to those equations already presented, namely (2.59) & (3.18). Thus

$$\bar{\varepsilon}_{yx} = \bar{\nu}_t = \bar{\tau}_a / (\rho \partial U_d / \partial y) \tag{6.93}$$

(Note the dual notation here, since $\bar{\varepsilon}_{yx} = \bar{\nu}_t$)

Figure 6.54 shows that the values of the dimensionless eddy viscosity λ, inferred from the boundary shear stress and velocity data, lie between 0.1 and 2.0, with a mean of around 0.45, for all the channels used in those experiments. As might be expected from the increased roughness, this is considerably greater than the smooth boundary value of 0.07, as explained in Section 6.2.3 and adopted in (6.12) & (6.55), and also different from the value of 0.24 given in Table 3.4 for the main channels in rivers. These differences are not surprising, given the smaller aspect ratios and the larger differential in roughness between the bed and banks used in the experiments by Alhamid (1991).

Because the SKM was developed as a tool for the predicting the lateral variations of both the depth-averaged velocity and boundary shear stress, it is particularly suited to solving certain types of problem in hydraulic engineering. Three examples are now briefly outlined to illustrate the range of applications. Some of these will be returned to in Chapter 7, where further details given.

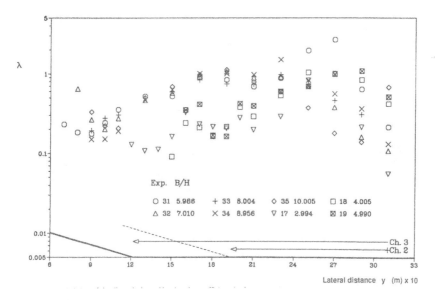

Figure 6.54 Lateral variations of the dimensionless eddy viscosity, λ, for trapezoidal channels with roughened walls and a smooth bed from Knight, Yuen & Alhamid (1994).

6.5.3   Predicting boundary shear stress using the SKM

a.   Case 1 Rectangular channels

The first application is the simplest: to determine the influence of sidewalls in any work involving rectangular flumes. This application is also useful in illustrating the calibration procedure for the 3 key parameters in the SKM. One approach to calibration of any numerical model is to use a trial and error procedure. Another might be to use a more advanced method using a multi-objective evolutionary algorithm (NSGA-II) and a Pareto based approach. Both methods have been used with the SKM, and details concerning these methodologies and application may be found in Chlebek & Knight (2006) and Sharifi et al. (2009).

Using the first approach, based on trial and error, Table 6.2 illustrates the general influence of the 3 calibration parameters (f, λ & Γ) used within each panel of the SKM. For example, in the third row, for a given depth, H, when λ increases, with fixed f & Γ, it causes $\%SF_w$ to increase and Q to decrease.

Using the 152 mm (= $2b$) flume data as an example, listed in Knight, Demetriou & Hamed (1984) and as explained in Chlebek & Knight (2006), a single panel was used to model half the channel, but sub-divided further into 20 slices. Initial values of λ and Γ were assumed to be 0.02 and 0.05 respectively, and f values were then varied from 0.010 to 0.046 in increments of 0.002. In some cases the most suitable value for a parameter, f for example, will differ depending on which target variable is being compared, i.e. discharge, Q, or percentage shear force on the walls, $\%SF_w$. This is illustrated in Figure 6.55 for two particular tests. Experiment 2 had a low depth, with an aspect ratio, Asp (= $2b/h$), equal to 1.77, whereas Experiment 7 had a higher depth and an aspect ratio of 0.99. In Figure 6.55(a) it is seen that the zero error point associated with discharge coincides with the zero error point for $\%SF_w$ is at around $f = 0.021$ and $\lambda = 0.020$ for both Q and $\%SF_w$. But in Figure 6.55(b) it is seen that the optimum f values are different for each variable ($f \approx 0.020$ for Q and ≈ 0.040 for $\%SF_w$). Similar problems existed with the calibration of λ but since this parameter had a lower impact on the overall results it was less evident. The trial and error approach is useful to see the effects of individual parameters on the results during the calibration process, but it is tedious. Some results of this procedure are shown in Figure 7.3 in Chapter 7 for Case R1. Having obtained the optimum values for the 3 calibration parameters that give the best overall discharge and shear force results, the modelled lateral distributions of velocity and boundary shear stress still need to be compared with the measured ones.

Table 6.2 Effect of varying calibration parameters in SKM.

Vary	Fixed	+/–	Effect on %SF$_w$	Effect on Q
f	λ & Γ	increase f	down	down
		decrease f	up	up
λ	f & Γ	increase λ	up	down
		decrease λ	down	up
Γ	f & λ	increase Γ	up	down
		decrease Γ	down	up

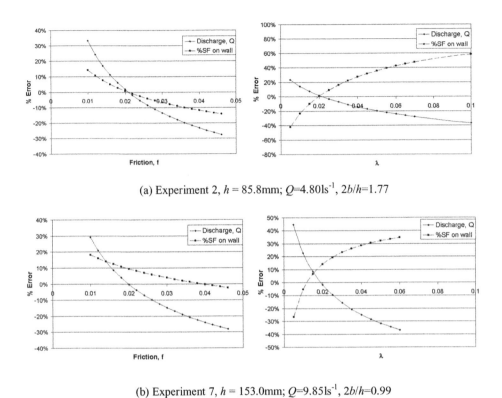

(a) Experiment 2, $h = 85.8$mm; $Q=4.80$ls^{-1}, $2b/h=1.77$

(b) Experiment 7, $h = 153.0$mm; $Q=9.85$ls^{-1}, $2b/h=0.99$

Figure 6.55 Errors in modelled Q and $\%SF_w$ associated with calibrating f and λ from Chlebek & Knight (2006).

In the second more rigorous approach, highlighted by Sharifi *et al.* (2009), use is made of a multi-objective evolutionary algorithm. Four objective functions were specified, (6.94) & (6.95), to measure the difference between the observed and model generated results. As the mean streamwise velocity and local boundary shear stress distributions for each case consisted of many experimental points, the sum of squared errors (SSE) was selected as the goodness-of-fit measure. In contrast, the absolute percentage error (APE) was selected as the performance measure for the single measured and calculated values of discharge (Q) and the percentage of shear force on the walls of the channel ($\%SF_w$):

$$f_1(X) = \sum_i \left\{ (U_d)_{SKM} - (U_d)_{exp} \right\}^2 \quad f_2(X) = \sum_i \left\{ (\tau_b)_{SKM} - (\tau_b)_{exp} \right\}^2 \quad (6.94)$$

$$f_3(X) = \left| \frac{Q_t - Q_{SKM}}{Q_t} \right| \times 100 \quad f_4(X) = \left| \frac{\%(SF_w)_t - \%(SF_w)_{SKM}}{\%(SF_w)_t} \right| \times 100 \quad (6.95)$$

where $X = (f_1, \lambda_1, \Gamma_1, \cdots f_N, \lambda_N, \Gamma_N)$ is the variable vector in the design domain search space, Ω. The subscripts *SKM* and *exp* refer to the predictions obtained using the

Figure 6.56 Experimental flumes with uniform and differential roughness (from Alhamid, 1991).

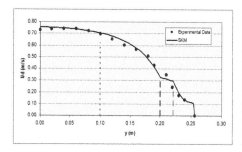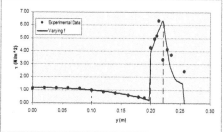

Figure 6.57 Experimental and predicted velocity and boundary shear stress distributions using SKM for Experiment 27, Alhamid, with aspect ratio = 7.03, smooth bed and R1 on the walls (from Sharifi, Sterling & Knight, 2009).

SKM model and experimental data, respectively. In $f_3(X)$ and $f_4(X)$ the subscript t is used to denote the global value of either Q or $\%SF_w$ and indicates that for these two functions the channel is considered as a whole, i.e. with the panels 'removed'.

A more sophisticated schematization was used, with up to 4 panels, again each sub-divided into smaller elements. This illustrates both a weakness and strength of the SKM, since some knowledge or experience of the number and location of secondary flow cells is required for very accurate simulations. This issue is addressed by Knight *et al.* (2007) and Omran *et al.* (2008) where detailed simulations are presented. Data from Alhamid's (1991) experiments were used as these were undertaken with uniform and differential roughness, as shown in Figures 6.56 & 6.57. Depending on the available data, any combination of the above objectives can be minimized simultaneously, or additional objective functions added, e.g. one involving the friction factor. However, it is felt that those listed above made use of the best available data and enabled a good comparison with previously published experimental results. For these, see Sharifi *et al.* (2009).

b. Case 2 Compound channels

The second application is for overbank flow in compound channels. Figure 6.58 shows both zonal discharge and shear force simulations over relative depths from 0.05 to 0.5 for two different data sets, one from the FCF and the other from earlier experiments by Knight *et al.* (1981 & 1984). The calculated $\%Q_{mc}$ and $\%SF_{mc}$ values

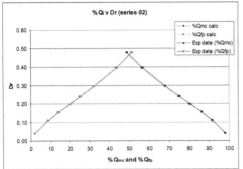 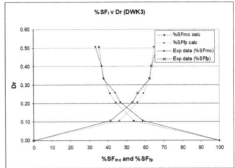

Figure 6.58 Comparison of measured and computed %Q and %SF in main channel and floodplains (after Knight 2013).

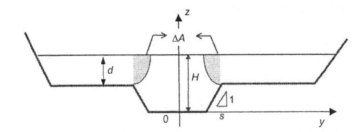

Figure 6.59 A compound channel showing typical zero shear stress lines (from Knight & Tang, 2008).

agree reasonably well with the experimental data over the entire range of *Dr* tested. Further details are available in Knight (2013a).

c.   Case 3 Link between the area method and the SKM

The third application illustrates a link between the SKM and the so-called 'area-method' for predicting the zonal discharges in channels, illustrated in Figure 6.59. In the latter, the interaction between zones is allowed for by adjusting the flow areas in each zone, based on the 'apparent shear stress' (as yet unknown) acting on each vertical division line between the main river channel and its associated floodplain, as suggested by Knight & Tang (2008). Hence:

$$A_C' = A_C - (\Delta A_L + \Delta A_R) \tag{6.96}$$

$$A_L' = A_L + \Delta A_L \tag{6.97}$$

$$A_R' = A_R + \Delta A_R \tag{6.98}$$

where A_C = main channel area, A_L & A_R = left and right floodplain areas, all based on vertical division lines and thus requiring the apparent shear stresses to be known on these interfaces.

The corresponding modified areas, A_C', A_L' & A_R', are those areas that when multiplied by the zonal velocity, give the correct zonal discharge, thus taking into account the position of the zero-shear stress interfaces. The areas ΔA_L & ΔA_R can be obtained by considering the equilibrium of forces acting on each floodplain, as follows:

Region left: $\sum F_L - d\overline{\tau}_{aL} = \rho g A_L S$ (6.99)

Region right: $\sum F_R - d\overline{\tau}_{aR} = \rho g A_R S$ (6.100)

in which F_L, F_R = total shear force on the left and right floodplains, $\overline{\tau}_{aL}$, $\overline{\tau}_{aR}$ = depth-averaged apparent shear stress on the left and right hand vertical interfaces between the main channel and the floodplain, and d = flow depth over the floodplain at the interface. If the arbitrary interface with zero shear stress is taken into account, then the equilibrium of forces acting on each floodplain is given by:

Region left: $F_L = \rho g \, (A_L + \Delta A_L) \, S$ (6.101)
Region right: $F_R = \rho g \, (A_R + \Delta A_R) \, S$ (6.102)

Combining (6.99)–(6.102) and rearranging gives

$$\Delta A_L = (\overline{\tau}_{aL} d)/(\rho g S) \qquad (6.103)$$

$$\Delta A_R = (\overline{\tau}_{aR} d)/(\rho g S) \qquad (6.104)$$

Thus, if the depth-mean 'apparent shear stresses' on each vertical interface $(\overline{\tau}_{aL}, \overline{\tau}_{aR})$ are known, then the discharge in each region can be obtained from the Manning or Darcy-Weisbach resistance laws

$$Q_R = A_R U'_R = \frac{A_R A_R'^{2/3}}{n_R P_R^{2/3}} \sqrt{S} \qquad (6.105)$$

$$Q_R = A_R U_R' = A_R \sqrt{\frac{8gS}{f_R} \left(\frac{A'_R}{P_R} \right)} \qquad (6.106)$$

where P is wetted perimeter of channel boundary, n & f are the Manning and Darcy-Weisbach resistance coefficients respectively, and Q is the discharge. The total discharge is then

$$Q = Q_L + Q_C + Q_R \qquad (6.107)$$

For the symmetric trapezoidal compound channel shown in Figure 6.60, divided into 3 regions, the main channel (panels 1–3) and two floodplains (panels 4 & 6 and 5 & 7). In this case each panel consists of either a flat bed or a transverse sloping bed region. Applying (6.108) to one half of the main channel, and noting that $\overline{\tau}_{aL} = \overline{\tau}_{aR} = \tau_a$, then the apparent shear stress on a vertical interface is given by

$$-\tau_a = \frac{1}{H_2} \int_0^{b1} [\rho \frac{f_1}{8} (U_d^{(1)})^2 - \rho g H S] dy + \frac{1}{H_2} \int_{b1}^{b2} [\rho \frac{f_2}{8} (U_d^{(2)})^2 \sqrt{1 + 1/s_2{}^2} - \rho g \xi_2 S] dy \qquad (6.108)$$

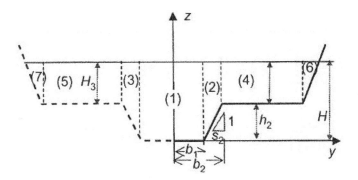

Figure 6.60 A schematised trapezoidal compound channel with 2 floodplains and 7 panels.

where U_d is given by the appropriate velocity distribution equations (6.14) & (6.19). Substituting the appropriate equations for U_d into (6.108) gives:

$$-\tau_a = I + II \tag{6.109}$$

in which

$$I = \frac{1}{H_2} \int_0^{b1} [\rho \frac{f_1}{8}(A_1 e^{\gamma_1 y} + A_2 e^{-\gamma_1 y} + k_1) - \rho g H S] dy$$

$$= \frac{\rho f_1}{8 H_2 \gamma_1}[A_1(e^{\gamma_1 b_1} - 1) - A_2(e^{-\gamma_1 b_1} - 1) + k_1 \gamma_1 b_1] - \rho g b_1 S H / H_2 \tag{6.110}$$

$$II = \frac{s_2}{H_2} \int_{H_2}^{H} \frac{\rho f_2}{8}\left[(A_3 \xi_2^{\alpha_2} + A_4 \xi_2^{-(\alpha_2+1)} + \omega_2 \xi_2 + \eta_2)\sqrt{1 + \frac{1}{s_2^2}} - \rho g \xi_2 S\right] d\xi_2 = \frac{\rho f_2 \sqrt{1 + s_2^2}}{8 H_2}$$

$$\left[\frac{A_3}{\alpha_2 + 1}(H^{\alpha_2+1} - H_2^{\alpha_2+1}) - \frac{A_4}{\alpha_2}(H^{-\alpha_2} - H_2^{-\alpha_2}) + \frac{1}{2}\omega_2(H^2 - H_2^2) + \eta_2(H - H_2)\right] -$$

$$\frac{\rho g s_2 S}{2 H_2}(H^2 - H_2^2) \tag{6.111}$$

Since $\bar{\tau}_{aL} = \bar{\tau}_{aR} = \tau_a$, it follows that by inserting τ_a into (6.103) & (6.104) and using (6.96)–(6.98) gives the modified areas A_C', A_R' & A_L'. Hence applying (6.105) gives the zonal discharge for the right-hand floodplain, Q_R and applying a similar procedure gives equations for Q_L & Q_C. Summing these individual zonal discharges via (6.107) gives the total discharge for a given stage. Repeated application, using different depths, will yield the stage-discharge relationship, as illustrated elsewhere by Mc Gahey *et al.* (2006) and Abril & Knight (2004).

In addition to the stage-discharge relationship, it should be noted that this methodology also gives the boundary shear forces acting on different elements of the wetted perimeter, since

$$\tau_b = \rho \frac{f}{8} U_d^2 \tag{6.112}$$

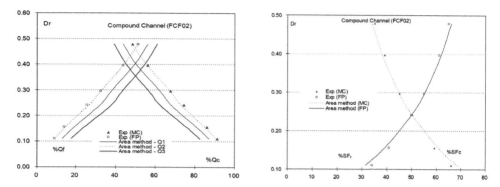

Figure 6.61 Zonal discharges and shear forces for Series 02 FCF (from Knight & Tang, 2008).

Thus for two of the panels in the main channel, the shear forces are obtained by integration as:

$$SF_1 = \frac{\rho f_1}{8\gamma_1}\left[A_1(e^{\gamma_1 b_1} - 1) - A_2(e^{-\gamma_1 b_1} - 1) + k_1\gamma_1 b_1\right] \tag{6.113}$$

$$SF_2 = \frac{\rho f_2\sqrt{1+s_2^2}}{8}\left[\frac{A_3}{(\alpha_2+1)}(H^{\alpha_2+1} - H_2^{\alpha_2+1}) - \frac{A_4}{\alpha_2}(H^{-\alpha_2} - H_2^{-\alpha_2})\right.$$
$$\left. + \frac{\omega_2}{2}(H^2 - H_2^2) + \eta_2(H - H_2)\right] \tag{6.114}$$

The coefficients $A_1 - A_4$ have already been given for this case, so (6.113) & (6.114) illustrate how shear forces on particular elements may be determined analytically also using the SKM. Comparisons between predicted and measured values are shown in Figure 6.61. The area method using option Q2 (= $U'A$) gives the best results, as would be expected. For further details see Knight & Tang (2008).

6.6 RESISTANCE EQUATIONS FOR SURFACES, SHAPE EFFECTS AND TREES

Standard resistance equations can become so familiar that it is easy to forget either the assumptions implicit in their derivation or their limits of application. Since the Colebrook-White equation is often used for flow calculations in pipes and open channels, as in the CES & SKM, it is important to consider how it should be applied to rivers for which it was not originally conceived. Four issues are therefore now examined, namely: the links between velocity distribution laws and resistance formulation, the effects of cross-sectional shape on resistance, the development of an equation for the drag force caused by trees, and the effects of flow-dependency on resistance which might vary with time, such as when sediment or vegetation are present.

6.6.1 Velocity distribution laws and resistance formulation

Various forms of the Colebrook-White equation, such as (2.26), (2.27) & (3.12), have been given in Sections 2.2.3 and 3.2.2 and Table 3.3 has indicated the range of coefficients that have been adopted by researchers for rigid impervious boundaries. It is instructive to understand the origin of some of these coefficients, and how this might affect a particular application. Starting with the velocity distribution laws for rough and smooth surfaces, given earlier as (2.48) & (2.51), and noting the uncertainty about the empirical coefficients mentioned there, these are now replaced by the more general forms:

$$\frac{u}{u_*} = A \ln\left(\frac{u_* z}{\nu}\right) + B \quad \text{(smooth surfaces, z being normal to the surface)} \quad (6.115)$$

$$\frac{u}{u_*} = A \ln\left(\frac{z}{k_s}\right) + B \quad \text{(rough surfaces, z being normal to the surface)}$$

$$(6.116) \text{ \& } (2.49)$$

where A & B are coefficients that are subject to debate and therefore need to be chosen carefully. Nikuradse and Prandtl originally took them to be $A = 2.50$ ($= 1/\kappa$, i.e. $\kappa = 0.40$) and $B = 5.50$, but the findings of the Stanford Conference (1968) and Patel (1965) are now taken as the most authoritative, as listed in Table 6.3. The Stanford conference (1968) was convened to decide on the range of coefficients that should be used in general fluid flow computations, and agreed that it is better to take $A = 2.44$ (i.e. $\kappa = 0.410$) and $B = 5.0$ as this virtually unifies the Stanford, Patel and Clauser velocity distribution results, setting them apart from the original Nikuradse results. See Kline *et al.* (1969).

Prandtl's logarithmic velocity distribution laws, (2.48) & (2.51), clearly needs some modification close to the wall to ensure that $u = 0$ for $z = 0$. Furthermore, experiments indicate that turbulent boundary layers have several distinct zones, as shown in Figure 6.62. Different velocity distributions exist in each zone and affect the way the shear stresses are transferred from the fluid to the wall surface. These are particularly important in the lower two layers where the Preston tube is normally placed to obtain

Table 6.3 Values of the coefficients A & B in the velocity distribution laws and values of the coefficients C, D, E & κ in the resistance laws for smooth surfaces.

Source	A	B	C	D	E	κ
Stanford (1968)	2.44	5.00	1.986	−1.021	0.863	0.410
Patel (1965)	2.39	5.45	1.946	−0.805	0.845	0.418
Nikuradse (1932)	2.50	5.50	2.035	−0.913	0.884	0.400
Clauser (1954)	2.49	4.90	2.027	−1.114	0.880	0.402
Prandtl (1925)	2.50	5.50	2.000	−0.800	0.869	0.400

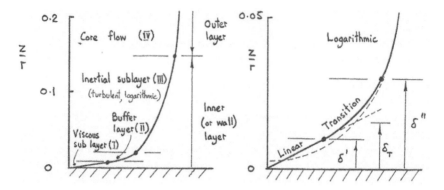

Figure 6.62 Various zones within a turbulent boundary layer.

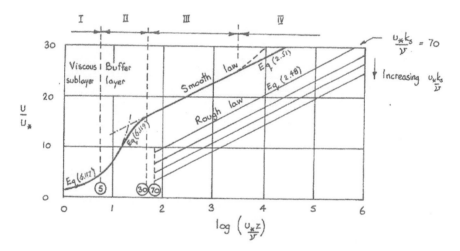

Figure 6.63 Velocity distributions over smooth and rough surfaces (U^+ versus Z^+ format).

the boundary shear stress. In general, there are 4 distinct zones (I – IV), beginning with the viscous sub-layer close to the wall. Even in turbulent flow, the flow very close to a solid surface reverts back to laminar flow for smooth surfaces and is known as the '*viscous sub-layer*' (I). Outside this zone the flow is turbulent (III) & (IV), albeit with a thin '*buffer layer*' (II) in between. The zone within which the velocity is logarithmic and exactly described by (6.115) & (6.116) above, is known as the '*inertial sub-layer*' (III), which may extend throughout the remainder of the boundary layer if it is 'bounded'. In an 'unbounded' boundary layer, then a region known as the '*core flow*' (IV) may develop, extending out indefinitely from the solid surface. Zone IV may be deemed not to exist in most river flows.

It is now possible to relate these distribution laws for both smooth and rough surfaces, as shown schematically in Figure 6.63 between the relationship between u/u_* and u_*z/v for a wide range of u_*z/v values. Figure 6.64 shows some actual data from Nezu & Nakagawa, (1993) plotted in a similar format to Figure 6.63.

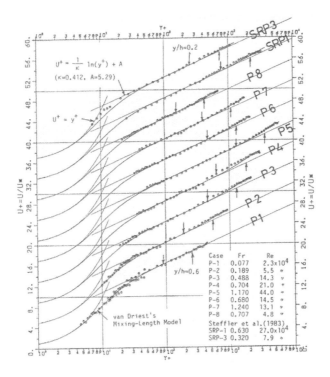

Figure 6.64 Experimental data on the relationship between U^+ and Y^+ for a smooth surface (From Nezu & Nakagawa, 1993).

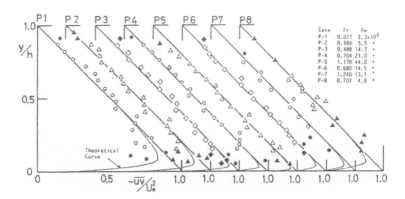

Figure 6.65 Experimental data for shear stress measured in open channel flow (From Nezu & Nakagawa, 1986).

Figure 6.65 shows the corresponding linear distributions of shear stress over the depth in the inertial layer and how the stresses revert to a laminar condition very close to the bed. The dimensional variables, u/u_* and $u_* z/v$, are often given the symbols U^+ and Z^+ (or Y^+ in 2D flow where the y axis is normal to the boundary). Figure 6.63 shows the

4 zones of the boundary layer, I-IV, and how the relevant equations, (6.117), (6.118), (6.115) & (6.116), apply in each region. The limiting values of $u_* z / v = 5$, 30 & 70 are also marked, corresponding to the ranges for each of the appropriate equations. The velocity distribution in the viscous sub-layer and its thickness δ', are given by

$$\frac{u}{u_*} = \frac{u_* z}{v} \qquad\qquad (6.117)$$

(valid for $0 < u_* z / v < 5$)

$$\delta' = \frac{5v}{u_*} \qquad\qquad (6.118)$$

Various equations have been proposed for the velocity distribution in the buffer layer, between δ' and δ'', notably by van Driest, von Karman and others. The simple von Karman equations for the velocity distribution and outer limit are

$$\frac{u}{u_*} = 5.0 \ln\left(\frac{u_* z}{v}\right) - 3.05 \qquad\qquad (6.119)$$

(valid for $5 < u_* z / v < 30$)

$$\delta'' = \frac{30v}{u_*} \qquad\qquad (6.120)$$

These velocity distribution laws are important as they have a direct influence on any resistance relationship that uses the mean velocity of the cross section in its formulation, as indicated by (2.35), where $f = 8(u_* / \bar{u})^2$. It therefore becomes necessary to compare not only velocity distribution equations with data from flow in pipes (e.g. Nikuradse, Clauser, Patel and Stanford), examining the u/u_* versus $u_* z / v$ relationships in the 4 main boundary layer regions, but also to integrate the distributions across the pipe diameter to obtain the section-mean velocity, $\bar{u}$, as this gives a corresponding theoretical equation for the resistance coefficient f. The integration usually ignores the buffer and viscous sub-layers, which are small in smooth pipes, and simply non-existent in rough pipes, where they are then subsumed into the roughening excrescences. The wake zone may also be ignored since the flow in most pipes is fully developed. By integrating the discharge through an annulus within a pipe with respect to the radius, r, as shown in Figure 6.66, and noting that $r_o = d/2$ gives Q as

$$Q = 2\pi u_* \int_0^{r_o} \left[A \ln\left(\frac{u_* z}{v}\right) + B \right] (r_o - z) \, dz$$

which after changing variables and integrating leads to $\dfrac{\bar{u}}{u_*} = A \ln\left(\dfrac{u_* r_o}{v}\right) + \left(B - \dfrac{3A}{2}\right)$

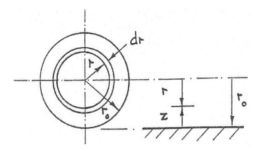

Figure 6.66 Notation for circular pipe.

When this is inserted into (2.35) it gives a 'universal' resistance law for a smooth surface as

$$\frac{1}{\sqrt{f}} = \frac{1}{\sqrt{8}}\left[A\ln\left(\frac{\overline{u}d}{\nu}\sqrt{f}\right) + \left(B - \frac{3A}{2}\right) + A\ln\left(\frac{1}{2\sqrt{8}}\right) \right] \tag{6.121}$$

The values of A & B may be taken from any of the velocity laws given earlier so that, for example, with $A = 2.5$ and $B = 5.5$ (after Nikuradse), this will result in the following resistance equation:

$$\frac{1}{\sqrt{f}} = 0.884\ln\left(\frac{\overline{u}d}{\nu}\sqrt{f}\right) - 0.91 \quad \text{(for a smooth surface)} \tag{6.122}$$

Using actual resistance data from pipe flow experiments, Prandtl suggested a formula which is very similar in both form and with similar numerical coefficients, given earlier as (2.28) and repeated here as Prandtl's law in either of two common formulations, (6.123) or (6.124):

$$\frac{1}{\sqrt{f}} = 0.869\ln\left(\frac{\overline{u}d}{\nu}\sqrt{f}\right) - 0.8 \tag{6.123}$$

or

$$\frac{1}{\sqrt{f}} = 2.0\log_{10}\left(\frac{\overline{u}d}{\nu}\sqrt{f}\right) - 0.8 \tag{6.124}$$

Prandtl's law is valid for Re up to 3.4×10^6, and can be written more generally as

$$\frac{1}{\sqrt{f}} = C\log_{10}\left(Re\sqrt{f}\right) - D \tag{6.125}$$

where C & D are the coefficients in any resistance equation that is based on a velocity equation with coefficients A & B. These are also listed in Table 6.3. A similar analysis involving the integration of the velocity distribution law for rough surfaces may also be undertaken, to give the corresponding resistance equation for rough surfaces as:

$$\frac{1}{\sqrt{f}} = \frac{1}{\sqrt{8}}\left[A\ln\left(\frac{d}{2k_s}\right) + \left(B - \frac{3A}{2}\right)\right] \tag{6.126}$$

Again, using Nikuradse's values ($A = 2.5$ & $B = 8.5$), by way of example, gives equation (6.127):

$$\frac{1}{\sqrt{f}} = 0.884\ln\left(\frac{d}{2k_s}\right) + 1.68 \tag{6.127}$$

(based on the velocity distribution law)

Using actual resistance data from pipe flow experiments, Prandtl suggested a formula which again is very similar in both form and numerical values, which is now traditionally used for rough surfaces:

$$\frac{1}{\sqrt{f}} = 0.869\ln\left(\frac{d}{2k_s}\right) + 1.74 \tag{6.128}$$

or

$$\frac{1}{\sqrt{f}} = 2.0\log_{10}\left(\frac{d}{2k_s}\right) + 1.74 \tag{6.129}$$

For open channel flow problems, the hydraulic radius, R ($d = 4R$) traditionally replaces the pipe diameter, d, and equation (6.129) then becomes

$$\frac{1}{\sqrt{f}} = 2.0\log_{10}\left(\frac{2R}{k_s}\right) + 2\log_{10}(7.413) \tag{6.130}$$

or

$$\frac{1}{\sqrt{f}} = 2.0\log_{10}\left(\frac{14.84R}{k_s}\right) \tag{6.131}$$

as expressed by (2.29). This is of necessity a somewhat simplified summary, but is probably sufficient for the purposes of this book. For those interested in further details, see Schlichting (1979), Reynolds (1974) and Nezu & Nakagawa (1993).

6.6.2   Effects of cross-sectional shape on resistance

Having established suitable resistance equations for smooth and rough surfaces in turbulent flow, some researchers have attempted to introduce a 'shape factor', a, in order to apply them to ducts and channels of different shapes, or even to account for any differences between flows in pipes and open channels, expressing (6.131) in a more 'general' form

$$\frac{1}{\sqrt{f}} = 2.0 \log_{10}\left(\frac{aR}{k_s}\right) \tag{6.132}$$

Ignoring turbulent flow for the moment, some clues about shape effects are to be found from laminar flow where resistance coefficients may be derived from first principles for certain shapes. For example, solving the Navier-Stokes equations gives $f = 64/Re$ for flow in a circular pipe, $f = 53.3/Re$ for flow in a triangular section $(\theta = 60°)$, $f = 96/Re$ for flow between parallel plates and $f = 56.9/Re$ for flow in a square duct. This indicates that for laminar flows there is an almost two-fold variation in f values due to the effects of shape alone. Turbulent flow data follows a similar trend, but with a smaller variation in f values.

Expressing laminar flow resistances in general terms, as $f = K/Re$, Figure 6.67 shows some theoretical values of K for 'overbank flow' calculated for compound duct flows by Lai (1987) for different floodplain widths $(Wr = B/b)$, symmetries and relative depths $(Dr = (H-h)/H$. The main channel semi-width aspect ratio $(ASP = b/h)$ was held constant at 1.0 throughout to generate the maximum secondary flow and for comparison with flow in square ducts. As can be seen, K values vary significantly with the shape of the cross-section. Figure 6.68 shows a test comparison, taken from Lai (1987), between numerical and measured distributions of velocity and boundary shear stress using the laminar flow experiments by Allen & Ullah (1967) in a T-shaped narrow channel. Having found good agreement, Figures 6.69–6.74 show further numerical results from Lai (1987) for laminar flow in compound ducts, together with the corresponding turbulent flow measurements in the same shaped ducts. In many cases the similarities are obvious, as in the variation of kinetic energy correction coefficients α (CORE in Figures 6.69 & 6.70) with Dr, and $\%Q_{mc}$ (Figures 6.73 & 6.74). There is less similarity in others, such as for the apparent shear forces on the vertical interface between the main channel and its floodplain, $\%ASF_v$, in Figures 6.71 & 6.72, which indicate a significant difference between the laminar and turbulent flow fields at low Dr values.

Alternatively, Prandtl's early work is of use as it enabled researchers to investigate the variation of mixing length and eddy viscosity within a cross section through what he termed the 'velocity defect' law. Originally he assumed that near a solid boundary the fluid shear stress was constant within what he deemed to be a very thin layer and therefore equal to the boundary shear stress $(\tau = \tau_o)$, and that the mixing length, ℓ_m, varied linearly with distance, z, from any surface $(\ell_m = \kappa z)$, where κ is a constant, yet to be determined empirically. Inserting these into (6.47) with a notation change gives the velocity distribution in the inertial sub-layer, since

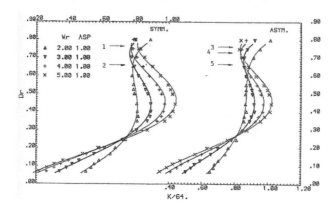

Figure 6.67 Laminar flow coefficients (numerical values of K in $f = K/Re$ for different duct shapes), from Lai (1987).

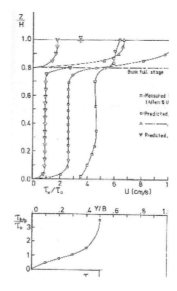

Figure 6.68 Laminar flow velocity and boundary shear stresses (measured by Allen & Ullah, 1967 and predicted by Lai, 1987).

$$\tau_{zx} = -\overline{\rho u'w'} = \rho \ell_m^2 \left(\frac{\partial u}{\partial z}\right)^2 \quad \text{and} \quad \tau_o = \rho \kappa^2 z^2 \left(\frac{\partial u}{\partial z}\right)^2 \tag{6.133}$$

and $\dfrac{du}{dz} = \dfrac{\sqrt{(\tau_o/\rho)}}{\kappa z} = \dfrac{u_*}{\kappa z}$ which when integrated yields $u = \dfrac{u_*}{\kappa}\ln z + C$, or

$$\frac{u}{u_*} = \frac{1}{\kappa}\ln\left(\frac{z}{z_o}\right) \tag{6.134}$$

where $C = $ constant $= -(u_*/\kappa)\ln z_o$ and $u = 0$ when $z = z_o$, and the constants κ, C or z_o have to be determined experimentally. Prandtl further assumed that

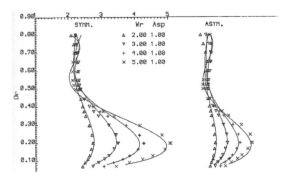

Figure 6.69 Laminar duct flow α versus *Dr* and *B/b* (numerical) from Lai (1987).

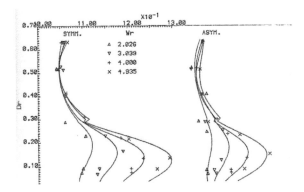

Figure 6.70 Turbulent duct flow α versus *Dr* and *B/b* (experimental) from Lai (1987).

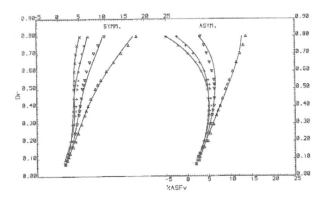

Figure 6.71 Laminar duct flow %ASF_v v Dr and B/b (numerical) from Lai (1987).

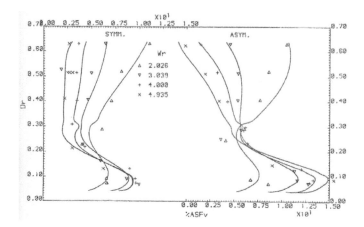

Figure 6.72 Turbulent duct flow %*ASF*ᵥ v *Dr* and *B/b* (experimental) from Lai (1987).

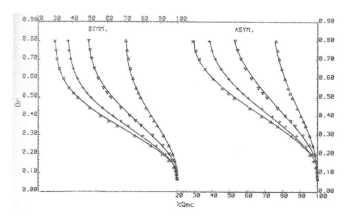

Figure 6.73 Laminar duct flow %*Q*ₘ𝒸 v *Dr* and *B/b* (numerical) from Lai (1987).

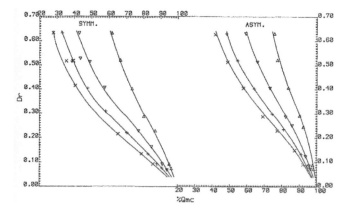

Figure 6.74 Turbulent duct flow %*Q*ₘ𝒸 v *Dr* and *B/b* (experimental) from Lai (1987).

$$z_o = \frac{\beta v}{u_*} \tag{6.135}$$

where β = constant such that $\dfrac{u}{u_*} = \dfrac{1}{\kappa}\ln\left(\dfrac{u_*z}{v}\right) - \dfrac{1}{\kappa}\ln\beta$, which is of the general form

$$\frac{u}{u_*} = A\ln\left(\frac{u_*z}{v}\right) + B \quad \text{(smooth pipe law)} \tag{6.136}$$

where A & B are given by

$$A = 1/\kappa; \quad B = -(1/\kappa)\ln\beta \tag{6.137}$$

Early experiments gave $\kappa = 0.40$ and $\beta = 0.111$, i.e. $A = 2.5$ and $B = 5.5$, thus giving the velocity distribution law for the inertial sub-layer as given by (6.115). However, as noted in Figures 6.65 & 6.66, this is only valid for $u_*z/v > 30$, and that the distance, z_o, at which $u = 0$ is given by

$$z_o = 0.111\frac{v}{u_*} = 0.111\frac{\delta'}{5} = \frac{\delta'}{45},$$

i.e. is very small. The distance δ_T can be found by ignoring the buffer layer entirely and equating the velocities at $z = \delta_T$ from buffer and inertial layers

$$\frac{u}{u_*} = \frac{u_*\delta_T}{v} = 2.5\log_{10}\left(\frac{u_*\delta_T}{v}\right) + 5.5 \quad \Rightarrow \quad \frac{u_*\delta_T}{v} = 11.6$$

Hence the distance δ_T may be given by $\delta_T = 11.6v/u_*$ (i.e. about $2\delta'$), as indicated in Figure 6.62.

For rough surfaces, Prandtl's logarithmic velocity law, (6.134), is still applicable and he assumed that that z_o was proportional to the size of the Nikuradse equivalent sand roughness size, k_s, giving

$$z_o = \gamma k_s \tag{6.138}$$

where γ = constant. Hence

$$\frac{u}{u_*} = \frac{1}{\kappa}\ln\left(\frac{z}{k_s}\right) - \frac{1}{\kappa}\ln\gamma \tag{6.139}$$

Early experiments gave values of $\gamma = 1/30$ and $\kappa = 0.40$. Thus the velocity distribution law for rough surfaces became (6.116), where A & B are constants. Experiments show that B is a function of (u_*k_s/v), as shown in Schlichting (1979).

Fortuitously, Prandtl's formulation of the logarithmic velocity distribution law has subsequently been found to be widely applicable, even when the two initial assumptions are invalid. In many flows it is more realistic to assume

$$\tau = \tau_o \left(1 - \frac{z}{\delta}\right)$$ (6.140)

and

$$\ell_m = \kappa z \left(1 - \frac{z}{\delta}\right)^{1/2}$$ (6.141)

where δ = boundary layer thickness.

This pair of assumptions will lead to the same logarithmic law, (6.134), as may be seen by inserting them into (6.133). The reason as to why this pair of assumptions is better is now explained with reference to Figure 6.75, which is related to Figures 6.6 to 6.8 in Section 6.2. For flows in pipes in which the walls are parallel, or in open channels in which the flow is uniform and not changing with x, any further growth in the boundary layer is inhibited, since for fully developed flow $\delta = r$ (radius of pipe) and $\delta = h$ (depth of flow in channel).

From the Navier-Stokes equations with $u\partial u/\partial x = 0$, then $0 = -\frac{1}{\rho}\frac{\partial p}{\partial x} + \frac{1}{\rho}\frac{\partial \tau}{\partial z}$. Since $\partial p/\partial x$ is constant in uniform or fully developed flow, integrating over the boundary layer gives

$$\int_0^z \frac{1}{\rho}\frac{dp}{dx}dz = \int_{\tau_o}^\tau \frac{1}{\rho}\frac{d\tau}{dz}dz, \quad \text{hence} \quad \frac{1}{\rho}\frac{dp}{dx}z = \frac{\tau}{\rho} - \frac{\tau_o}{\rho}$$

Since for $z = \delta$, $u = u_{max}$ and $\tau = 0$, this yields $\dfrac{dp}{dx} = -\dfrac{\tau_o}{\delta}$ and hence

$$\tau = \tau_o \left(1 - \frac{z}{\delta}\right)$$ (6.142)

which is linear, as assumed in (6.140). Applying the logarithmic law to the whole flow, (6.139) gives the maximum velocity as $\dfrac{u_{max}}{u_*} = \dfrac{1}{\kappa}\ln\left(\dfrac{\delta}{z_o}\right)$. Subtracting (6.139) gives the 'velocity defect law' as

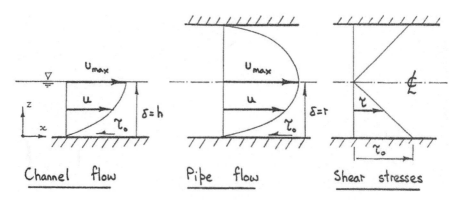

Figure 6.75 Velocity and shear stress distributions in channel and pipe flow.

$$\frac{u_{max} - u}{u_*} = \frac{1}{\kappa}\ln\left(\frac{\delta}{z}\right) \qquad (6.143)$$

Experiments by Nikuradse in smooth and rough pipes in the 1930s confirmed this 'velocity defect law', which in a pipe of radius r the term δ in (6.139) is replaced by r to give

$$\frac{u_{max} - u}{u_*} = \frac{1}{\kappa}\ln\left(\frac{r}{z}\right) \qquad (6.144)$$

The same experiments also gave information about the distribution of Prandtl's mixing length and Boussinesq's eddy viscosity in pipe flow. Assuming a linear shear stress distribution, then $\tau_o\left(1 - \frac{z}{r}\right) = \rho \ell_m^2 \left(\frac{\partial u}{\partial z}\right)^2$. Hence by measuring $u = f\{z\}$, and knowing r and τ_o (from pressure drop readings, $\tau_o = -(d/4)(\partial p/\partial x)$), the mixing length, ℓ_m, may be determined and also the turbulent eddy viscosity, v_t, from (6.48). Some values of ℓ_m and v_t for pipe flow are shown in Figures 6.76 & 6.77, which may be compared with the corresponding values for open channel flow in Figures 6.7 & 6.6. Figure 6.76 indicate that ℓ_m varies as a power law, often given by the following equation, which when expanded by the binomial theorem gives

$$\frac{\ell_m}{r} = 0.14 - 0.08\left(1 - \frac{z}{r}\right)^2 - 0.06\left(1 - \frac{z}{r}\right)^4 \quad \Rightarrow \quad \frac{\ell_m}{r} = 0.4\frac{z}{r} - 0.44\left(\frac{z}{r}\right)^2 + ... \;(6.145)$$

Thus near the wall, $\ell_m = 0.4z$ [Prandtl's original hypothesis, but as also assumed in (6.141)] or $\ell_m = \kappa z$ [where $\kappa = 0.4$ (von Karman's constant)]

This indicates that very close to the wall Prandtl's original idea that the mixing length varied linearly was correct, as shown by the line (2) in Figure 6.76, which is valid for say $z/R < \sim 0.05$.

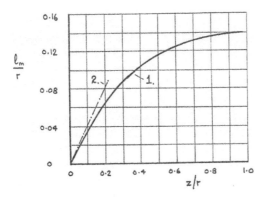

Figure 6.76 Variation of mixing length across pipe radius (Nikuradse, 1932) (after Schlichting, 1979).

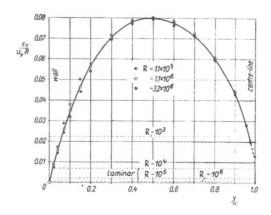

Figure 6.77 Variation of eddy viscosity with radius for turbulent flow in a pipe (after Schlichting, 1979).

Using this value of κ, and inserting it into (6.144) gives the universal velocity defect distribution law

$$\frac{u_{max} - u}{u_*} = 2.5\ln\left(\frac{r}{z}\right) \quad \text{or} \quad \frac{u_{max} - u}{u_*} = 5.75\log_{10}\left(\frac{r}{z}\right) \tag{6.146}$$

This law was found to be valid for all Reynolds numbers $> 4 \times 10^3$ and for types of roughness. The universal velocity distribution law, or defect law, is known to fit the data well for smooth and rough pipes but requires knowledge of the maximum velocity, u_{max}, which itself is an unknown quantity and must be determined by other means. Alternatively, a more general distribution law must be developed, that does not use u_{max} at all, and instead uses the section-mean velocity or shear velocity that most engineers prefer to work with. This is the reason for the depth-mean or area-mean value has been used throughout this book.

One of the purposes in going back to some of the basic and early work on turbulent flow in simple shapes such as pipes and ducts, is to illustrate how little we know about key turbulent parameters in open channel flow, as noted by Lighthill (1970). For example, the relationships between turbulent intensities, Reynolds stresses, secondary currents, flow structures, and boundary shear stress have not yet been comprehensively measured enough for flows in channels with simple trapezoidal shapes, let alone for prismatic compound shapes and non-prismatic shapes such as in natural rivers. This is why we are still so heavily reliant on decades of work on 2-D wall-governed turbulence and boundary layers. What is needed now is a similar programme of work specifically for 3-D flows in open channels, as highlighted by Knight (2013). These data could then become the basic building blocks for future engineers to benefit from, without having to rely on partial data sets from individual experiments, written up in various papers, often conducted with differing objectives, and not necessarily aimed at this fundamental task.

As already highlighted in (2.61) and Section 2.5.1, resistance coefficients need to be carefully defined, especially when interpreting resistance data or using data to calibrate a model. As indicated for a typical river cross section, such as in Figures 2.44–2.48, the wetted perimeter, P, increases rapidly once the flow goes overbank, whereas the cross-sectional area, A, increases very little, thus leading to a marked decrease in the hydraulic radius ($R = A/P$). This effect changes the normal relationship between the global or 1-D friction factor, f, and the Reynolds number, Re, shown in Figure 2.10, and instead creates a discontinuity at the bankfull stage, as shown by Figures 6.78 & 6.79. The compound channel laboratory data in Figure 6.78 indicates that as the flow increases, the f values decrease with increasing Re for all inbank flows until the bankfull discharge is reached, at which point the f values suddenly drop to around half their bankfull value before increasing again along a particular line dependent of the ratio of floodplain to main channel width, B/b. The same effect is seen in Figure 6.79 for the River Severn data although the discontinuity is not as marked due to the floodplain edges being not as sharp as in the laboratory channels. A discontinuity may also arise in the stage-discharge relationship, as shown in Figure 6.80,

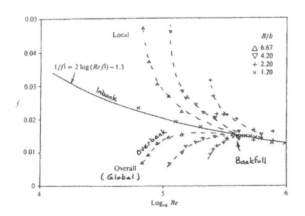

Figure 6.78 f versus Re, FCF laboratory data from FCF, Series 01–04 (from Knight, 2006).

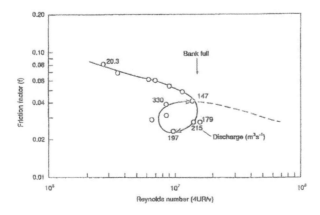

Figure 6.79 f versus Re, FCF field data, from River Severn at Montford (from Knight, 2006).

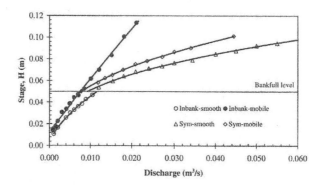

Figure 6.80 H versus Q laboratory data for rigid and mobile beds (from Atabay & Knight, 2006).

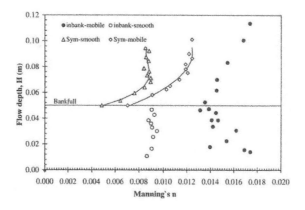

Figure 6.81 H versus n laboratory data for rigid and mobile beds (from Atabay & Knight, 2006).

where for a rigid bed channel the discharge drops at the bankfull stage to below the bankfull discharge. Thus the channel may flow at either of two depths in this region, creating oscillations in numerical models or hysteresis between rising and falling limbs of a flood hydrograph. A marked discontinuity may also be seen in values of Manning's n, as shown in Figure 6.81. Further details are given by Atabay & Knight (2006).

Other types of surface roughness types that are important to mention are those of strip roughness and sediment bed forms. Strip roughness is used frequently to enhance mixing and where sediment is present river beds will develop bed features such as ripples and dunes, as shown in Figure 5.2. Both of these greatly enhance surface roughness and therefore specialist texts should be consulted, such as ESDU (1979), Knight & Macdonald (1979), White et al. (1980), Chang (1988), Knight (2006b) and Morvan et al. (2008). One remaining type of roughness that is not commonly dealt with in textbooks on river engineering is that of trees, despite being frequently encountered in river modelling. It is therefore considered in some detail in the next section, and how the drag force caused by trees may be incorporated into the modelling system of either SKM or CES.

6.6.3   Drag force caused by trees

Trees are common features alongside rivers and on their floodplains, and Figure 6.82 shows a typical river channel with a line of trees either side of the main channel at the edges of the floodplain. Figure 6.83 shows the effect of these on the lateral distribution of depth-averaged velocity, taken from experiments undertaken by Sun (2007) and reported by Shiono *et al.* (2009). Also shown in Figure 6.83 is the theoretical distribution, predicted by the SKM with a component added that allows for the effects of drag forces on trees. This section explains the background to this additional resistance term and how it might be applied in practice within the SKM or CES.

In order to take into account the effect of vegetation or trees within the SKM, an additional drag force term, F_d needs to be added to the governing equation (2.58) and (6.7). The drag force per unit fluid volume for any such obstacles per unit volume of flow may be expressed as

$$F_d = \frac{1}{2}\rho\sum_i\left(C_dS_fA_p\right)_i U_d^2 \tag{6.147}$$

where i is the type of vegetation and C_d is the drag coefficient.

For an array of rods, the wake formed by the upstream rods can diminish the drag coefficient C_d on the downstream rods. S_F is the shading factor and A_p (1/m) is the projected area of i plants in unit volume. In the case of dense vegetation, the porosity also needs to be considered. The porosity, α, accounts for the blockage effects on the flow by the vegetation and is defined as

Figure 6.82 Trees bordering a river channel at the floodplain edges (From Shiono).

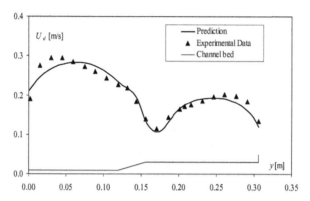

Figure 6.83 Simulation of single tree line by SKM (experimental study on tree lines at floodplain edges) (From Sun, 2007).

$$\alpha = 1-\left(N_v A_v\right)_i \tag{6.148}$$

where A_v is the average cross-sectional area of i vegetation stems and N_v is the averaged i vegetation density per unit area. Adding the drag force per unit area to (6.7) then gives

$$\alpha g H S_o - \alpha \frac{f}{8} U_d^2 \left(1+\frac{1}{s^2}\right)^{1/2} + \frac{\partial}{\partial y}\left\{\alpha\lambda H^2 \left(\frac{f}{8}\right)^{1/2} U_d \frac{\partial U_d}{\partial y}\right\}$$
$$+\sum_i \frac{1}{2}(C_d S_f A_p H)_i U_d^2 = \alpha \frac{\partial}{\partial y}\left[H(\rho U V)_d\right] \tag{6.149}$$

This can be solved numerically, but when the water depth is constant (6.149) may be solved analytically. The formulation of the drag force term is now considered, using numerical results with experimental data to establish the correct formulation, and then secondly using analytical results, based on a new analytical solution, to illustrate how it can be used within the SKM for a constant depth domain.

1. Formulation of the drag force term using numerical results with experimental data

As mentioned previously, prior to applying the SKM to open channel flow, the friction factor, dimensionless eddy viscosity and secondary flow term should all be calibrated. Additionally, the drag coefficient, shading factor and porosity are now also required in order to solve the newly formulated equation, (6.149). Rameshwaran and Shiono (2007) used SKM numerically to predict the FCF data and the following is extracted from their paper to demonstrate how to model all the input parameters for SKM and to illustrate some of the results.

Consider first the input parameter of friction factor for two sets of experimental FCF data. Case A is for a smooth bed and Case B is for cylindrical rods on the floodplains. The experimental set up and programme for the FCF series A is briefly described here, but further details may be found in Knight & Shiono (1990) and Wormleaton & Merrett (1990). The FCF flume was 60 m long and 10 m wide, had a maximum discharge of 1.1 m³s⁻¹, and built so that a variety of a channel geometries could be constructed within it. A typical cross-section of a straight compound channel is shown in Figure 6.84. Case A was a 'smooth floodplain' experiment and Cases B and C were the 'vegetated floodplain' experiments with different geometric parameters. The channel surfaces in all experiments were composed of smooth cement mortar. The vegetated floodplain was roughened by vertical rods placed in a triangular array of 60° on the smooth surface of the floodplain as shown in Figure 6.84. The rods had a diameter of 25 mm and were spaced to give a density of 12 rods per m². The lateral spacing between rods was 310 mm and the longitudinal spacing of the rows was 268.5 mm. The flow rate was measured using calibrated orifice plates and the water surface elevations by using digital pointer gauges. Detailed measurements of primary velocity and bed shear stress were carried out in the main channel and on the floodplain for cases A and B, and in the main channel only for case C, at different flow depths under steady uniform flow conditions. The primary velocity was measured using miniature propeller meters and the bed shear stress by a Preston tube. Turbulence measurements were undertaken using a Laser Doppler Anemometer (LDA) system at relative depths $Dr \approx 0.10, 0.15, 0.20$ & 0.25 in Case A.

The formulation for estimating the friction factor across the channel is given below using the roughness height, k_s, or skin friction of the channel surface. The skin

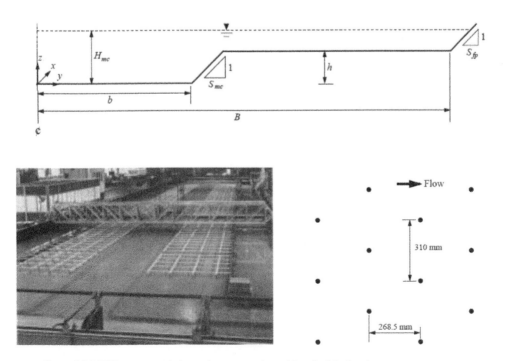

Figure 6.84 FCF compound channel cross-section with cylindrical rod array arrangement.

friction of the smooth channel surface was obtained from experimental data from the straight wide trapezoidal channel. Figure 6.85 shows that the Manning coefficient n took an average value of 0.009 for all flow depths.

The equivalent sand grain roughness height, k_s, was estimated to be 0.16 mm, calculated from the relationship $n = k_s^{1/6} / (8.25\sqrt{g})$ by Ackers (1991). With $k_s = 0.16$ mm, the Colebrook–White equation (C–W) was then used to calculate a local friction factor f for the smooth surface (bed) at any location in a cross-section with water depth H, from

$$\frac{1}{\sqrt{f}} = -2\log\left(\frac{3.02}{R_e\sqrt{f}} + \frac{k_s}{12.3H}\right) \tag{6.150}$$

where Re is the local Reynolds number, defined as $Re = 4U_d H / \nu$. Using $U_* = \sqrt{gS_0 H}$ and $f = 8U^2/U_*^2$, (6.150) can be re-arranged as

$$f = \frac{1}{\left[-2\log\left(\dfrac{3.02\nu}{\sqrt{128gH^3 S_0}} + \dfrac{k_s}{12.3H}\right)\right]^2} \tag{6.151}$$

For the vegetated floodplain, the wake formed by the rods generates a complex flow pattern on the floodplain and distorts the vertical profile of velocity from the 2D logarithmic profile which was assumed in the Colebrook–White equation. Therefore, a calibrated version of the Colebrook–White equation with a different coefficient in the k_s term was used

$$f = \frac{1}{\left[-2\log\left(\dfrac{3.02\nu}{\sqrt{128gH^3 S_0}} + \dfrac{k_s}{1.20H}\right)\right]^2} \tag{6.152}$$

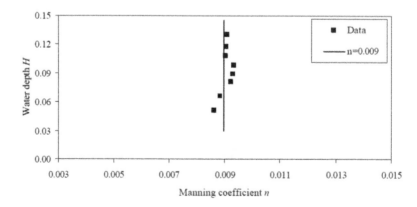

Figure 6.85 Manning coefficient for FCF data (Case A), Rameshwaran & Shiono (2007).

The validity of these equations is assessed by comparison with the experimental data. Figure 6.86 shows the experimentally calculated values of the averaged friction factor, the main channel and on the floodplain, based on the measured bed shear stress and depth-averaged velocity, and the calculated values using (6.151) & (6.152) for all the depths in cases A and B. Figure 6.87 shows the measured and calculated values of the local friction factor across the section for a relative flow depth of $Dr \approx$ 0.2 in cases A and B. From these Figures, it is seen that the friction factors calculated with (6.151) & (6.152) agree reasonably well with the experimental data.

i. *Eddy viscosity vt*

Several eddy viscosity models have been used in the literature to predict transverse momentum exchange in the mixing layer. The range of models used in this paper, are listed in Table 6.4. Model 1 is the constant eddy viscosity model where the coefficient λ is constant across the section. In Model 2 (Shiono & Knight,1991) and Model 3 (Abril & Knight, 2004), the coefficient λ is constant in the main channel but a function of the relative depth on the floodplain. These models were all derived from experimental data. van Prooijen *et al.* (2005) derived Model 4 by introducing a combination of the bed-generated and shear-generated turbulence. Model 5 is a similar form to Model 4, but the mixing layer approach proposed by Alavian & Chu (1985)

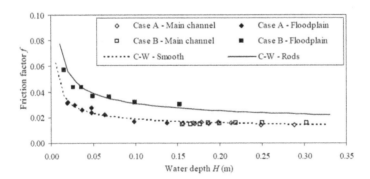

Figure 6.86 Measured and reproduced average friction factor for cases A and B (Rameshwaran & Shiono, 2007).

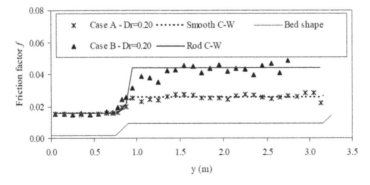

Figure 6.87 Measured and predicted friction factor f across a section ($Dr \approx 0.20$) (Rameshwaran & Shiono, 2007).

Table 6.4 Models for eddy viscosity.

Models	Eddy viscosity	Coefficients
Model 1	$$v_t = \lambda\sqrt{\frac{f}{8}\bar{U}_d H}$$	$\lambda_{mc} = \lambda_{fp}$
Model 2 (Shiono and Knight, 1991)	$$v_t = \lambda\sqrt{\frac{f}{8}\bar{U}_d H}$$	$\lambda_{fp} = \lambda_{mc}\,(2Dr)^{-4}$
Model 3 (Abril and Knight, 2004)	$$v_t = \lambda\sqrt{\frac{f}{8}\bar{U}_d H}$$	$\lambda_{fp} = \lambda_{mc}\,(-2.0 + 1.20\,Dr^{-1.44})$
Model 4 (van Prooijen et al., 2005)	$$v_t = \lambda\sqrt{\frac{f}{8}\bar{U}_d H + (\beta\delta)^2 \left|\frac{\partial\bar{U}_d}{\partial y}\right|}$$	$\lambda_{mc} = \lambda_{fp},\ \beta_{mc} = \beta_{fp}$
Model 5 (Alavian and Chu, 1985)	$$v_t = \lambda\sqrt{\frac{f}{8}\bar{U}_d H + \beta\delta\Delta\bar{U}_d}$$	$\lambda_{mc} = \lambda_{fp},\ \beta_{mc} = \beta_{fp}$

is adopted, which uses the velocity difference between the main channel and flood-plain $(U_{dmc} - U_{dfp})$ as the velocity scale. In Models 4 & 5, β is a coefficient and δ is the width of the mixing layer, as defined by van Prooijen *et al.* (2005).

$$\delta = 2\ (y75\% - y25\%)$$

where $y75\%$ is where $U_d 75\%$ [$= U_{dfp} + 0.75(U_{dmc} - U_{dfp})$], $y25\%$ is where $U_d 25\%$ [$= Udfp + 0.25(U_{dmc} - U_{dfp})$] and U_{dmc} & U_{dfp} are the depth-averaged velocities at the centre of main channel and floodplain respectively (see Figure 3 in van Prooijen *et al.*, 2005).

ii. Advection term (secondary flow)

The experimental data from Shiono & Knight (1991) showed that the Reynolds shear stress term is insignificant outside the lateral shear layer region. Thus at the centre of the main channel and on the floodplain outside the shear layer region, the advection term can be estimated using (6.149) with the measured bed shear stress and velocity distributions across the channel, and ignoring the transverse shear stress term. The calculated values of $\Gamma/\alpha\rho gHS_0$ for cases A and B are shown in Figure 6.88. From Figure 6.88(a), the main channel and floodplain $\Gamma/\alpha\rho gHS_0$ values for case A are about 0.2 and −0.2 respectively. For the vegetated floodplain in case B, Figure 6.88(b) shows that the main channel and floodplain $\Gamma/\alpha\rho gHS_0$ values are 1.2 Dr and 0.6 Dr respectively. It should be noted that these values were worked out on the assumption of there being no gradient in the transverse shear stress τ_{yx}. The values are relatively high when the water depth is large and there is also the question of the existence of second-ary flow on the floodplain with rods. This is discussed later.

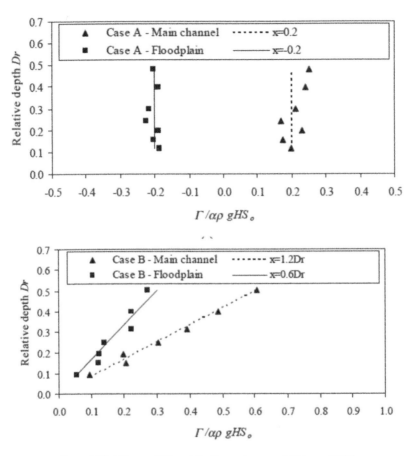

Figure 6.88 Values of $\Gamma/\alpha\rho g H S_0$ (Rameshwaran & Shiono, 2007).

iii. Porosity α

In these experiments, the rod diameter D was 25 mm and the rod density N_v was 12 per m² giving a porosity of 0.9941 from (6.148).

iv. Drag coefficient C_d

In the FCF experiments, vegetation on the floodplain was idealised by vertical smooth rods placed in a triangular array of 60°. For a smooth rod (i.e. cylinder), the experimental drag coefficient C_d can be determined from Schlichting (1962), as a function of the rod Reynolds number Re_{rod} ($= U_d D/v$). For the range of rod Reynolds numbers in these experiments ($10^3 \leq Re_{rod} \leq 10^4$), the drag coefficient C_d is about 1.0.

v. Shading factor

Nepf (1999) showed that this effect increases as the lateral and longitudinal array spacing between rods decreases. Using the wake interference model given by Nepf (1999), aligned rods with a ratio of 0.87 between the longitudinal and lateral rows and a fractional volume of 0.0075 gives a bulk drag coefficient $(C_d \times S_f)$ of 1.0. Thus, $S_f = 1.0$ when $C_d = 1.0$.

Once setting all the input data to SKM, one can use a finite difference method to solve (6.149). Details of the finite difference method can be found in Chapter 7, together with examples of input data and output results using the Fortran program source code given in Appendix 6. A Fortran compiler can be found in the following website *http://www.silverfrost.com/32/ftn95/ftn95_personal_edition.aspx* or and freely downloaded from *http://www.cs.yorku.ca/~roumani/fortran/ftn.htm*. Some results produced by the program are shown in Figure 6.89.

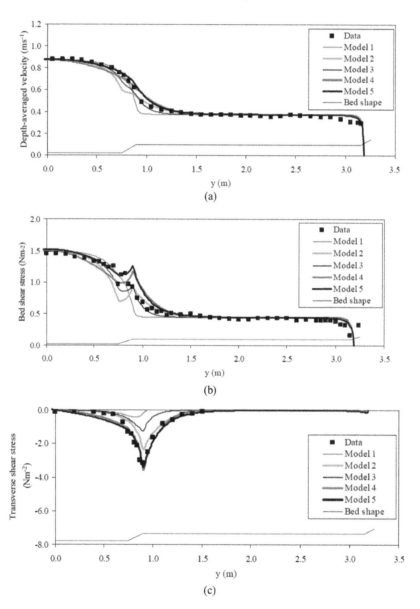

Figure 6.89 Comparison of flow field for case A (*Dr* ≈ 0.20). (a) Depth averaged velocity, U_d b. Bed shear stress τ_b (c) Transverse shear stress (Rameshwaran & Shiono, 2007).

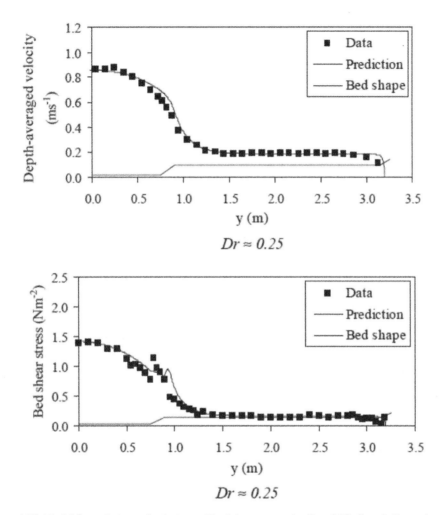

Figure 6.90 Model 5 predictions of velocity and bed shear stress for *Dr* = 0.25, Case B (Rameshwaran & Shiono, 2007).

Since Model 5 appeared to be the best for predicting the depth averaged velocity and bed shear stress, it was used for predicting the data for Case B, as shown in Figure 6.90.

It should be noted that the lateral distributions of both U_d and τ_b should be predicted simultaneously as it is frequently possible to get good agreement with one parameter but not the other, as shown earlier in Section 6.5, Case 1, when simulating both %SF_w & %Q_{mc}. The use of a calibrated friction factor or Manning coefficient, based on simulating velocity distributions, may not be adequate for predicting boundary shear stress distributions where the roughness is dominated by drag forces and vegetation. This is illustrated in Figure 6.91, where U_d is predicted well by the Conveyance Estimation System (CES), but not τ_b. Analytical solutions for predicting

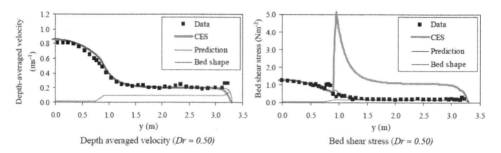

Figure 6.91 Predictions of velocity and bed shear stress for Dr = 0.50 (Rameshwaran & Shiono, 2007).

velocity and bed shear stress in the same data are described in Shiono *et al.* (2009) and dealt with next in more detail.

2.   Comparison of analytical results for a constant depth domain with experimental data

i. *Analytical solution*

The analytical solution for (6.149) for a flat bed with vegetation is given in Appendix 5. The FCF compound channel had rods arranged on a staggered grid on the flood-plains and therefore applying the analytical solution in Appendix 5 gives

$$F = U_d^2 = A_1 e^{\gamma y} + A_2 e^{-\gamma y} + \frac{gHS_0 (1 - \beta)}{\dfrac{f}{8} + \dfrac{1}{2} C_d N A_p H} \tag{6.153}$$

where N is a number of rods per unit area,

$$\gamma = \left(\frac{\dfrac{f}{8} + \dfrac{1}{2} C_d N A_p H}{\dfrac{\lambda H^2}{2} \left(\dfrac{f}{8} \right)^{1/2}} \right)^{\!1/2}$$

$$\Gamma = \frac{\partial}{\partial y} \left\{ \rho H (UV)_d \right\}$$

$$\beta = \frac{\Gamma}{\rho g H S_0}$$

For a linear side slope of 1 in s ($s = L/h$: L = slope width and h = height), the solution in Appendix 5 is

$$U_d = \left[A_3 \xi^{\alpha_1} + A_4 \xi^{-(\alpha_1 + 1)} + \omega \xi + \eta \right]^{1/2}$$

where

$$\alpha_1 = -\frac{1}{2} + \frac{1}{2}\left\{1 + \frac{s(1+s^2)^{1/2}}{\lambda}(8f)^{1/2}\right\}^{1/2}$$

$$\alpha_2 = -\frac{1}{2} - \frac{1}{2}\left\{1 + \frac{s(1+s^2)^{1/2}}{\lambda}(8f)^{1/2}\right\}^{1/2}$$

$$\omega = \frac{gS_0}{\dfrac{(1+s^2)^{1/2}}{s}\dfrac{f}{8} - \dfrac{\lambda}{s^2}\left(\dfrac{f}{8}\right)^{1/2}}$$

$$\eta = -\frac{\Gamma}{\dfrac{\rho f}{8}\left(1+\dfrac{1}{s^2}\right)^{1/2}}$$

These solutions are now applied to the following sub-sections: the main channel flat bed and side sloping bank domains, and the floodplain flat bed and side sloping walls of the FCF compound channel, together with the boundary conditions at the subsections listed in 6.3. As the channel is symmetric, one can use just half of the channel and four sub-sections: 1) main channel flat bed, 2) main channel side bank, 3) floodplain bed and 4) floodplain side bank.

The boundary condition at the centre of the main channel is

$$\frac{\partial U_{d1}}{\partial y} = 0$$

Between the main channel and its sloping bank, assuming a smooth transition of depth-averaged velocity,

$$U_{d1} = U_{d2} \quad \text{and} \quad \frac{\partial U_{d1}}{\partial y} = \frac{\partial U_{d2}}{\partial y}$$

Similarly, at the edge of the floodplain, assuming a smooth transition,

$$U_{d2} = U_{d3} \quad \text{and} \quad \frac{\partial U_{d2}}{\partial y} = \frac{\partial U_{d3}}{\partial y}$$

and also at the bottom of the floodplain side bank,

$$U_{d3} = U_{d4} \quad \text{and} \quad \frac{\partial U_{d3}}{\partial y} = \frac{\partial U_{d4}}{\partial y}$$

Finally, at the water surface of the floodplain side bank,

$$U_{d4} = 0$$

This leads to $A_8 = 0$ and $U_d = \left[A_7 \xi^{\alpha_1} + \omega \xi + \eta \right]^{1/2}$ in the 4th sub-section.

In Section 6.3, on boundary conditions, the Reynolds stress is assumed to be equal at all boundary interfaces, but here it is assumed that the depth-averaged velocity is continuous and that there is a smooth transition between subsections, even if f and λ differ in each sub-section. When different f and λ are applied in adjacent sub-sections and the Reynolds stresses are equal at the boundaries, when Section 6.3 is applied the velocities are not smooth at the boundaries. There are now 7 unknown coefficients, A_1, A_2 ....A_7 with 7 boundary conditions. In order to obtain predictions of velocity and bed shear stress FCF data, appropriate values of f, λ and the advection term, Γ are required, together with an appropriate value for C_d for the rods on the floodplain.

In previous sections, a description of all the parameters adopted for the FCF channel flow has been given. However, it is noticeable that in order to predict the bed shear stress correctly the secondary flow terms in the main channel and floodplain are rather large. The magnitude of the secondary flow was in some cases 60% of the weight component in the main channel and 30% on the floodplain for large relative depths. On the floodplain in particular, it is not likely that one would expect one large secondary flow cell over the whole floodplain, since the rods would destroy its structure. Furthermore, high values are somewhat unrealistic as for typical secondary flows in compound channels. It should also be noted that a drag coefficient of 1.0 was used in the simulations. However, the measured drag coefficient C_d was estimated by the force balance using the measured bed shear stress outside of the shear layer on the floodplain and the weight component, giving $C_d = 1.3$. Shiono & Rameshwaran (2015) used $C_d = 1.3$ to predict bed shear stress well. The reasons for this may be caused by the measurement location of the velocity and boundary shear stress, which was at the midpoint between two rods in the longitudinal direction. This location might be affected by other phenomena related to rods, such as wakes produced behind each rod, non-uniformity, etc.

In the main channel, the structure of secondary flows in a compound channel with cylindrical rods on the floodplain was recently observed by Dupuis *et al.* (2017). Two distinct secondary cells in the upper and lower layers in the main channel were observed and the magnitude of the force induced by the depth averaged secondary flow was of the same order of that induced by the Reynolds stress and also one order larger than that due to bed shear stress. It is therefore understandable that the contribution of 60% of the weight component by secondary flow is a realistic value. Refer to Figure 6.92.

ii. *Trees or vegetation planted in a single line.*
In (i) it was shown how the velocity and boundary shear stress distributions could be predicted by using either numerical methods or analytical solutions, for the case of a compound channel with vegetation planted uniformly on the floodplains. However, as already indicated by Figure 6.82, trees can often be seen in a single line on the edge of floodplains and river channels. The analytical solution has recently been applied to such a case, to simulate one line of square wooden rods at the edge of a floodplain

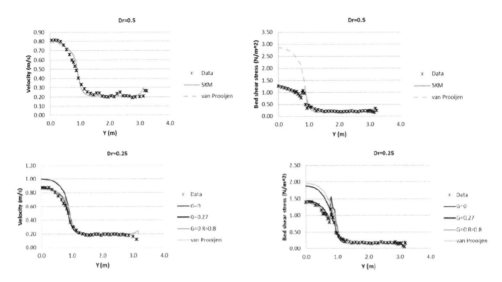

Figure 6.92 Predicted velocity and boundary shear stresses.

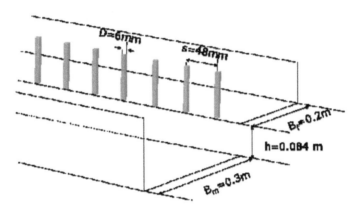

Figure 6.93 Experimental compound channel used at Kansai University.

of an asymmetric compound channel. The experimental data were obtained from the laboratory of Kansai University in Japan, using a channel with the dimensions shown in Figure 6.93.

The compound channel was constructed in a 14 m long and 0.5 m wide Perspex tilting flume with one line of emergent rods along the floodplain edge. The channel geometrical parameters were $B_m = 0.3$ m, $B_f = 0.2$ m, and $h = 0.084$ m and the energy gradient was set to be 0.00123. The square wooden rods used to simulate emergent vegetation had dimensions of 6 mm × 6 mm and were 100 mm high and spaced at 48 mm intervals. The one relative water depth that was used was 0.5. The longitudinal and transverse velocities were measured at the midpoint between the rods and are shown in Figure 6.94.

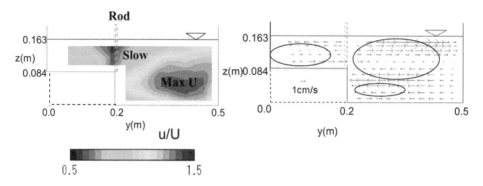

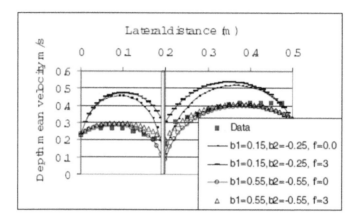

Figure 6.94 Longitudinal and transverse velocities in an asymmetric compound channel.

Figure 6.95 Simulation of depth mean velocities in the Kansai flume.

It is interesting to observe two distinct secondary flow cells in the upper and lower layers in the main channel. These cells are similarly observed in the main channel for rods on the floodplains by Dupuis *et al.* (2017) and contribute a large momentum transfer within the main channel. It can also be seen that there is one large secondary cell on the floodplain. The β and f values were calibrated from the data. The advection term β was 0.55, the same as the ratio of the drag force to the weight component. A drag coefficient C_d of 2.1 was used for a squared rod and S_f was set to 1.0 in this study. Some velocity results are shown in Figure 6.95. The extra eddy f' is an additional non dimensional eddy viscosity which does not much change the velocity profile, even if the value given is large. This is because the parameter is of 2nd order in the differential term, but Γ, which is of first order term, does reduces the velocity significantly, as seen in the simulations in Figure 6.95 (b = β in the Figure).

iii. *Drag force and advection term relationship*
The force balance method can be used to calculate the total drag force in a vegetated compound channel. Under quasi uniform flow condition, the sum of the total boundary

shear force per unit length and the total drag force is equal to the weight component per unit length, which is expressed by the following equation

$$\int_{P} \tau_b dp + F_d = \rho g A S_0 \qquad (6.154)$$

where P is the wetted parameter and A is the channel cross-section area.

From this equation, the total drag force F_d per unit channel length was calculated using the measured boundary shear stress along the wetted perimeter. The drag force per unit length was 0.43 N/m, 55% of the weight component. The advection term was also calibrated using the experimental data. The β values [$\beta = \Gamma / \rho g H' S_0$ with $H' = H$ in the main channel, and $H' = (H - h)$ on the floodplain] appear to be equal to the drag force in the SKM model. Sun and Shiono (2009) performed similar experiments using cylindrical rods and also demonstrated that the drag force and β have a linear relationship.

iv. *Rod spacing*
An investigation of the effect of rod spacing on discharge was carried out by undertaking 3 numerical tests. The first was with $\beta = 0.0$ in the main channel and on the floodplain; the second was with $\beta = 0.55$ for both the main channel and floodplain, and the third was with β equal to the ratio of drag force to the weight component. The percentage changes in discharge are shown in Figure 6.96, where s is the distance between rods and D the rod dimension. Thus for Q = 0.042 m³s⁻¹ and b1 = b2 = 0 (b = β in the Figure), the % change is zero for s/D = 10,000. For b1 = b2 = 0.6, then for s/D > 1000, the discharges are almost constant, meaning no effect of rods on discharge, whereas for s/D < 32, the discharge reduction is considerable. According to the observation of s/D along three rivers, the most frequent range is 8 < s/D < 20, indicating that the% change in discharge is more than 25%. The b values shown in Figure 6.96 become significant as the rod spacing decreases. Without the advection term equal to drag force, i.e. b1 = b2 = 0, the discharge and boundary shear stress could be overestimated (see Figure 6.93). This result indicates that the design of plant

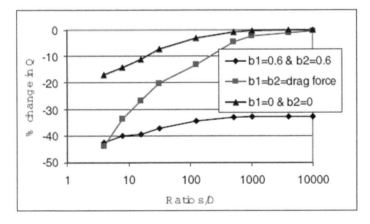

Figure 6.96 Percentage change in discharge.

spacing along the edge of a river is crucially important in flood management. It should be noted that the results are for a channel with a relatively small aspect ratio.

v. *The effect of drag force to the total flow resistance*

The effect of the distance between two rods in the longitudinal direction for one line of rods at the edge of the floodplain on the discharge was investigated in the last section. The next consideration is obviously the effect of the change of width of the channel on the flow and the total resistance. This can be seen by the change of bed shear stress at the centre of the main channel changing from $\rho g H S_o$ (bed shear stress for a wide channel). The distance between two rods was set to 16 times the rod diameter ($s/D = 16$). The diameter of the rod was 9 mm and floodplain height was 36 mm and the floodplain width was 150 mm. The main channel water depth was 75 mm. The SKM was then used to simulate the bed shear stress at the centre of the main channel as the main channel width was varied, keeping the water depth the same. The results are shown in Figure 6.97 for different aspect ratios, together with a Bristle case and the no rod case (Fa = no rod case, Fa + Fd = rod case and Bristle case respectively in the Figure). For a bed width to water depth of 10 there is a 20% ~40% reduction from that of 2D flow, and for an aspect ratio of 40 there is still a 10% reduction. Obviously, as the water depth decreases, less reduction would be expected because the drag force decreases.

vi. *Lateral variations in the friction factor and non-dimensional eddy viscosity*

The friction factor and the non-dimensional eddy viscosity are assumed to be constant in each sub-section for the SKM solutions. When applying the SKM with these constants in each sub-section the solutions give some discrepancy in the shear layer in the vicinity of vegetation. White and Nepf (2008) show that the measured non dimensional eddy viscosity in an open channel flow with vegetation tends to vary in the shear layer. The friction factor also appears not to be constant in the shear layer region caused by vegetal drag, as shown by Xin and Shiono (2009). Further measurements were carried out by Shiono *et al.* (2012) to confirm this, using turbulence measurement in an open channel with one line of rods placed in the centre of the channel. The results in Figure 6.98 show that both the friction factor and non-dimensional eddy viscosity are not constant in the vicinity of rods.

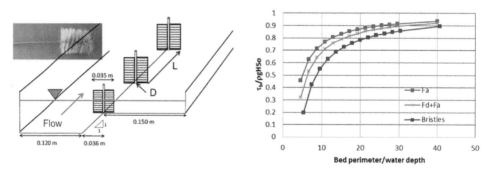

Figure 6.97 Dimensionless bed shear stress.

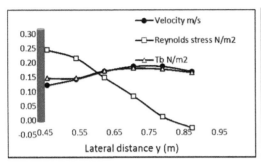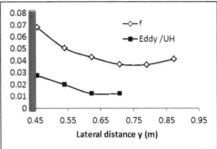

Figure 6.98 Variation of certain parameters in the vicinity of rods.

It therefore follows that the eddy viscosity and friction factor need to be refined in such region in SKM in order to predict accurately such distributions which are particularly important for transport mechanisms of sediment and pollutants. New analytical solutions were derived based on having a variation of friction factor and non-dimensional eddy viscosity, as given in Appendix 5. A brief illustration of the solution that includes the variations of these parameters is now given, based on the notation indicated in Figure 6.99.

The friction factor is now set to a linear variation as a first approximation, in the form $f = f_0\xi$ where

$$\xi = 1 - \left(\frac{y-y_0}{l}\right)\left(1 - \frac{f_{max}}{f_0}\right) \tag{6.155}$$

l is shear layer width, y_0, is at the outside of the shear layer, f_0, is at $y = y_0$, f_{max} is at the rod, $y_R = y_0 + l$. This distribution of the friction factor is shown in Figure 6.100, together with analytical and experimental data. Also shown are the corresponding eddy viscosity values.

In order to have an analytical solution with friction, the non-dimensional eddy viscosity in the SKM has to have the form

$$v_t = \lambda\sqrt{\frac{f_0}{8}}\xi^3 HU \tag{6.156}$$

Through the calibration of λ undertaken using the data, λ was found to be 0.1 in the shear layer and 0.03 in the outside of shear layer. The distribution of the eddy viscosity of (6.156) divided by UH, is also shown in Figure 6.100, which shows that the experimental data and new analytical solution are in reasonable agreement in the shear layer region. The analytical solution for a constant water depth, given in Appendix 5, is therefore confirmed as

$$U^2 = A_1 \zeta^{\alpha 1} + A_2 \zeta^{-\alpha 2} + \frac{gHS_0}{\frac{f_0}{8} + D} \zeta^{-1}(1-\beta) \tag{6.157}$$

where

$$\alpha_1 = -1 + \sqrt{1 + \frac{f_0}{8D}} \qquad \alpha_2 = -1 - \sqrt{1 + \frac{f_0}{8D}}$$

$$D = \lambda \sqrt{\frac{f_0}{8}} \frac{J^2 H^2}{2} \qquad J = -\left(\frac{1}{l}\right)\left(1 - \frac{f_{max}}{f_0}\right) \tag{6.158}$$

The analytical predictions of velocity and boundary shear stress in the vicinity of the rod are shown in Figure 6.101. The prediction of the velocity distribution appears to be reasonable, but the prediction of boundary shear stress distribution is good near the rod but less so further away. This solution was applied to the compound channel flow data obtained from Kansai University. The velocity and boundary shear stress were predicted and compared with the solution using constant friction factor and non-dimensional eddy viscosity. The results are shown in Figure 6.102 where it can

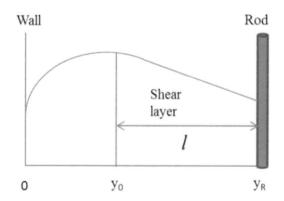

Figure 6.99 Schematic shear layer close to a roughness element.

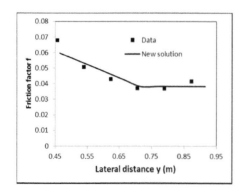
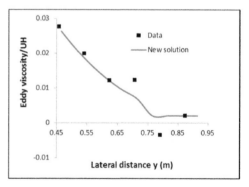

Figure 6.100 Analytical and experimental results.

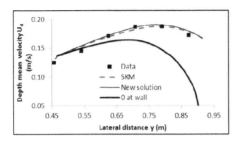
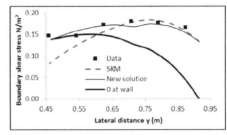

Figure 6.101 Analytical predictions of velocity and boundary shear stress in the vicinity of a rod.

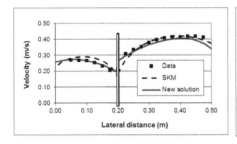
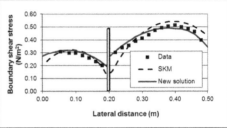

Figure 6.102 Analytical predictions of velocity and boundary shear stress for Kansai flume.

be seen that the new solution improves the prediction. It should noted that with the boundary condition of velocity = 0 at the wall, a large discrepancy occurs near the wall. When the velocity at the wall is set to $U^2 = 0.75RgS_0/(f/8)$, which is 75% of the bed shear stress, the Figure shows that both the SKM and the new solution give better predictions near the wall.

vii. *Drag coefficient of rod with any shape*
The distribution of velocity in a flow and the projected area of the object both affect the drag coefficient, C_d, when using any drag formula. In this section, the drag force on objects different from rods are examined, using a triangular plate and Pandtl's 7th power law for velocity as it approximates the logarithmic velocity profile well in a wide open channel. Theoretically and experimentally this demonstrates that the drag force of a triangle with its apex pointing downwards differs from its value when pointing upwards. Experiments in a wide open channel flow were therefore carried out using triangular and rectangular plates to confirm these changes in drag force. A force measuring device (FMD) was designed to measure the drag forces directly and the approach velocities were measured using Acoustic Doppler Velocimetry (ADV).

A drag formula is commonly used in 1-D, 2-D and 3-D numerical and mathematical models to solve for the flow resistance due to vegetation in river channels or two stage channels, e.g. Rameshwaran and Shiono (2007), Shiono *et al.* (2009), Tan, *et al.* (2010), Sun *et al.* (2011), Sanjou & Nezu, (2011), Shiono, *et al.* (2012), Kawahara *et al.* (2013) and Shiono & Rameshwaran (2015). The drag formula consists of water density, ρ, a drag coefficient, C_d, a project area, A_p, and an approach velocity, U. to give drag force F_d:

$$F_d = \frac{1}{2}\rho C_d A_p U^2 \tag{6.159}$$

The drag force of a horizontal thin layer of an object, ΔA_p, at any vertical location in a 2 D flow can be expressed by equation (6.159) in the form

$$\Delta F_d = \frac{1}{2}\rho C_d U^2 \, \Delta A_p \tag{6.160}$$

The drag force on a cylindrical rod in flow can be calculated as indicated in Figure 6.103. Using the 1/7th power law, as given by (2.55), the local streamwise velocity at any location can be expressed by

$$u = U_{max}\left(z/h\right)^{1/7} \tag{6.161}$$

where U_{max} is the maximum velocity at the water surface, h is the water depth and z is the vertical distance from the bed. Integrating equation (6.160) over the water depth gives the total drag force due to the object. In this case it is assumed for simplicity

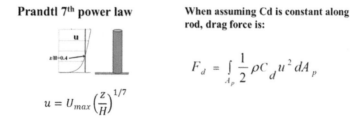

Figure 6.103 Drag force formula for a cylindrical rod using Prandtl's 7th power law.

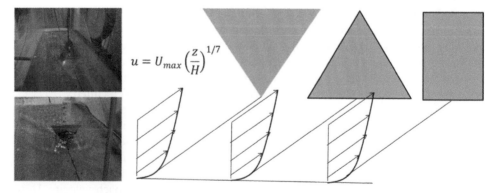

Figure 6.104 Drag force due to triangular and rectangular plates in a uniform flow.

that C_d is constant, although it could be varied if required, and that there is uniform flow in a wide open channel. In order to investigate the effect of velocity distribution and the shape of the projected area, some simple calculations and flume experiments were conducted on triangular and rectangular plates having the same projected area. Figure 6.104 illustrates a triangular plate in position in a laboratory flume, with the apex pointing either upwards or downwards. Before undertaking the experiments, the drag force on both shapes, and the corresponding drag force on a rectangular plate of the same projected area, were all calculated as follows.

Case 1 Drag force on a triangular plate with its apex pointing downwards
The projected area of a horizontal strip, dz, of a triangular plate with its apex pointing downwards at any vertical location is

$$dA = 2z \tan\frac{\theta}{2} dz$$

where θ is the angle of the triangle and h is the height of the triangle. The integration of equation (6.160) for the triangular plate with its apex pointing downwards gives

$$F_d = \int_0^h \frac{1}{2}\rho C_d 2U_{max}^2 \tan(\theta/2)z\left(\frac{z}{h}\right)^{2/7} dz = \frac{7}{16}\rho C_d U_{max}^2 h^2 \tan(\theta/2) \qquad (6.162)$$

Case 2 Drag force on a triangular plate with its apex pointing upwards
The projected area of a horizontal strip, dz, of a triangular plate with its apex pointing upwards at any vertical location is

$$dA = 2(h-z)\tan\frac{\theta}{2} dz$$

The integration of equation (6.160) for the triangular plate with its apex pointing upwards gives

$$F_d = \int_0^h \frac{1}{2}\rho C_d 2U_{max}^2 \tan(\theta/2)\left(\frac{z}{h}\right)^{2/7}(h-z)dz = \frac{49}{144}\rho C_d U_{max}^2 h^2 \tan(\theta/2)$$
$$(6.163)$$

The ratio of the two drag forces, one with the apex downwards and the other upwards is 1.286.

Case 3 Drag force on a rectangular plate with the same projected area
For a rectangular plate having the same projected area as the triangular plate, the projected area of a horizontal strip, dz, at any vertical location is

$$dA = h\tan\left(\frac{\theta}{2}\right)dz$$

The drag force is then given by

$$F_d = \int_0^h \frac{1}{2}\rho C_d U^2_{max} h\tan(\theta/2)\left(\frac{z}{h}\right)^{2/7} dz = \frac{7}{18}\rho C_d U^2_{max} h^2 \tan(\theta/2) \qquad (6.164)$$

The ratios of the rectangular drag force to the drag force on the triangular plate in the downwards and upwards orientation are 0.888 and 1.143 respectively. These calculations were then repeated using the mean velocity over each plate and the total projected area.

Case 4 Drag force on a triangular plate, but using the mean velocity and whole projected area
Since Case 2 is the most rudimentary shape of a tree, this case is repeated, but using the depth averaged velocity and the whole projected area, to give

$$u_{mean} = \frac{7}{8}U_{max}; \quad A = h^2 \tan(\theta/2)$$

The drag force is then

$$F_d = \frac{49}{128}\rho C_d U^2_{max} h^2 \tan(\theta/2) \qquad (6.165)$$

Comparing this with (6.164) shows that the drag force for the triangular shape with its apex pointing upwards is only slightly different from the rectangular case (1.6%).

The above analysis suggests that the velocity distribution and the object shape significantly affect the drag force compared to when using the mean velocity and the whole projected area of the object. In order to adjust for this difference, either the value of Cd needs changing or separate factors need to be introduced to account for the object shape and velocity distribution. In the examples above, the factors for a triangle with the apex pointing upwards or downwards are 0.888 and 1.143 respectively. The tree shape might be simulated by using the average velocity and the same C_d as for the cylindrical rod, with a shape factor to adjust for the difference between the drag force on a tree and a cylindrical rod. In the case of multiple trees, in-line alongside a river, the wakes of individual trees will distort the velocity distribution within the area of study, causing it to be no longer 2-D, thus making the drag force differ from the single tree case. It is therefore better to introduce a velocity distribution factor, as well as a shape factor, in any drag force formula. Further research is needed to determine what kind of factors are needed.

Case 5 Theoretical moments of overturning based on drag forces
It is important to know the overturning moment on an object, such as a tree, and these may be readily calculated in a similar way to the drag forces. For the triangular plate with its apex pointing downwards, the overturning moment is

$$M = \int_0^b z dF = \int_0^b \frac{1}{2}\rho C_d 2U^2_{max} 2\tan(\theta/2) z^2 \left(\frac{z}{b}\right)^{2/7} dz = \frac{7}{23}\rho C_d 2U^2_{max} \tan(\theta/2) b^3$$

(6.166)

For the triangular plate with its apex pointing upwards, the overturning moment is

$$M = \int_0^b z dF = \int_0^b \frac{1}{2}\rho C_d 2U^2_{max} \tan(\theta/2) \left(\frac{z}{b}\right)^{2/7} (b-z) dz = \frac{49}{368}\rho C_d U^2_{max} \tan(\theta/2) b^3$$

(6.167)

The ratio of the downwards to the upwards triangle is 1.39. The moment on the rectangular with the same area as the triangle is

$$M = \int_0^b z dF = \int_0^b \frac{1}{2}\rho C_d U^2_{max} b z \left(\frac{z}{b}\right)^{2/7} dz = \frac{7}{32}\rho C_d U^2_{max} b^3$$

(6.168)

The ratios of the rectangular one to the triangular ones are 1.64 for upwards and 2.285 for downwards. As can be seen from these ratios, the effects of the distribution of velocity and projected area on both the drag force and the overturning moment are significant.

Since the drag coefficient, C_d, is one of key parameters in the formula for drag force, many researchers have investigated it by measuring the drag force on an object, either directly with a measuring device or indirectly using a momentum balance. Many studies have therefore been conducted in laboratory open channels and wind tunnels on flows with vegetation, e.g. Armanini et al. (2005), Kohyari et al. (2009), Tanaka et al. (2011), Shucksmith et al. (2011), Jalonen & Jaevela (2013), Wunder et al. (2013) and Whittaker et al. (2013). Armanini et al. (2005) used 8 strain gauges to measure the drag forces on willow trees in a relatively large flume and calculated the drag coefficient, C_d, for various mean velocities and different projected areas of the trees. They showed C_d varying from 0.9 to 2.0 over a large range of the stem Reynolds numbers for the emergent cases and C_d varying from 1.5 to 3.0 for the submerged cases. They also measured the drag force on the trees without leaves and found that the drag reduced by up to 40%, with C_d values approaching 0.9~1.2, which are similar values to those for a cylindrical rod. Kothyari et al. (2009) measured the drag force on a cylindrical rod using a load cell under nearly uniform flow conditions, giving a C_d value of ~0.9 that was also almost independent of the rod Reynolds number. The C_d value was worked out using the depth averaged velocity and the projected area of the cylinder. They also produced an equation for the drag on an array of cylinders. Wunder et al. (2013) also measured the drag force on a willow tree by direct means, using a force measuring device. They estimated the projected area of the tree by taking pictures with a waterproof camera in an unstressed condition, with no flow, and a stressed condition, with flow. The C_d values, worked out using the mean velocity and the measured projected areas were ~0.5 for leafy willows in the unstressed condition and ~0.85 in the stressed condition. Without leaves, the C_d values increased with velocity from 0.5 to 2.0, with a scattering around the cylindrical rod C_d value of 1.0~1.2. From these studies it is clear that C_d varies considerably according to different flow and vegetation conditions. All these C_d values were generally calculated using the depth averaged velocity and the whole projected area over the water depth.

viii. *Experimental work to ascertain actual drag forces*
In order to investigate the effect of velocity distribution and object shape on C_d values, experiments were conducted in an open channel to measure the drag force on an equilateral triangular plate with a triangle height of 80 mm. The approach velocity was measured by ADV at the centre of the triangle upstream over the entire water depth, using a sampling rate of 50 Hz over a period of 60 s. The drag force was measured directly by a force measuring device (FMD), which was designed and developed at Loughborough University. Any object, such as a triangular plate, a cylindrical rod or a tree was inserted in a tube attached to the FMD. This device consisted of a load cell from a kitchen electronic weighing scale with a range from 1 g to 5 kg (note ones with 0.01 g to 500 g are also available), a SY034 USB digital strain gauge amplifier (ABS $18 \times 60 \times 49$ mm) with plug-in connector $+/-6$ mV/V which was connected to the load cell with a 10 m long cable. The FMD was built in a box to avoid any influence from water movement and then fixed under the flume bed, with the chosen object poking out above through the channel floor.

The triangle was completely submerged and the measured velocity distribution checked to ensure that it followed the assumed 7th power law closely. The drag forces on the triangular plate with the apex pointing downwards and then upwards were 0.633 N and 0.526 N respectively, giving a ratio of up to down of 1.203, which is relatively close to the theoretical value of 1.286. The drag forces calculated from (6.162) – (6.163) using the measured velocity with $C_d = 1.0$ were 0.255 N and 0.211 N. The measured drag forces gave $C_d = 2.49$ (0.633/0.255) and 2.49 (0.526/0.211) respectively which is slightly higher than the value of $C_d = 2.1$ for an emergent rectangular plate. This discrepancy is only for the submerged triangle rather than for the emergent condition. However, the drag force calculated using the depth averaged velocity and the projected area with $C_d = 1.0$, was 0.214 N, giving C_d values for the triangular plate with its apex up and then down of 2.96 (0.633/0.214) and 2.46 (0.526/0.214) respectively. For the apex pointing upwards, the drag is about 20% higher than that obtained by the integration of velocity and project area. This shows that when using $C_d = 2.1$, as given in some hydraulic textbooks, the drag force would be underestimated.

ix. *Drag force due to a line of trees*
When there are multiple rods or trees in a single line in a channel, the velocity in between them would be expected to vary spatially due to the wake zone. The wake decay rate is known to depend on the ratio of a rod diameter to distance, according to theoretical wake decay equation, (6.169), given by Schlichting (1979).

$$\frac{u}{U} = 1 - \frac{\sqrt{10}}{18\beta}\left(\frac{x}{C_D d}\right)^{-(1/2)}\left\{1-\left(\frac{y}{b}\right)^{3/2}\right\}^2 \tag{6.169}$$

where b = shear layer width, C_D = drag coefficient, d = effective tree diameter, u = depth mean velocity, U = maximum velocity, x = distance from the tree, y = lateral distance and β = mixing length coefficient. Ozaki *et al.* (2013) used real trees in the centre of the open channel to demonstrate that this equation fits the measured velocity distribution. For example, with b = 0.18, C_d = 1.0 and x/d = 100, then u/U = 0.884, which was close to his observations, as shown in Figure 6.105. It should be noted that the recovery of the flow is not complete even for x/d = 100, suggesting that the approach velocity in front of an object varies with x/d and will not be necessarily fully recovered unless the spacing between objects is very large.

In case of multiple rods in one line, the approach velocity also varies for the first rod, second, etc. In order to find when the approach velocity becomes constant after a certain number of rods, an experiment was carried out to measure the lateral velocity distribution in front of the tree, again using ADV. In all, 10 trees were set in one line in the centre of a relatively wide channel and it was found that the velocity distribution suffers little change after the 6th tree (see Figures 106 & 107). It is therefore expected that continuous tree lines alongside river banks will behave in a similar manner.

6.7   FLOW DEPENDENT RESISTANCE ISSUES

In rivers and estuaries, it should be remembered that resistance does not vary only with stage, but possibly also with time, for example, where sediments are transported and vegetation is present. Figure 6.108 shows a typical looped stage-discharge and looped resistance relationship for the River Blackwater and Figure 6.109 the causative flood hydrograph. Rivers are by their nature subject to short term (floods) and long term (seasonal) changes. Pictures of the seasonal variation in roughness for this river and its floodplain are shown in Chapter 2, Figures 2.8 & 2.9. Figure 6.108 indicates that the resistance is less during the flood recession (falling limb), due to the vegetation being flattened by the flood in its first progress overbank when water initially inundates the floodplain (rising limb). See Sellin & van Beesten (2004), Knight (1981) and Gunawan *et al.* (2010) for further details.

Figure 6.110, taken from Knight (1981), shows how resistance data might be collected under unsteady flow conditions by measuring values of the various terms in the

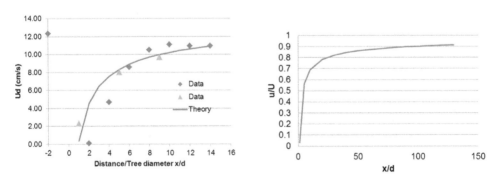

Figure 6.105 Measured wake decay rates.

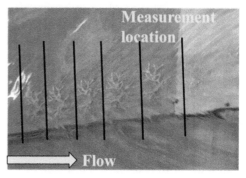

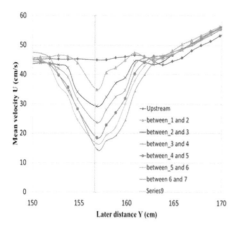

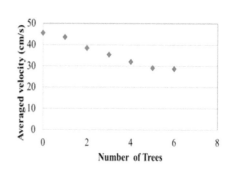

Figure 6.106 Measured wakes behind model trees.

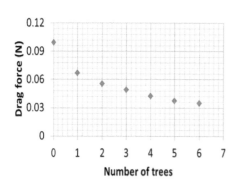

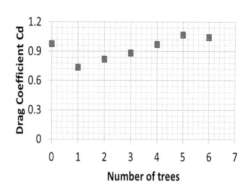

Figure 6.107 Measured drag force of model tree at different location and Cd.

1-D St Venant equations. A subsequent model calibration study based on these data is described by Wallis & Knight (1984).

Unsteady flow is also a feature of flood routing and validation issues in 1-D flood routing models, using either the St Venant or Variable Parameter Muskingum-Cunge

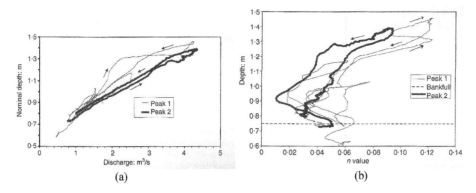

Figure 6.108 Looped stage-discharge and looped resistance relationships for a compound two-stage channel with vegetated floodplains (after Sellin & van Beesten, 2004).

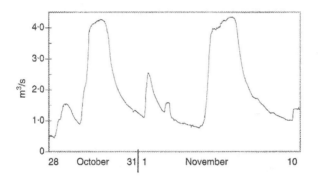

Figure 6.109 Unsteady flow hydrograph for river Blackwater, 2000 (after Sellin & van Beesten, 2004).

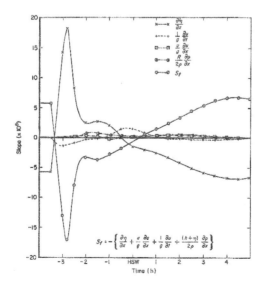

Figure 6.110 Unsteady flow friction slope for a reach of the Conwy estuary (after Knight, 1981).

(VPMC) equations, are dealt with by Knight (2006b) and Tang *et al.* (2001). Related to this, it may be helpful to know that the SKM may be used to determine flood wave speed, c, for any discharge, since c is related to the stage-discharge relationship and channel geometry by the following equation

$$c = \left(\frac{1}{B}\right)_s \left(\frac{dQ}{dh}\right)_c \qquad (6.170)$$

where the subscripts s and c refer to storage and conveyance respectively. This recognizes that part of the cross-section of a river may have a 'dead' zone in which there is little or no discharge and water is effectively simply stored, as for example where there is a bankside region with thick vegetation, and another 'active' zone in which water is flowing, known as a conveyance zone. In flood routing models it is important to distinguish between the two and to model the conveyance accordingly, as shown by Knight (2006). Where the SKM is used to obtain the h v Q relationship for a particular river reach, then knowing the floodplain geometry the c v Q relationship may be calculated from (6.170). This is particularly useful in extending the c v Q relationship to discharges higher than observed, as shown by Tang (1999), Tang, Knight & Samuels (1999a&b & 2001).

Chapter 7

Worked examples using the Shiono & Knight Method (SKM)

ABSTRACT

This chapter illustrates the use of the Shiono & Knight Method (SKM) in analyzing flows in different shaped prismatic channels as well as rivers. Spreadsheet programs using Microsoft EXCEL are the basis for numerical and analytical tools dealing with the lateral distributions of depth-averaged velocity and boundary shear stress, U_d & τ_b. A dedicated SKM FORTRAN program is also given in Appendix 6 which, together with CES, provides the river engineer with powerful tools and options for flow analysis. The governing SKM equations and the algebraic expressions for the A_i coefficients are provided for a number of the Cases and Examples presented in Chapter 6, for both inbank and overbank flows. Particular consideration is given to model validation, based on other numerical and experimental studies, including those for drag forces on trees, leading to a new analytical solution for inclusion in the SKM. These examples are aimed at giving users confidence in the use of these tools and to enable them to develop the SKM further for their own particular purposes.

7.1 USING EXCEL TO SOLVE THE SKM EQUATIONS TOGETHER WITH ANALYTICAL EXPRESSIONS FOR THE A$_i$ COEFFICIENTS

7.1.1 Introduction and explanatory notes

EXCEL is ideally suited for solving the SKM equations by either matrix methods or by direct solution using algebraic expressions for the A_i coefficients to give the distributions of U_d and τ_b within each panel. If programs are set up with a general philosophy in mind then repeated runs will give the stage discharge relationship and all the ancillary information that may be needed to calculate the boundary shear forces on particular boundary elements, the zonal discharges in sub-areas, such as floodplains, and other hydraulic details. Boundary shear stress distributions are often needed as inputs to sediment or pollution models. It is not the aim of this chapter to give specific programs suited for EXCEL but to give the reader enough information to help them get started on writing their own, to their own specification, which tends to be the best way of learning. For the uninitiated, there are many books on EXCEL and other resources available, for example those by Bloch (2000), Korol (2000) and Walkenbach (1999).

It is wise not to start with too ambitious a target, but instead to begin by solving some simple inbank flow cases in order to appreciate how to define symbols and obtain analytical expressions for the A_i coefficients, as already indicated in Chapter 6. It is sensible to formulate the relevant basic equations that arise from the various boundary conditions pertaining to each panel, then solve them by using the readily available EXCEL tools, such as MINVERSE and MMULT. The same basic equations should then be solved algebraically to give expressions for the various A_i coefficients, which should be the same as those obtained by matrix inversion. If there is agreement, then the lateral distributions of U_d and τ_b can be listed, plotted and integrated to give the total discharge and shear forces. The experience gained on exploring inbank flows in simple rectangular and trapezoidal channels will be repaid when it comes to embarking on more complex cases, such as the one indicated in Figure 7.1. It is simply not worth tackling by algebraic means more than a 9 node case, which is generally sufficient for defining many river cross sections with two adjacent floodplains. It should also be noted that in the example shown, the thalweg (deepest point in the cross section), is used as a local origin, so that the panels have either positive of negative side slopes, either side of the thalweg. Reference to Case T2 and equations (15) & (16) in Chapter 6 should make the reader aware of the importance of +/− for coding details. Alternatively, when dealing with large rivers it makes sense to choose a local origin to one side, as done in the FORTRAN program listed in Appendix 6 and in the CES software.

In the interest of brevity, and to satisfy the strict demands of the analytical solution in SKM, the "μ" factors, which were introduced in Chapter 6 to account for where the parameters f, λ & Γ differ markedly between panels, are not considered in most of the subsequent algebraic expressions for the A_i coefficients, but may be added by the user if felt appropriate. The examples now given start with some simple problems to build confidence in those new to the SKM.

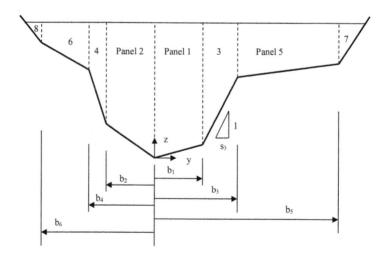

Figure 7.1 Generalised compound river cross section with 9 nodes and 8 panels.

7.1.2   Inbank flows in rectangular channels

Rather than give extended lists of code or data validation sources, Tables will provide summaries of input and output parameters for those Cases now considered. Note that the equations referred to now are numbered as in Chapter 6 [(1) to (n)] for each particular case, without a prefix 7.

Case R1   (based on Example 6, in Chapter 2, Section 2.3.1)

In order to appreciate the link between SKM parameters on the lateral distributions of depth-averaged velocity, U_d, and local boundary shear stress, τ_b, consider a 20 m wide rectangular channel. Referring back to Section 6.4.1 and Case R1 in Chapter 6, use equations (13) & (16), with $b = 10$ m, to determine the distributions of both U_d and τ_b for a 20 m wide rectangular channel with a bed slope of 1.0×10^{-3} and a depth $H = 5.0$ m. It is suggested that you begin by using values of $f = 0.020$, $\lambda = 0.070$ and $\Gamma = 0.15$, as the f value corresponds with a Manning n value of around 0.020 and a Nikuradse k_s value of 0.015 m. This should lead to $\beta_1 = 0.003059$, $k_1 = 19.554$ and $\gamma_1 = 0.23905$ with $A_1 = 1.77601$, $C_1 = 3.5520$ & $C_2 = 19.554$. It is generally helpful to include a calculation for the overall mean shear stress, as given by (2.32) in Chapter 2, and the conversion equations (2.29) & (2.30), or those in Chapter 3 as (3.14) – (3.16), so that one can check the overall output values and change the input values of f, n or k_s easily, as required.

  If you select a single panel to simulate half the cross section, as in Case R1, then the solution is straightforward since the analytical distributions involve only the hyperbolic functions, cosh and sinh and no matrix algebra is required. It is suggested that you calculate U_d values at intervals of $b/20$ (i.e. 0.5 m in this case), calculating τ_b, values via (15) and plot your results as shown in Figure 7.2, These may be compared with those in Chapter 2, Figure 2.25. Note that in Figure 2.25, Δy changes beyond $y = 9.5$ m, reducing to 0.1 m, then down to 0.005 m in order to illustrate the steep changes in values close to the wall, whereas in Figure 7.2, Δy has a fixed value. Check also that the tabulated value of the bed shear stress at the centerline ($y = 0$) is in agreement with that given by (18). Determine the mean shear stress on half the bed, using equation (20), and compare it with the average value from your tabulated results. Note that the mean shear force on any bed element Δy long is the average

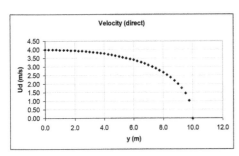

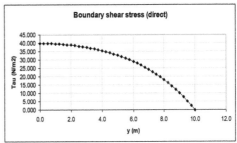

Figure 7.2 Distributions of U_d and τ_b for Case R1, based on Example 6 (for comparison with those in Figure 2.25).

of the two stresses times Δy only if there is no side slope and the bed is horizontal. Depending on the value of Δy, small differences may emerge with the overall mean values, so it is always advisable to divide any panel into a number of sub-elements for calculation purposes, 20 being often sufficient, as suggested here. Having now calculated the shear forces, dF, on each sub-element, the total bed shear force, SF_b, may be calculated by summation, giving 575 N, as shown by the output data in Table 7.1. The overall mean bed shear stress may be determined 28.775 Nm^{-2}, $(SF_b = \overline{\tau}_b \times 2b)$. Since the total shear force, SF_T, is 980.7 N $(= \tau_o P = \rho g R S_o P = \rho g A S_o)$ it follows that $SF_w = 405.197$ N (or for one wall 202.6 N). Expressing this in percentage terms, using (6.80), gives $\%SF_w = 41.32\%$ (both walls) and $\%SF_b$ (entire bed) $= 58.68\%$. For this particular aspect ratio $(2 \times 10/5)$ of 4.0, this corresponds to a wetted perimeter ratio, P_b/P_w of 2.00. Comparing this with the value of $\%SF_w$ shown in Figure 6.47, it is evident that your calculated value is considerably higher than the 30.47% expected. It may of course be increased by lowering λ, keeping other values the

Table 7.1 Some tabulated results for Example 6 (Case RI).

y (m)	Ud (ms^{-1})	y (m)	Tau (Nm^{-2})	dQ (m^{3}s^{-1})	Bed dF (N)
0	4.000247	0	40.00495	9.9966	19.9866
0.5	3.997071	0.5	39.94145	9.9807	19.92287
1	3.987482	1	39.75002	9.9485	19.79449
1.5	3.971294	1.5	39.42795	9.8994	19.59964
2	3.948195	2	38.97061	9.8324	19.33552
2.5	3.917727	2.5	38.37147	9.7463	18.99836
3	3.879276	3	37.62195	9.6391	18.58333
3.5	3.832041	3.5	36.71135	9.5088	18.08449
4	3.775003	4	35.62663	9.3523	17.49472
4.5	3.706873	4.5	34.35227	9.1661	16.80558
5	3.62602	5	32.87006	8.9455	16.00721
5.5	3.530371	5.5	31.15879	8.6845	15.0882
6	3.41725	6	29.19399	8.3755	14.03539
6.5	3.283142	6.5	26.94755	8.0080	12.83373
7	3.12329	7	24.38735	7.5679	11.46603
7.5	2.930991	7.5	21.47677	7.0340	9.912734
8	2.696232	8	18.17417	6.3737	8.151623
8.5	2.402692	8.5	14.43232	5.5280	6.157508
9	2.019674	9	10.19771	4.3634	3.901868
9.5	1.471022	9.5	5.409762	0.6987	0.489483
9.6	1.323616	9.6	4.379896	0.6192	0.385225
9.7	1.153186	9.7	3.324597	0.5251	0.278393
9.8	0.947262	9.8	2.243263	0.4053	0.168927
9.9	0.673877	9.9	1.135275	0.1998	0.056252
9.99	0.214265	9.99	0.114774	0.0046	0.00043
9.995	0.151554	9.995	0.057422	0.0019	0.000144
10	0	10	0		
			Sum =	164.4052	287.5387
			Q total =	328.8104	

same, or by increasing f, but some thought needs to be given before simply adjusting one coefficient arbitrarily.

If you have included $\%SF_w$ as an output on your spreadsheet, try varying λ and observe the changes to both your distributions of U_d and τ_b as well as with other key output values, like discharge and SF values. You might also like to only consider a constant interval for Δy of 0.5 m and observe the slight changes in $\%SF_w$. Although this is a simple example to calculate, it illustrates a significant issue concerning model simulations. Firstly, that although it may be straightforward to obtain the total discharge, Q, it is not necessarily as straightforward to obtain the correct percentage values of SF on either the bed or the walls. This arises for a number of reasons, firstly because there is no experimental data on the distributions or mean values of boundary shear stress or depth-averaged velocity for this particular hypothetical example and, secondly because using just one panel and hyperbolic functions does not necessarily mimic the true lateral distribution of U_d well enough over the entire width. More panels should improve the representation, as already shown in Figures 2.32 & 2.34, but this is not necessarily needed, as shown in the next example.

Case R1   (based on Example 7 in Chapter 2, Section 2.3.1)

Refer to example 7 in Chapter 2, and repeat Case R1, but this time using some of the data referred to in Section 6.5.3 and in the papers by Knight et al. (1982 & 1984), Knight, Yuen & Alhamid (1994) and Chlebek & Knight (2006). Consider a small smooth 152 mm wide laboratory channel, with $S_o = 9.66 \times 10^{-4}$, Manning's n of around 0.010 (approximately equivalent to a k_s value of 0.00025 m). Taking $b = 0.076$ m and using values of $f = 0.0203$, $\lambda = 0.011$ and $\Gamma = 0.047$, gives $\beta_1 = 0.024806$, $k_1 = 7.28163$ and $\gamma_1 = 15.13179$ with $A_1 = 1.77601$, $C_1 = 3.5520$ & $C_2 = 19.554$. This gives $Q = 0.138$ m³s⁻¹ and $\%SF_w = 73\%$ for $2b/H = 0.76$ or $P_b/P_w = 0.38$. Check that for the suggested input Nikuradse k_s value of 0.25 mm this corresponds approximately with the f and Manning n values selected. Figure 7.3 illustrates how this procedure may be repeated to simulate the $\%SF_w$ & mean τ_b values over a wide range of aspect ratios using just a single panel. Despite simulating the experimental data quite well, there are clearly limits on the accuracy of the results imposed by choosing only a single panel to represent half the cross section. As also indicated in Section 6.5.3, further

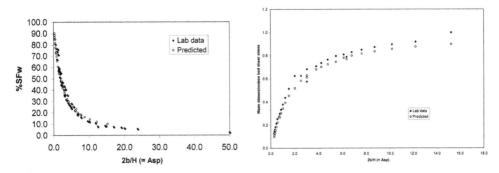

Figure 7.3 Simulated $\%SF_w$ & mean τ_b values versus aspect ratio, $2b/H$, by SKM (from Chlebek & Knight, 2006).

improvements may be made by choosing 4 objective functions, in addition to using more panels to capture the physical processes more accurately, as shown by the results in Figures 2.33 & 2.34 for a trapezoidal channel case.

Case R2 *(similar to Example 8 in Chapter 2, but with H = 5.0 m)*

Refer back to the 2 panel simulation in Chapter 2, Example 8. The lateral distributions of U_d in the two panels are

$$U_d^{(1)} = \left[A_1 e^{\gamma_1 y} + A_2 e^{-\gamma_1 y} + k_1 \right]^{1/2} \qquad U_d^{(2)} = \left[A_3 e^{\gamma_2 y} + A_4 e^{-\gamma_2 y} + k_2 \right]^{1/2}$$

and the 4 A_i coefficients are given in Section 6.4.1 by

$$A_1 = \cfrac{\gamma_2 \left(k_2 - k_1 \right) \left[e^{\gamma_2 (b-b_1)} + e^{-\gamma_2 (b-b_1)} \right] - 2\gamma_2 k_2}{\left[e^{\gamma_2 (b-b_1)} + e^{-\gamma_2 (b-b_1)} \right] \left\{ \left[\gamma_2 + \gamma_1 \left(\dfrac{\mu_1}{\mu_2} \right) \right] e^{\gamma_1 b_1} + \left[\gamma_2 - \gamma_1 \left(\dfrac{\mu_1}{\mu_2} \right) \right] e^{-\gamma_1 b_1} \right\}}$$
$$- 2\gamma_1 \left(\dfrac{\mu_1}{\mu_2} \right) \left[e^{\gamma_1 b_1 - \gamma_2 (b-b_1)} - e^{-\gamma_1 b_1 - \gamma_2 (b-b_1)} \right]$$

$$A_2 = A_1$$

$$A_3 = \frac{1}{2\gamma_2 e^{\gamma_2 b_1}} \left\{ \left[\gamma_2 + \gamma_1 \left(\frac{\mu_1}{\mu_2} \right) \right] e^{\gamma_1 b_1} A_1 + \left[\gamma_2 - \gamma_1 \left(\frac{\mu_1}{\mu_2} \right) \right] e^{-\gamma_1 b_1} A_2 - \gamma_2 \left(k_2 - k_1 \right) \right\}$$

$$A_4 = -e^{-\gamma_2 b} \left[A_3 e^{\gamma_2 b} + k_2 \right]$$

The governing SKM equations in matrix form are:

$$\begin{bmatrix} 1 & -1 & 0 & 0 \\ e^{\gamma_1 b_1} & e^{-\gamma_1 b_1} & -e^{\gamma_2 b_1} & -e^{-\gamma_2 b_1} \\ \left(\dfrac{\mu_1}{\mu_2} \right) \gamma_1 e^{\gamma_1 b_1} & -\left(\dfrac{\mu_1}{\mu_2} \right) \gamma_1 e^{-\gamma_1 b_1} & -\gamma_2 e^{\gamma_2 b_1} & \gamma_2 e^{-\gamma_2 b_1} \\ 0 & 0 & e^{\gamma_2 b} & e^{-\gamma_2 b} \end{bmatrix} \times \begin{bmatrix} A_1 \\ A_2 \\ A_3 \\ A_4 \end{bmatrix} = \begin{bmatrix} 0 \\ k_2 - k_1 \\ 0 \\ -k_2 \end{bmatrix}$$

Now model a rectangular channel using two panels, as Case R2 in Section 6.4.1, but with $H = 5.0$ m. Allow for using two panels of unequal width, as indicated in Figure R2, but input the data initially so that $b_1 = b_2 = 5.0$ m. Some suggested input values and the corresponding output values are given in Tables 7.2 & 7.3 for these hypothetical cases. See results in Figure 7.4 and in 2nd row of Table 7.3. This will allow you to check that the output values are the same as those in row 1 of Table 7.3 for Case R1. As expected they are much the same. The highlighted numbers in Table 7.2 indicate where certain parameters are changed from the row above, so that the inputs for multiple runs of the same program may be identified easily. The lists also enable comparisons to be made for the effects of different inputs on outputs quickly.

Table 7.2 Input parameters for rectangular channel cases.

Case	b (m)	b_1 (m)	b_2 (m)	f_1	f_2	λ_1	λ_2	Γ_1	Γ_2	k_s (m)
R1	10	*	*	0.020	0.020	0.070	0.070	0.150	0.150	0.015
R2	10	5.0	5.0	0.020	0.020	0.070	0.070	0.150	0.150	0.015
R2	10	5.0	5.0	**0.040**	0.020	0.070	0.070	0.150	0.150	**0.500**
R2	10	**9.0**	**1.0**	**0.040**	0.020	0.070	0.070	0.150	0.150	**0.500**
R2	10	**3.0**	5.0	**0.01**	**0.01**	**0.10**	**0.10**	0	0	0.015
R2	10	**3.0**	5.0	0.01	0.01	**0.03**	**0.03**	0	0	0.015
R4	20	10	10.0	5.0	0.020	0.020	0.070	0.070	0.150	0.150
R4	20	**5.0**	**15.0**	5.0	0.020	0.020	0.070	0.070	0.150	0.150
R4	20	10	10.0	5.0	**0.040**	0.020	0.070	0.070	0.150	0.150
R4	20	**5.0**	**15.0**	5.0	0.040	0.020	0.070	0.070	0.150	0.150

Table 7.3 Output results for rectangular channel cases.

Case	H (m)	Q_{total} (m³s⁻¹)	Q_1 (m³s⁻¹)	Q_2 (m³s⁻¹)	SF_{total} (N)	SFb (N)	SFb_1 (N)	SFb_2 (N)	$\%SF_w$ (%)	$\%SF_w$ eq 6.82
R1	5.0	328.8	*	*	980.7	575.1	*	*	41.32	30.47
R2	5.0	328.6	97.08	67.21	980.7	575.4	188.7	99.0	41.32	30.47
R2	5.0	273.4	76.97	59.75	980.7	634.80	236.99	80.41	35.27	30.47
R2	5.0	252.1	119.97	6.07	980.7	657.4	324.38	4.33	32.97	30.47
R2	3.0	227.0	66.50	47.01	588.4	379.95	122.91	67.07	35.43	19.37
R2	3.0	255.6	71.67	56.16	588.4	473.29	142.7	93.9	19.57	19.37
R4	5.0	327.6	163.4	163.4	980.7	575.02	287.5	287.5	41.37	30.47
R4	5.0	327.4	260.2	67.2	980.7	592.1	116.4	475.7	39.62	30.47
R4	5.0	282.1	129.5	152.6	980.7	622.0	260.9	361.1	36.58	*
R4	5.0	258.2	198.7	59.5	980.7	660.6	105.6	555.1	32.64	*

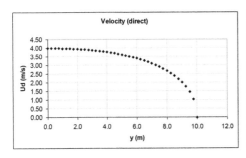

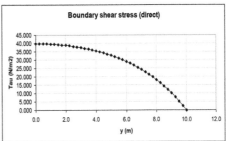

Figure 7.4 Case R2 $b_1 = b_2 = 5$ m; $f_1 = f_2 = 0.02$; $k_s = 0.015$ m (cp with Figure 2.25 in Chapter 2).

Because the $\%SF_w$ value is ~10% too high, try increasing the bed roughness in one panel, e.g. make $f_1 = 0.04$, leaving all other parameter values the same. See results in Figure 7.5 and row 3 in Table 7.3.

It is apparent from comparing Figures 7.4 & 7.5 that the central velocities have decreased from around 4.0 ms⁻¹ to 3.0 ms⁻¹, and that the bed shear stresses in panel 1 have increased from ~40 Nm⁻² to ~50 Nm⁻². In panel 2 they have decreased, thus the

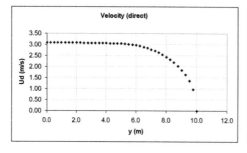
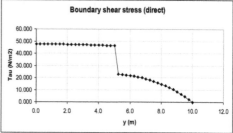

Figure 7.5 Case R2 $b_1 = b_2 = 5$ m; $f_1 = 0.04, f_2 = 0.02; k_s = 0.05$ m.

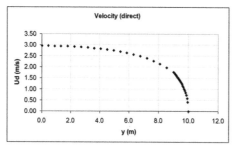
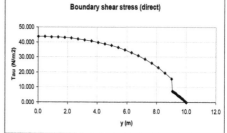

Figure 7.6 Case R2, $b_1 = 9$ m, $b_2 = 1.0$ m; $f_1 = 0.02, f_2 = 0.04; k_s = 0.05$ m.

overall SF_b should be larger making %SF_w smaller as intended. See Table 7.3, which shows that %SF_w is now 35.27% in row 3, reduced from 41.32% in row 2 above. Note the sudden and physically questionable change in τ_b at the interface between panels 1 & 2, arising from the sharp change in f values and the consequence of defining τ_b in terms of (2.61). There is a paucity of data on this issue, apart from that presented by Tominaga & Nezu (1991), referred to in Section 6.3.2. However, this issue may be overcome in SKM by varying friction coefficents in a transition region between panels, as outlined in section 2.3.2 and illustrated in Figure 2.34. In order to reduce the value of the calculated %SF_w further, extend panel 1 to 9.0 m and rerun your spreadsheet keeping everything else the same. The results are as shown in Figure 7.6 and Table 7.3, row 4. This helps to see more detail.

   You may now wish to repeat some runs for the same depth but with $b_1 = 9.0$ m & 9.9 m, with the corresponding values for b_2 of 1.0 m & 0.1 m respectively, and compare your results with those in Figure 2.29. You may wish also to later reproduce the results in Figure 2.25 by altering the Γ values.

Case R2   (Example 8 in Chapter 2)

Using same spreadsheet, alter the depth to $H = 3.0$ m and try the same problem as Example 8 in Chapter 2. Input the data shown in Table 7.2 and results are shown in row 5 of Table 7.3. Note that with $f_1 = f_2 = 0.01$ and $\lambda_1 = \lambda_2 = 0.10$ leads to

$Q = 227.0 \text{ m}^3\text{s}^{-1}$ & $SF_T = 588.4$ N. The calculated %SF_w value is 35.43%, which is ~16% higher than the correct value of 19.37%, according to (6.82) and Figure 6.47. Now decrease just the input values of λ to $\lambda_1 = \lambda_2 = 0.030$ and repeat your calculation. This should increase the Q value to 255.6 m^3s^{-1} and reduce the %SF_w value to 19.57%, which is now much closer to the desired value of 19.37%. This is not to say that these are the 'correct' values, but illustrates how the SKM can easily give distributions of U_d and τ_b leading to discharge and shear force values that may be compared with either known data or theoretically determined values. It also illustrates the logic encapsulated in Table 6.2 earlier, and illustrates that even a 'simple' problem like flow in a rectangular channel has many complexities that make it not a 'simple' problem at all to deal with. Figures 6.10–6.13 indicate some of these complexities, that even for the most advanced CFD models find challenging.

Case R4 (based on Example 8 in Chapter 2)

Alternatively, construct a new spreadsheet for Case R4 for a rectangular with two panels of unequal width offset from the centerline. Check it initially with $b_1 = b_2 = 10$ m, for $H = 5.0$ m, a similar problem as Example 8 in Chapter 2 with $H = 3.0$ m, and compare you results shown in Figure 7.7 with those in Figure 2.25. The output results are listed in the first row of the R4 series in Table 7.3, which should be compared with those in the first two rows of the R2 results. Slight differences arise from the having plots for each side, different Δy values and panel definitions. The distributions in Figures 7.7 are however essentially identical to those in Figures 7.4 & 2.25. By repeating this calculation with $b_1 = 5$ m and $b_2 = 15$ m will lead to the curves shown in Figure 7.8, with the origin displaced and two panels of unequal size. The flow is still symmetric, with the centre of the channel, placed at $y = -5.0$ m.

Now create some asymmetric flow by altering one f value, such as making $f_2 = 0.04$, firstly keeping the two panels of equal size of 10 m, before then making them different sizes of 15 m & 5 m again. The resulting distributions of U_d and τ_b are shown in Figures 7.9 & 7.10. Some of the key output values for the parameters are again tabulated in Table 7.3 in order to assist in checking your results and to illustrate the changes that occur when certain input parameters are changed.

Before tackling more complex examples in SKM, it is sensible to explore another simple channel shape that features in canal and urban drainage problems, and examine

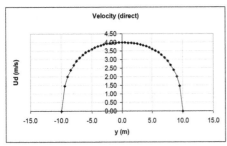
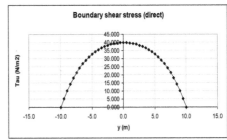

Figure 7.7 Case R4 $b_1 = b_2 = 10$ m; $f_1 = f_2 = 0.020$; $k_s = 0.015$ m (cp plots in Chapter 2).

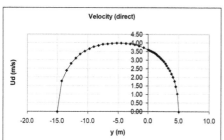
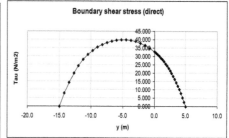

Figure 7.8 Case R4 $b_1 = 5$ m, $b_2 = 15$ m; $f_1 = f_2 = 0.020$; $k_s = 0.015$ m.

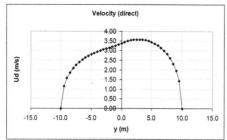
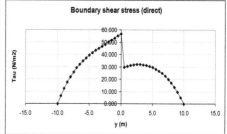

Figure 7.9 Case R4 with $b_1 = b_2 = 10$ m; $f_1 = 0.02$, $f_2 = 0.04$; $k_s = 0.05$ m.

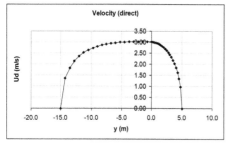
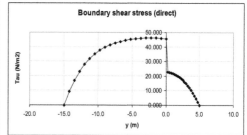

Figure 7.10 Case R4 with $b_1 = 5$ m, $b_2 = 15$ m; $f_1 = 0.02$, $f_2 = 0.04$; $k_s = 0.05$ m.

flows in trapezoidal; channels. These are particularly good cases to consider, as there is a considerable amount of experimental data on the distributions of turbulence, velocity and boundary shear stresses for such cases, allowing for comparison with the model simulations. It also introduces panels with sloping sides for the first time.

7.1.3 Inbank flows in trapezoidal channels

Of the 7 trapezoidal channels cases dealt with in Chapter 6, only numerical answers to Cases T1 & T2 are presented herein in order to avoid too much repetition, given their similarity with the rectangular cases already dealt with. Now refer back to

Section 6.4.2 and Case T1 which relates to symmetric flow in a simple trapezoidal channel. Use two panels to represent half the cross section, as indicated in Figure T1. The lateral distributions of U_d in the two panels are

$$U_d^{(1)} = \left[A_1 e^{\gamma_1 y} + A_2 e^{-\gamma_1 y} + k_1\right]^{1/2} \quad \text{and} \quad U_d^{(2)} = \left[A_3 \xi_2^{\alpha_2} + A_4 \xi_2^{-(\alpha_2+1)} + \omega_2 \xi_2 + \eta_2\right]^{1/2}$$

and the 4 A_i coefficients are given by

$$A_1 = \frac{(\alpha_2 - 1)\omega_2 H - k_1 \alpha_2 + \alpha_2 \eta_2}{\left[\left\{\alpha_2 + \left(\frac{\mu_1}{\mu_2}\right)s_2 \gamma_1 H\right\}e^{\gamma_1 b}\right] + \left[\left\{\alpha_2 - \left(\frac{\mu_1}{\mu_2}\right)s_2 \gamma_1 H\right\}e^{-\gamma_1 b}\right]}$$

$$A_2 = A_1$$

$$A_3 = \frac{1}{H^{\alpha_2}}\left[A_1 e^{\gamma_1 b} + A_2 e^{-\gamma_1 b} + k_1 - \omega_2 H - \eta_2\right]$$

In a similar manner to the rectangular cases, the inputs and outputs are summarized in Tables 7.4 & 7.5, respectively, allowing the reader to check their own spreadsheets.

Case T1   (based on Examples 1–3 in Chapter 2, Section 2.1)

Create a spreadsheet based on Figure T1 in Chapter 6, allowing for input variables of H, S_o, b & s (side slope) and input parameters of f, λ & Γ for each of two panels. For Example 1 in Chapter 2, set $H = 1.50$ m, $S_o = 0.0003$, $b = 7.5$ m and $s = 1.0$. It is suggested that you begin by using values of $f = 0.060$, $\lambda = 0.070$ and $\Gamma = 0.15$, as this f value corresponds with a Manning n value of around 0.031, and a Nikuradse k_s value of around 0.20 m. This should lead to $\beta_1 = 0.03399$, $k_1 = 0.5684$ and $\gamma_1 = 1.04867$ with $A_1 = -5.2696\text{E-}05$, $A_2 = -5.2696\text{E-}05$ and $A_3 = -0.2937$. These values should give $Q = 17.96$ m³s⁻¹ and $SF_b = 61.96$ N, which for $SF_T = 72.82$ N gives $\%SF_w = 20.81\%$ & $\%SF_b = 79.19\%$. These are all slightly in error, as they differ from the desired discharge of $Q = 16.90$ m³s⁻¹ and the $\%SF_w$ & $\%SF_b$ values of 18.29% & 81.71% respectively given by (6.82) or Figure 6.47. These outputs are tabulated in Table 7.5 for the inputs tabulated in Table 7.4.

Now progressively change the f_1 and f_2 values, together with k_s with the aim of achieving the twin goals of $Q = 16.90$ m³s⁻¹ and the $\%SF_w$ & $\%SF_b$ values of 18.29%

Table 7.4 Input parameters for trapezoidal channel cases.

Case T1	H (m)	b (m)	f_1	f_2	λ_1	λ_2	Γ_1	Γ_2	k_s (m)
Example 1	1.50	7.5	0.060	0.060	0.070	0.070	0.150	0.150	0.20
Example 1	1.50	7.5	0.070	0.060	0.070	0.070	0.150	0.150	0.20
Example 1	1.50	7.5	0.065	0.070	0.070	0.070	0.150	0.150	0.30
Example 2	1.095	7.5	0.064	0.080	0.070	0.070	0.150	0.150	0.150
Example 3	1.50	7.5	0.050	0.060	0.070	0.070	0.150	0.150	0.050
Example 4	1.65	7.5	0.065	0.060	0.070	0.070	0.150	0.150	0.15
T1 Ex	20	7.5	5.0	0.020	0.020	0.070	0.070	0.150	0.150

Table 7.5 Ouput parameters for trapezoidal channel cases.

Case TI	H (m)	Q (m³s⁻¹)	SF_T (N)	SF_b (N)	$\%SF_w$ %	$\%SF_b$ %	$\%SF_w$ By (6.82)	$\%SF_b$ By (6.82)
Example 1	1.50	17.96	72.82	61.96	20.81	79.19	18.29	81.71
Example 1	1.50	16.74	72.82	62.44	22.26	77.74	18.29	81.71
Example 1	1.50	17.20	72.82	61.81	19.94	80.06	18.29	81.71
Example 2	1.095	10.59	51.86	44.73	13.95	86.05	13.23	86.77
Example 3	1.50	29.29	159.36	136.87	19.36	80.64	18.29	81.71
Example 4	1.65	30.17	176.88	152.47	23.50	76.50	20.06	79.94
TI Ex	20	7.5	5.0	0.020	0.020	0.070	0.070	

& 81.71 given by (6.82). Row 3 of Table 7.4 will yield row 3 of Table 7.5, giving $Q = 17.20$ m³s⁻¹ and the $\%SF_w$ & $\%SF_b$ values of 19.94% & 80.06%, which differ from the target values by 1.8%, –2.0% & 9.0% respectively. The larger difference in the $\%SF_w$ value is less important given the smaller contribution of SF_w to the overall total SF. Furthermore errors of <2% in the two dominant figures are acceptable without examining the distribution plots of U_d and τ_b in more detail, a suggested by (6.94) & (6.95) in Section 6.5.3. The reader is encouraged to explore this further by reference to Knight *et al.* (1994) and subsequent papers by Knight *et al.* (2007), Omran & Knight (2010), Sharifi *et al.* (2008, 2009 & 2010). It should be evident by now that detailed assessment of any model output is predicated on the availability of good quality assured experimental data, as emphasized by Knight (2013a&b) and indicated in the comparisons drawn here from Example T1.

Examples 2–4 for trapezoidal channels may be readily solved, or explored, using the same spreadsheet, noting the changes in key parameters where appropriate. In Example 2, use an input $H = 1.095$ m, in Example 3, use an input $S_o = 0.000657$ and in Example 3b use an input $H = 1.65$ m.

7.1.4   Inbank flows in urban drainage channels

Refer back to Section 6.4.2 and Case U1 which is for symmetric flow in an urban drainage channel or a bridge deck drainage channel. Use one panel to represent half the cross section, as indicated in Figure U1. The lateral distribution of U_d in the single panel is

$$U_d^{(2)} = \left[A_3 \xi_2^{\alpha_2} + A_4 \xi_2^{-(\alpha_2+1)} + \omega_2 \xi_2 + \eta_2 \right]^{1/2}$$

and the 2 A_i coefficients are given by

$$A_3 = -\frac{\omega(H-h)^{-(\alpha+1)} + \omega(\alpha+1)(H-h)H^{-(\alpha+2)} + \eta(\alpha+1)H^{-(\alpha+2)}}{\alpha(H-h)^{-(\alpha+1)}H^{(\alpha-1)} + (\alpha+1)(H-h)^\alpha H^{-(\alpha+2)}}$$

$$A_4 = \frac{A_3\alpha H^{(\alpha-1)} + \omega}{(\alpha+1)H^{-(\alpha+2)}}$$

Create a spreadsheet based on Figure U1 in Chapter 6, allowing for input variables of H, S_o, b, h & s (side slope) and input parameters of f, λ & Γ for a single panel. For Example U1, set $H = 0.05$ m, $S_o = 0.001$, $b = 0.15$ m and $s = 7.5$, making $h = 0.02$ m, as these represent one type of urban drainage unit tested by Mohammadi (1998) as part of a study into gradually varied and spatially varied flow in bridge deck units, as featured in Figure 7.11. Additional experiments were also undertaken in uniform flow conditions in another flume for both types of unit shown in Figure 7.11. Some further details are given by Mohammadi & Knight (1999). It is suggested that you begin by using values of $f = 0.015$, $\lambda = 0.200$ and $\Gamma = 0.150$, as this f value corresponds with a Manning n value of around 0.031, and a Nikuradse k_s value of around 0.20 m. This should lead to $A_3 = 4.68997\text{E+04}$ & $A_4 = -3.71878\text{E-10}$. For the particular depth chosen, these values should give $Q = 0.00263$ m³s⁻¹ (or 2.6 ls⁻¹) and $SF_b = 0.4146$ N, which for $SF_T = 0.649$ N gives $\%SF_w = 36.12\%$ & $\%SF_b = 63.88\%$. Now change the water depth to 0.07 m and re-calculate the discharge. Repeat the calculations for the same two depths and compare your answers with the stage-discharge curves given in Figure 7.12. The power law equations for these curves are given in Table 7.6 to make it easier to draw comparisons with your output.

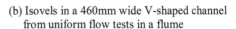

(a) Spatially varied flow unit for testing 300mm wide bridge deck drainage units

(b) Isovels in a 460mm wide V-shaped channel from uniform flow tests in a flume

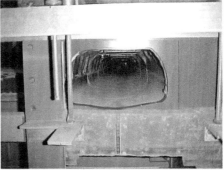

(c) Kerb side drainage channel on a bridge deck

(d) Bridge deck drainage channel (300mm wide)

Figure 7.11 Urban drainage units, 300 & 460 mm wide, being tested in two flumes.

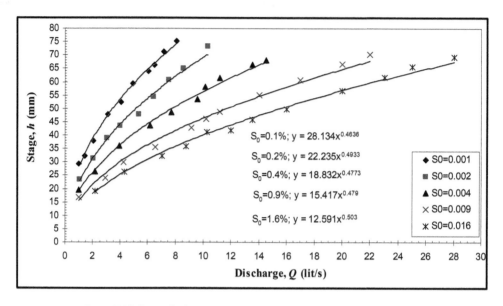

Figure 7.12 Stage-discharge relationships for 300 mm wide CIS channel.

Table 7.6 Stage-discharge relationships for 300 mm wide CIS units.

Equation	Slope	k	n
	0.001	28.129	0.4613
$H = kQ^n$	0.002	22.077	0.5024
where	0.004	18.832	0.4773
H is in mm	0.009	15.414	0.4788
Q is in ls^{-1}	0.016	12.591	0.5030

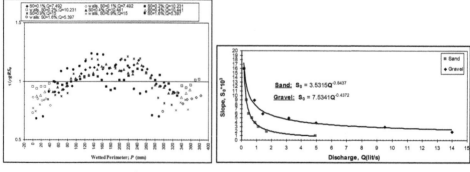

(a) Lateral distributions of stress   (b) Threshold conditions for sediments

Figure 7.13 Boundary shear stress data for 450 mm wide units.

The 450 mm wide V-shaped channel shown in Figure 7.11(b) was studied in more detail with extensive measurements of boundary shear stress and velocity being undertaken under uniform flow conditions. Typical lateral distributions of local boundary shear stress are shown in Figure 7.13(a) for selected slopes and a discharge of around $10 \, \text{ls}^{-1}$. As can be seen, for mild slopes ($S_o = 0.1\%$) the distribution is very uniform, whereas for steeper channels at higher flow rates some perturbations were observed. The threshold conditions for sand and gravel were also explored under uniform flow conditions and some test results are shown in Figure 7.13(b).

Finally, use your U1 spreadsheet with $H = 5.1$ m, $S_o = 0.001$, $b = 10$ m, $s = 100$, $f = 0.020$, $\lambda = 0.070$ and $\Gamma = 0.15$ to make an approximation to a rectangular channel and compare your answer with those in Figure 7.2. Other shaped urban drainage channels in which boundary shear stresses have been studied are referred to in Knight & Sterling (2000) and Sterling & Knight (2000 & 2002). Circular pipes running part full are commonly used and their sediment carrying capacity is closely related to the shear stresses. Circular pipes also afford a simple way of measuring flow by the end depth method, as shown by Sterling & Knight (2001). Having now illustrated how the SKM is used for studying inbank flows in channels of various shapes, attention is now turned to overbank flows in both prismatic channels and in natural rivers.

7.1.5 Overbank flow in compound trapezoidal channels

As outlined in Section 1.5 in Chapter 1 and Section 2.7.1 in Chapter 2, the scientific basis of both the SKM and CES methods arose from a decade long research project involving the Flood Channel Facility, built at HR Wallingford. One aim of this project was the acquisition of high quality data on overbank flows, so that river models could be better validated. One experiment, FCF 020301, from Series 2 is therefore selected here, as an exemplary exercise, from the many that could have been chosen. There is an extensive literature on the analysis of these experimental studies, such as in Knight & Shiono (1990 & the Discussion in 1991a), Shiono & Knight (1991b), Atabay *et al.* (2006) and in the related website *www.flowdata.bham.ac.uk*, which may be referred to for more information.

Refer back to Section 6.4.3 and Case C1 which relates to symmetric flow in a trapezoidal compound channel. Create your own spreadsheet, initially using four panels, as indicated in Figure C1. The lateral distributions of U_d in the four panels are then given by:

$$U_d^{(1)} = \left[A_1 e^{\gamma_1 y} + A_2 e^{-\gamma_1 y} + k_1 \right]^{1/2} \qquad \text{for } 0 \le y \le b$$

$$U_d^{(2)} = \left[A_3 \xi_2^{\alpha_2} + A_4 \xi_2^{-(\alpha_2 + 1)} + \omega_2 \xi_2 + \eta_2 \right]^{1/2} \qquad \text{for } 0 \le y \le b + s_2 h \quad \text{or} \quad H \ge \xi_2 \ge (H - h)$$

$$U_d^{(3)} = \left[A_5 e^{\gamma_3 y} + A_6 e^{-\gamma_3 y} + k_3 \right]^{1/2} \qquad \text{for } (b + s_2 h) \le y \le B \quad \text{or} \quad \xi_3 = (H - h)$$

$$U_d^{(4)} = \left[A_7 \xi_4^{\alpha_4} + A_8 \xi_4^{-(\alpha_4 + 1)} + \omega_4 \xi_4 + \eta_4 \right]^{1/2} \qquad \text{for } B \le y \le B + s_4(H - h) \quad \text{or} \quad (H - h) \le \xi_4 \le 0$$

and the 8 A_i coefficients are presented below in 3 alternative versions for ease of use.

a With no μ factors, as in the original SKM

$$A_1 = \frac{\left[(\alpha_2 - 1)\omega_2 H + (\eta_2 - k_1)\alpha_2\right]m_2 + (2\alpha_2 + 1)m_3}{\left[(\alpha_2 + \gamma_1 s_2 H)m_2 + (2\alpha_2 + 1)m_1\right]e^{\eta b} + \left[(\alpha_2 - \gamma_1 s_2 H)m_2 + (2\alpha_2 + 1)m_1\right]e^{-\eta b}}$$

$$A_2 = A_1$$

$$A_3 = $$
$$\frac{\left[(\alpha_2 + 1) - \gamma_1 s_2 H\right]e^{\eta b}A_1 + \left[(\alpha_2 + 1) + \gamma_1 s_2 H\right]e^{-\eta b}A_2 - (\alpha_2 + 2)\omega_2 H - (\alpha_2 + 1)(\eta_2 - k_1)}{(2\alpha_2 + 1)H^{\alpha_2}}$$

$$A_4 = \frac{\left[(\alpha_2 + \gamma_1 s_2 H)e^{\eta b} + (\alpha_2 - \gamma_1 s_2 H)e^{-\eta b}\right]A_1 - (\alpha_2 - 1)\omega_2 H - (\eta_2 - k_1)\alpha_2}{(2\alpha_2 + 1)H^{-(\alpha_2 + 1)}}$$

$$A_5 = \frac{(\alpha_4 - 1)\omega_4 H_3 - (k_3 - \eta_4)\alpha_4 - (\alpha_4 - \gamma_3 s_4 H_3)e^{-\gamma_3 B}A_6}{(\alpha_4 + \gamma_3 s_4 H_3)e^{\gamma_3 B}}$$

$$A_6 = $$
$$\frac{\left[(\alpha_2 + \gamma_3 s_2 H_3)H_3^{\alpha_2 - 1}\right]A_3 - \left[(\alpha_2 + 1) - \gamma_3 s_2 H_3\right]H_3^{-(\alpha_2 + 2)}A_4 - \gamma_3 s_2\left[k_3 - (\omega_2 H_3 + \eta_2)\right] + \omega_2}{2\gamma_3 s_2 e^{-\gamma_3 b_1}}$$

$$A_7 = H_3^{-\alpha_4}\left[A_5 e^{\gamma_3 B} + A_6 e^{-\gamma_3 B} + k_3 - \omega_4 H_3 - \eta_4\right]$$

$$A_8 = 0$$

b With individual μ factors

$$A_1 = \frac{\left[(\alpha_2 - 1)\mu_2 \omega_2 H + (\eta_2 - k_1)\mu_2 \alpha_2\right]m_2 + \mu_2(2\alpha_2 + 1)m_3}{\left[(\mu_2 \alpha_2 + \mu_1 \gamma_1 s_2 H)m_2 + \mu_2(2\alpha_2 + 1)m_1\right]e^{\eta b} + \left[(\mu_2 \alpha_2 - \mu_1 \gamma_1 s_2 H)m_2 + \mu_2(2\alpha_2 + 1)m_1\right]e^{-\eta b}}$$

$$A_2 = A_1$$

$$A_3 = \frac{\left[\mu_2(\alpha_2 + 1) - \mu_1 \gamma_1 s_2 H\right]e^{\eta b}A_1 + \left[\mu_2(\alpha_2 + 1) + \mu_1 \gamma_1 s_2 H\right]e^{-\eta b}A_2 - \mu_2(\alpha_2 + 2)\omega_2 H - \mu_2(\alpha_2 + 1)(\eta_2 - k_1)}{\mu_2(2\alpha_2 + 1)H^{\alpha_2}}$$

$$A_4 = \frac{\left[(\mu_2 \alpha_2 + \mu_1 \gamma_1 s_2 H)e^{\eta b} + (\mu_2 \alpha_2 - \mu_1 \gamma_1 s_2 H)e^{-\eta b}\right]A_1 - (\alpha_2 - 1)\mu_2 \omega_2 H + (\eta_2 - k_1)\mu_2 \alpha_2}{\mu_2(2\alpha_2 + 1)H^{-(\alpha_2 + 1)}}$$

$$A_5 = \frac{(\alpha_4 - 1)\mu_4 \omega_4 H_3 - (k_3 - \eta_4)\mu_4 \alpha_4 - (\mu_4 \alpha_4 - \mu_3 \gamma_3 s_4 H_3)e^{-\gamma_3 B}A_6}{(\mu_4 \alpha_4 + \mu_3 \gamma_3 s_4 H_3)e^{\gamma_3 B}}$$

$$A_6 = \frac{\left[(\mu_2 \alpha_2 + \mu_3 \gamma_3 s_2 H_3)H_3^{\alpha_2 - 1}\right]A_3 - \left[(\mu_2(\alpha_2 + 1) - \mu_3 \gamma_3 s_2 H_3)\right]H_3^{-(\alpha_2 + 2)}A_4 - \mu_3 \gamma_3 s_2\left[k_3 - (\omega_2 H_3 + \eta_2)\right] + \mu_2 \omega_2}{2\mu_3 \gamma_3 s_2 e^{-\gamma_3 b_1}}$$

$$A_7 = H_3^{-\alpha_4}\left[A_5 e^{\gamma_3 B} + A_6 e^{-\gamma_3 B} + k_3 - \omega_4 H_3 - \eta_4\right]$$

$$A_8 = 0$$

c   With ratios of μ factors. Note that by setting $(\mu_1/\mu_2) = 1.0$, etc., gives version (a)

$$A_1 = \cfrac{\left[(\alpha_2 - 1)\omega_2 H + (\eta_2 - k_1)\alpha_2\right]m_2 + (2\alpha_2 + 1)m_3}{\left[\left(\alpha_2 + \left(\dfrac{\mu_1}{\mu_2}\right)\gamma_1 s_2 H\right)m_2 + (2\alpha_2 + 1)m_1\right]e^{\eta_1 b} + \left[\left(\alpha_2 - \left(\dfrac{\mu_1}{\mu_2}\right)\gamma_1 s_2 H\right)m_2 + (2\alpha_2 + 1)m_1\right]e^{-\eta_1 b}}$$

$$A_2 = A_1$$

$$A_3 = \cfrac{\left[(\alpha_2 + 1) - \left(\dfrac{\mu_1}{\mu_2}\right)\gamma_1 s_2 H\right]e^{\eta_1 b} A_1 + \left[(\alpha_2 + 1) + \left(\dfrac{\mu_1}{\mu_2}\right)\gamma_1 s_2 H\right]e^{-\eta_1 b} A_2 - (\alpha_2 + 2)\omega_2 H - (\alpha_2 + 1)(\eta_2 - k_1)}{(2\alpha_2 + 1)H^{\alpha_2}}$$

$$A_4 = \cfrac{\left[\left(\alpha_2 + \left(\dfrac{\mu_1}{\mu_2}\right)\gamma_1 s_2 H\right)e^{\eta_1 b} + \left(\alpha_2 - \left(\dfrac{\mu_1}{\mu_2}\right)\gamma_1 s_2 H\right)e^{-\eta_1 b}\right]A_1 - (\alpha_2 - 1)\omega_2 H - (\eta_2 - k_1)\alpha_2}{(2\alpha_2 + 1)H^{-(\alpha_2 + 1)}}$$

$$A_5 = \cfrac{(\alpha_4 - 1)\omega_4 H_3 - (k_3 - \eta_4)\alpha_4 - \left(\alpha_4 - \left(\dfrac{\mu_3}{\mu_4}\right)\gamma_3 s_4 H_3\right)e^{-\gamma_3 B} A_6}{\left(\alpha_4 + \left(\dfrac{\mu_3}{\mu_4}\right)\gamma_3 s_4 H_3\right)e^{\gamma_3 B}}$$

$$A_6 = \cfrac{\left[\left(\alpha_2 + \left(\dfrac{\mu_3}{\mu_2}\right)\gamma_3 s_2 H_3\right)H_3^{\alpha_2 - 1}\right]A_3 - \left[(\alpha_2 + 1) - \left(\dfrac{\mu_3}{\mu_2}\right)\gamma_3 s_2 H_3\right]H_3^{-(\alpha_2 + 2)} A_4 - \left(\dfrac{\mu_3}{\mu_2}\right)\gamma_3 s_2\left[k_3 - (\omega_2 H_3 + \eta_2)\right] + \omega_2}{2\left(\dfrac{\mu_3}{\mu_2}\right)\gamma_3 s_2 e^{-\gamma_3 b_1}}$$

$$A_7 = H_3^{-\alpha_4}\left[A_5 e^{\gamma_3 B} + A_6 e^{-\gamma_3 B} + k_3 - \omega_4 H_3 - \eta_4\right]$$

$$A_8 = 0$$

In matrix form the initial equations based on versions (a) & (b) may be written as:

Version (a)

$$
\begin{bmatrix}
1 & -1 & 0 & 0 & 0 & 0 & 0 & 0 \\
e^{\gamma_1 b} & e^{-\gamma_1 b} & -H^{\alpha_2} & -H^{-(\alpha_2+1)} & 0 & 0 & 0 & 0 \\
\gamma_1 s_2 e^{\gamma_1 b} & -\gamma_1 s_2 e^{-\gamma_1 b} & \alpha_2 H^{\alpha_2-1} & -(\alpha_2+1)H^{-(\alpha_2+2)} & 0 & 0 & 0 & 0 \\
0 & 0 & H_3^{\alpha_2} & H_3^{-(\alpha_2+1)} & -e^{\gamma_3 b_1} & -e^{-\gamma_3 b_1} & 0 & 0 \\
0 & 0 & \alpha_2 H_3^{\alpha_2-1} & -(\alpha_2+1)H_3^{-(\alpha_2+2)} & \gamma_3 s_2 e^{\gamma_3 b_1} & -\gamma_3 s_2 e^{-\gamma_3 b_1} & 0 & 0 \\
0 & 0 & 0 & 0 & e^{\gamma_3 B} & e^{-\gamma_3 B} & -H_3^{\alpha_4} & 0 \\
0 & 0 & 0 & 0 & \gamma_3 s_4 e^{\gamma_3 B} & -\gamma_3 s_4 e^{-\gamma_3 B} & \alpha_4 H_3^{\alpha_4-1} & 0 \\
0 & 0 & 0 & 0 & 0 & 0 & 0 & 1
\end{bmatrix}
$$

$$
\times
\begin{bmatrix}
A_1 \\ A_2 \\ A_3 \\ A_4 \\ A_5 \\ A_6 \\ A_7 \\ A_8
\end{bmatrix}
=
\begin{bmatrix}
0 \\
\omega_2 H + \eta_2 - k_1 \\
-\omega_2 \\
k_3 - \omega_2 H_3 - \eta_2 \\
-\omega_2 \\
\omega_4 H_3 + \eta_4 - k_3 \\
-\omega_4 \\
0
\end{bmatrix}
$$

Version (b)

$$
\begin{bmatrix}
1 & -1 & 0 & 0 & 0 & 0 & 0 & 0 \\
e^{\gamma_1 b} & e^{-\gamma_1 b} & -H^{\alpha_2} & -H^{-(\alpha_2+1)} & 0 & 0 & 0 & 0 \\
\mu_1 \gamma_1 s_2 e^{\gamma_1 b} & -\mu_1 \gamma_1 s_2 e^{-\gamma_1 b} & \mu_2 \alpha_2 H^{\alpha_2-1} & -\mu_2(\alpha_2+1)H^{-(\alpha_2+2)} & 0 & 0 & 0 & 0 \\
0 & 0 & H_3^{\alpha_2} & H_3^{-(\alpha_2+1)} & -e^{\gamma_3 b_1} & -e^{-\gamma_3 b_1} & 0 & 0 \\
0 & 0 & \mu_2 \alpha_2 H_3^{\alpha_2-1} & -\mu_2(\alpha_2+1)H_3^{-(\alpha_2+2)} & \mu_3 \gamma_3 s_2 e^{\gamma_3 b_1} & -\mu_3 \gamma_3 s_2 e^{-\gamma_3 b_1} & 0 & 0 \\
0 & 0 & 0 & 0 & e^{\gamma_3 B} & e^{-\gamma_3 B} & -H_3^{\alpha_4} & 0 \\
0 & 0 & 0 & 0 & \mu_3 \gamma_3 s_4 e^{\gamma_3 B} & -\mu_3 \gamma_3 s_4 e^{-\gamma_3 B} & \mu_4 \alpha_4 H_3^{\alpha_4-1} & 0 \\
0 & 0 & 0 & 0 & 0 & 0 & 0 & 1
\end{bmatrix}
$$

$$
\times
\begin{bmatrix}
A_1 \\ A_2 \\ A_3 \\ A_4 \\ A_5 \\ A_6 \\ A_7 \\ A_8
\end{bmatrix}
=
\begin{bmatrix}
0 \\
\omega_2 H + \eta_2 - k_1 \\
-\mu_2 \omega_2 \\
k_3 - \omega_2 H_3 - \eta_2 \\
-\mu_2 \omega_2 \\
\omega_4 H_3 + \eta_4 - k_3 \\
-\mu_4 \omega_4 \\
0
\end{bmatrix}
$$

Using FCF experiment 020301 ($b = 0.75$ m, $B = 3.15$ m, $s_2 = s_4 = 1.0$, $h = 0.150$ m, $H = 0.177$ m and $S_o = 1.029 \times 10^{-3}$) together with the input parameters suggested in Table 7.7, should give the associated parameters shown in Table 7.8 and produce the results shown in Table 7.9 and Figure 7.14 for the relative depth, Dr, of 0.157. The f values in panels 1 & 3 are based on Abril & Knight's (2004) equation relating floodplain (fp) and main channel (mc) friction factors across the entire FCF series,

Table 7.7 Input parameters for FCF simulation 020301.

Panel	b	s	f	lambda	GAMMA
1	0.75000	***	0.01597	0.0700	0.2687
2	0.15000	1.000	0.02335	0.6347	0.2687
3	2.25000	***	0.03073	1.1994	−0.0701
4	0.02784	1.000	0.03073	1.1994	−0.0701

Top width = 3.17784.

Table 7.8 Associated parameters for FCF simulation 020301.

mu	alpha	eta	omega	beta	k	gamma
0.0088461				0.150002767	0.7627	6.353166839
0.0969851	0.2005416	−0.06509	−0.333926642			
0.2102510				−0.250001243	0.0912	11.54743454
0.2102510	0.1294113	0.01290	−0.146174278			

A1 = −3.46119E-03
A2 = −3.46119E-03
A3 = 6.84949E-01
A4 = −4.268136E-04
A5 = −2.333262E-18
A6 = 4.46666E+03
A7 = 1.07746E-01
A8 = 0.

as illustrated in Figure 7.17 and discussed later in this section. Figure 7.14 illustrates the computed results compared with the experimentally measured U_d & τ_b values. It should be noted that individual μ factors were included in this case, in order to illustrate them in a spreadsheet format and their influence.

Now change the depth to 0.198 m, making a relative value of 0.197, which corresponds to a different FCF experiment, 020501. Using the input parameters suggested in Table 7.10 should give the associated parameters shown in Table 7.11 and produce the results shown in Figure 7.15.

Although a thorough investigation of errors is beyond the scope of this book, Table 7.12 shows some values for a single set of FCF experiments (Series 2), covering a range of relative depths, Dr (= $(H-b)/H$), from 0.05 to 0.50. The overall discharge and mean shear force errors are within 2%, and the average values for the main channel (mc = panels 1&2) and floodplain (fp = panels 3&4) are also in single figures. The individual errors in each panel or zone are also indicated, and are generally acceptable, except in the two smallest zones, 2 & 4, as might be expected. It was not possible to make measurements of U & τ_b in zone 4 for the lowest depth in the series (020101). Zone 4 contributes very little to the overall discharge anyway and zone 2 is represented by just a single panel, despite the complexity of the flow in that region. The average errors indicate a reasonably good first fit, which can of course be improved on by either fine tuning the parameters for an individual panel further or by increasing the number of panels in those regions where precision is required. An example of this

Table 7.9 Output values of U_d & τ_b (FCF Experiment 020301).

y (m)	Depth (m)	Ud (m/s)	Tau (N/m²)	dQ (m³s⁻¹)	Zonal Q (m³s⁻¹)
0.000	0.178	0.8693	1.5087		
0.075	0.178	0.8689	1.5071	0.01159	
0.150	0.178	0.8674	1.5019	0.01158	
0.225	0.178	0.8645	1.4920	0.01155	
0.300	0.178	0.8596	1.4750	0.01150	
0.375	0.178	0.8514	1.4470	0.01141	
0.450	0.178	0.8379	1.4016	0.01127	
0.525	0.178	0.8157	1.3282	0.01103	
0.600	0.178	0.7785	1.2098	0.01063	
0.675	0.178	0.7145	1.0191	0.00996	
0.750	0.178	0.5972	0.7118	0.00875	0.10926
0.765	0.163	0.5939	1.0295	0.00152	
0.780	0.148	0.5900	1.0161	0.00138	
0.795	0.133	0.5854	1.0001	0.00124	
0.810	0.118	0.5797	0.9809	0.00110	
0.825	0.103	0.5728	0.9576	0.00095	
0.840	0.088	0.5641	0.9288	0.00081	
0.855	0.073	0.5529	0.8924	0.00067	
0.870	0.058	0.5378	0.8443	0.00053	
0.885	0.043	0.5158	0.7764	0.00040	
0.900	0.028	0.4777	0.6660	0.00026	0.00887
1.125	0.028	0.3185	0.3896	0.00249	
1.350	0.028	0.3033	0.3534	0.00195	
1.575	0.028	0.3022	0.3507	0.00190	
1.800	0.028	0.3021	0.3505	0.00189	
2.025	0.028	0.3021	0.3505	0.00189	
2.250	0.028	0.3021	0.3505	0.00189	
2.475	0.028	0.3021	0.3505	0.00189	
2.700	0.028	0.3019	0.3502	0.00189	
2.925	0.028	0.3003	0.3463	0.00189	
3.150	0.028	0.2768	0.2943	0.00181	0.01949
3.153	0.025	0.2759	0.2923	0.00002	
3.156	0.022	0.2748	0.2900	0.00002	
3.158	0.019	0.2735	0.2873	0.00002	
3.161	0.017	0.2719	0.2839	0.00001	
3.164	0.014	0.2699	0.2798	0.00001	
3.167	0.011	0.2674	0.2746	0.00001	
3.169	0.008	0.2640	0.2677	0.00001	
3.172	0.006	0.2591	0.2579	0.00001	
3.175	0.003	0.2506	0.2413	0.00000	
3.1778	0.000	0	0	0.00000	0.00010
			Q =	**0.13772**	**0.13772**
			main channel =	**Zones 1 & 2 =**	**0.11813**
			floodplain =	**Zones 3 & 4 =**	0.01960

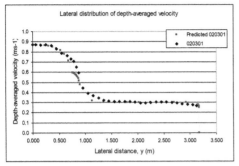
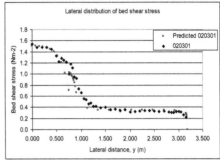

Figure 7.14 Comparison of experimental and model results for FCF simulation 020301 ($Dr = 0.157$).

Table 7.10 Input parameters for FCF simulation 020501.

Panel	b	s	f	lambda	GAMMA
1	0.75000	***	0.01612	0.0700	0.2991
2	0.15000	1.000	0.02085	0.3515	0.2991
3	2.25000	***	0.02558	0.6330	−0.1208
4	0.04796	1.000	0.02558	0.6330	−0.1208

Table 7.11 Associated parameters for FCF simulation 020501.

mu	alpha	eta	omega	beta	k	gamma
0.0088875				0.149999146	0.8411	5.720806985
0.0507535	0.3129017	−0.08114	−0.706381856			
0.1012356				−0.249998447	0.1888	8.813469345
0.1012356	0.2089978	0.02671	−0.322088504			

A1 = −5.18932E-03
A2 = −5.18932E-03
A3 = 1.13516E+00
A4 = −8.918837E-05
A5 = −3.007495E-14
A6 = 3.62567E+02
A7 = 2.70320E-01
A8 = 0.

is shown later for these two FCF cases in Figure 7.21. These simulations also serve to indicate the simplicity and effectiveness and of using the SKM for open channel flow analysis.

Determining zonal discharges and percentage shear forces on selected boundary elements may be obtained by other means, as indicated in Section 6.5.3(c), where a link is made between the SKM and the 'area method'. The latter method can provide a direct check on the values in your spreadsheet, as indicated by Knight & Tang (2008). Another check on any output or original data is to compare the variation of individual parameters. There have been many guidance 'rules' for all parameters within the SKM

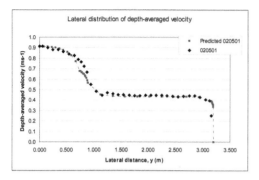
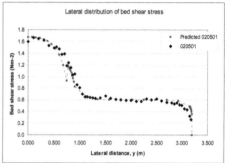

Figure 7.15 Comparison of predicted and experimental distributions of U_d & τ_b (Experiment FCF 020501, $Dr = 0.197$).

Table 7.12 Errors in simulations of discharge and boundary shear forces for FCF Series 2.

Experiment =	020101	020203	020301	020401	020501	020601	020701	020801	Average
Q total =	−6.12	−3.03	−2.36	1.51	−1.68	−0.11	0.02	0.51	−1.41
SF total =	−11.33	−6.89	−4.30	−2.25	−0.31	1.80	4.91	7.06	−1.41
Q_1	−3.85	−1.29	−1.35	2.07	−0.46	0.43	1.77	2.58	−0.01
Q_2	−30.52	−18.77	−14.31	−9.03	−11.51	−7.40	−5.50	−5.23	−12.78
Q_3	−15.20	−6.25	−1.76	3.20	−2.51	−0.02	−1.51	−0.89	−3.12
Q_4	−	84.60	−13.07	13.69	50.16	51.98	18.46	14.40	
Q_{mc}	−5.85	−2.74	−2.45	1.08	−1.51	−0.35	0.97	1.63	−1.15
Q_{fp}	−15.10	−6.08	−1.82	3.27	−2.20	0.39	−1.21	−0.55	−2.91
SF_1	−8.92	−3.38	−2.78	−7.26	−1.76	0.89	3.75	5.05	−1.80
SF_2	2.29	−20.56	−12.94	−5.84	−6.43	−3.69	−6.07	−4.47	−7.21
SF_3	−25.21	−7.53	−3.85	4.81	0.89	1.63	5.32	8.23	−1.96
SF_4	−	132.22	9.17	−24.36	169.24	194.78	80.14	34.98	
SF_{mc}	−7.19	−6.63	−4.57	−7.03	−2.56	0.09	1.83	3.09	−2.87
SF_{fp}	−25.02	−7.07	−3.71	4.17	2.18	3.40	7.05	9.38	−1.20

system, based on measured data obtained both from the field and the laboratory. Figure 7.16 shows how the FCF experimental data for the local f values on the floodplain remain more or less constant with y, if based on (2.61), involving the depth-averaged velocity, U_d, and the local boundary shear stress, τ_b. This led to the equation $f_{fp}/f_{mc} = 0.669 + 0.331Dr^{-0.719}$ and similar equations for the other key parameters, λ & Γ, some of which are presented either in Table 3.4 or elsewhere by Shiono & Knight (1991), Knight & Abril (1996) and Abril & Knight (2004). Figure 7.17 shows the trend line for Series 01–03 and how the floodplain roughness increases with decreasing floodplain depth (= $H-h$) or relative depth, Dr (= $(H-h)/H$). The same feature of constancy in floodplain f value may be seen in the rod roughness data in Figure 6.87. You may observe the same feature, using your calculated U_d & τ_b results for FCF 020301, and cross check that the laterally integrated value agrees with your input value for panel 3.

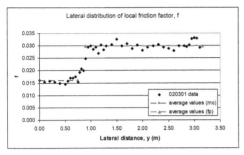
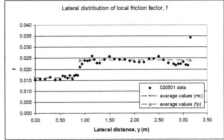

Figure 7.16 Measured f values, based on U_d & τ_b values (FCF 020301 & 020501).

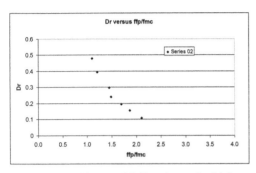
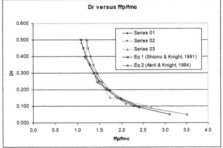

Figure 7.17 Measured f_{fp}/f_{mc} values v Dr; (a) Series 02: (b) All experiments in FCF series 01–03.

7.1.6 Overbank flow in compound rectangular channels

Refer back to Section 6.4.3 and Case C2 which relates to symmetric flow in a rectangular compound channel. Use two panels, as indicated in Figure C2. The lateral distributions of U_d in the two panels are given by:

$$U_d^{(1)} = \left[A_1 e^{\gamma_1 y} + A_2 e^{-\gamma_1 y} + k_1\right]^{1/2} \quad \& \quad U_d^{(2)} = \left[A_3 e^{\gamma_2 y} + A_4 e^{-\gamma_2 y} + k_2\right]^{1/2}$$

and the 4 A_i coefficients are given by

$$A_1 = \frac{\phi_2 \gamma_2 \left(e^{\gamma_2 b} + e^{\gamma_2 (2B-b)}\right)\left[k_2\left(1 - e^{\gamma_2(B-b)}\right) - k_1\right]}{-\left(e^{\gamma_2 b} - e^{\gamma_2 (2B-b)}\right)\left[\phi_2 \gamma_2 k_2 e^{\gamma_2 (B-b)} - \rho \frac{f_w}{8} h k_1\right]}{\phi_2 \gamma_2 \left(e^{\gamma_1 b} + e^{-\gamma_1 b}\right)\left(e^{\gamma_2 b} + e^{\gamma_2(2B-b)}\right) - \phi_1 \gamma_1\left(e^{\gamma_1 b} - e^{-\gamma_1 b}\right)\left(e^{\gamma_2 b} - e^{\gamma_2(2B-b)}\right)}{-\rho \frac{f_w}{8} h\left(e^{\gamma_1 b} + e^{-\gamma_1 b}\right)\left(e^{\gamma_2 b} - e^{\gamma_2(2B-b)}\right)}$$

$$A_2 = A_1$$

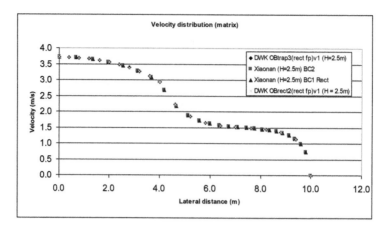

Figure 7.18 Comparison between U_d distributions for a trapezoidal compound channel with a nearly vertical internal wall with that for a rectangular compound channel for $H = 2.5$ m ($S_o = 0.001, b = 4$ m, $B = 10$ m, $h = 2$ m; $f_1 = f_2 = 0.01$ & $f_3 = 0.02$; $\lambda_1 = \lambda_2 = 0.1$ & $\lambda_3 = 0.2$; $\Gamma_1 = \Gamma_2 = 1.0$ & $\Gamma_3 = -0.75$) after Knight (2013b).

$$A_3 = \frac{\left[\phi_1 \gamma_1 \left(e^{\gamma_1 b} - e^{-\gamma_1 b}\right) + \rho \frac{f_w}{8} b \left(e^{\gamma_1 b} + e^{-\gamma_1 b}\right)\right] A_1 - \phi_2 \gamma_2 k_2 e^{\gamma_2 (B-b)} + \rho \frac{f_w}{8} b k_1}{\phi_2 \gamma_2 \left(e^{\gamma_2 b} + e^{\gamma_2 (2B-b)}\right)}$$

$$A_4 = -A_3 e^{2\gamma_2 B} - k_2 e^{\gamma_2 B}$$

These were used to compare simulations of the U_d distributions for a trapezoidal compound channel with a nearly vertical internal wall with that for a rectangular compound channel, i.e. Cases C1 & C2, as shown in Figure 7.18. Different spreadsheets and boundary conditions were used, as illustrated in Figure 7.18 and explained further in Tang & Knight (2008) and Knight (2013b). See also Yang *et al.* (2013).

7.1.7   Overbank flow in the River Severn

One of the earliest applications of SKM to an actual river was that by Knight, Shiono & Pirt (1989) as part of a larger study into resistance coefficients and stage-discharge relationships at three gauging stations in what was then the Severn-Trent Water Authority. These sites were at Montford Bridge on the River Severn and at North Muskham and Yoxall on the River Trent, sites that have featured already in Chapter 2 and in the CES simulations in Chapter 4.

Refer back to Sections 2.5.2, 4.2.4 & 6.4.3 which relate to overbank flow in natural rivers and create your own spreadsheet, initially using 7 panels, as indicated in Figure 2.45, and set $S_o = 2.0 \times 10^{-4}$. Try to reproduce the velocity distributions in Figure 7 of Knight *et al.* (1989), for $H = 4.77, 6.45, 6.92$ and 7.81 m to correspond with the gaugings shown. Set $\Gamma = 0$ as a first attempt, before adjusting f & λ values progressively for each panel. It should be noted that for a bed level of 50.282 m OD the local stage datum was set at 52.000, making a difference of 1.718 m between the two. Further guidance on choice of coefficients is given in Section 4.2.4 (CES simulation) and 7.2.4 (SKM simulation).

7.1.8 Flows in regime channels of lenticular shape with sediment

The transport of sediment in rivers or canals under regime conditions, when there is a balance between sediment load, discharge, bed slope and geometry of the channel (width, depth & shape), is of particular interest to engineers and environmentalists concerned with water supply, rehabilitation of rivers and flood risk assessment. Since the majority of sediment transport formulae are based on the boundary shear stress, and the SKM was developed with this in mind, it is an obvious candidate for estimating sediment transport in prismatic channels, since the SKM determines the lateral distribution of τ_b in complex flow fields. Because the SKM & CES rely on the cross section being built up of linear elements, it is suggested that you create a spreadsheet of initially just 4 panels, before attempting one with many more panels, to represent a regime channel. Figure 7.19 indicates an attempt at simulating small scale regime channels using SKM and experimental data. Further details are available in Ayyoubzadeh (1997) and Cao (1995). Dealing with this topic is beyond the scope of this book and reference should therefore be made to papers by Knight & Yu (1995 & 1998), Cao & Knight (1995, 1996 & 1998), as well as the fundamental papers on analytical regime theory, by Bettess et al. (1988), White, Bettess & Paris (1982), the Wallingford Tables for the design of stable alluvial channels by White, Bettess & Paris (1981), and the ASCE task force papers on river width adjustment by Thorne et al. (1998) and Darby et al. (1998).

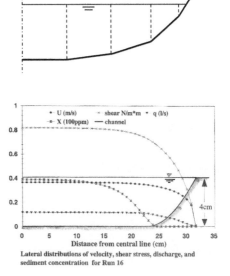

Figure 7.19 Studies in regime channels and SKM simulations using schematized cross sections.

7.2   USING EXCEL TO SOLVE THE SKM EQUATIONS NUMERICALLY

7.2.1   Introduction and explanatory notes

Examples of how to set the SKM in EXCEL in order to solve the $n \times n$ matrix for predicting the lateral distributions of depth-averaged velocity and boundary shear stress across a river are now described. Some of the case studies referred to in Chapters 2 & 4 are deliberately used again for two reasons; firstly because much of the data have already been illustrated in earlier chapters and, secondly, because they illustrate many of practical issues that need to be considered when modelling rivers.

7.2.2   The Flood Channel Facility (FCF), Series 02

Beginning with Experiment 020301 from the FCF Series 02, a trapezoidal compound channel with a symmetric cross section, only half the cross section is required, because of the flow and geometrical symmetry. Using 4 panels, as done in Case C1 in Section 7.1.5, and the notation defined in Figure 7.20, the calculation processes as are described below. Some procedures are repeated for clarity.

The SKM equations below have the term Γ' added to (2.58a) for a constant water depth domain with parameters f, λ and Γ' as before, where

$$\frac{\partial}{\partial y}\left[H(\rho UV)_d\right] = \Gamma = \rho\Gamma',$$

Let $s = \infty$, so that $1/s = 0$, then the basic equation becomes

$$\rho g H S_0 - \rho\frac{f}{8}U_d^2 + \frac{\partial}{\partial y}\left\{\rho\lambda H^2\left(\frac{f}{8}\right)^{1/2}U_d\frac{\partial U_d}{\partial y}\right\} - \frac{1}{2}\rho C_d A_p U_d^2 = \rho\Gamma' \qquad (A5.65)$$

$$\rho g H S_0 - \left(\rho\frac{f}{8} + \frac{1}{2}\rho C_d A_p\right)U_d^2 + \frac{\partial}{\partial y}\left\{\rho\lambda H^2\left(\frac{f}{8}\right)^{1/2}U_d\frac{\partial U_d}{\partial y}\right\} = \rho\Gamma'$$

$$E_1 - E_2 U_d^2 + E_3\frac{\partial}{\partial y}\left\{\frac{\partial U_d^2}{\partial y}\right\} = \Gamma' \qquad (A5.66)$$

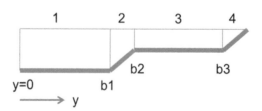

Figure 7.20 Cross section of the FCF.

where

$$E_1 = gHS_0,$$

$$E_2 = \frac{f}{8} + \frac{1}{2}C_dA_p$$

$$E_3 = \frac{\lambda H^2}{2}\left(\frac{f}{8}\right)^{1/2}$$

The analytical solution of SKM for a constant water depth, as in subsections 1 & 3 is

$$U^2 = A_1 e^{\gamma y} + A_2 e^{-\gamma y} + k$$

$$\gamma = \pm\sqrt{\frac{E_2}{E_3}}$$

$$k = \frac{E_1 - \Gamma}{E_2} = \frac{E_1'}{E_2}(1 - \Gamma'') \quad \text{where} \quad \Gamma'' = \frac{\Gamma'}{E_1}$$

$$k = \frac{E_1 - \Gamma'}{E_2}$$

For flow in a varying depth domain with a transverse bed slope, the momentum equation is

$$\rho g S_0 \xi - \left(\rho \frac{f}{8}\psi + \rho \frac{1}{2}C_dA_p\right)F + \frac{\rho\lambda}{2s^2}\left(\frac{f}{8}\right)^{1/2}\left[2\xi\frac{\partial F}{\partial \xi} + \xi^2\frac{\partial^2 F}{\partial \xi^2}\right] = \rho\Gamma' \qquad \text{(A5.74)}$$

Eliminating density ρ, out, then the equation becomes:

$$E_1\xi - E_2 F + E_3\left[2\xi\frac{\partial F}{\partial \xi} + \xi^2\frac{\partial^2 F}{\partial \xi^2}\right] = \Gamma'$$

where

$$E_1' = gS_0 \qquad\qquad\qquad\qquad\qquad\qquad\qquad\qquad \text{(A5.77)}$$

$$E_2 = \left(\frac{f}{8}\psi + \frac{1}{2}C_dA_p\right) \qquad\qquad\qquad\qquad \text{(A5.78)}$$

$$E_3 = \frac{\lambda}{2s^2}\left(\frac{f}{8}\right)^{1/2} \qquad\qquad\qquad\qquad\qquad \text{(A5.79)}$$

where $\psi = \left(1 + \frac{1}{s^2}\right)^{1/2}$ and $\xi = H \pm \frac{(y-b)}{s}$, when the coordinate $y = 0$ is at the centre of the main channel. In this particular example, the local depth will be $\xi = H - \frac{(y-b)}{s}$, where $b1$ and $b3$ replace b when $y > b1$ and $y > b3$ for subsections 2 and 4 respectively. The analytical solution is

$$U^2 = A_3 \xi^{\alpha_1} + A_4 \xi^{-(\alpha_1+1)} + \omega \xi + \eta$$

$$\alpha_1 = -\frac{1}{2} + \frac{1}{2}\sqrt{1 + 4\frac{E_2}{E_3}} \tag{A5.81}$$

$$\alpha_2 = -\frac{1}{2} - \frac{1}{2}\sqrt{1 + 4\frac{E_2}{E_3}} = -(\alpha_1 + 1) \tag{A5.82}$$

$$\omega = \frac{E_1'}{E_2 - 2E_3}$$

$$\eta = \frac{\Gamma'}{E_2} = \frac{E_1'}{E_2}\Gamma'' \quad \text{where} \quad \Gamma'' = \frac{\Gamma'}{E_1'}$$

The key unknown parameters in the solutions are f, λ & Γ, which must now be determined. First, the most significant one is f, the resistance coefficient, which depends on the bed material and the type of vegetation on the boundary. Bed materials may be expressed by the equivalent sand grain roughness height, k_s, and then the friction factor can be determined from the Colebrook–White equation (C–W)

$$\frac{1}{\sqrt{f}} = -2\log\left(\frac{3.02}{R_e\sqrt{f}} + \frac{k_s}{12.3H}\right) \quad \text{and} \quad f = \frac{1}{\left[-2\log\left(\frac{3.02\upsilon}{\sqrt{128gH^3S_0}} + \frac{k_s}{12.3H}\right)\right]^2}$$

The equivalent sand grain roughness height, k_s, can give a Manning coefficient, n, by calculating from the relationship $n = k_s^{1/6}/(8.25\sqrt{g})$ (Ackers, 1991). Sometimes a Manning coefficient, n, is given, and the friction factor can be estimated by the following equation:

$$f = \frac{8gn^2}{H^{1/3}}$$

Second, a value for λ is required. The standard value is 0.07 for a smooth bed and 0.16 for a mixing zone, (Elder's value). This value can be varied between 0.07 and 1.0, and is best calibrated from measured velocity data. Finally, Γ, the value of the secondary flow is required. This is often the most difficult parameter to determine, due to paucity of measured data. It is usually obtained by calibration, using measured velocity data initially and setting Γ to zero. Thereafter Γ may be calibrated more closely using both measured velocity and boundary shear stress data.

Once the values of the parameters f, λ & Γ are set, then a set of simultaneous equations or the matrix equation ($\mathbf{Ax} = \mathbf{B}$) is obtained using the appropriate boundary conditions. First, setting $y = 0$ at the centre of the main channel, the boundary conditions (BCs) can be taken at $y = 0$, b1, b2 and b3. The boundary conditions at the interfaces between the subsections are as follows:

1   **BC at $y = 0$,** the centre of the channel: $\frac{\partial U_1^2}{\partial y} = 0$, since $\frac{\partial U_1}{\partial y} = 0$ for smooth transition and symmetry of velocity at the centre of the channel, i.e. a symmetric condition.

Analytical solution: $U_1^2 = A_1 e^{\eta_1 y} + A_2 e^{-\eta_1 y} + k_1$

At $y = 0$ $\frac{\partial U_1^2}{\partial y} = 0$, arranging the unknown terms A_1 and A_2 for this boundary condition gives $A_1 \gamma_1 e^0 - A_2 \gamma_1 e^0 = 0$

The left hand side terms except A_1 and A_2 can be determined for all conditions at $y = 0$. Entries for the 1st row in the Matrix are:

1st row $[\gamma_1 e^0, -\gamma_1 e^0, 0, 0.....0]$ $B: [0]$

where γ_1 is for subsection 1.

2   **BC at $y = b1$,** the edge of the main channel bed: $U_1 = U_2$ and $\frac{\partial U_1^2}{\partial y} = \frac{\partial U_2^2}{\partial y}$ and $\xi_1 = H$

From $U_1 = U_2$ (the same as $U_1^2 = U_2^2$), because of it being continuous, arrange unknown terms, A_1, A_2, A_3 and A_4 on the left hand side of the equation and known terms on the right hand side.

$$A_1 e^{\eta_1 b1} + A_2 e^{-\eta_1 b1} + k_1 = A_3 \xi^{\alpha_{1,2}} + A_4 \xi_1^{-(\alpha_{1,2}+1)} + \omega_2 \xi_1 + \eta_2$$

$$A_1 e^{\eta_1 b1} + A_2 e^{-\eta_1 b1} - A_3 \xi_1^{\alpha_{1,2}} - A_4 \xi_1^{-(\alpha_{1,2}+1)} = -k_1 + \omega_2 \xi_1 + \eta_2$$

The left and right hand side terms are determined by all conditions at $y = b1$. For example, $\xi_1 = H$ etc., γ and k in subsection 1, and α_1, ω, η and s are in subsection 2. Entries of the 2nd row in the Matrix are then:

2nd row $\left[e^{\eta_1 b1}, e^{-\eta_1 b1}, -\xi_1^{\alpha_{1,2}}, -\xi_1^{-(\alpha_{1,2}+1)}, 0, 0.... \right]$ $B: [-k_1 + \omega_2 \xi_1 + \eta_2]$

From $\frac{\partial U_1^2}{\partial y} = \frac{\partial U_2^2}{\partial y}$, $\frac{\partial U_2^2}{\partial y} = \frac{\partial \xi}{\partial y} \frac{\partial U_2^2}{\partial \xi}$, $\xi_1 = H$, $\frac{\partial \xi}{\partial y} = -\frac{1}{s_2}$ and $s = 1.0$, arranging unknown terms, A_1, A_2, A_3 and A_4

$$A_1 \gamma_1 e^{\eta_1 b1} - A_2 \gamma_1 e^{-\eta_1 b1} = A_3 \alpha_{1,2} \left(\frac{-1}{s_2}\right) \xi_1^{\alpha_{1,2}-1} - A_4 (\alpha_{1,2}+1) \left(\frac{-1}{s_2}\right) \xi_1^{-(\alpha_{1,2}+1)-1} + \omega_2 \left(\frac{-1}{s_2}\right)$$

$$A_1 \gamma_1 e^{\eta_1 b1} - A_2 \gamma_1 e^{-\eta_1 b1} - A_3 \alpha_{1,2} \left(\frac{-1}{s_2}\right) \xi_1^{\alpha_{1,2}-1} + A_4 (\alpha_{1,2}+1) \left(\frac{-1}{s_2}\right) \xi_1^{-(\alpha_{1,2}+1)-1} = + \omega_2 \left(\frac{-1}{s_2}\right)$$

where γ in subsection 1, and α_1, ω and s are in subsection 2. Entries of the 3rd row in the Matrix are then:

3rd row $\left[\gamma e^{\gamma b1}, -\gamma e^{-\gamma b1}, -\alpha_{1,2} \left(\frac{-1}{s_2}\right) \xi_1^{\alpha_{1,2}-1}, (\alpha_{1,2}+1) \left(\frac{-1}{s_2}\right) \xi_1^{-(\alpha_{1,2}+1)-1}, 0, 0.....0 \right]$

$B: \left[\omega_2 \frac{-1}{s_2} \right]$

3   BC at $y = b2$, the start of floodplain, $U_2 = U_3$ and $\frac{\partial U_2^2}{\partial y} = \frac{\partial U_3^2}{\partial y}$ and $\xi_2 = H - h$ (h = floodplain height)

From $U_2^2 = U_3^2$ arranging unknown terms, A_3, A_4, A_5 and A_6 by substituting $H - h$ into ξ_2, and b1 into y

$$A_3 \xi_2^{\alpha_{1,2}} + A_4 \xi_2^{-(\alpha_{1,2}+1)} + \omega_2 \xi_2 + \eta_2 = A_5 e^{\gamma_3 b2} + A_6 e^{-\gamma_3 b2} + k_3$$

$$A_3 \xi_2^{\alpha_{1,2}} + A_4 \xi_2^{-(\alpha_{1,2}+1)} - A_5 e^{\gamma_3 b2} - A_6 e^{-\gamma_3 b2} = k_3 - \omega_2 \xi_2 - \eta_2$$

where γ and k in subsection 3, and α_1, ω, η and s are in subsection 2. Entries of the 4th row in the Matrix are then:

4th row $\left[0, 0, \xi_2^{\alpha_{1,2}}, \xi_2^{-(\alpha_{1,2}+1)}, -e^{\gamma_3 b2}, -e^{-\gamma_3 b2}, 0, 0.. \right]$ $B: \left[k_3 - \omega_2 \xi_2 - \eta_2 \right]$

From $\frac{\partial U_2^2}{\partial y} = \frac{\partial U_3^2}{\partial y}$, $\frac{\partial U_2^2}{\partial y} = \frac{\partial \xi}{\partial y} \frac{\partial U_2^2}{\partial \xi}$, $\xi_2 = H - h$, $\frac{\partial \xi}{\partial y} = -\frac{1}{s_2}$ and $s_2 = 1.0$, arranging unknown terms, A_3, A_4, A_5 and A_6

$$A_3 \alpha_{1,2} \frac{-1}{s_2} \xi_2^{\alpha_{1,2}-1} - A_4 \left(\alpha_{1,2} + 1 \right) \frac{-1}{s_2} \xi_2^{-(\alpha_{1,2}+1)-1} + \omega_2 \frac{-1}{s_2} = A_5 \gamma_3 e^{\gamma_3 b2} - A_6 \gamma_3 e^{-\gamma_3 b2}$$

$$A_3 \alpha_{1,2} \frac{-1}{s_2} \xi_2^{\alpha_{1,2}-1} - A_4 \left(\alpha_{1,2} + 1 \right) \frac{-1}{s_2} \xi_2^{-(\alpha_{1,2}+1)-1} - A_5 \gamma_3 e^{\gamma_3 b2} + A_6 \gamma_3 e^{-\gamma_3 b2} = -\omega_2 \frac{-1}{s_2}$$

where α_1, ω, and s are in subsection 2 and γ in subsection 3. Entries of the 5th row in the Matrix are,

5th row $\left[0, 0, \alpha_{1,2} \frac{-1}{s_2} \xi_2^{\alpha_{1,2}-1}, -(\alpha_{1,2}+1) \frac{-1}{s_2} \xi_2^{-(\alpha_{1,2}+1)-1}, -\gamma_3 e^{\gamma_3 b2}, \gamma_3 e^{-\gamma_3 b2}, 0, .0 \right]$

$B: \left[-\omega_2 \frac{-1}{s_2} \right]$

4   BC at $y = b3$, the end of floodplain, the solution in subsection 4 are $U_3 = U_4$ and $\frac{\partial U_3^2}{\partial y} = \frac{\partial U_4^2}{\partial y}$ and $\xi_3 = H - h$

The analytical solution in subsection 4 should be: $U_4^2 = A_7 \xi^{\alpha_1} + \omega_4 \xi + \eta_4$ since $A_8 \xi^{-(\alpha_1+1)} \to \infty$ as $\xi \to 0$ or $U = 0$ at the water edge ($\xi = 0$) in subsection 4, thus A_8 should be set to zero.

From $U_3^2 = U_4^2$, arranging unknown terms, A_5, A_6 and A_7

$$A_5 e^{\gamma_3 b3} + A_6 e^{-\gamma_3 b3} - A_7 \xi_3^{\alpha_{1,4}} = +\omega_4 \xi_3 + \eta_4 - k_3$$

where γ and k in subsection 3, and α_1, ω, η and s are in subsection 4. Entries of the 6th row in the Matrix are,

6th row $\left[0, 0, 0, 0, e^{\gamma_3 b3}, e^{-\gamma_3 b3}, -\xi_3^{\alpha_{1,4}} \right]$ $B: \left[\omega_4 \xi_3 + \eta_4 - k_3 \right]$

From $\frac{\partial U_3^2}{\partial y} = \frac{\partial U_4^2}{\partial y}$, $\frac{\partial U_4^2}{\partial y} = \frac{\partial \xi}{\partial y}\frac{\partial U_4^2}{\partial \xi}$, $\xi_3 = H - b$, $\frac{\partial \xi}{\partial y} = -\frac{1}{s_4}$ and $s_4 = 1.0$, arranging unknown terms, A_5, A_6 and A_7

$$A_5\gamma_3 e^{\gamma_3 b3} + A_6\gamma_3 e^{-\gamma_3 b3} = A_7\frac{-1}{s_4}\alpha_{1,4}\xi_3^{\alpha_{1,4}-1} + \omega_4\frac{-1}{s_4}$$

$$A_5\gamma_3 e^{\gamma_3 b3} - A_6\gamma_3 e^{-\gamma_3 b3} - A_7\frac{-1}{s_4}\alpha_{1,4}\xi_3^{\alpha_{1,4}-1} = +\omega_4\frac{-1}{s_4}$$

where γ in subsection 3, and α_1, ω, η and s are in subsection 4. Entries of the 7th row in the Matrix are,

$$\text{7th row}\left[0,0,0,0,\gamma_3 e^{\gamma_3 b3},-\gamma_3 e^{-\gamma_3 b3},-\frac{-1}{s_4}\alpha_{1,4}\xi_3^{\alpha_{1,4}-1}\right] \quad B:\left[\omega_4\frac{-1}{s_4}\right]$$

Therefore the matrix, A is 7×7 with 7 unknowns of x with B known. To solve Ax = B, it is required to form the matrix, A as follows (see Table 7.14):

1st row $\left[\gamma_1 e^0, -\gamma_1 e^0, 0,0,0,0,0\right]$ $B:[0]$

2nd row $\left[e^{\gamma_1 b1}, e^{-\gamma_1 b1}, -H^{\alpha_{1,2}}, -H^{-(\alpha_{1,2}+1)}, 0,0,....\right]$ $B:\left[-k_1 + \omega_2 H + \eta_2\right]$

3rd row $\left[\gamma_1 e^{\gamma_1 b1}, -\gamma_1 e^{-\gamma_1 b1}, -\alpha_{1,2}\left(\frac{-1}{s_2}\right)H^{\alpha_{1,2}-1}, (\alpha_{1,2}+1)\left(\frac{-1}{s_2}\right)H^{-(\alpha_{1,2}+1)-1}, 0,0.....0\right]$
$B:\left[\omega_2\frac{-1}{s_2}\right]$

4th row $\left[0,0,(H-b)^{\alpha_{1,2}}, (H-b)^{-(\alpha_{1,2}+1)}, -e^{\gamma_3 b2}, -e^{-\gamma_3 b2}, 0,0..\right]$
$B:\left[k_3 - \omega_2(H-b) - \eta_2\right]$

5th row $\left[0,0,\alpha_{1,2}\frac{-1}{s_2}(H-b)^{\alpha_{1,2}-1}, -(\alpha_{1,2}+1)\frac{-1}{s_2}(H-b)^{-(\alpha_{1,2}+1)-1},\right.$
$\left.-\gamma_3 e^{\gamma_3 b2}, \gamma_3 e^{-\gamma_3 b2}, 0,.0\right]$ $B:\left[-\omega_2\frac{-1}{s_2}\right]$

6th row $\left[0,0,0,0,e^{\gamma_3 b3}, e^{-\gamma_3 b3}, -(H-b)^{\alpha_{1,4}}\right]$ $B:\left[\omega_4(H-b) + \eta_4 - k_3\right]$

7th row $\left[0,0,0,0,\gamma_3 e^{\gamma_3 b3}, -\gamma_3 e^{-\gamma_3 b3}, -\frac{-1}{s_4}\alpha_{1,4}(H-b)^{\alpha_{1,4}-1}\right]$ $B:\left[\omega_4\frac{-1}{s_4}\right]$

All compound channel parameters are set out in Excel to look like those below in Table 7.13.

Table 7.13 Input values for FCF 020301.

		Section 1	Section 2	Section 3	Section 4
Width (m)		0–0.75	0.75–0.9	0.9–3.03	3.05–
Water	H	0.1778	0.1778	0.0278	0.0278
Gravity	g	9.81	9.81	9.81	9.81
Slope	So	0.001027	0.001027	0.001027	0.001027
Slope	S		1		1
Cd					
d	Rod dia	0	0	0	0
Rod No	No (N)	0	0	0	0
Ap	Ndbh/(b*1)	0	0	0	0
Drag	1/2CdAp	0	0	0	0
ks		0.00016	0.00016	0.00016	0.00016
friction	f	0.01591	0.022777	0.029645	0.029645
dim eddy	lamda	0.6	0.2	3.234897	3.264897
Γ/pgHSo	Γ''	0	0	–0.2	0
	E1	0.001791	0.010075	0.00028	0.010075
	E2	0.001989	0.004027	0.003706	0.005241
	E3	0.000423	0.005336	7.61E-05	0.098461
coeff	γ_1	2.168459		6.978403	
coeff	γ_2	–2.16846		–6.9784	
coeff	α_1		0.502303		0.050659
coeff	α_2		–1.5023		–1.05066
	k	0.900734		0.090699	
const Mat	ω		–1.51609		–0.05256
	η		0		0

Table 7.14 Matrix A values (FCF Experiment 020301).

Matrix A is:

y	H				A matrix				
b0 0	0.1778	2.168459	–2.16846	0	0	0	0	0
b1 0.75	0.1778	5.085252	0.196647	–0.41999	–13.3915	0	0	0
b1 0.75	0.1778	11.02716	–0.42642	1.186513	–113.15	0	0	0
b2 0.9	0.0278	0	0	0.165363	217.5283	–534.089	–0.00187	0
b2 0.9	0.0278	0	0	–2.98786	11755.16	–3727.09	0.013066	0
b3 3.15	0.0278	0	0	0	0	3.52E+09	2.84E-10	–0.83402
b3 3.15	0.0278	0	0	0	0	2.46E+10	–2E-09	1.519803

The unknown coefficients, A_1, A_2...A_7 can be obtained using the Excel function: highlight like green 7 column of A_1, A_2...A_7 and then write [=mmult(minverse(A),B)], (where A is highlighted and B is highlighted) in A1 column and press the keys [ctr+shift+enter] together.

All coefficients are calculated now, then substituting them into the solutions Subsection 1, for 0 < y < 0.75 m

Table 7.15 Matrix B values (FCF Experiment 020301).

Matrix B is:
B
0
−1.1703
1.516095
0.132846
−1.51609
−0.09216
0.052561

Table 7.16 Coefficients for FCF Experiment 020301.

A	
A1	−0.06884
A2	−0.06884
A3	1.913766
A4	0.00022
A5	−3.7E-12
A6	123.6337
A7	0.094781

$$U_1^2 = A_1 e^{\gamma_1 y} + A_2 e^{-\gamma_1 y} + k_1$$

Subsection 2, for $0.75 < y < 0.9$ m, with $\xi = H - \dfrac{(y - b1)}{s}$

$$U_2{}^2 = A_3 \xi^{\alpha_{1,2}} + A_4 \xi^{-(\alpha_{1,2}+1)} + \omega_2 \xi + \eta_2$$

Subsection 3, $0.9 < y < 3.15$ m,

$$U_3^2 = A_5 e^{\gamma_3 y} + A_6 e^{-\gamma_3 y} + k_3$$

Finally subsection 4, for $3.15 < y < Ym$ at $H = 0$, with $\xi = (H - h) - \dfrac{(y - b3)}{s}$

$$U_4^2 = A_7 \xi^{\alpha_{1,4}} + \omega_4 \xi + \eta_4$$

Taking square root gives the velocity, U. Calculated values of U are shown in Table 7.17.

A plot of U_d v y is shown in Figure 7.21.

Table 7.17 Calculated values of U_d for FCF Experiment 020301.

y	H	U^2	U	dQ	To
0	0.1778	0.763061	0.873534		1.517519
0.1	0.1778	0.759812	0.871672	0.015515	1.511056
0.2	0.1778	0.74991	0.865973	0.015448	1.491364
0.3	0.1778	0.732888	0.856089	0.015309	1.457512
0.4	0.1778	0.707942	0.841393	0.015091	1.407903
0.5	0.1778	0.673896	0.820912	0.014778	1.340193
0.6	0.1778	0.629141	0.793184	0.014349	1.251189
0.65	0.1778	0.60211	0.775958	0.006975	1.197432
0.7	0.1778	0.571565	0.756019	0.00681	1.136686
0.75	0.1778	0.537147	0.732903	0.006618	1.068238
0.8	0.1278	0.492003	0.70143	0.005479	1.400826
0.85	0.0778	0.422918	0.650321	0.003474	1.204126
0.9	0.0278	0.322184	0.567612	0.001608	0.917318
0.95	0.0278	0.254	0.503984	0.000745	0.941234
1	0.0278	0.205899	0.453761	0.000666	0.762991
1.05	0.0278	0.171967	0.414689	0.000604	0.637249
1.1	0.0278	0.148029	0.384746	0.000556	0.548545
1.15	0.0278	0.131143	0.362136	0.000519	0.485968
1.2	0.0278	0.11923	0.345297	0.000492	0.441824
1.3	0.0278	0.104897	0.323879	0.00093	0.388713
1.5	0.0278	0.094215	0.306945	0.001754	0.349128
1.7	0.0278	0.091569	0.302604	0.001695	0.339323
1.9	0.0278	0.090912	0.301517	0.001679	0.336889
2.1	0.0278	0.090744	0.301237	0.001676	0.336264
2.3	0.0278	0.090677	0.301127	0.001675	0.336018
2.5	0.0278	0.090562	0.300934	0.001674	0.335589
2.7	0.0278	0.090132	0.30022	0.001671	0.333999
2.75	0.0278	0.089895	0.299825	0.000417	0.33312
2.8	0.0278	0.089559	0.299264	0.000416	0.331875
2.85	0.0278	0.089083	0.298468	0.000415	0.330111
2.9	0.0278	0.088408	0.297336	0.000414	0.32761
2.95	0.0278	0.087452	0.295723	0.000412	0.324066
3	0.0278	0.086096	0.293422	0.000409	0.319042
3.05	0.0278	0.084174	0.290128	0.000406	0.311921
3.07	0.0278	0.083197	0.288439	0.000161	0.3083
3.15	0.0278	0.077589	0.278547	0.00063	0.287516
3.16	0.0178	0.076349	0.276313	4.94E-05	0.282922
3.17	0.0078	0.073711	0.271497	2.14E-05	0.273146

7.2.3 Overbank flow in the River Main

The first river case considered is now the River Main, which has previously been studied in Section 4.2.3. First, input the cross section details, as shown in Figure 7.22, using 7 panels. Setting $y = 0$ at the centre of the main channel for convenience, but it does not matter where $y = 0$ is, as long as it has a definite reference, such as the right hand edge of the water surface on the left hand floodplain bank. In this example, y = 0

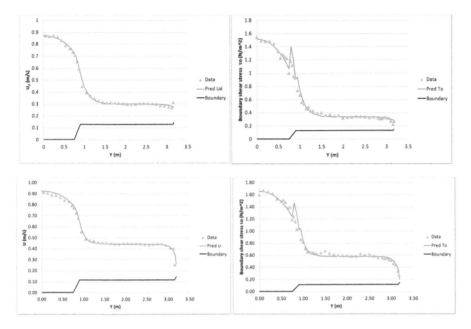

Figure 7.21 Predicted velocity distribution and bed shear stress for FCF 020301 and 020501 (Dr = 0.15 & 0.25 respectively).

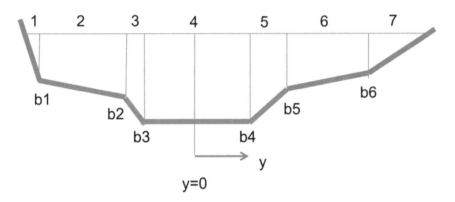

Figure 7.22 Schematic cross section of River Main.

is set at the centre of the main channel. There are then 6 panel interfaces at b1, b2....
b6 for the boundary conditions, giving a 12×12 matrix.

Start building a matrix using the boundary conditions at b1 to b6, in a similar
manner to the previous example for the FCF. Define water depth in subsections 1,2..,7
as follows:

Subsection 1: $\xi = H - ((b2 - b3) / s_3 + (b1 - b2) / s_2) - (y - b1) / s_1$

Subsection 2: $\xi = H - (b3 - b2) / s_3 - (y - b2) / s_2$

Subsection 3: $\xi = H - (y - b2)/s_2$

Subsection 4: $\xi = H$

Subsection 5: $\xi = H - (y - b4)/s_5$

Subsection 6: $\xi = H - ((b5 - b4)/s_5) - (y - b5)/s_6$

Subsection 7: $\xi = H - ((b5 - b4)/s_5 + (b6 - b5)/s_6) - (y - b6)/s_7$

1   BC at y = b1: $U_1 = U_2$ and $\frac{\partial U_1^2}{\partial y} = \frac{\partial U_2^2}{\partial y}$ with the water depth:

$$\xi_1 = H - ((b2 - b3)/s_3 + (b1 - b2)/s_2)$$

The term $A\xi^{-(\alpha_1 + 1)} = 0$ at the water surface since $\xi^{-(\alpha_1 + 1)} \to \infty$ as $\xi \to 0$ or $U = 0$ at $\xi = 0$.

Thus the analytical solution in subsections 1 and 7 is in the form: $U^2 = A\xi^{\alpha_1} + \omega\xi + \eta$

For subsection 1, the solution is

$$U_1^2 = A_1\xi_1^{\alpha_{1,1}} + \omega_1\xi_1 + \eta_1$$

For $U_1 = U_2$, $(U_1^2 = U_2^2)$, arranging the unknown terms, A_1, A_2 and A_3 gives

$$A_1\xi_1^{\alpha_{1,1}} + \omega_1\xi_1 + \eta_1 = A_2\xi_1^{\alpha_{1,2}} + A_3\xi_1^{-(\alpha_{1,2}+1)} + \omega_2\xi_1 + \eta_2$$

$$A_1\xi_1^{\alpha_{1,1}} - \left(A_2\xi_1^{\alpha_{1,2}} + A_3\xi_1^{-(\alpha_{1,2}+1)}\right) = \omega_2\xi_1 + \eta_2 - (\omega_1\xi_1 + \eta_1)$$

Entries of the 1st row in the Matrix are then:

1st row $\left[\xi_1^{\alpha_{1,1}}, -\xi_1^{\alpha_{1,2}} - \xi_1^{-(\alpha_{1,2}+1)}, 0, 0, ...\right]$ B: $\left[\omega_2\xi_1 + \eta_2 - (\omega_1\xi_1 + \eta_1)\right]$

For $\frac{\partial U_1^2}{\partial y} = \frac{\partial U_2^2}{\partial y}, \frac{\partial U^2}{\partial y} = \frac{\partial \xi}{\partial y}\frac{\partial U^2}{\partial \xi}, \frac{\partial \xi}{\partial y} = -\frac{1}{s}$, arranging the unknown terms, A_1, A_2 and A_3 gives

$$\left[A_1\alpha_1\xi_1^{\alpha_{1,1}-1} + \omega_1\right]\left(\frac{-1}{s_1}\right) = \left[A_2\alpha_{1,2}\xi_1^{\alpha_{1,2}-1} - A_3(\alpha_{1,2}+1)\xi_1^{-(\alpha_{1,2}+1)-1} + \omega_2\right]\left(\frac{-1}{s_2}\right)$$

$$A_1\alpha_1\xi_1^{\alpha_{1,1}-1}\left(\frac{-1}{s_1}\right) - \left(\frac{-1}{s_2}\right)\left(A_2\alpha_{1,2}\xi_1^{\alpha_{1,2}-1} - A_3(\alpha_{1,2}+1)\xi_1^{-(\alpha_{1,2}+1)-1}\right) =$$

$$+ \omega_2\left(\frac{-1}{s_2}\right) - \omega_1\left(\frac{-1}{s_1}\right)$$

Entries of the 2nd row in the Matrix are then:

$$
\text{2nd row} \quad \left[\alpha_{1,1}\xi_1^{\alpha_{1,1}-1}\left(\frac{-1}{s_1}\right),\, -\alpha_{1,2}\xi_1^{\alpha_{1,2}-1}\left(\frac{-1}{s_2}\right),\, \left(\alpha_{1,2}+1\right)\xi^{-\left(\alpha_{1,2}+1\right)-1}\left(\frac{-1}{s_2}\right),\, 0, 0, \dots \right]
$$

$$
B: \left[\omega_2 \frac{-1}{s_2} - \left(\omega_1 \frac{-1}{s_1}\right) \right]
$$

2   BC at y = b2: $U_2 = U_3$ and $\frac{\partial U_2^2}{\partial y} = \frac{\partial U_3^2}{\partial y}$ and $\xi_2 = H - (b2-b3)/s_3$,
For $U_2^2 = U_3^2$, arranging unknown terms, A_2, A_3, A_4 and A_5

$$
A_2\xi_2^{\alpha_{1,2}} + A_3\xi_2^{-\left(\alpha_{1,2}+1\right)} + \omega_2\xi_2 + \eta_{12} = A_4\xi_2^{\alpha_{1,3}} + A_5\xi_2^{-\left(\alpha_{1,3}+1\right)} + \omega_3\xi_2 + \eta_3
$$

$$
A_2\xi_2^{\alpha_{1,2}} + A_3\xi_2^{-\left(\alpha_{1,2}+1\right)} - \left(A_4\xi_2^{\alpha_{1,3}} + A_5\xi_2^{-\left(\alpha_{1,3}+1\right)}\right) = +\omega_3\xi_2 + \eta_3 - \left(\omega_2\xi_2 + \eta_2\right)
$$

Entries of the 3rd row in the Matrix are,

$$
\text{3rd row} \quad \left[\xi_2^{\alpha_{1,2}}, \xi_2^{-\left(\alpha_{1,2}+1\right)}, -\xi_2^{\alpha_{1,3}}, -\xi_2^{-\left(\alpha_{1,3}+1\right)}, 0, 0, \dots \right] \quad B: \left[\omega_3\xi_2 + \eta_3 - \left(\omega_2\xi_2 + \eta_2\right) \right]
$$

For $\frac{\partial U_2^2}{\partial y} = \frac{\partial U_3^2}{\partial y}$, $\frac{\partial U^2}{\partial y} = \frac{\partial \xi}{\partial y}\frac{\partial U^2}{\partial \xi}, \frac{\partial \xi}{\partial y} = -\frac{1}{s}$, arranging unknown terms, A_2, A_3, A_4 and A_5

$$
\left[A_2\alpha_{1,2}\xi_2^{\alpha_{1,2}-1} - A_3\left(\alpha_{1,2}+1\right)\xi_2^{-\left(\alpha_{1,2}+1\right)-1} + \omega_2 \right]\left(\frac{-1}{s_2}\right)
$$

$$
= \left[A_4\alpha_{1,3}\xi_2^{\alpha_{1,3}-1} - A_5\left(\alpha_{1,3}+1\right)\xi_2^{-\left(\alpha_{1,3}+1\right)-1} + \omega_3 \right]\left(\frac{-1}{s_3}\right)
$$

$$
\left[A_2\alpha_{1,2}\xi_2^{\alpha_{1,2}-1} - A_3\left(\alpha_{1,2}+1\right)\xi_2^{-\left(\alpha_{1,2}+1\right)-1} \right]\left(\frac{-1}{s_2}\right)
$$

$$
- \left[A_4\alpha_{1,3}\xi_2^{\alpha_{1,3}-1} - A_5\left(\alpha_{1,3}+1\right)\xi_2^{-\left(\alpha_{1,3}+1\right)-1} \right]\left(\frac{-1}{s_3}\right)
$$

$$
= \omega_3\left(\frac{-1}{s_3}\right) - \omega_2\left(\frac{-1}{s_2}\right)
$$

Entries of the 4th row in the Matrix are then:

$$
\text{4th row} \quad \left[\alpha_{1,2}\xi_2^{\alpha_{1,2}-1}\left(\frac{-1}{s_2}\right), -\left(\alpha_{1,2}+1\right)\xi_2^{-\left(\alpha_{1,2}+1\right)-1}\left(\frac{-1}{s_2}\right), \right.
$$

$$
\left. -\alpha_{1,3}\xi_2^{\alpha_{1,3}-1}\left(\frac{-1}{s_3}\right), \left(\alpha_{1,3}+1\right)\xi_2^{-\left(\alpha_{1,3}+1\right)-1}\left(\frac{-1}{s_3}\right), 0, 0, \dots \right]
$$

$$
B: \left[\omega_3\left(\frac{-1}{s_3}\right) - \omega_2\left(\frac{-1}{s_2}\right) \right]
$$

3   BC at y = b3, $U_3 = U_4$ and $\frac{\partial U_3^2}{\partial y} = \frac{\partial U_4^2}{\partial y}$ and $\xi_3 = H$,

For $U_3^2 = U_4^2$ arranging unknown terms, A_4, A_5, A_6 and A_7

$$A_4 \xi_3^{\alpha_{1,3}} + A_5 \xi_3^{\alpha_{1,3}-(+1)} + \omega_3 \xi_3 + \eta_3 = A_6 e^{\gamma_4 b3} + A_7 e^{-\gamma_4 b3} + k_4$$

$$A_4 \xi_3^{\alpha_{1,3}} + A_5 \xi_3^{-(\alpha_{1,3}+1)} - \left(A_6 e^{\gamma_4 b3} + A_7 e^{-\gamma_4 b3} \right) = k_4 - \left(\omega_3 \xi_3 + \eta_3 \right)$$

Entries of the 4th row in the Matrix are then:

5th row $\quad \left[\xi_3^{\alpha_{1,3}}, \xi_3^{-(\alpha_{1,3}+1)}, -e^{\gamma_4 b3}, -e^{-\gamma_4 b3}, 0, 0, \ldots \right] \quad B: \left[-\left(\omega_3 \xi_3 + \eta_3 \right) + k_4 \right]$

For $\frac{\partial U_3^2}{\partial y} = \frac{\partial U_4^2}{\partial y}, \frac{\partial U^2}{\partial y} = \frac{\partial \xi}{\partial y}\frac{\partial U^2}{\partial \xi}, \frac{\partial \xi}{\partial y} = -\frac{1}{s}$, arranging the unknown terms, A_4, A_5, A_6 and A_7

$$\left(A_4 \alpha_{1,3} \xi_3^{\alpha_{1,3}} - A_5 \left(\alpha_{1,3}+1 \right) \xi_3^{-(\alpha_{1,3}+1)} + \omega_3 \right)\left(\frac{-1}{s_3} \right) = A_6 \gamma_4 e^{\gamma_4 b3} - A_7 \gamma_4 e^{-\gamma_4 b3}$$

$$\left(A_4 \alpha_{1,3} \xi_3^{\alpha_{1,3}} - A_5 \left(\alpha_{1,3}+1 \right) \xi_3^{-(\alpha_{1,3}+1)} \right)\left(\frac{-1}{s_3} \right) - \left(A_6 \gamma_4 e^{\gamma_4 b3} - A_7 \gamma_4 e^{-\gamma_4 b3} \right) = -\omega_3 \left(\frac{-1}{s_3} \right)$$

Entries of the 6th row in the Matrix are then:

6th row $\quad \left[\alpha_{1,3}\xi_3^{\alpha_{1,3}}\left(\frac{-1}{s_3} \right), -\left(\alpha_{1,3}+1 \right)\xi_3^{-(\alpha_{1,3}+1)}\left(\frac{-1}{s_3} \right), -\gamma_4 e^{\gamma_4 b3}, \gamma_4 e^{-\gamma_4 b3}, 0, 0, \ldots \right]$

$\quad B: \left[-\omega_3 \left(\frac{-1}{s_3} \right) \right]$

4   BC at y = b4: $U_4 = U_5$ and $\frac{\partial U_4^2}{\partial y} = \frac{\partial U_5^2}{\partial y}$ and $\xi_4 = H$,

For $U_4^2 = U_5^2$, arranging unknown terms, A_4, A_5, A_6 and A_7

$$A_6 e^{\gamma_4 b4} + A_7 e^{-\gamma_4 b4} + k_4 = A_8 \xi_4^{\alpha_{1,5}} + A_9 \xi_4^{-(\alpha_{1,5}+1)} + \omega_5 \xi_4 + \eta_5$$

$$A_6 e^{\gamma_4 b4} + A_7 e^{-\gamma_4 b4} - A_8 \xi_4^{\alpha_{1,5}} - A_9 \xi_4^{-(\alpha_{1,5}+1)} = -k_4 + \omega_5 \xi_4 + \eta_5$$

Entries of the 7th row in the Matrix are then:

7th row $\quad \left[e^{\gamma_4 b4}, e^{-\gamma_4 b4}, -\xi_4^{\alpha_{1,5}}, -\xi_4^{-(\alpha_{1,5}+1)}, 0, 0, \ldots \right] \quad B: \left[-k_4 + \omega_5 \xi_4 + \eta_5 \right]$

For $\frac{\partial U_1^2}{\partial y} = \frac{\partial U_2^2}{\partial y}, \frac{\partial U^2}{\partial y} = \frac{\partial \xi}{\partial y}\frac{\partial U^2}{\partial \xi}, \frac{\partial \xi}{\partial y} = -\frac{1}{s}, \xi_4 = H$, arranging unknown terms, A_6, A_7, A_8 and A_9

$$A_6 \gamma_4 e^{\gamma_4 b4} - A_7 \gamma_4 e^{-\gamma_4 b4} = A_8 \alpha_{1,5} \left(\frac{-1}{s_5}\right) \xi_4^{\alpha_{1,5}-1}$$

$$- A_9 \left(\alpha_{1,5} + 1\right)\left(\frac{-1}{s_5}\right) \xi_4^{-(\alpha_{1,5}+1)-1} + \omega_5 \left(\frac{-1}{s_5}\right)$$

$$A_6 \gamma_4 e^{\gamma_4 b4} - A_7 \gamma_4 e^{-\gamma_4 b4} - A_8 \alpha_{1,5} \left(\frac{-1}{s_5}\right) \xi_4^{\alpha_{1,5}-1}$$

$$+ A_9 \left(\alpha_{1,5} + 1\right)\left(\frac{-1}{s_5}\right) \xi_4^{-(\alpha_{1,5}+1)-1} = +\omega_5 \left(\frac{-1}{s_5}\right)$$

Entries of the 8th row in the Matrix are then:

8th row $\left[\gamma_4 e^{\lambda_4 b4}, -\gamma_4 e^{-\lambda_4 b4}, -\alpha_{1,5}\left(\frac{-1}{s_5}\right)\xi_4^{\alpha_{1,5}-1}, (\alpha_{1,5}+1)\left(\frac{-1}{s_5}\right)\xi_4^{-(\alpha_{1,5}+1)-1}, 0,0..... \right]$

$B: \left[\omega_5 \left(\frac{-1}{s_5}\right) \right]$

5   BC at y = b5: $U_5 = U_6$ and $\frac{\partial U_5^2}{\partial y} = \frac{\partial U_6^2}{\partial y}$ and $\xi_5 = H - (b5 - b4)/s_5$,
For $U_5^2 = U_6^2$, arranging unknown terms, A_8, A_9, A_{10} and A_{11}

$$A_8 \xi_5^{\alpha_{1,5}} + A_9 \xi_5^{-(\alpha_{1,5}+1)} + \omega_5 \xi_5 + \eta_5 = A_{10}\xi_5^{\alpha_{1,6}} + A_{11}\xi_5^{-(\alpha_{1,6}+1)} + \omega_6 \xi_5 + \eta_6$$

$$A_8 \xi_5^{\alpha_{1,5}} + A_9 \xi_5^{-(\alpha_{1,5}+1)} - \left(A_{10}\xi_5^{\alpha_{1,6}} + A_{11}\xi_5^{-(\alpha_{1,6}+1)}\right) = +\omega_6 \xi_5 + \eta_6 - \left(\omega_5 \xi_5 + \eta_5\right)$$

Entries of the 9th row in the Matrix are then:

9th row $\left[\xi_5^{\alpha_{1,5}}, \xi_5^{-(\alpha_{1,5}+1)}, -\xi_5^{\alpha_{1,6}} - \xi_5^{-(\alpha_{1,6}+1)}, 0,0,.... \right]$ $B: \left[\omega_6 \xi_5 + \eta_6 - \left(\omega_5 \xi_5 + \eta_5\right) \right]$

For $\frac{\partial U_5^2}{\partial y} = \frac{\partial U_6^2}{\partial y}$, $\frac{\partial U^2}{\partial y} = \frac{\partial \xi}{\partial y}\frac{\partial U^2}{\partial \xi}$, $\frac{\partial \xi}{\partial y} = -\frac{1}{s}$, arranging unknown terms, A_8, A_9, A_{10} and A_{11}

$$\left[A_8 \alpha_{1,5}\xi_5^{\alpha_{1,5}} - A_9 \left(\alpha_{1,5}+1\right)\xi_5^{-(\alpha_{1,5}+1)} + \omega_5 \right]\left(\frac{-1}{s_5}\right)$$

$$= \left[A_{10}\alpha_{1,6}\xi_5^{\alpha_{1,6}} - A_{11}\left(\alpha_{1,6}+1\right)\xi_5^{-(\alpha_{1,6}+1)} + \omega_6 \right]\left(\frac{-1}{s_6}\right)$$

$$\left[A_8 \alpha_{1,5}\xi_5^{\alpha_{1,5}} - A_9 \left(\alpha_{1,5}+1\right)\xi_5^{-(\alpha_{1,5}+1)} \right]\left(\frac{-1}{s_5}\right)$$

$$- \left[A_{10}\alpha_{1,6}\xi_5^{\alpha_{1,6}} - A_{11}\left(\alpha_{1,6}+1\right)\xi_5^{-(\alpha_{1,6}+1)} \right]\left(\frac{-1}{s_6}\right) = \omega_6 \left(\frac{-1}{s_6}\right) - \omega_5 \left(\frac{-1}{s_5}\right)$$

Entries of the 10th row in the Matrix are then:

$$10\text{th row} \quad \left[\alpha_{1,5}\xi_5^{\xi\,\alpha_{1,5}}\left(\frac{-1}{s_5}\right), -\left(\alpha_{1,5}+1\right)\xi_5^{-(\alpha_{1,5}+1)}\left(\frac{-1}{s_5}\right), \right.$$

$$\left. -\alpha_{1,6}\xi_5^{\xi\,\alpha_{1,6}}\left(\frac{-1}{s_6}\right), \left(\alpha_{1,6}+1\right)\xi_5^{-(\alpha_{1,6}+1)}\left(\frac{-1}{s_6}\right), 0, 0, ... \right]$$

$$B: \left[\omega_6\left(\frac{-1}{s_6}\right) - \omega_5\left(\frac{-1}{s_5}\right) \right]$$

6   BC at y = b6: $U_6 = U_7$ and $\frac{\partial U_6^2}{\partial y} = \frac{\partial U_7^2}{\partial y}$ and $\xi_6 = H - \left((b5-b4)/s_5 + (b6-b5)/s_6\right)$

For $U_6^2 = U_7^2$, arranging unknown terms, A_{10}, A_{11} and A_{12}

$$A_{10}\xi_6^{\xi\,\alpha_{1,6}} + A_{11}\xi_6^{-(\alpha_{1,6}+1)} + \omega_6\xi_6 + \eta_6 = A_{12}\xi_6^{\xi\,\alpha_{1,7}} + \omega_7\xi_6 + \eta_7$$

$$A_{10}\xi_6^{\xi\,\alpha_{1,6}} + A_{11}\xi_6^{-(\alpha_{1,6}+1)} - A_{12}\xi_6^{\xi\,\alpha_{1,7}} = \omega_7\xi_6 + \eta_7 - \left(\omega_6\xi_6 + \eta_6\right)$$

Entries of the 11th row in the Matrix are then:

$$11\text{th row} \quad \left[\xi_6^{\xi\,\alpha_{1,6}}, -\xi_6^{\xi\,\alpha_{1,6}}, -\xi_6^{-(\alpha_{1,7}+1)}, 0, 0, ... \right] \quad B:\left[\omega_6\xi_6 + \eta_6 - (\omega_7\xi_6 + \eta_7) \right]$$

For $\frac{\partial U_6^2}{\partial y} = \frac{\partial U_7^2}{\partial y}, \frac{\partial U_6^2}{\partial y} = \frac{\partial\xi}{\partial y}\frac{\partial U^2}{\partial\xi}, \frac{\partial\xi}{\partial y} = -\frac{1}{s}$, arranging unknown terms, A_{10}, A_{11} and A_{12}

$$\left[A_{10}\alpha_{1,6}\xi_6^{\xi\,\alpha_{1,6}-1} - A_{11}\left(\alpha_{1,6}+1\right)\xi_6^{-(\alpha_{1,6}+1)-1} + \omega_6 \right]\left(\frac{-1}{s_6}\right) = \left[A_{12}\alpha_{1,7}\xi_6^{\xi\,\alpha_{1,7}-1} + \omega_7 \right]\left(\frac{-1}{s_7}\right)$$

$$\left[A_{10}\alpha_{1,6}\xi_6^{\xi\,\alpha_{1,6}-1} - A_{11}\left(\alpha_{1,6}+1\right)\xi_6^{-(\alpha_{1,6}+1)-1} \right]\left(\frac{-1}{s_6}\right) - A_{12}\alpha_{1}\xi_6^{\xi\,\alpha_{1,7}-1}\left(\frac{-1}{s_7}\right) =$$

$$\omega_7\left(\frac{-1}{s_7}\right) - \omega_6\left(\frac{-1}{s_7}\right)$$

Entries of the 12th row in the Matrix are then:

$$12\text{th row} \quad \left[\alpha_{1,6}\xi_6^{\xi\,\alpha_{1,6}-1}\left(\frac{-1}{s_6}\right), \left(\alpha_{1,6}+1\right)\xi_6^{-(\alpha_{1,6}+1)-1}\left(\frac{-1}{s_6}\right), -\alpha_{1,7}\xi_6^{\xi\,\alpha_{1,7}-1}\left(\frac{-1}{s_7}\right), 0, 0, ... \right]$$

$$B:\left[\omega_7\frac{-1}{s_7} - \omega_6\frac{-1}{s_6} \right]$$

All 12 rows for the matrix are made, the matrix (see Table 7.19):

$$1\text{st row} \quad \left[\xi_1^{\xi\,\alpha_{1,1}}, -\xi_1^{\xi\,\alpha_{1,2}}, -\xi_1^{-(\alpha_{1,2}+1)}, 0, 0, ...0 \right]$$

$$2\text{nd row} \quad \left[\alpha_{1,1}\xi_1^{\xi\,\alpha_{1,1}-1}\left(\frac{-1}{s_1}\right), -\alpha_{1,2}\xi_1^{\xi\,\alpha_{1,2}-1}\left(\frac{-1}{s_2}\right), \left(\alpha_{1,2}+1\right)\xi_1^{-(\alpha_{1,2}+1)-1}\left(\frac{-1}{s_2}\right), 0, 0, ... \right]$$

3rd row $\left[\xi_2^{\alpha_{1,2}}, \xi_2^{-(\alpha_{1,2}+1)}, -\xi_2^{\alpha_{1,3}} - \xi_2^{-(\alpha_{1,3}+1)}, 0, 0, ... \right]$

4th row $\left[0, ..\alpha_{1,2}\xi_2^{\alpha_{1,2}-1}\left(\dfrac{-1}{s_2}\right), -(\alpha_{1,2}+1)\xi_2^{-(\alpha_{1,2}+1)-1}\left(\dfrac{-1}{s_2}\right), \right.$

$\left. -\alpha_{1,3}\xi_2^{\alpha_{1,3}-1}\left(\dfrac{-1}{s_3}\right), (\alpha_{1,3}+1)\xi_2^{-(\alpha_{1,3}+1)-1}\left(\dfrac{-1}{s_3}\right), 0, 0, ... \right]$

5th row $\left[0, ..\xi_3^{\alpha_{1,3}}, \xi_3^{-(\alpha_{1,3}+1)}, -e^{\gamma_4 b3}, -e^{-\gamma_4 b3}, 0, 0, ... \right]$

6th row $\left[0, ..\alpha_{1,3}\xi_3^{\alpha_{1,3}}\left(\dfrac{-1}{s_3}\right), -(\alpha_{1,3}+1)\xi_3^{-(\alpha_{1,3}+1)}\left(\dfrac{-1}{s_3}\right), -\gamma_4 e^{\gamma_4 b3}, \gamma_4 e^{-\gamma_4 b3}, 0, 0, ... \right]$

7th row $\left[0, e^{\lambda_4 b4}, e^{-\lambda_4 b4}, -\xi_4^{\alpha_{1,5}}, -\xi_4^{-(\alpha_{1,5}+1)}, 0, 0, ... \right]$

8th row $\left[0, ...\gamma_4 e^{\lambda_4 b4}, -\gamma_4 e^{-\lambda_4 b4}, -\alpha_{1,5}\left(\dfrac{-1}{s_5}\right)\xi_4^{\alpha_{1,5}-1}, \right.$

$\left. (\alpha_{1,5}+1)\left(\dfrac{-1}{s_5}\right)\xi_4^{-(\alpha_{1,5}+1)-1}, 0, 0.. \right]$

9th row $\left[0, ..\xi_5^{\alpha_{1,5}}, \xi_5^{-(\alpha_{1,5}+1)}, -\xi_5^{\alpha_{1,6}}, -\xi_5^{-(\alpha_{1,6}+1)}, 0 \right]$

10th row $\left[0, ..\alpha_{1,5}\xi_5^{\alpha_{1,5}}\left(\dfrac{-1}{s_5}\right), -(\alpha_{1,5}+1)\xi_5^{-(\alpha_{1,5}+1)}\left(\dfrac{-1}{s_5}\right), \right.$

$\left. -\alpha_{1,6}\xi_5^{\alpha_{1,6}}\left(\dfrac{-1}{s_6}\right), (\alpha_{1,6}+1)\xi_5^{-(\alpha_{1,6}+1)}\left(\dfrac{-1}{s_6}\right), 0 \right]$

11th row $\left[0, ..\xi_6^{\alpha_{1,6}}, -\xi_6^{\alpha_{1,6}}, -\xi_6^{-(\alpha_{1,7}+1)} \right]$

12th row $\left[0, ...\alpha_{1,6}\xi_6^{\alpha_{1,6}-1}\left(\dfrac{-1}{s_6}\right), (\alpha_{1,6}+1)\xi_6^{-(\alpha_{1,6}+1)-1}\left(\dfrac{-1}{s_6}\right), -\alpha_{1,7}\xi_6^{\alpha_{1,7}-1}\left(\dfrac{-1}{s_7}\right) \right]$

$Ax = B$

B column is:

1st row B: $\left[\omega_2\xi_1 + \eta_2 - (\omega_1\xi_1 + \eta_1) \right]$

2nd row B: $\left[\omega_2\dfrac{-1}{s_2} - \left(\omega_1\dfrac{-1}{s_1}\right) \right]$

3rd row B: $\left[\omega_3\xi_2 + \eta_3 - (\omega_2\xi_2 + \eta_2) \right]$

4th row B: $\left[\omega_3\left(\dfrac{-1}{s_3}\right) - \omega_2\left(\dfrac{-1}{s_2}\right) \right]$

5th row B: $\left[-(\omega_3\xi_3 + \eta_3) + k_4 \right]$

6th row B: $\left[-\omega_3\left(\dfrac{-1}{s_3}\right)\right]$

7th row B: $\left[-k_4 + \omega_5\xi_4 + \eta_5\right]$

8th row B: $\left[\omega_5\left(\dfrac{-1}{s_5}\right)\right]$

9th row B: $\left[\omega_6\xi_5 + \eta_6 - \left(\omega_5\xi_5 + \eta_5\right)\right]$

10th row B: $\left[\omega_6\left(\dfrac{-1}{s_6}\right) - \omega_5\left(\dfrac{-1}{s_5}\right)\right]$

11th row B: $\left[\omega_6\xi_6 + \eta_6 - \left(\omega_7\xi_6 + \eta_6\right)\right]$

12th row B: $\left[\omega_7\dfrac{-1}{s_7} - \left(\omega_6\dfrac{-1}{s_6}\right)\right]$

The River Main parameters are shown in Table 7.18.

Table 7.18 River Main parameters.

		Centre				20.5		
		Slope	Slope	Slope	Flat	Slope	Slope	Slope
		Section 1	Section 2	Section 3	Section 4	Section 5	Section 6	Section 7
Water	H	0.405	0.865	1.785	1.785	1.785	0.815	0.355
Gravity	g	9.81	9.81	9.81	9.81	9.81	9.81	9.81
Slope	So	0.00297	0.00297	0.00297	0.00297	0.00297	0.00297	0.00297
Slope	s	−1.69492	−16.6667	−0.98039	0	1.030927835	19.60784314	1.449275
Cd								
d	Rod dia	0	0	0	0	0	0	0
Rod No	No (N)	0	0	0	0	0	0	0
Ap	Ndbh/(b°l)	0	0	0	0	0	0	0
Drag	1/2CdAp	0	0	0	0	0	0	0
ks								
friction	f	0.389707	0.266249	0.391277	0.079252994	0.415341246	0.273628434	0.407206
Manning coeff	n	0.054	0.054	0.074	0.035	0.076	0.054	0.054
dim eddy	lamda	0.07	0.07	0.07	0.15	0.07	0.07	0.07
G/pgHSo	Γ				0		0	
	E1	0.029136	0.029136	0.029136	0.052007225	0.0291357	0.0291357	0.029136
	E2	0.05656	0.033341	0.069864	0.009906624	0.072329698	0.034248007	0.061842
	E3	0.002689	2.3E−05	0.008053	0.023784857	0.007503589	1.68362E−05	0.003759
coeff	γ_1					0.645375642		
coeff	γ_2					−0.64537564		
coeff	α_1	4.113412	37.58833	2.48753		2.644733181	44.60475522	3.586506
coeff	α_2	−5.11341	−38.5883	−3.48753		−3.644733181	−45.6047552	−4.58651
	k				5.249742332			
const Mat	ω	0.569258	0.875076	0.541985		0.508276677	0.851564125	0.536344
	η	0	0	0	0	0	0	0

Table 7.19 Matrix A for the River Main.

Creating matrix A

Matrix Y (m)		Y (m)	H (m)	1	2	3	4	5	6	7	8	9	10	11	12	B
5.3	b1	-15.2	0.42	0.341678	-2.3259E-08	-102366339.4	0	0	0	0	0	0	0	0	0	-0.57321
5.3	b1	-15.2	0.42	0.594167	-6.7322E-08	310918162.3	0	0	0	0	0	0	0	0	0	-1.45146
13.5	b2	-7	0.88	0	0.075016108	15.14826167	-0.87671194	-1.29617	0	0	0	0	0	0	0	17.18786
13.5	b2	-7	0.88	0	0.103630458	-21.95929209	-1.0459466	3.048742	0	0	0	0	0	0	0	21.09284
14.4	b3	-6.1	1.8	0	0	0	1.831250802	0.303375	-0.05143	-19.4423	0	0	0	0	0	-28.4312
14.4	b3	-6.1	1.8	0	0	0	1.068096738	-0.34886	-0.02502	9.458061	0	0	0	0	0	-21.166
26.6	b4	6.1	1.8	0	0	0	0	0	19.44232	0.051434	-1.89131	-0.29374	0	0	0	5.145635
26.6	b4	6.1	1.8	0	0	0	0	0	9.458061	-0.02502	1.105014	-0.32991	0	0	0	-7.58012
27.6	b5	7.1	0.83	0	0	0	0	0	0	0	0.817082	1.47454	-0.0845141	-14.25583677	0	-5.49428
27.6	b5	7.1	0.83	0	0	0	0	0	0	0	-1.03529	3.591593	0.06886262	-12.49170595	0	7.519176
35.7	b6	15.2	0.37	0	0	0	0	0	0	0	0	0	1.87978E-06	1437773.689	-0.17329	-0.02888
35.7	b6	15.2	0.37	0	0	0	0	0	0	0	0	0	-3.4359E-06	2826154.75	0.569717	-0.70971

Table 7.20 Values of the A_i coefficients for the River Main.

	ctl-shift-enter mmult (minverse(A),b) *Coeff* A
1	−1.95623
2	21.41244
3	−9.3E-10
4	−17.8643
5	0.061917
6	−1E-05
7	−4688.01
8	−4.96133
9	0.252246
10	21.44517
11	−1.1E-07
12	−0.71708

Now solve the matrix with EXCEL, using [=mmult(minverse(A),B)] with ctl-shift-enter keys together to obtain the values shown in Table 7.20.

Now all unknown coefficients are obtained, substituting A's coefficients in the analytical solution in each subsection, then varying y for the flat bed subsection or ξ sloping bed subsection gives velocities. The output values are shown in Table 7.21, with different colours highlighting the various subsections.

Subsection 1: $U_1^2 = A_1 \xi^{\alpha_{1,1}} + \omega_1 \xi + \eta_1$

Subsection 2: $U_2^2 = A_3 \xi^{\alpha_{1,2}} + A_4 \xi^{-(\alpha_{1,2}+1)} + \omega_2 \xi + \eta_2$

Subsection 3: $U_3^2 = A_5 \xi^{\alpha_{1,3}} + A_6 \xi^{-(\alpha_{1,3}+1)} + \omega_3 \xi + \eta_3$

Subsection 4: $U_4^2 = A_7 e^{\gamma_4 y} + A_8 e^{-\gamma_4 y} + k_4$

Subsection 5: $U_5^2 = A_9 \xi^{\alpha_{1,5}} + A_{10} \xi^{-(\alpha_{1,5}+1)} + \omega_5 \xi + \eta_5$

Subsection 6: $U_6^2 = A_{11} \xi^{\alpha_{1,6}} + A_{12} \xi^{-(\alpha_{1,6}+1)} + \omega_6 \xi + \eta_6$

Subsection 7: $U_7^2 = A_{13} \xi^{\alpha_{1,7}} + A_{14} \xi^{-(\alpha_{1,7}+1)} + \omega_7 \xi + \eta_7$

The calculated U_d & τ_b values for the River Main are shown plotted in Figure 7.23. The lateral distribution of U_d is relatively smooth, but the boundary shear stress has distinct peaks on the both river banks. This arises because the Manning roughness coefficients or friction factors used on the river banks are significantly larger than those on either the floodplains or the main channel, despite being calibrated with field data. The large Manning coefficient could be due to bushes and trees on the river banks, but the boundary shear stresses on the

Table 7.21 Output values of U_d & τ_b for the River Main.

Solution y	H	U²	U	dq	Tb
−15.7	0.125	0.173913	0.417028869		6.07356621
−15.2	0.42	0.416923	0.645695501	0.072398098	14.56019172
−15.2	0.42	0.416923	0.645695501	0	9.990849253
−14.2	0.48	0.579716	0.761390845	0.316594428	13.89191719
−13.2	0.54	0.658061	0.811209691	0.401013137	15.76932652
−12.2	0.6	0.732231	0.855705159	0.475070732	17.54668928
−11.2	0.66	0.809473	0.899707123	0.552954869	19.39765373
−10.2	0.72	0.905458	0.951555827	0.638685718	21.69778648
−9.2	0.78	1.09051	1.044274743	0.748436464	26.13222771
−7	0.88	2.679283	1.636851654	2.447868401	64.20450891
−7	0.88	2.679283	1.636851654	0	62.0097152
−6.1	1.8	4.656558	2.157905916	2.288238815	107.7720402
−6.1	1.8	4.656558	2.157905916	0	27.37604536
−5.1	1.8	6.287963	2.507581039	4.198938259	36.96712321
−4.1	1.8	7.283977	2.698884408	4.685818902	42.82272198
−3.1	1.8	7.884994	2.808023113	4.956216769	46.35611771
−2.1	1.8	8.236071	2.869855593	5.110090835	48.42011193
−1.1	1.8	8.421943	2.902058435	5.194722625	49.51285926
−0.1	1.8	8.487471	2.913326447	5.233846394	49.89809936
0.9	1.8	8.44847	2.906625201	5.237956484	49.66881172
1.9	1.8	8.295527	2.880195709	5.208138819	48.76965671
2.9	1.8	7.991729	2.826964683	5.136444353	46.98361906
3.9	1.8	7.463753	2.731987007	5.003056521	43.87963029
4.9	1.8	6.584169	2.565963528	4.768155481	38.70852822
6.1	1.8	4.756859	2.181022577	5.126744993	27.96572117
6.1	1.8	4.756859	2.181022577	0	110.7867353
6.6	1.315	3.742382	1.934523639	1.602490808	87.15965925
7.1	0.83	2.804215	1.674579011	0.967690648	65.30985643
7.1	0.83	2.804215	1.674579011	0	69.01532556
8.1	0.779	1.712577	1.308654589	1.200005715	42.14871304
9.1	0.728	1.188406	1.090140451	0.903746031	29.24820135
10.1	0.677	0.930518	0.964633351	0.721739298	22.90122955
11.2	0.626	0.790977	0.88936893	0.603941243	19.46696 11
15.2	0.37	0.288983	0.537571443	1.456763428	7.112243786
15.2	0.37	0.288983	0.537571443	0	10.5276913
15.6	0.094	0.093888	0.306412052	0.039160834	3.420364963

banks are unrealistic if based on the bed material alone. In order to obtain sensible boundary shear stress values, it is necessary to separate out the resistance due to bed materials from the resistance due to any vegetation such as bushes and trees, as the latter should be treated as producing form drag, unlike the bed materials which produce largely skin resistance. A drag force formula therefore needs to be introduced into SKM, which is the topic of Section 7.2.4 for dealing with such issues as illustrated in Figure 7.24.

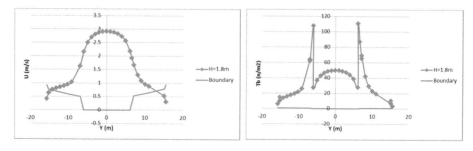

Figure 7.23 Lateral variation of U_d & τ_b values for River Main, $H = 1.8$ m.

Figure 7.24 River Severn, showing vegetation on banks.

7.2.4 Overbank flow in the River Severn with trees on the floodplain

The second river example is that of the River Severn at Montford. This site has been extensively studied, and featured already in Chapter 2, Section 2.5.2, in the CES simulations in Chapter 4, Section 4.2.4, as well as in the analytical SKM modelling section in Chapter 6, Section 6.6.3 and in Chapter 7, Section 7.1.7. An equation for the drag force caused by trees was developed in Chapter 6, Section 6.6.3, for specific inclusion in the mathematical modelling approach to rivers adopted by the SKM. The River Severn is a particularly a good case to illustrate its use.

The geometric configuration of the channel cross section is similar to the River Main, having a flat bed in the main channel and sloping banks and floodplains.

Exactly the same EXCEL program used for the River Main can therefore be used, the only differences being in the friction factors, lamda values, and the side slope angles of the river banks, floodplains and floodplain side banks. After inputting these particular parameters in the program, the velocities and boundary shear stresses can be obtained.

The various values for the geometry which set the locations for the boundary conditions are given in Table 7.22, together with the corresponding bed slopes. Table 7.23 shows the input values for each panel and Table 7.24 the corresponding

Table 7.22 Nodal geometry used for panels in River Severn.

Panel nodes	y (m)	z (m)	s
LHS	−0.85	8.375	−4
b1	−79.5	7	−63
b2	−16.5	6	−1.3333
b3	−8.5	0	0
b4	8.5	0	0
b5	17.5	6	1.5
b6	40.5	7	23
RHS	50	8.33803	7.1

Table 7.23 Input values for the 7 panels used for River Severn at Montford Bridge.

		Slope	Slope	Slope	Flat	Slope	Slope	Slope
		Section 1	Section 2	Section 3	Section 4	Section 5	Section 6	Section 7
Water	H	0.8	1.8	7.8	7.8	7.8	1.8	0.8
Gravity	g	9.81	9.81	9.81	9.81	9.81	9.81	9.81
Slope	So	0.0002	0.0002	0.0002	0.0002	0.0002	0.0002	0.0002
Slope	s	−4	−63	−1.33333	0	1.5	23	7.1
Cd		0	0	0	0	0	0	0
Ap	bh/(bL)	0	0	0.2405	0	0.2405	0	0
L		0	0	20	0	20	0	0
Drag	1/2CdAp	0	0	0	0	0	0	0
friction	f	0.038	0.038	0.312828	0.015829023	0.352141401	0.1	0.1
Manning coeff	n	0.029	0.029	0.082	0.02	0.087	0.029	0.029
dim eddy	lamda	0.2	0.2	0.2	0.3	0.2	0.2	0.2
Γ/pgHSo	Γ	0	0	0	0	0	0	0
	E1	0.001962	0.001962	0.001962	0.0153036	0.001962	0.001962	0.001962
	E2	0.004896	0.004751	0.048879	0.001978628	0.052902661	0.012511809	0.012623
	E3	0.000431	1.74E-06	0.011123	0.405940638	0.009324618	2.11349E-05	0.000222
coeff	γ_1				0.069815331			
coeff	γ_2				−0.06981533			
coeff	α_1	2.908316	51.80716	1.655078		1.933811888	23.83616102	7.060844
coeff	α_2	−3.90832	−52.8072	−2.65508		−2.933811888	−24.836161	−8.06084
	k				7.734450682			
const Mat	ω	0.486283	0.413303	0.073668		0.057278942	0.157343421	0.161086
	η	0	0	0	0	0	0	0

Table 7.24 Matrix A for River Severn at Montford Bridge.

Matrix Y (m)		Y (m)	H (m)	1	2	3	4	5	6	7	8	9	10	11	12	B
5.3	b1	−79.5	0.8	0.522583	−9.536E-06	−131081.8689	0	0	0	0	0	0	0	0	0	−0.05838
5.3	b1	−79.5	0.8	0.474949	−9.8023E-06	137342.4863	0	0	0	0	0	0	0	0	0	−0.11501
13.5	b2	−16.5	1.8	0	1.67859E+13	3.3.966E-14	−2.64542735	−0.21001	0	0	0	0	0	0	0	−0.61134
13.5	b2	−16.5	1.8	0	7.668868E+12	−1.54122E-14	−1.82432829	0.232326	0	0	0	0	0	0	0	0.048691
14.4	b3	−8.5	7.8	0	0	0	29.95614417	0.00428	−0.55243	−1.81019	0	0	0	0	0	7.15984
14.4	b3	−8.5	7.8	0	0	0	4.767283382	−0.00109	−0.03857	0.126379	0	0	0	0	0	−0.05525
26.6	b4	8.5	7.8	0	0	0	0	0	1.810187	0.552429	−53.1059	−0.00241	0	0	0	−7.28767
26.6	b4	8.5	7.8	0	0	0	0	0	0.126379	−0.03857	8.777514	−0.00061	0	0	0	−0.03819
27.6	b5	17.5	1.8	0	0	0	0	0	0	0	3.11637	0.17827	−12153392.45	−4.571E-07	0	0.180116
27.6	b5	17.5	1.8	0	0	0	0	0	0	0	−2.23203	0.193708	699765.4639	−2.74217E-07	0	0.031345
35.7	b6	40.5	0.8	0	0	0	0	0	0	0	0	0	0.004898209	255.1953093	−0.20689	0.00294
35.7	b6	40.5	0.8	0	0	0	0	0	0	0	0	0	−0.00634535	344.4604235	0.257183	−0.01585

values for the matrix A coefficients. Table 7.25 gives the values of the A_i coefficients used to determine the velocities in each panel.

The calculated velocities are compared with measured data in Figure 7.25 along with the calculated boundary shear stresses. The distribution of velocity is well simulated by adjusting the friction factors, but again two unrealistic spikes in the boundary shear stress distribution are apparent. Since such spikes may indicate the presence of some vegetation on the river banks (see Figure 7.24), the skin friction arising from the bed material and the possibility of extra drag forces arising from vegetation are now considered.

In this second simulation, the drag force formula is now introduced for the river banks in order to demonstrate that it will reduce these peak boundary shear stresses. The friction factor on the river banks in the previous calculation and was calibrated and found to be 0.087. It is now assumed that the river banks would be composed of the same bed materials as the river bed, so are now set to the same friction factor of 0.038 as for the left hand floodplain. The fact that there are trees covering the whole of the river banks is now taken into account, by assuming that the trees have a projected area per unit bank area of 0.2405 and are 20 m apart with

Table 7.25 Values for the A_i coefficients for River Severn at Montford Bridge.

Coeff A	ctl-shift-enter mmult (minverse(A),b)
	Coeff A
1	−0.17231
2	−1.7E-14
3	−2.4E-07
4	0.0357
5	1.066102
6	−2.60067
7	−2.56833
8	0.021843
9	0.487972
10	−2.1E-08
11	−1.8E-05
12	−0.03708

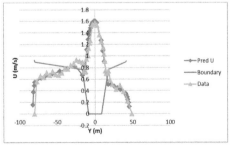
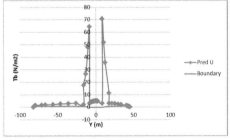

Figure 7.25 Lateral variations of U_d & τ_b for the River Severn, $H = 7.8$ m.

Table 7.26 Revised input values for the 7 panels used for River Severn at Montford Bridge.

		Slope	Slope	Slope	Flat	Slope	Slope	Slope
		Section 1	Section 2	Section 3	Section 4	Section 5	Section 6	Section 7
Water	H	0.8	1.8	7.8	7.8	7.8	1.8	0.8
Gravity	g	9.81	9.81	9.81	9.81	9.81	9.81	9.81
Slope	So	0.0002	0.0002	0.0002	0.0002	0.0002	0.0002	0.0002
Slope	s	−4	−63	−1.33333	0	1.5	23	7.1
Cd		0	0	1	0	1	0	0
Ap	bh/(bL)	0	0	0.2405	0	0.2405	0	0
L		0	0	20	0	20	0	0
Drag	1/2CdAp	0	0	0.012025	0	0.012025	0	0
friction	f	0.038	0.038	0.038	0.015829023	0.038	0.1	0.1
Manning coeff	n	0.029	0.029	0.029	0.02	0.029	0.029	0.029
dim eddy	lamda	0.2	0.2	0.2	0.3	0.2	0.2	0.2
Γ/pgHSo	Γ	0	0	0	0	0	0	0
	E1	0.001962	0.001962	0.001962	0.0153036	0.001962	0.001962	0.001962
	E2	0.004896	0.004751	0.017963	0.001978628	0.01773379	0.012511809	0.012623
	E3	0.000431	1.74E-06	0.003877	0.405940638	0.003063122	2.11349E-05	0.000222
coeff	γ_1				0.069815331			
coeff	γ_2				−0.06981533			
coeff	α_1	2.908316	51.80716	1.709836		1.957529126	23.83616102	7.060844
coeff	α_2	−3.90832	−52.8072	−2.65508		−2.957529126	−24.836161	−8.06084
	k				7.734450682			
const Mat	ω	0.486283	0.413303	0.192184		0.169027981	0.157343421	0.161086
	η	0	0	0	0	0	0	0

a porosity of 0.9. The drag coefficient, Cd, is assumed to take the standard value of 1.0. Table 7.26 shows the revised input values for each panel and Table 7.27 the corresponding values for the matrix A coefficients. Table 7.28 gives the values of the A_i coefficients used to determine the velocities in each panel.

The revised values of the velocity and boundary shear stress are shown tabulated in Table 7.29 and plotted in Figure 7.26. As may be seen by comparing Figures 7.25 & 7.26, the boundary shear stresses on the river banks are substantially reduced despite the velocity prediction being the same as before.

7.2.5 Overbank flow in a compound trapezoidal channel with rod roughness

The scientific background to the equation representing the drag force on trees was discussed in Chapter 6, Section 6.6.3. The formulation of this equation was based on experimental evidence, one element of which was the data from Series 07 of the FCF research programme. Figure 6.84 illustrates a typical array of the 25 mm diameter rod roughness that was used in this series, roughening both floodplains to a density of 12 per m². FCF Experiment 070601 is now modelled using the same program as for the Rivers Main and Severn.

Table 7.30 shows the input values for each panel and Table 7.31 the corresponding values for the matrix A coefficients. Table 7.32 gives the values of the A_i coefficients

Table 7.27 Revised matrix A for River Severn at Montford Bridge.

Matrix Y (m)	Matrix	Y (m)	H (m)	1	2	3	4	5	6	7	8	9	10	11	12	B
5.3	b1	−79.5	0.8	0.522583	−9.536E-06	−131081.8689	0	0	0	0	0	0	0	0	0	−0.05838
5.3	b1	−79.5	0.8	0.474949	−9.8023E-06	137342.4863	0	0	0	0	0	0	0	0	0	−0.11501
13.5	b2	−16.5	1.8	0	1.67859E+13	3.3966E-14	−2.73195857	−0.20335	0	0	0	0	0	0	0	−0.39801
13.5	b2	−16.5	1.8	0	7.668868E+12	−1.54122E-14	−1.94633376	0.229607	0	0	0	0	0	0	0	0.137578
14.4	b3	−8.5	7.8	0	0	0	33.5224255	0.003824	−0.55243	−1.81019	0	0	0	0	0	6.235416
14.4	b3	−8.5	7.8	0	0	0	5.511331604	−0.001	−0.03857	0.126379	0	0	0	0	0	−0.14414
26.6	b4	8.5	7.8	0	0	0	0	0	1.810187	0.552429	−55.7572	−0.0023	0	0	0	−6.41603
26.6	b4	8.5	7.8	0	0	0	0	0	0.126379	−0.03857	9.328752	−0.00058	0	0	0	−0.11269
27.6	b5	17.5	1.8	0	0	0	0	0	0	0	3.160118	0.175802	−121539245	−4.571E-07	0	−0.02103
27.6	b5	17.5	1.8	0	0	0	0	0	0	0	−2.29112	0.19257	699765.4639	−2.74217E-07	0	0.105844
35.7	b6	40.5	0.8	0	0	0	0	0	0	0	0	0	0.004898209	255.1953093	−0.20689	0.002994
35.7	b6	40.5	0.8	0	0	0	0	0	0	0	0	0	−0.00634535	344.460235	0.257183	−0.01585

Table 7.28 Revised values for the A_i coefficients for River Severn at Montford Bridge.

Coeff A	ctl-shift-enter mmult(minverse(A),b) Coeff A
1	−0.17231
2	−9.2E-15
3	−2.4E-07
4	0.013062
5	1.018672
6	−2.45503
7	−2.45135
8	0.011066
9	0.336821
10	9.48E-08
11	−1.8E-05
12	−0.03708

Table 7.29 Revised SKM U_d & τ_b output values for the River Severn, $H = 7.8$ m.

y	Solution y	H	U²	U	dq	Tb	H
−84.5	−84.5	−0.4.5					
−83.5	−83.5	−0.2					
−82.5	−82.5	0.05	0.024286	0.155839084		0.115357645	
−81.5	−81.5	0.3	0.14069	0.375086278	0.046455969	0.668276152	
−80.5	−80.5	0.55	0.237173	0.487003966	0.183194177	1.126571097	0.929487
−79.5	−79.5	0.8	0.298981	0.546791921	0.348906112	1.420161675	0.897436
−79.5	−79.5	0.8	0.298981	0.546791921	0	1.420161675	0.897436
−71.5	−71.5	0.926984	0.383112	0.6189603	4.026471165	1.819781303	0.881156
−66.5	−66.5	1.006349	0.415927	0.644923812	3.054386605	1.975651936	0.870981
−56.5	−56.5	1.165079	0.481531	0.693923993	7.268030939	2.287269911	0.850631
−46.5	−46.5	1.32381	0.547134	0.739685136	8.920234576	2.598886975	0.830281
−36.5	−36.5	1.48254	0.612731	0.782771386	10.68136163	2.910472456	0.809931
−26.5	−26.5	1.64127	0.677042	0.822825576	12.5389477	3.215949159	0.789581
−16.5	−16.5	1.8	0.588768	0.76731229	13.6802337	2.796648711	0.769231
−16.5	−16.5	1.8	0.588768	0.76731229	0	2.796648711	0.769231
−13.5	−13.5	4.05	0.944136	0.971666725	7.629770425	4.484647063	0.480769
−10.5	−10.5	6.3	1.52163	1.233543727	17.11794613	7.2277431	0.192308
−8.5	−8.5	7.8	1.940812	1.393130169	41.4804952	9.218855419	0
−8.5	−8.5	7.8	1.940812	1.393130169	0	3.840144135	0
−7.5	−7.5	7.8	2.141975	1.463548708	11.14104762	4.238171167	0
−6	−6	7.8	2.392859	1.546886781	17.6104761	4.734677041	0
−4.5	−4.5	7.8	2.585108	1.607827203	18.45507681	5.114967463	0
−3	−3	7.8	2.720834	1.649495053	19.0553352	5.383517956	0
−1.5	−1.5	7.8	2.801525	1.673775796	19.44113446	5.543176379	0
0	0	7.8	2.828069	1.681686211	19.62945274	5.595695292	0

(Continued)

Table 7.29 (Continued).

y	Solution y	H	U²	U	dq	Tb	H
1.5	1.5	7.8	2.800755	1.673545513	19.62810558	5.541651191	0
3	3	7.8	2.719284	1.649025108	19.43703813	5.380450839	0
4.5	4.5	7.8	2.582762	1.607097347	19.04831636	5.110324749	0
6	6	7.8	2.38969	1.54586229	18.44481387	4.728307766	0
7	7	7.8	2.228539	1.492829149	11.85089661	4.409449209	0
8.5	8.5	7.8	1.936191	1.391470858	16.87315504	3.831001847	0
8.5	8.5	7.8	1.936191	1.391470858	0	9.196907953	0
10.5	10.5	6.466667	1.521874	1.233642445	18.72580823	7.228899994	0.17094
12.5	12.5	5.133333	1.142372	1.068818128	31.81973755	5.42626791	0.34188
17.5	17.5	1.8	0.398434	0.631215919	14.73362841	1.892559299	0.769231
17.5	17.5	1.8	0.398434	0.631215919	0	4.980419207	0.769231
21.5	21.5	1.626087	0.266079	0.515828255	3.929873083	3.325984857	0.791527
25.5	25.5	1.452174	0.22918	0.478727237	3.061501252	2.864747089	0.813824
29.5	29.5	1.278261	0.201159	0.448507384	2.531753659	2.514485913	0.83612
33.5	33.5	1.104348	0.17376	0.416846876	2.061800583	2.172016471	0.858417
40.5	40.5	0.8	0.121199	0.34813583	2.549388017	1.514981954	0.897436
40.5	40.5	0.8	0.121199	0.34813583	0	1.514981954	0.897436
41.5	41.5	0.659155	0.104227	0.322841437	0.244764947	1.302832419	0.915493
42.5	42.5	0.51831	0.083135	0.288330884	0.179908472	1.039183736	0.93355
43.5	43.5	0.377465	0.060766	0.246508214	0.119773826	0.759578744	0.951607
44.5	44.5	0.23662	0.038115	0.195230149	0.067816171	0.476435139	0.969664
45.5	45.5	0.095775	0.015428	0.124209468	0.026544982	0.192849901	0.987721

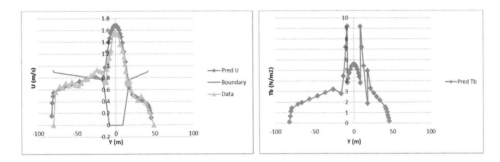

Figure 7.26 Revised lateral variations of U_d & τ_b for the River Severn, $H = 7.8$ m.

used to determine the velocities in each panel. Table 7.33 gives the SKM output values for U_d & τ_b for this particular depth, H, of 0.215 m, or relative depth, Dr, of 0.304.

The necessity of validating models against good quality data is illustrated again in Figures 7.27 and 7.28, where a comparison is made between output values for U_d & τ_b for $Dr = 0.50$ and a model by van Prooijen, taken from Shiono and Ramesh (2015). The prediction of velocity and bed shear stress using SKM for the rods on the flood-plain appears to agree well with the data. However, it is apparent that van Prooijen's model predicts the velocities well but not the boundary shear stresses, caused by not

Table 7.30 Input values for the 47 panels used for FCF Experiment 070501.

		Section 1	Section 2	Section 3	Section 4
Width (m)		0–0.75	0.75–0.9	0.9–3.03	3.05–
Water	H	0.2	0.2	0.05	0.05
Gravity	g	9.81	9.81	9.81	9.81
Slope	So	0.001027	0.001027	0.001027	0.001027
Slope	S		1	1	1
Cd		0	0	1.3	0
d	Rod dia			0.025	
Rod No	No (N)	0	0	12.4	0
Ap	Ndbh/(b*1)	0	0	0.0155	0
Drag	1/2CdAp	0	0	0.010075	0
ks		0.00016			
friction	f	0.015444	0.023722	0.032	0.032
dim eddy	lamda	0.75	0.07	1.913464	0.07
Γ/pgHSo	Γ'''	0	0	0	0
	E1	0.002015	0.010075	0.000504	0.010075
	E2	0.001931	0.004194	0.014075	0.005657
	E3	0.000659	0.001906	0.000151	0.002214
coeff	γ_1	1.711492		9.645928	
coeff	γ_2	-1.71149		-9.64593	
coeff	α_1		1.06534		1.174965
coeff	α_2		-2.06534		-2.17496
	k	1.043727		0.03579	
const Mat	ω		26.39261		8.193179
	η		0		0

Table 7.31 Matrix A for FCF Experiment 070501.

y		H	A matrix						B
b0	0	0.2	1.711492	-1.71149	0	0	0	0	0
b1	0.75	0.2	3.609681	0.277033	-0.18004	-27.7722	0	0	4.234794
b1	0.75	0.2	6.177941	-0.47414	0.958998	-286.795	0	0	-26.3926
b2	0.9	0.05	0	0	0.041111	486.4845	-5891.91	-0.00017 0	-1.28384
b2	0.9	0.05	0	0	-0.87595	20095.12	-56832.9	0.001637 0	26.39261
b3	3.15	0.05	0	0	0	0	1.57E+13	6.37E-14 -0.0296	0.373869
b3	3.15	0.05	0	0	0	0	1.51E+14	-6.1E-13 0.69565	-8.19318

Table 7.32 Values for the A_i coefficients for FCF Experiment 070501.

A	
A1	-0.14441
A2	-0.14441
A3	-26.6472
A4	4.98E-05
A5	1.14E-15
A6	1252.399
A7	-12.0256

Table 7.33 SKM values for U_d & τ_b (FCF Experiment 070501, $H = 0.20$ m).

y	H	U²	U	dQ	To
0	0	0.754905	0.868853		1.457386
0.1	0.2	0.750665	0.866409	0.017353	1.4492
0.2	0.2	0.737819	0.858964	0.017254	1.424401
0.3	0.2	0.71599	0.846162	0.017051	1.38226
0.4	0.2	0.684538	0.827368	0.016735	1.32154
0.5	0.2	0.642539	0.801586	0.01629	1.240458
0.6	0.2	0.58876	0.767307	0.015689	1.136634
0.65	0.2	0.556974	0.746307	0.007568	1.075269
0.7	0.2	0.521621	0.722233	0.007343	1.007019
0.75	0.2	0.482443	0.694581	0.007084	0.931383
0.8	0.15	0.430302	0.655974	0.005909	1.275965
0.85	0.1	0.352533	0.593745	0.003905	1.045359
0.9	0.05	0.248353	0.49835	0.002048	0.736435
0.95	0.05	0.167018	0.408679	0.001134	0.668074
1	0.05	0.116806	0.341769	0.000938	0.467223
1.05	0.05	0.085806	0.292927	0.000793	0.343225
1.1	0.05	0.066668	0.258202	0.000689	0.266673
1.15	0.05	0.054853	0.234207	0.000616	0.219412
1.2	0.05	0.047559	0.21808	0.000565	0.190235
1.3	0.05	0.040276	0.200688	0.001047	0.161102
1.5	0.05	0.036442	0.190897	0.001958	0.145766
1.7	0.05	0.035885	0.189432	0.001902	0.143538
1.9	0.05	0.035804	0.189219	0.001893	0.143215
2.1	0.05	0.035793	0.189189	0.001892	0.143171
2.3	0.05	0.035795	0.189196	0.001892	0.143181
2.5	0.05	0.035824	0.189272	0.001892	0.143295
2.7	0.05	0.036023	0.189797	0.001895	0.144091
2.75	0.05	0.036167	0.190177	0.000475	0.144669
2.8	0.05	0.036401	019079	0.000476	0.145604
2.85	0.05	0.03678	0.19178	0.000478	0.147119
2.9	0.05	0.037393	0.193373	0.000481	0.149572
2.95	0.05	0.038387	0.195925	0.000487	0.153546
3	0.05	0.039996	0.19999	0.000495	0.159984
3.05	0.05	0.042603	0.206405	0.000508	0.170411
3.07	0.05	0.044053	0.209887	0.000208	0.17621
3.15	0.05	0.053665	0.231657	0.000883	0.214661
3.16	0.04	0.053837	0.232028	9.27E-05	0.215348
3.17	0.03	0.050461	0.224636	6.85E-05	0.201845
3.18	0.02	0.042559	0.206298	4.31E-05	0.170236

considering the rod drag force separately from the bed friction. The same conclusion may be drawn from Figure 6.92, in which the earlier version of the CES model did not allow for the additional drag force arising from the rod roughness. This brief analysis shows the importance of introducing a drag force formula into the SKM. The new analytical form for such a drag force, presented in Section 6.6.3, appears to offer a useful addition to the science of modelling vegetation.

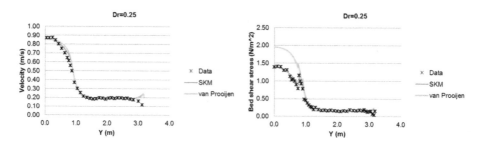

Figure 7.27 Comparison of SKM values for U_d & τ_b with those by van Prooijen. (FCF Experiment 070501, $H = 0.20$ m, $Dr = 0.25$) (from Shiono and Ramesh, 2015).

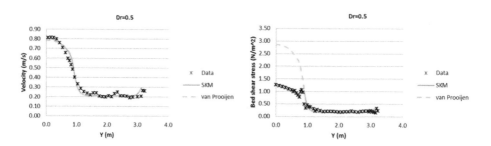

Figure 7.28 Comparison of SKM values for U_d & τ_b with those by van Prooijen. (FCF Experiment 070801, $H = 0.303$ m, $Dr = 0.5$) (from Shiono and Ramesh, 2015).

7.2.6   Flow in the River Rhône, France

An example of an approximately straight trapezoidal compound river channel with floodplains that are relatively uniformly covered with of mature trees is the River Rhône in France. Recent studies by Terrier *et al.* (2008) have compared velocity measurements by Acoustic Doppler Current Profiler (ADCP) with those given by the SKM for a flood in April 2006. These are shown in Figure 7.29 for a discharge of 2,400 m³s⁻¹. The modified version of the SKM which incorporates the drag force on trees is seen to give satisfactory results.

7.2.7   Flow in the Yangtze River, China

The largest river to which the SKM has been applied to is the Yangtze River in China for which the measured discharge was $Q = 23,127$ m³s⁻¹ at the Zhu Tuo gauging station. The cross section was simulated by 7 panels, as shown in Figure 7.30, and appropriate parameters chosen for the river. The measured depth-averaged velocity data is shown in Figure 7.31, together with the SKM simulation. Integration of the velocity data gave a discharge of $Q = 22,924$ m³s⁻¹, a difference of just −0.9%.

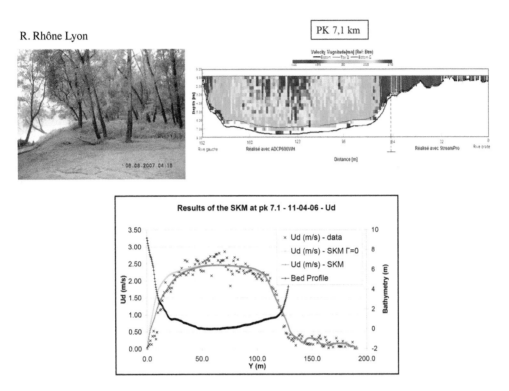

Figure 7.29 River Rhône, France, showing trees, ADCP velocity data and predicted values by SKM.

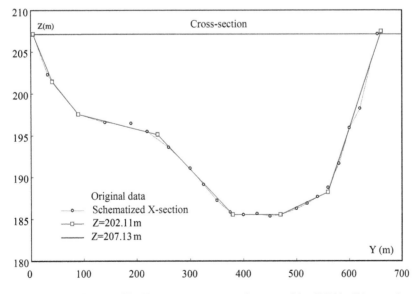

Figure 7.30 Yangtze River at Zhu Tuo gauging station schematized for SKM by 7 linear elements.

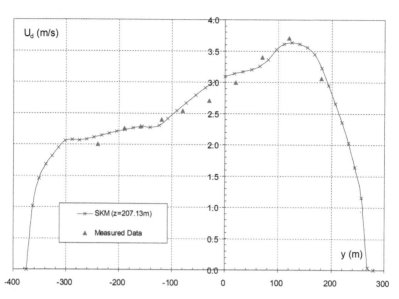

Figure 7.31 Predicted and measured velocity distribution in the Yangtze River (Zhu Tuo gauging station for a water level of z = 207.13 m).

7.2.8　Flow in the Paute River, Ecuador

An unusual application of the SKM involved modelling sediment transport and the regrading of 12 km of the Paute River in Ecuador. Following a landslide in 1993, 30 million m³ of earth and rock slid and blocked the river, creating a lake upstream. Eventually, a month later, the dam burst, releasing 200 million m³ of water, causing a disastrous flood downstream with an estimated peak discharge of 8,300 m³s⁻¹, as illustrated in Figure 7.32, and also 10 million m³ of debris to move downstream. The geomorphology of the whole system was thus substantially changed, with river bed levels rising by up to 30 m. The hydraulic master plan involved river bed surveys and modelling studies using the SKM, as described by Abril & Knight (2004), leading to the stabilization of the river by the means of a series of drop structures.

7.2.9　Flow in other world rivers

The SKM has now been used on many rivers since its development in the late 1990s, as indicated by some of the examples selected for inclusion in this book. Further examples are described in Abril & Knight (2002 & 2012) and in Mc Gahey *et al.* (2008), from which Figure 7.33 is reproduced. International Standards are now beginning to cite commercially available software for use in evaluating stage-discharge relationships, as for example in ISO 18320, in which both the CES and SKM are referred to in some detail.

　　In summary, this Chapter has demonstrated that the SKM is capable of analyzing flows in a variety of different shaped channels, over a wide range of discharges, from

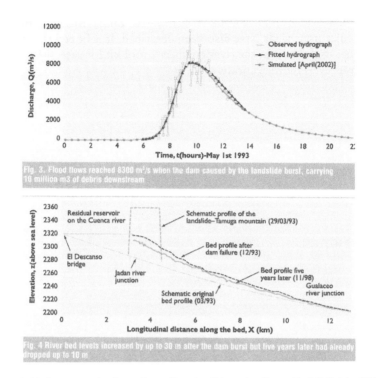

Figure 7.32 Stabilizing the Paute River, Ecuador, S America (from Abril & Knight, 2004).

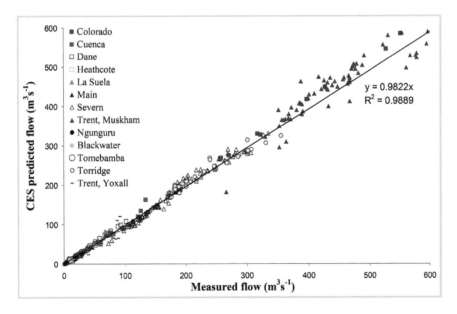

Figure 7.33 Predicted and measured discharges on several rivers, world-wide (from Mc Gahey et al., 2008).

small scale laboratory flumes ($Q = 2.0$ ls^{-1}) to European rivers (~2,000 m^{3}s^{-1}), and even large-scale world rivers (>23,000 m^{3}s^{-1}), a ~10^7 range. Methods for adapting it to meet particular user needs have also been described. It is hoped that the SKM will provide another useful tool in the river engineer's tool kit for assessing flood risk and estimating river levels more accurately.

Chapter 8

Further examples – estimating flow, level and velocity in practice

ABSTRACT

This chapter provides further examples of the use of the CES-AES methods through a series of practical applications involving stage-discharge, stage-conveyance and velocity prediction for a range of channel types (size, shape, cover, sinuosity). Sites from rivers in the UK, France, China, New Zealand and Ecuador are considered.

The survey data for the examples are provided in Appendix 3.

8.1 ESTIMATING AND USING STAGE-DISCHARGE RELATIONSHIPS

8.1.1 Stage-discharge prediction for the River Dane at Rudheath, Cheshire, UK

Aim

To predict the stage-discharge and upper and lower uncertainty scenarios for the River Dane site using the CES-AES software and available site information. The outputs are compared with measured data and then calibrated based on updating the unit roughness, to provide improved confidence in the results.

The Rudheath site on the River Dane

The River Dane rises in the Peak District, flowing west to join River Wheelock, and then north-west towards Northwick in Cheshire. The upstream catchment area is 407 km² with a mean annual flood discharge of approximately 78 m³s⁻¹ and a 1:100 year flood of the order 161 m³s⁻¹. The 6 km study reach is shown in Figure 8.1.

The river channel is strongly meandering with a sinuosity of 1.8 however the measurements were taken in a straight portion of the reach (cross-section 30, Figure 8.2a), 10 m downstream of the road bridge. The gauging station, which has been in existence since 1949, was originally based on the velocity area approach. Around 1980, a flat V-weir (Figure 8.2b) for low flows and a current metering cableway for high flows was built. Flow measurements prior to 1980 should be used with caution (Environment Agency, 2005). The data measurements include discharge and water level taken

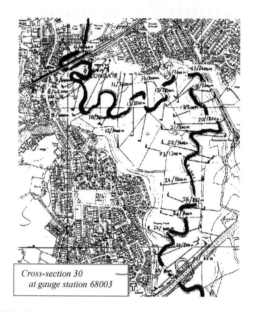

Figure 8.1 The River Dane study reach (Ervine & MacLeod, 1999).

(a)	(b)

Figure 8.2 Gauging station on the River Dane at Rudheath, Cheshire, showing (a) the cableway and (b) the flat v-weir (courtesy Environment Agency).

from January through to July, 1995. The maximum recorded flood during this period was 108 m³s⁻¹, and the corresponding flood level was 16.66 m ODN.

The channel widths are typically 25 m and the bankfull depth is 3.6 m. The river cross-sections are trapezoidal in shape with side slopes of 30°–40°. The longitudinal floodplain slope is in the range 0.00111 to 0.00077 and the river channel has an average bed slope of 0.0005. The vegetation consists of thick growth over the channel edges, which is less dense over the floodplains (Figure 8.3).

Roughness and cross-section definition

Based on the description and photographs, the Roughness Advisor values include (Figure 8.4):

- *Channel bed* – The RA sand ($n_{sur} = 0.02$) and trailing bankside plants ($n_{veg} = 0.05$) are adopted giving $n_l = 0.054$. The trailing bankside plants have a wide range, (0.0 to 0.10). As these are not present everywhere, a value on the lower side is adopted, 0.03. This gives $n_l = 0.036$.
- *Berms* – The RA emergent reeds ($n_{veg} = 0.15$) and sand ($n_{sur} = 0.02$) are adopted giving the total unit roughness $n_l = 0.151$. The reasoning for selecting emergent reeds is that the berm vegetation appears thicker and taller than the floodplain cover.
- *Floodplain* – The RA medium grass ($n_{veg} = 0.08$) and sand ($n_{sur} = 0.02$) are adopted giving $n_l = 0.082$.

Results for the River Dane site

Figure 8.5 provides the CES stage-discharge prediction using these RA roughness values and the measured data. The flows are over-predicted at lower flows (less than 20 m³s⁻¹) and under predicted at higher lows. All data fall within the upper and lower uncertainty scenarios, which provide a wide range due to the seasonal variations

Figure 8.3 Typical vegetation on the River Dane.

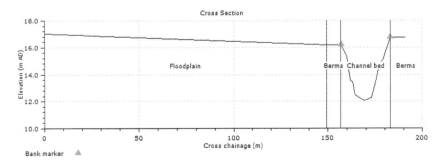

Figure 8.4 River Dane cross-section, roughness zones and top-of-bank marker locations.

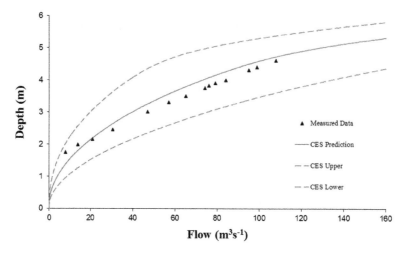

Figure 8.5. Stage-discharge prediction for the River Dane at Rudheath.

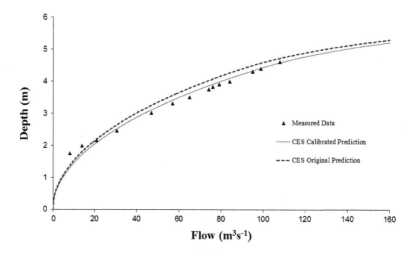

Figure 8.6 Calibrated stage-discharge prediction for the River Dane at Rudheath.

for the vegetation. The reason for the over-prediction at lower flows may be partly related to the dense seasonal vegetation, which is abundant in summer when the low flows are measured, resulting in increased flow depths. A further reason is that the V-shaped weir immediately downstream of the site affects the ratings at low flows. There may also be variation in the longitudinal water surface slope, as it is subject to local topographic variations at low flows.

Further calibration for the River Dane site

Whilst the under-prediction at lower flows can be explained, it is possible to improve the predictions at higher flows through refining the roughness estimates. Most of the data measurements are for inbank flow, as bankfull depth is close to 4 m. It is therefore reasonable to assume that the channel roughness estimate could be improved. The bed material is sand and, as such, cannot be changed. However, the trailing bankside plants had a wide range of possible roughness values (illustrated by the wide upper and lower scenario curves in Figure 8.5), and previously $n_{veg} = 0.03$ was adopted. Thus, to increase the flow predictions for higher flows, a value of $n_{veg} = 0.025$ is adopted. Figure 8.6 provides the CES stage-discharge prediction using the original and revised RA roughness values. The result is that the refined predicted flows follow the measured data more closely.

8.1.2   Stage-discharge prediction for the river blackwater, Coleford Bridge, North-East Hampshire, UK

Aim

To predict the stage-conveyance and upper and lower uncertainty scenarios for the River Blackwater site using the CES-AES software and available site information. The outputs are compared with measured data for the different seasons (winter and summer), to demonstrate the utility of adopting different roughness values for different seasons.

The Coleford Bridge site on the River Blackwater

The River Blackwater is situated in north-east Hampshire in the south of England. It serves a catchment area of approximately 40 km² that responds to local storms. In 1993, the construction of a major trunk road along the River Blackwater valley required the relocation of the river. This provided the opportunity to the then UK National Rivers Authority to design the new channel and undertake a multidisciplinary study of an environmentally acceptable channel design. This 3 km reach of the River Blackwater has therefore been extensively monitored in terms of flow parameters, vegetation growth patterns and channel deformation.

The channel was reconstructed as a doubly meandering two-stage trapezoidal channel, with a main channel and floodplain sinuosity of 1.18 and 1.06 respectively. It is capable of carrying the estimated 100 year flood, 4.3 m³s⁻¹, and an upstream spillway ensures that the flow does not exceed 4 m³s⁻¹. The study reach is 520 m long, the main channel width is 5.75 m and the average berm width is 4.9 m (Sellin & van Beesten, 2002; Sellin & van Beesten, 2004). The reach-averaged longitudinal bed slope is 0.001, whereas the average measured water surface slope has been as low as 0.00044. The inbank vegetation builds up in the summer (Chapter 2, Figures 2.8 and 2.9; Figure 8.7) and is washed out in Autumn, while the out-of-bank vegetation is cut in October. The channel bank vegetation is emergent reeds with a silt substrate and the bed consists of silt and fine gravel. The berms are covered in heavy grass and weed growth.

The measured flow parameters available for calibration are stage (pressure transducers), discharge (electromagnetic gauge) and change in Manning's n with depth and season. These measurements achieve good accuracy for high flows, but are less

Figure 8.7 River Blackwater vegetation in (a) May 1996 downstream from Coleford Bridge (b) March 1995 downstream from Coleford Bridge and (c) May 1996 upstream from the footbridge (Courtesy of Dr Robert Sellin, personal communication).

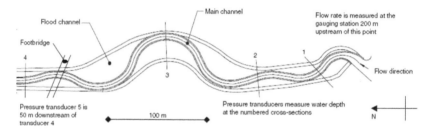

Figure 8.8 Plan form layout of the River Blackwater (Sellin & van Beesten, 2004).

accurate at low flows. The relevant data is that of cross-section 3 (Figure 8.8) which coincides with gauge 3 and is located at the bend apex, with a main channel sinuosity of 1.18. Stage-discharge measurements for three events are available: mid-winter, late winter and mid-summer.

Roughness and cross-section definition

Based on the description and photographs, the Roughness Advisor values include (Figure 8.9):

- *Channel bed* – The RA silt (n_{sur} = 0.018) and submerged fine-leaved plants (n_{veg} = 0.1) are adopted giving n_l = 0.102. The 0.6–1.2 m plants are selected, as this corresponds to the typical flow depths that are observed in the channel. The time-series should be activated for the submerged fine-leaved plants, as this will provide the seasonal distribution of roughness.

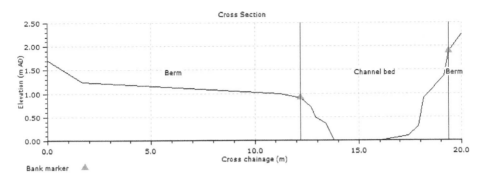

Figure 8.9 River Backwater cross-section, roughness zones and top-of-bank marker locations.

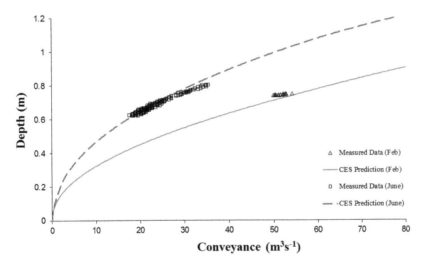

Figure 8.10 Stage-conveyance prediction for the River Blackwater at Coleford Bridge.

- *Channel berms* – The RA silt (n_{sur} = 0.018) and emergent reeds (n_{veg} = 0.15) are adopted giving the total unit roughness n_l = 0.151. The time-series should also be activated for the emergent reeds, as this will provide the seasonal distribution of roughness.

Results for the River Blackwater site

The flow measurements for the River Blackwater are closely linked to the water surface slope, and this varies with depth of flow, with rising and falling longitudinal levels as the event hydrograph passes through and the vegetation bends or washes out. The CES is therefore used to assess conveyance rather than flow rate, given this sensitivity to water surface slope. Figure 8.10 provides the CES stage-conveyance prediction using the RA roughness values and the measured data for February and June. The CES predicts the data reasonable well, capturing the seasonal variation in roughness, the greater flow capacity in February and the significantly reduced flow capacity in June. The upper and lower uncertainty scenarios are not plotted here, as they are broad due to large variation in seasonal roughness.

8.1.3 Stage-discharge prediction for the River Torridge, Torrington, North Devon, UK

Aim

To predict the stage-discharge and upper and lower uncertainty scenarios for the River Torridge site using the CES-AES software and available site information. The outputs are compared with measured data, to provide improved confidence in the results.

The River Torridge at Torrington, Devon, UK

The River Torridge rises near Meddon, flows 93 km through the Devon farming country and ultimately flows into the Bristol Channel. At the Torrington site, gauged water level and discharge measurements are available. The gauging station is located in a straight reach, with a 28 m wide main channel and a bankfull stage of 3.2 m. This is a natural section with a flood plain on the left bank. All measured floods are contained by a flood bank 30 m from the river bank, but larger floods overtop this embankment. The cableway spans 57 m across the river and flood plain to the embankment. The reach-average longitudinal bed slope is 0.00145. The ground material consists of gravels and the berm with trees is shown in Figure 8.11.

Roughness and cross-section definition

Based on the description and photographs, the Roughness Advisor values include (Figure 8.12):

Figure 8.11 Vegetation near the River Torridge gauging station at Torrington (courtesy of the Environment Agency).

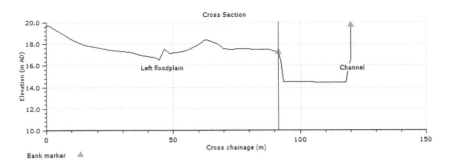

Figure 8.12 River Torridge cross-section, roughness zones and top-of-bank marker locations.

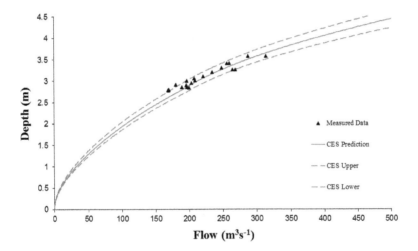

Figure 8.13 Stage-discharge prediction for the River Torridge at Torrington.

- *Channel* – The RA coarse gravels are adopted for the channel ($n_{sur} = n_l = 0.027$).
- *Left floodplain* – The RA coarse gravels ($n_{sur} = 0.027$) and light brush/trees ($n_{veg} = 0.06$) are adopted for the left floodplain, giving the total unit roughness $n_l = 0.065$.

Results for the River Torridge site

Figure 8.13 provides the CES stage-discharge prediction using these RA roughness values and the measured data. Data measurements are only available for a small range of depths, however these are useful in that they cover both inbank and out-of-bank flows. The CES predicts the data reasonably well and, although the data are scattered, they largely falls within the upper and lower uncertainty scenarios. The scatter in the data may reflect the accuracy of the gauging, the seasonable variability in vegetation and/or how well the floodplain topography is captured.

8.1.4 Stage-discharge prediction for the River Blackwater at Ower, Hampshire and Wiltshire, UK

Aim

To predict the stage-discharge and upper and lower uncertainty scenarios for the River Blackwater site at Ower using the CES-AES software and available site information. The outputs are compared with measured data, to provide improved confidence in the results.

River Blackwater at Ower, Hampshire and Wiltshire, UK

The River Blackwater is a river in the English counties of Hampshire and Wiltshire, which is a tributary of the River Test. It should not be confused with the River Blackwater in north-east Hampshire (Section 8.1.2). The Ower gauging station (Figure 8.14) is located along the River Blackwater, Hampshire, about 10 km upstream of the River Test confluence. The section has a Crump profile weir, with a crest width of 6.16 m, which measures low flows reasonably well. For medium to high flows (up to 10 m^3s^{-1}) a velocity area calibration is used as well as current metering (Acoustic Doppler Current Profiler, ADCP). For very high flows, an ultrasonic gauge has been installed however there are no records to date. Although there is a small footbridge 1.5 m upstream of the site, water spills out of bank long before its soffit level of 9.83 m is reached (1.31 m stage). The bankfull stage is about 1.9 m and the reach-averaged longitudinal bed slope is 0.0016. Station measurements include water level and discharge. Although the weir may raise water levels above normal depth for low flows, the measurements used are all taken in the depth range well above the crest of the weir, near bankfull (>5 times the weir height), and thus the effect of the weir is drowned. The channel in the vicinity of the gauge has vertical sheet piling, with vegetation growing out from the top of it. In places, trees have fallen or are growing in the channel and silt islands with vegetation are also present.

Figure 8.14 Gauging station at Ower on the River Blackwater, Hampshire (courtesy of the Environment Agency).

Roughness and cross-section definition

Based on the description and photographs, the Roughness Advisor values include (Figure 8.15):

- *Channel* – The RA silt ($n_{sur} = 0.018$) and light brush and trees ($n_{veg} = 0.06$) are adopted for the channel, giving the total unit roughness $n_l = 0.063$.
- *Banks* – The RA sheet pile ($n_{sur} = 0.025$) and height varying grass ($n_{veg} = 0.041$) are adopted for the banks, giving the total unit roughness $n_l = 0.048$.
- *Floodplain* – The RA silt ($n_{sur} = 0.018$) and medium brush and trees ($n_{veg} = 0.07$) are adopted for the left floodplain, giving the total unit roughness $n_l = 0.072$.

Results for the River Blackwater site at Ower

Figure 8.16 provides the CES stage-discharge prediction using these RA roughness values and the measured data. The measured data cover inbank and low out-of-bank flows. The CES predicts the data reasonably well, other than one outlier where the CES under-predicts the flow. This data point is also an outlier in terms of the flow measurement trend. All data, including the outlier, fall within the upper and lower uncertainty scenarios.

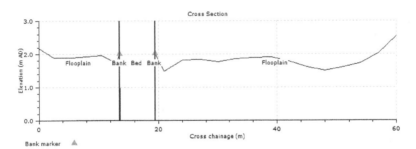

Figure 8.15 River Blackwater cross-section, roughness zones and top-of-bank marker locations.

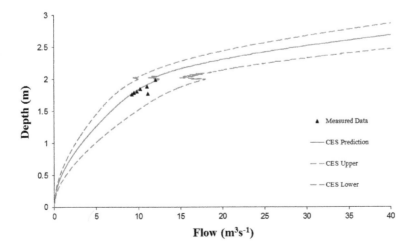

Figure 8.16 Stage-discharge prediction for the River Blackwater at Ower.

8.1.5   Stage-discharge prediction for the River Heathcote, Sloan Terrace, NZ

Aim

To predict the stage-discharge, stage conveyance and upper and lower uncertainty scenarios for the River Heathcote at Sloan Terrace using the CES-AES software and available site information. The outputs are compared with measured data, to provide improved confidence in the results. The CES is also adopted to estimate the Manning *n* values.

River Heathcote at Sloan Terrace, New Zealand (Hicks & Mason, 1998)

The River Heathcote serves a catchment area of 62.5 km², has a mean annual flood of 7 m³s⁻¹ and an average flow rate of 1.6 m³s⁻¹. Measurements were taken along the River Heathcote at Sloan Terrace from July to August 1989. Water level and discharge were measured at two simple channel cross-sections along a 115 m straight reach of constant width, ~8 m. The water surface slope varied from 0.00062 at low depths to 0.00035 at large depths. The river bed consists of angular cobbles and gravel intermixed with mud and silt and the banks are characterised by steep sides with long grass (Figure 8.17).

Roughness and cross-section definition

Based on the description and photographs, the Roughness Advisor values include (Figure 8.18):

* *Channel* – The RA coarse gravel (n_{sur} = 0.027) and mosses (n_{veg} = 0.015) are adopted for the channel, giving the total unit roughness n_l = 0.0309.
* *Banks* – The RA fine gravel (n_{sur} = 0.022) and height varying grass (n_{veg} = 0.041) are adopted for the banks, giving the total unit roughness n_l = 0.0465.

Results for the River Heathcote site

Figure 8.19 provides the CES stage-discharge prediction using these RA roughness values and the measured data, where all the flow measurements are made inbank.

(a)   (b)

Figure 8.17 River Heathcote at Sloan Terrace looking (a) downstream and (b) upstream along the reach (Hicks & Mason, 1998).

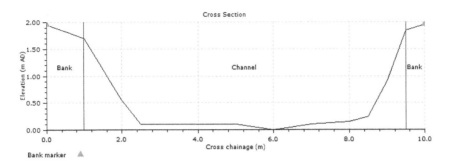

Figure 8.18 River Heathcote cross-section, roughness zones and top-of-bank marker locations.

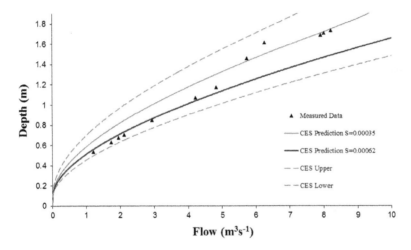

Figure 8.19 Stage-discharge prediction for the River Heathcote at Sloan Terrace.

As the water surface slope varies throughout the flow depth, the CES simulations have been undertaken for two slopes to represent the lower and higher depth ranges, 0.00062 and 0.00035 respectively. Here, the larger slope is in the lower depth range, which may be attributed to less bed vegetation at Sloan Terrace (angular cobbles and gravel) and hence lower depths near the channel beds, compared to the larger depths, where bank side vegetation is present. The results show that the CES predicts the data reasonably well for the lower and higher depths ranges and they illustrate the importance of water surface slope in estimating discharge. The measured data all falls within the upper and lower uncertainty scenarios.

Given the variability in the water surface slope, the CES can also be used to estimate stage-conveyance, which is less slope dependent and can therefore be estimated with a single curve. Figure 8.20 provides the CES conveyance, $K (= Q/S_f^{1/2})$ prediction, which compares reasonably well to the measured data, although the shape of the curve is not identical.

The CES enables users to back-calculate an equivalent Manning n resistance parameter. In modelling practice, a Manning n value is assigned to the channel bed,

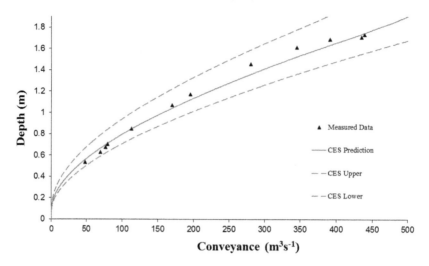

Figure 8.20 Stage-conveyance prediction for the River Heathcote at Sloan Terrace.

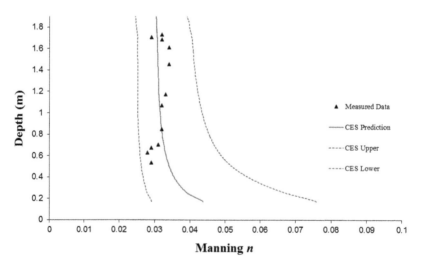

Figure 8.21 Back-calculated Manning *n* values for the River Heathcote at Sloan Terrace.

banks and berms, which is constant for all flow depths. In the field, measurements show how Manning's *n* varies with flow depth (e.g. Chow, 1959), with a greater resistance at low depths where roughness size is comparable with flow depth. The observed Manning *n* values for the River Heathcote site are fairly constant with depth and fall within the range 0.028 to 0.034 (Hicks and Mason, 1998). Figure 8.21 provides the CES back-calculated and observed Manning *n* values with depth. These show a reasonable correlation throughout the measured depth range, with the CES average curve falling through the center of the data points, and all data fall within the upper and lower uncertainty scenarios.

8.1.6 Stage discharge prediction for the River Cuenca, Ucubamba, Ecuador

Aim

To predict the stage-discharge, stage-conveyance and upper and lower uncertainty scenarios for the River Cuenca at Ucubamba, using the CES-AES software and available site information. The outputs are compared with measured data, to provide improved confidence in the results.

River Cuenca at Ucubamba, Ecuador (Abril & Knight, 2004)

The River Cuenca is a natural mountain river located in the Southern Andean region of Ecuador, and it is a tributary to the Paute River. Data measurements, taken along the River Cuenca at Ucubamba, are available from the Hydrologic Institute of Ecuador. These cover 16 discharges over the two year period 1998 to 2000, and include water levels, discharge, velocity and roughness. The channel is 40 m wide and the reach-averaged longitudinal bed slope is 0.0150. The river bed at Ucubamba consists of large boulders, approximately 1 m in diameter, and hence the Manning n values, back-calculated from discharge and water level measurements, range from 0.05 at large depths through to 0.30 at shallow depths (Abril & Knight, 2004).

Roughness and cross-section definition

Based on the description and photographs, the Roughness Advisor values include (Figure 8.22):

- *Channel bed* – The RA cobbles (n_{sur} = 0.035), boulders >50% (n_{irr} = 0.045) and mosses (n_{veg} = 0.015) are adopted for the channel, giving the total unit roughness n_l = 0.059.

Results for the River Cuenca site

Figure 8.23 provides the CES stage-discharge prediction using these RA roughness values and the measured data, where all the flow measurements are made inbank. Although the data is fairly scattered, the trend is reasonably predicted by the CES

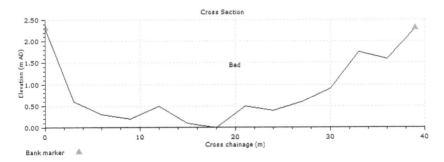

Figure 8.22 River Cuenca cross-section, roughness zone and top-of-bank marker locations.

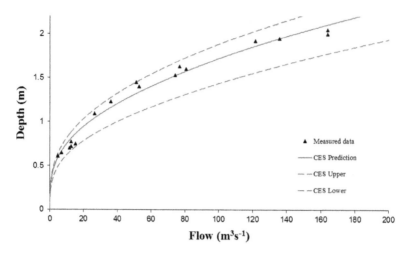

Figure 8.23 Stage-discharge prediction for the River Cuenca at Ucubamba.

and all data fall within the upper and lower uncertainty scenarios. At low and high depths, the curve tends to under-predict the discharge capacity, whereas for medium depths the curve over-predicts the discharge capacity. This is likely to reflect that the roughness variation with depth due to the boulders is not fully captured by the CES roughness formulae (see also Chapter 4, Section 4.2.6).

8.2   ESTIMATING AND USING SPATIAL VELOCITIES

8.2.1   Velocity prediction for the River Severn at Shrewsbury

Aim

To predict the depth-averaged velocity and upper and lower uncertainty scenarios for the River Severn site at Llandrinio, using the CES-AES software and available site information. The outputs are compared with measured data, to provide improved confidence in the results.

River Severn at Llandrinio, Shrewsbury, UK (Babaeyan-Koopaei et al, 2002)

A reach of the River Severn, 20 km east of Shrewsbury, has been monitored for research purposes. The 600 m long study reach consists of a single meander, located south of Llandrinio near Lower Farm (Figure 8.24). The main channel is 30 m wide, 6–7 m deep relative to the right floodplain (180 m wide) and 9 m deep relative to the left floodplain. The longitudinal bedslope is 0.000146. Figure 8.25 shows the vegetation near Llandrinio. Detailed measurements of velocity, using both an Acoustic Doppler Velocity (ADV) and cable supported directional current meter, and turbulence using 3D ADV were recorded immediately downstream of the single meander.

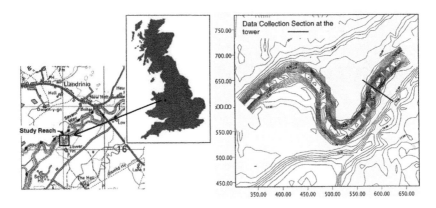

Figure 8.24 Location of the River Severn site in Shrewsbury (Babaeyan-Koopaei *et al.* 2002).

Figure 8.25 River Severn vegetation near Llandrinio.

Roughness and cross-section definition

Based on the site description, the RA values include (Figure 8.26):

- *Channel bed* – The RA sand ($n_{sur} = n_l = 0.02$) is adopted as the bed material.
- *Channel banks* – The RA sand ($n_{sur} = 0.02$) and medium grass 0.75–1.0 m ($n_{veg} = 0.08$) are adopted giving $n_l = 0.082$.
- *Floodplain* – The RA sand ($n_{sur} = 0.02$) and turf ($n_{veg} = 0.021$) are adopted giving $n_l = 0.029$.

Results for River Severn at Llandrinio

This asymmetrical compound section is located on a relatively straight reach downstream of a large meander (Figure 8.24). As the meandering flow patterns and momentum effects are still likely to be present in this downstream reach, a low sinuosity value of 1.1 is assumed. For a flow depth of 7.62 m, the CES predicts a discharge of

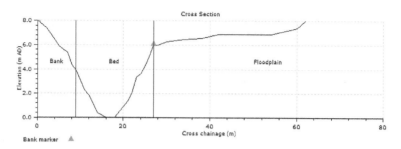

Figure 8.26 River Severn cross-section, roughness zones and top-of-bank marker locations.

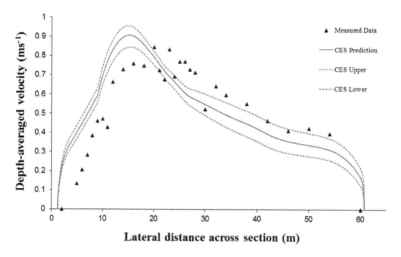

Figure 8.27 Depth-averaged velocity prediction for River Severn, Llandrinio, for depth 7.62 m.

101 m³s⁻¹ and the measured flow is 103 m³s⁻¹ (within 2%). Figure 8.27 provides the CES depth-averaged velocity for this depth. The velocity distribution is reasonably predicted in terms of velocity magnitude, however the skewness of the data is not fully captured, suggesting that the upstream bend effects are affecting the local velocities at the site more than was accounted for by the CES.

8.2.2   Velocity prediction for the River Frome at East Burton, Dorset, UK

Aim

To explore the effect of re-introducing gravels on the local flow conditions at a site along the River Frome. The CES-AES software is used to estimate stage-discharge, bankfull flow and depth-averaged velocities for the existing situation (*scenario: no gravel*) and for a scenario where gravel is introduced, altering the cross-section shape and roughness (*scenario: gravel*). The approach assumes that the gravel

(a)   (b)

Figure 8.28 River Frome at East Burton looking (a) upstream and (b) downstream along the reach (courtesy of Centre of Ecology and Hydrology).

which is introduced does not move. No flow measurement data is available for this site.

River Frome at East Burton, Dorset, UK

The River Frome in the south of England is 48 km long, serves a catchment of 454 km² and has an average annual discharge of 6.38 m³s⁻¹. The river runs over sands, clays and gravels which overlie the chalk in the region. At the site in East Burton, the river was dredged for flood alleviation in 1950s and, as a result, gravels are no longer available for over-wintering juvenile salmon to burrow into for cover. The site is characterised by sand and silt, with ranunculus plants covering large portions of the channel (Figure 8.28). The longitudinal bed slope is 0.0016 and the section is located in a gentle meander, with sinuosity 1.08.

Roughness and cross-section definition

Based on the site description, the RA values include (Figure 8.29):

- *Channel bed* – The RA silt ($n_{sur} = n_l = 0.018$) is adopted as the bed material.
- *Channel bed with vegetation* – The RA silt ($n_{sur} = 0.018$) and submerged fine-leaved plants ($n_{veg} = 0.10$) are adopted giving $n_l = 0.102$.

For the proposed gravel scenario, whereby the channel bed is raised be the introduction of the gravel, the RA values include (Figure 8.30):

- *Channel bed* – The RA gravel ($n_{sur} = n_l = 0.025$) is adopted as the bed material.
- *Channel bed with vegetation* – The RA silt ($n_{sur} = 0.025$) and submerged fine-leaved plants ($n_{veg} = 0.10$) are adopted giving $n_l = 0.103$.

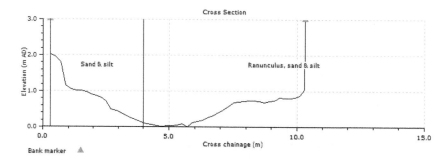

Figure 8.29 River Frome cross-section, roughness zones and top-of-bank marker locations for the *no gravel* scenario.

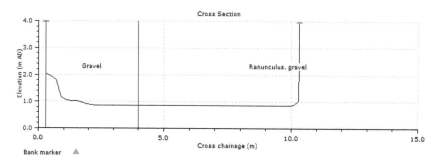

Figure 8.30 River Frome cross-section, roughness zones and top-of-bank marker locations for the *gravel* scenario.

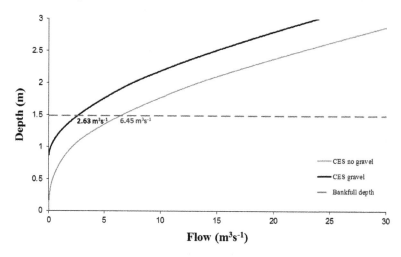

Figure 8.31 Stage-discharge prediction for the River Frome, East Burton.

Results for the River Frome at East Burton

Figure 8.31 provides the CES stage-discharge predictions for the two scenarios using these RA roughness values. This illustrates the large reduction in channel capacity

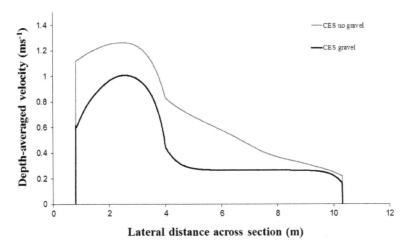

Figure 8.32 Depth-averaged velocity predictions for River Frome for bankfull depth 1.49 m.

with the introduction of the gravels, assuming these do not move, with bankfull discharge reducing from 6.45 m³s⁻¹ to 2.63 m³s⁻¹. Figure 8.32 provides the corresponding depth-averaged velocity profiles for the two scenarios, where the velocity is reduced throughout the section with the introduction of the gravels. This type of information can be used to help design the optimum channel shape and cover for promoting habitat conditions of different species at low flows. It can also be used to consider if reductions in channel capacity pose a concern for flooding, particularly, for high flows.

8.2.3   Velocity prediction for the River Rhône, Lyon, France

Aim

To predict the depth-averaged velocity for two river sections along the River Rhône, using the CES-AES software and available site information. The outputs are compared with measured data, to provide improved confidence in the results.

River Rhône, Lyon, France

The River Rhône is one of the major rivers in France, rising in the Swiss Alps, passing through Lake Geneva and running through south-eastern France to its mouth on the Mediterranean Sea. It is 813 km long, serves a catchment of 98,000 km² and the average annual discharge at Arles, near the mouth, is 1,710 m³s⁻¹. Data, measured with an Acoustic Doppler Current Profiler (ADCP), are available for a reach south of Lyon. This reach has a longitudinal bed slope of 0.0005 and is characterised by sand and variable grass cover, with shrubs and trees in the floodplain (see Chapter 7, Figure 7.27).

Roughness and cross-section definition

Based on the site description, the RA values include (Figures 8.33 and 8.34):

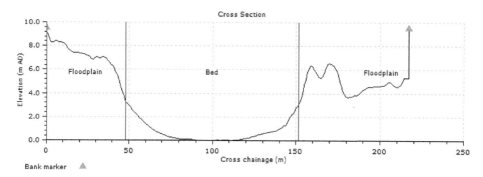

Figure 8.33 River Rhône, cross-section PK7-85, roughness zones and top-of-bank marker locations.

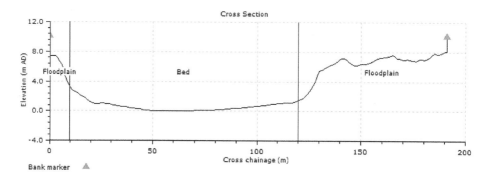

Figure 8.34 River Rhône, cross-section PK7-1, roughness zones and top-of-bank marker locations.

- *Channel bed* – The RA sand (n_{sur} = 0.02) and height varying grass (n_{veg} = 0.041) are adopted, giving n_l = 0.0456.
- *Floodplain* – The RA sand (n_{sur} = 0.02) and heavy stand of trees (n_{veg} = 0.15) are adopted giving n_l = 0.151.

Results for River Rhône, Lyon

Figure 8.35 provides the CES depth-averaged velocity for cross-section PK7-85 at a depth of 9.2 m. The CES predicts a flow of 2,330 m³s⁻¹ and the measured flow is 2,380 m³s⁻¹, which is within 2.1%. The velocity distribution is reasonably well predicted across the section, other than the right floodplain, where the CES over-predicts the flows. This may be that the impact of the vegetation is greater than the RA roughness values suggest, or that there are localized features on the floodplain that have been missed.

Figure 8.36 provides the CES depth-averaged velocity for cross-section PK7-1 at a depth of 9.21 m. The CES predicts a flow of 2,385 m³s⁻¹ and the measured flow is 2,380 m³s⁻¹, which is within 0.2%. The velocity distribution is reasonably well predicted across the section, other than the floodplains, where the CES over-predicts the flows, but to a lesser degree than for cross-section PK7-85.

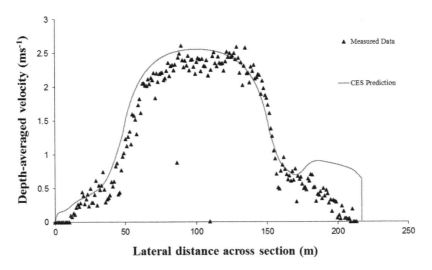

Figure 8.35 Depth-averaged velocity prediction for River Rhône at cross-section PK7-85, depth 9.2 m, CES predicted flow 2330 m³s⁻¹, measured flow 2,380 m³s⁻¹.

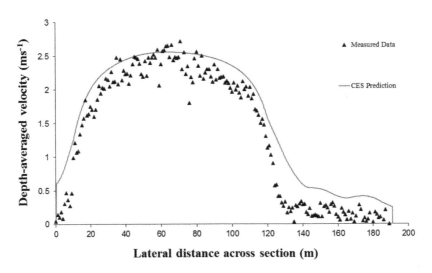

Figure 8.36 Depth-averaged velocity prediction for River Rhône at cross-section PK7-1, depth 9.21 m, CES predicted flow 2,385 m³s⁻¹, measured flow 2,380 m³s⁻¹.

8.2.4   Velocity prediction for the Yangtze River, China

Aim

To predict the depth-averaged velocity for a river section along the Yangtze River, using the CES-AES software and available site information. The outputs are compared with measured data, to provide improved confidence in the results.

Yangtze River, ZhuTuo, China

The Yantgze River is the longest river in Asia and the third-longest in the world. It drains one-fifth of the land area of the People's Republic of China and its river basin is home to one-third of the country's population. It is 6,300 km long, serves a catchment of 1,808,500 km² and the average annual discharge is 30,166 m³s⁻¹. The site is located at the ZhuTuo gauging station, in a low gradient reach with a longitudinal bed slope of 0.000065. The substrate of the Yangtze River basin consists of fine sand, silts and clay.

Roughness and cross-section definition

The roughness information for the site is limited. Thus, initially a run was undertaken with the unit roughness for the whole cross-section based on the RA silt ($n_{sur} = n_l = 0.018$). For a depth of 21.56 m (207.13 m above datum), this gives a flow rate of 22,691 m³s⁻¹, which is within 2% of the measured value 23,127 m³s⁻¹.

The velocity prediction was then compared to the measured data to refine the cross-section roughness information as follows (Figures 8.37):

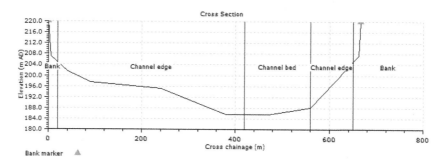

Figure 8.37 Yangtze River roughness zones and top-of-bank marker locations.

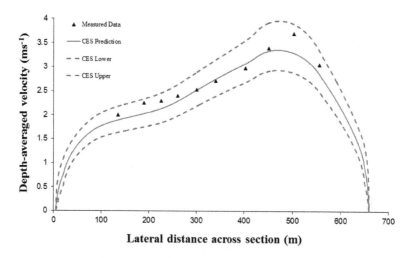

Figure 8.38 Depth-averaged velocity prediction for Yangtze River for a depth 21.56 m (207.13 m above datum), CES predicted flow 24,107 m³s⁻¹, measured flow 23,127 m³s⁻¹.

- *Channel bed* – The RA for very fine clay was adopted in the channel center i.e. similar to a clay lining ($n_{sur} = n_l = 0.014$)
- *Channel edge* – The RA silt ($n_{sur} = n_l = 0.018$) was adopted closer to the channel edges.
- *Channel banks* – The RA silt ($n_{sur} = 0.018$) and height varying grass ($n_{veg} = 0.041$) are adopted giving $n_l = 0.045$.

Results for Yangtze River

Figure 8.38 provides the CES depth-averaged velocity for a depth of 21.56 m (207.13 m above datum). The CES predicts a discharge of 24,107 m³s⁻¹ and the measured flow is 23,127 m³s⁻¹, which is within 4.1%. The velocity distribution is reasonably predicted across the section generally, however the highest velocities in the channel center are under-predicted. These under-predictions may arise from localized variations in channel roughness or features not captured in the analysis. All data fall within the upper and lower uncertainty scenarios.

Chapter 9

Concluding remarks

ABSTRACT

These concluding remarks draw together some of the key points made in earlier chapters and highlight the more important issues involved in modelling flows in rivers. Suggestions are made about future work that is required in the area of modelling as well as possible developments in the CES-AES software. The need for accurate measurements of certain hydraulic and turbulence parameters at full scale is especially highlighted. These should be carried out in conjunction with the development and rigorous testing of improved computer-based modelling tools, tested on a selected number of rivers and against benchmark laboratory data.

9.1 CONCLUDING REMARKS

As outlined in the Preface, the principal driver for the development of the Conveyance Estimation System and the Afflux Estimation System (CES-AES) software, as well as the underpinning research described throughout this book, was the need for better computer-based tools for flood risk management. Although predominately developed with the needs of the UK in mind, the subject matter is relevant to all rivers, watercourses and open channels, as shown by the application of the current software to a number of international rivers. These were carefully selected to extend both the range of discharge, as well as the type of river (mountain boulder to lowland) and substrate. Fuller accounts of the application of the CES component to a wide range of international rivers are given by Mc Gahey (2006) and Mc Gahey *et al.* (2006 & 2008).

The global scale of flooding, as outlined by Berz (2000), is highlighted in Chapter 1 and the examples drawn from Europe and Asia are used to illustrate the wide range of problems that flooding causes. Having established the international nature of floods, as well as the importance of making sound estimates of the conveyance capacity of channels and water levels under flood flow conditions, new drivers are described that impinge on the work of the hydraulics engineer. These derive from recent international legislation and planning directives, such as the European Water Framework Directive (2000) and the European Directive on the Assessment and Management of Flood Risks (2007), or similar legislation that has been enacted in Japan, the USA and other countries.

In the light of the European legislation, as well as the requirement for better flood risk mapping, the CES-AES software deliberately focuses on the determination of water level in rivers, rather than discharge, since it is the water level that relates particularly strongly to: 1) damage to infrastructure and property, 2) national and local economies, and 3) loss of life. Furthermore, since the water level in rivers and watercourses are predominately influenced by channel roughness and hydraulic structures, especially bridges and culverts, the two elements that feature strongly in the software are those of flow resistance and afflux. The development of ancillary tools, such as the Uncertainty Estimator (UE) and the Roughness Advisor (RA), are aimed at providing the best possible estimates of channel roughness and flow resistance (two related but separate concepts). The Roughness Advisor was developed from a very extensive knowledge of data world-wide, again making it suitable for application both within the UK and internationally. The development of this particular practical component of this toolset is described more fully elsewhere by Mc Gahey *et al.* (2009).

Another important focus and distinguishing feature of the CES-AES is that it deals with both inbank and overbank flows as a continuum, rather than as disparate types of flow. This is, after all, how rivers behave naturally, and one would therefore expect any modelling software to mimic this natural process. Although overbank flows introduce more complex flow structures, these are handled in a similar way as for inbank flows, through the use of the same three hydraulic calibration parameters for each panel, f, λ and Γ, representing local friction, eddy viscosity and secondary flows, thus unifying the methodology. Further parameters related to wave speed, sediment transport and dispersion may also be added, creating a holistic approach to modelling river flows, as illustrated in more detail elsewhere by Knight (2006c). Although the CES-AES is restricted to simulating steady flows with a fixed bed, this still allows many types of practical problems to be solved. Unsteady flow phenomena and flow over loose boundaries, such as sediments, may be readily treated by further development of the CES in the future, allowing alluvial resistance formulae to be used, together with channel adjustment algorithms for channel widening, deepening or slope changes.

One particularly useful feature, often not found in comparable software, is the ability of the CES-AES to provide lateral distributions of depth-averaged velocity, U_d, and boundary shear stress, τ_o, as well as conveyance and water levels. The lateral distributions of U_d are particularly useful for calibration purposes, since they mimic what is often measured when river gauging, as well as being useful in studies related to the maintenance of vegetation. The distribution of τ_o around the wetted perimeter is likewise particularly useful, as it is the most important parameter required for any studies into sediment behaviour and geomorphology.

The title of the book was deliberately chosen to highlight the practical emphasis given throughout, despite the significant scientific nature of some elements in certain chapters. The authors believe that it is no longer possible to treat even one-dimensional (1-D) flow as 'simple', since turbulent flows are inherently three-dimensional and should be recognised as such. Thus the examples in Chapter 2, although beginning with relatively straightforward hydraulics, based on Manning's equation, soon take the reader into the difficulties of modelling flows in rivers and watercourses using the Reynolds-Averaged Navier-Stokes equations, or RANS for brevity. At the same time, an attempt is made to illustrate the physical processes at work in river flows, without 'blinding the reader with too much science'. It is hoped that the judicious use of case studies throughout that chapter, without all the details of the RANS equations or the CES-AES methodology being explained, is helpful. It is often said that we learn by repetition, rather like peeling layers from an onion, each 'layer' adding to and building on our existing knowledge, taking us deeper into the subject matter. It is for this reason that the full description of the CES-AES methodology is delayed until Chapter 3, with detailed examples of the use of the software reserved until Chapter 4.

The use of examples throughout the book reflects this repetitive, 'layer' by 'layer' approach, beginning with some simple hydraulics calculations and concepts in Chapter 2 (Examples 1–5), followed by examples of flows in rectangular and trapezoidal channels, analysed by the Shiono & Knight method (SKM). The SKM is introduced at an early stage, as it is not only a pre-cursor to the CES approach, but also may be used at a spreadsheet level by the reader, as shown by the various examples given. These are followed by further 'layers', typically case studies of flows in engineered channels and natural rivers, selected to illustrate and aid understanding of flow structures and fluid flow behaviour, rather than just computational techniques. Particular emphasis is placed on understanding the physical processes and the role of the calibration parameters used within the computational model, a feature which is often overlooked in books on computational hydraulics and sometimes even by modellers alike. It is often said that river engineering is an 'art' as well as a 'science', and that modelling needs therefore to take into account two distinct elements: 'theoretical fluid mechanics' and a multitude of 'practical issues'.

This is illustrated by Figure 9.1, taken from Nakato & Ettema (1996), in which river engineering is envisaged as the joining together of two river banks, one named 'theoretical fluid mechanics' and the other 'practical problems'. On the left hand side (looking downstream), the shapes of the zones represent various practical river problems and their inter-related nature. To deal with a specific problem or practical issue, one has to understand its relationship with other problems, gain that knowledge, cross the river by one of the bridges, pick up the appropriate knowledge on the right hand bank (again looking downstream), return by another bridge and deal with the particular practical problem or issue in question. In doing so, one may have to repeat the journey several times, or recollect previous journeys, thus slowly building up a more complete picture of the river system itself. The two river banks should not therefore be regarded as opposites, but as complementary. Sometimes a false dichotomy develops between people who tend to inhabit one side of the river or the other, perhaps stereotyped by the outdated 'division' between practitioners and theoreticians, each regarding the other side as either aliens or at worst useless. However, rivers always have two banks, and the most fruitful advances in river engineering have occurred when the two sides meet, exchange views and deal with common issues. It should be noted that the various areas of theoretical knowledge also have various regions too. The river bank alongside the continuity and Navier-Stokes equations has sheet piling, indicating that the bank is firm and well established. It has been there since 1845. However, the sheet piling barely extends into the turbulence region, where the ground appears to be marshy and not well developed. The high rise buildings nearby indicate enormous growth, activity and development in this region, as typified by the growth in the subject of Computational Fluid Dynamics (CFD) and advanced experimental studies in turbulence. A complete understanding of turbulence is however for future generations to achieve.

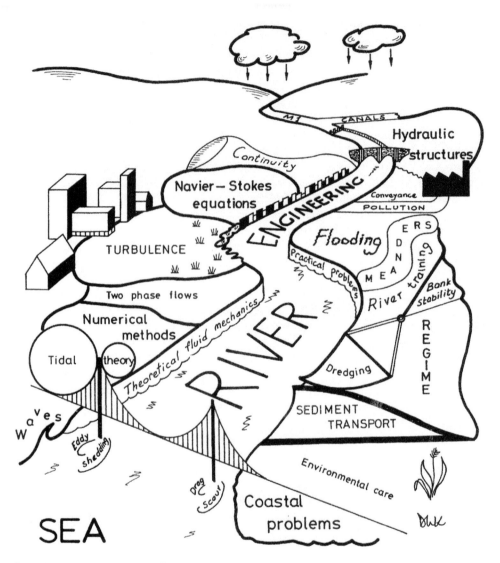

Figure 9.1 The art and science of river engineering (after Knight) [reproduced from Nakato & Ettema, (1996), page 448].

Despite our lack of knowledge of certain aspects of turbulence, the Reynolds-Averaged Navier-Stokes (RANS) equations have been used herein as the basis for the SKM as well as for the CES-AES software, as they are well established, even if the constituent equations for turbulence are not. Greater effort is needed in the area of turbulence measurements in open channels, especially in natural rivers at full scale. At present such data are relatively scarce, mainly due to cost considerations. Does an isometric view of the river between the banks in Figure 9.1 needs to be drawn, zoned and annotated by what is required below the free surface?

Figure 9.1 also reminds the reader that the effective use of any software in river engineering depends as much on the knowledge of the end-user as the software developer. Modelling should never be regarded as simply a matter of feeding numbers into a computer without fully understanding either the nature of the river flow one is attempting to model, or the simulation algorithms that approximate the turbulence. Furthermore it should be recognised that the particular practical problem or issue being investigated may be only one amongst many, as illustrated in Figure 9.1, where 'conveyance' is shown as just one relatively small region.

A common danger in interpreting model output is to believe that the visualisations (e.g. vector plots, velocity fields, etc.), are a true representation of the physical process. What is seen can be an artefact of the visualisation algorithms and not a true reflection of the solution of the numerical procedures in the hydrodynamic simulation. For example, post-processing of raw outputs may have smoothed results or introduced false flow structures. The numerical results themselves will also mask the true complexity of the turbulent flows in nature. There is sometimes a tendency for users to give greater weight to more detailed pictorial output from a 2-D or 3-D model, than from a 1-D model or measurements. A useful aid for understanding physical flow processes is the DVD from Cambridge University Press (for further information, see website: *http://www.cup.es/uk/catalogue.asp?isbn=0521721695*).

Having introduced the reader in Chapter 2 to some important fluid flow concepts, equations and a possible solution methodology via the SKM, as introductory 'layers', the worked examples may be considered as the next few 'layers'. These emphasise the distinction between roughness and fluid flow resistance, the various types of friction factor (global, zonal and local), the way that the SKM deals with both inbank and overbank flows within a unified methodology, the use of calibration parameters, the complex nature of flows through bridges and culverts, as well as some of the more common difficulties in modelling flows. After this introductory phase, a major series of 'layers' then follows in Chapter 3, beginning with a systematic exposition of flow structures in rivers and at bridges and culverts. At the same time the methodology behind the CES-AES is fully described, omitting unnecessary detailed application to actual rivers, bridges and culverts, to deepen understanding of the 3-D nature of various fluid flow structures, as well as highlighting how the different components are integrated within a single system to simulate water levels and flow behaviour.

One of these 'layers' is particularly important, namely Section 3.2.3, in which the CES methods, outputs and solution techniques are re-capitulated concisely. Tables 3.4 and 3.5 summarise the parameters used and the key assumptions, together with their implications and significance. Appendix 1 gives further details of all the equations used. Section 3.2.5 deals with the afflux methods employed within the AES in a similar way. These layers are the 'heart' of the onion, so to speak, and should be studied carefully. Software manuals and explanatory notes are sometimes notorious in not stating clearly and concisely the nature of the methodology used, together with any implicit assumptions. In addition to this, it is hoped that in due course an 'open-code' version of the CES-AES will be released so that engineers and research scientists can amend, replace or add in their own particular components. This attempt at 'transparency' is aimed at helping users to not only understand the CES-AES thoroughly, but also to use it more effectively and perhaps even to develop it further in the future.

Having now peeled the onion, so to speak, down to its heart, there are still six more Chapters that follow. Continuing with the food metaphor, one might therefore regard these as the more interesting chapters in a 'recipe book' in which, following the earlier chapters on basic cooking skills, kitchen equipment, nutrition and food types, one now begins to apply it and get some benefit. In the same manner, these later Chapters illustrate how the CES-AES and the SKM may be used in practice, beginning with an overview in Section 4.1, and followed by many specialist sections. These contain numerous examples, some relating to estimating stage-discharge relationships in different types of river, and others to estimating lateral distributions of velocity and boundary shear stress. Some of these examples show comparisons between the back-calculated Manning's n values with measured values. Others show the effect of changes in the stage-discharge relationships arising from changes in roughness, slope or sinuosity. The uses to which these are put, as well as the uncertainties are illustrated throughout the Chapter. Furthermore, the cross-section data for some representative rivers from the UK, South America and New Zealand are given in Appendix 3 for those interested in repeating the calculations themselves. Sections 4.3 and 4.4 deal with estimating afflux at bridges and culverts and illustrate the use of the AES by applying it to a wide range of cases, again some from the UK and some from overseas. This Chapter thus caters for those interested in international cuisine, and it is hoped that others may add their recipes to enrich the book in the future.

If Chapter 4 is regarded as illustrating 'established' recipes, based on experience, then Chapter 5 may be regarded as one of 'experimental' recipes, based on novel uses for the CES-AES. The two ideas or 'recipes' that have already been tried involve ecological and geomorphological issues, which serve to show the potentially wide range of application for the software. Section 4.6 has already indicated how the lateral distributions of depth-averaged velocity, U_d, can be used in assessing vegetation and the maintenance of weedy rivers, as well as cutting regimes, and the link between invertebrates and weed cover. These distributions might also be used to assess other ecological issues, such as balancing the conflicting requirements of flood risk management, when the maximum discharge capacity is usually needed, with that of biodiversity, when a more varied velocity distribution is required to support different flora and fauna. Habitat information such as response curves (velocity/depth preferences) and life scores can be used with the CES-AES outputs to identify preferred habitats. An example of both deepening and dredging has been given for the River Hooke in Section 4.6.3.

The second application is more strongly related to the lateral distributions of boundary shear stress, τ_o, as this parameter is frequently required in calculations involving sediment motion, as well as in many geomorphological assessments. Section 5.2 shows how both the stage-discharge and sediment rating curves may be estimated, for both inbank and overbank flows, again indicating a continuum. The CES approach may also be applied to other sediment issues, such as the determination of a channel shape for threshold conditions, the effects of widening or deepening the channel cross-section on the velocity field, as well as the design of stable channels and their correspondence with analytical regime formulae. See Cao & Knight (1996 & 1998), Knight & Yu (1995) and Yu & Knight (1998) for further details. The recent book by Ikeda & McEwan (2009) also contains useful information on the modelling of sediment behaviour in meandering rivers with overbank flow.

Chapters 6 & 7 may be regarded as the 'heart' of the explanation of the SKM, which was only briefly dealt with in Chapter 2, Section 2.3, as a necessary introduction to the CES. The SKM offers more possibilities in the description of flow in open channels, due to analytical solutions being obtained to the governing partial differential equation (pde), given as (2.58) or (3.4). These chapters, together with the 3 additional Appendices, 4–6, in this second edition, provide engineers and researchers with the necessary detail so that they can possibly adapt the SKM further to their own particular needs, or alternatively use it within Excel as they please. This extra level of detail will also be useful to those researchers who want to develop new analytical tools, such as the one between SKM and the 'area method', described in Section 6.5.3(c), or SKM and the analytical regime approach to sediment transport and morphology, as described in Section 7.1.8.

The worked examples in Chapter 7 illustrate the wide variety of engineering problems that the SKM is suited for. Since one of the objectives behind its development was to provide a tool for the prediction of boundary shear stress distributions in open channel flows, it is particularly suited for resistance and sediment studies. Since the method applies to both inbank and overbank flows, as well as the transition zone in between, it is especially suited for application to flood studies. It also has particular relevance to laboratory flume studies in which side-wall correction procedures are generally required.

Chapter 8 provides further worked examples for a range of channel types (size, shape, cover, sinuosity), based on the CES, illustrating its capabilities and providing comparative results with those in Chapters 6 & 7, based on the SKM. Sites from rivers in the UK, France, China, New Zealand and Ecuador are considered. In this respect, it should be noted that International Standards are now beginning to cite commercially available software for use in evaluating stage-discharge relationships, as for example ISO 18320 (2018), which refers to both the CES and SKM as one method of achieving this objective.

It is always important to appreciate what any software 'is' and what 'it is not'. It must be stressed that the CES-AES is not a universal tool aimed at solving all flow and water level problems in rivers and drainage channels and watercourses. It has been designed specifically as a relatively simple tool, dealing with practical issues of water level estimation in steady flow only, stage-discharge relationships, backwater effects, afflux at bridges and culverts, all of which are frequently encountered in engineering practice. It may be operated as a stand-alone piece of software, or embedded within other more complex river engineering modelling systems (e.g. as in ISIS and InfoWorks RS), allowing the non-expert and the experienced river engineer alike to use it, hopefully with relative ease. The limitations are listed in Sections 1.3 and 3.2.3, and the various examples illustrate its capabilities and potential, as well as some of its shortcomings and limitations.

Ancillary tools such as the Roughness Advisor (RA) and the Uncertainty Estimator (UE) are included in the book as they offer practical help in decision making over arguably the most important calibration parameter, the roughness. They also may be used in assessing the output from the combined influence of all three calibration parameters. The practical emphasis is highlighted again by the focus on stage-discharge relationships, since these are frequently used in making decisions in flood risk management, as well as in other areas of river engineering. The subject index highlights these, as do many of the worked examples.

In addition to a comprehensive list of references, mention should also be made of the data sources, described in Section 2.7 and also contained in the websites *www.flowdata.bham.ac.uk* and *www.river-conveyance.net*. It should be noted that the studies in the Flood Channel Facility (FCF) have featured significantly in the development of the CES-AES, as have other experimental studies. Without such data and their subsequent analysis, the concepts and equations presented herein would not have materialised. It is suggested that further progress in the ability to model flows in rivers successfully could therefore be made by designing carefully focussed experiments, some to be undertaken in the field and others in laboratories. These should be undertaken to specified standards of accuracy, and then combined with other quality-assured data from other researchers and practitioners, creating a resource that will benefit all. This re-iterates the comments made by Chezy in the 1760s, as noted by Mervyn Bramley in his Foreword to the first edition of this book!

9.2   CES AND SKM: DIFFERENCES AND SIMILARITIES

It is evident from the preceding chapters that there are some differences in approach between the SKM and the CES; these are now highlighted so that those developing these methods further may appreciate some of the subtle or less obvious ones. Firstly, a difference lies in the choice of main parameter, with the SKM using the depth-averaged velocity, U_d, and the CES using the depth integral of the local velocity which gives the discharge per unit width, q ($= U_d H$). The choice of q in preference to U_d, is made in order to deal with potential problems that may arise when modelling vertical internal boundary walls, as explained in Section 6.3. For all other types of boundary, horizontal or with a transverse slope, where H is continuous, the choice of U_d or q makes essentially no difference to the results, provided the boundary shear stresses are properly accounted for. For vertical walls, the SKM approach uses a force balance, as explained in detail in Section 6.3.3.

It should be noted that the coordinate axes in the SKM are in the streamwise direction, as explained in Section 6.1.1, whereas in the CES a global frame of reference is used with two axes in the horizontal plane and the third orientated vertically. The SKM choice derives from the fact that SKM was based initially on modelling the extensive velocity and boundary shear stress data from the Flood Channel facility (FCF), upon which the subsequent theoretical formulation of both the CES and the SKM were based. In the laboratory experiments great care was taken to ensure the flow in the flume was at normal depth and so the bed gradient, surface gradient and friction slope were identical; thus it is natural to base the analysis with axes inclined at this gradient to the horizontal. Later on, in the development of the CES which was intended for use in natural and artificial open watercourses, the global frame of reference was selected. The choice of the x-co-ordinate as being horizontal, facilitates the input of measured river survey data and application of the CES to cases where the bed gradient and hydraulic conditions are gradually varying.

There are subtle differences introduced in the formulation of the flow equations by the choice of axes. In the SKM the driving body force of gravity has a component in the streamwise coordinate direction, whereas in the CES the effect of gravity is introduced through the resolved component of hydrostatic pressure gradient in the horizontal direction since the inclined free surface is a plane of constant pressure. Since the streamwise gradient of most rivers is extremely small (under 1%) and the differences will be second-order in terms of the gradient, the choice of frame orientation between the SKM and the CES makes very little difference in most practical cases. However, care should be taken when applying either method to steep rivers and especially when formulating boundary shear stresses on river banks, as Equations (3.6) and (6.18) differ in this respect.

There are certain advantages in working with direct analytical solutions, as in the SKM approach, rather than relying solely on numerical procedures to solve the governing differential equations. Analytical solutions not only provide a simple check on any numerical solutions, but also allows one to obtain algebraic formulations for the individual A_i coefficients, enabling analytical expressions to be obtained for the percentage shear force on any boundary element, as illustrated by (6.113) & (6.114) for example. These should naturally agree with any values obtained by numerical integration laterally of local boundary shear stresses from an individual panel.

A further advantage of the SKM is that both the analytical and numerical approaches may be readily inserted and solved using MS EXCEL, making it easy for the user to adapt their programs without reference to, or the need of, any other software. Furthermore, key parameters, or alternative formulations for certain fluid processes may be readily altered to suit particular circumstances. This has been illustrated in Section 6.6.3, where the development of an equation for the drag force on trees (6.157) is described in detail.

One disadvantage of the SKM, highlighted in Section 6.4, is that of obtaining purely algebraic expressions for the individual A_i coefficients when the number of panels is large. The CES on the other hand can discretize any cross-section into as many panels as required (e.g. may be >1000 for large or complex channels) in order to best capture local variations in topography and hence flow parameters, making it easier to investigate curvilinear boundaries, such as alluvial channels illustrated in Figure 7.19. It is suggested that in the SKM, it is unrealistic to attempt to process more than 8 panels, as the algebra increases in complexity with the number of panels selected. Moreover, 8 panels are usually sufficient to represent most river cross-sections with two floodplains, as illustrated by Figure 7.1. Furthermore, each panel may be divided into dozens of smaller panels with the same 3 coefficients, f, λ & Γ, for greater accuracy. It has also been shown that where there are significant changes in parameters between panels or sub-panels, often resulting in sudden changes in boundary shear stress, the analytical solution may be 'adapted' to take this into account, as illustrated in Figure (2.34) and described at the end of Section 6.3.2. However, this approach should be undertaken carefully, as from a mathematical perspective it negates the variation of λ in the differential third term of (6.10).

One difficulty in seeking purely analytical solutions to the governing equation is that the algebraic expressions for the A_i coefficients and U_d, lead to elliptic integrals when any further integration is required to obtain purely analytical expressions for the stage-discharge relationship. Although it is possible to undertake these integrations, as shown by Liao & Knight (2007a&b), it is easier to undertake repeated simulations of U_d, at specific depths and build up the $H\,v\,Q$ relationship sequentially, as is done in the CES and SKM when used normally. However, the analytic $H\,v\,Q$ relationships do offer certain insights into the contributions of the individual parameters, f, λ & Γ to the stage-discharge relationship.

One significant advantage of using a mature, dedicated program, such as CES, is that other tools can be embedded within it, such as uncertainty limits, roughness estimators and ancillary data from comparative sites. In addition, the CES can be embedded within other hydrodynamic modelling tools. Further updates, developments and tools can then be easily added and mounted centrally, as in icloud computing. The CES is now available in this form in Flood Cloud, provided by CH2M Software. By contrast, the short Fortran program in Appendix 6 may be used by researchers who may wish to use the SKM for their own purposes, without having to use default parameters or restrictions placed on them by a commercial software provider. It has been demonstrated that MS EXCEL provides the user with a very flexible platform in which to explore and develop the SKM.

Although SKM and CES have incorporated certain key features of 3D flow, it should be remembered that it is still essentially a 1D streamwise program and should strictly only be used for this purpose, taking note of the limitations listed in Section 1.3. Despite this, many gauging stations are sited for technical reasons on predominately straight river reaches, where there are prismatic cross-sections, for which the SKM and the CES are eminently suitable. With increasing emphasis on the need to estimate river levels accurately, computational methods are becoming increasingly embedded in flood risk management procedures. This is also evident in International Standards, such as the one on 'Determination of the stage-discharge relationship' (ISO 18320), in which both the SKM and CES procedures are referenced. Both these methods provide engineers with well-researched tools which have been subsequently developed for practical use.

9.3   FUTURE DEVELOPMENTS

From the experience already gained from using the CES-AES, together with new science, a number of future developments and research areas have been recommended to ensure the scientific relevance of the CES-AES is maintained (EA/Defra, 2009). These recommendations are intended to deal with known deficiencies in elements of the theoretical background and methodology, incorporating emerging science, and developing related tools and aids. The review identifies upwards of 70 prioritised recommendations, prioritised following extensive consultation with a wide range of stakeholders in relation to benefits and value for money, which are categorised as either "enhancements to the science" or "enhancements to the software features". Some of the high priority aspects recommended for inclusion in the short-term are:

Updating the Roughness Advisor to reflect new science and roughness information – Roughness has the most significant contribution to the uncertainty in conveyance (Khatibi *et al.*, 1997; Latapie, 2003; Mc Gahey, 2006) and of the different roughness types, vegetation roughness includes the greatest natural variability (Sellin & van

Beesten, 2004; Defra/EA, 2003b). Much work (Dawson, 1978; Westlake & Dawson, 1982; Baattruo-Pedersen & Riis, 2004; Naden et al., 2006; O'Hare et al., 2008) has been undertaken on understanding this variability, including the seasonal variations and regrowth following cutting, through measurement of biomass, plan form cover and cross-sectional area occupied by plants. Available measurements are largely for two key species, emergent reeds (*Sparganium erectum*) and submerged fine-leaved plants (*Ranunculus pencillatus*), which are two of the most commonly occurring weeds in managed channels in England and Wales. There is information available to derive seasonal uncertainty distributions for vegetation roughness and to extend these to other vegetation morphotypes.

Other RA improvements include, for example, updating the photographic database to ensure each roughness type has ample photographic evidence; improving roughness values for pools and riffles based on recent findings at the River Restoration Centre (Janes et al., 2005); and updating the embedded River Habitat Survey information (Raven et al., 1998) to reflect the recently completed survey (2003–09 based on EA, 2003), including additional fields such as substrate type and trash present.

Developing a Channel Maintenance Module – There is a significant amount of literature and guidance available on channel maintenance (e.g. CAPM, 1997; EA, 1997; EA, 1998a&b; Fisher 2001; Buisson et al., 2009). Whilst the CES-AES provides a vital tool for supporting the exploration of "what-if" scenarios to investigate different channel management options, the process of investigating these is not straightforward. User experience is that navigation within the tool can be time-consuming, is not always intuitive and many of the outputs require further manipulation to provide them in a useful format for decision making. A CES-AES Channel Maintenance Module would improve the navigation of options within the current CES-AES tool to aid exploration of different "what-if" scenarios; process and summarise outputs to provide user-friendly graphs and tables which highlight the merits of the different scenarios; and incorporate a database of standard vegetation cuts (e.g. cut along one bank, cut 50% of inbank vegetation) with simple explanatory images.

Extending the existing capability to deal with different culvert inlet shapes and multiple barrels – There are a range of opening types available in the CES-AES for bridges and culverts, with some flexibility to define the shapes of the openings; however more complex types such as multiple culvert openings with different shapes and invert levels or bridges with relief culverts cannot be represented directly. In addition, the road or parapet level of bridges is assumed to be horizontal and approach road embankments cannot be represented. Whilst some of the issues are related to how the AES represents opening and cross section geometry, it is also becoming apparent that the coefficients used to compute energy losses for these configurations may require updating in light of more recent research (e.g. Hotchkiss et al., 2008; Kells et al., 2008; Tullis et al., 2008 for single inlets; Charbeneau, 2006 for multiple barrels). Thus, the CES-AES culvert inlet coefficients, which deal with idealised shapes and are based on the original US methods as adapted in the CIRIA guidance (Ramsbottom et al., 1997), could be improved to better represent inlets as they occur in nature. Accommodating more complex multi-barrel structures raises questions about the lateral distribution of velocity in the flow. A possible approach would be to use an approximate 2-D model such as the CES to determine the spatial velocity distribution of the undisturbed flow and hence to estimate the spatial velocity of the disturbed flow due to the presence of a structure using a simple adjustment, based on, for example, the observed flow behaviour (e.g. Charbeneau, 2006; Haderlie & Tullis, 2008).

Developing a blockage and trash screen module – The issues surrounding blockages and trash screens are detailed in Section 5.3. A CES-AES Blockage and Trash Screen Module would involve developing a hydraulic loss unit for dealing with trash screens (including percentage blocked) and more general channel blockage (e.g. woody debris) and a means to determine the impact on upstream water levels, building on current methods (e.g. EA, 2009) where they continue to offer utility and designed to link with ongoing research in this area (e.g. FRMRC2, see *gtr.rcuk.ac.uk*). Key challenges involve determining the distance upstream of a blocked culvert or trash screen at which the maximum afflux occurs and defining the point at which a channel is considered blocked.

Harmonisation of the CES-AES methods – The AES model for bridges and culverts incorporates a backwater calculation based on an energy balance through the structure to account for energy extracted from the main flow. This makes use of conveyance rating curves calculated both for the river channel as it would be in the absence of the structure (the 'undisturbed channel') and for the structure at its upstream and downstream faces. The AES computes the required physical parameters such as cross sectional flow area and wetted perimeter as functions of flow depth for a range of opening geometries. In particular, it works for closed shapes such as arches and pipes, which is essential when modelling the conveyance for a bridge or structure. This calculation is based on a Divided Channel Method using Manning n resistance parameters for three panels, consistent with the original experimental research (1950s). The Divided Channel Method and CES calculation typically provide different solutions at the transition section and the use of two distinct approaches has raised questions from users regarding the need for both. Harmonisation of the CES-AES methods will involve adapting and incorporating the CES calculation for the reaches with structures. An important change will be modifying the current CES algorithm for boundary treatment to deal with closed shapes such as circular culverts and arches and appropriate adaptation of the structure coefficients.

A targeted programme of data acquisition – is needed to ensure data on aquatic vegetation (e.g. seasonal measurements to build a long-term record), flow properties (e.g. flow, velocity, turbulence) and evidence of debris following storms (initiated in FRMRC2) is collated. The value of these measurements cannot be overstated as they are fundamental to method development, testing and validation, providing evidence of the true flood risk system behaviour. A philosophy of "measure more, model less" is essential to the ongoing development of tools such as the CES-AES.

Development of a habitat module – Section 5.1 highlights the importance of ecological functioning and habitat design when considering channel design and maintenance. A CES-AES Habitat Module would provide a CES-AES post-processing tool to enable users to make-use of detailed flow, velocity and shear stress information together with habitat data, for example, response curves or LIFE scores (Section 5.1) in support of identifying preferred channel management options. It would include a database of habitat information and facilitate processing of this, together with CES-AES outputs, into user-friendly graphs and tables.

Methods for extension of rating curves – Rating curves often do not cover high flood flows because of the difficulties in measuring flows under flood conditions and in some instances the gauge stations may be drowned. The extension of ratings is therefore of vital importance in planning and designing flood defence measures. Current best practice guidance (Ramsbottom & Whitlow, 2003) and tools provide a range of regression analysis techniques to fit polynomials to gauged data from within the measurable range, and hence extrapolation of these to predict water levels at flood flows. Emphasis is placed on selecting the most appropriate method for the site. For example, the hydraulic principles of extending inbank and out-of-bank rating curves are different and the site may be treated differently if a structure such as a weir is present. A further challenge in implementing the various approaches is that the surveyed cross-sections do not always extend across the entire inundated area, and there is therefore uncertainty about the true channel shape at large depths. This work would involve extending the data records at key sites to capture flood flows and using these to develop, test and validate methods consistent with the CES-AES for the extension of ratings.

Development of a sediment transport module – Section 5.2 introduces the importance of sediment and geomorphology issues. A CES-AES Sediment Transport Module would include a post-process tool offering a suite of analysis techniques based on, for example, methods of Du Boys (1879), Einstein (1942), Bagnold (1966), Engelund & Hansen (1966), Ackers & White (1973), Chang (1988) and Ackers (1990–93). These would make use of the CES-AES outputs such as the local velocity, shear stress and shear velocity to determine the sediment concentrations and loads. The user support in determining the preferred channel regime will build on work currently underway in the UK on Sediments and Habitats Phase II – Additional Studies and would link closely with the aforementioned Habitat and Channel Maintenance Modules.

Some of the medium-to-long term research areas include:

– Revamp the embedded CES conveyance calculation, for example, a 2-layer model for main channel and floodplain;
– Improved understanding of flow, vegetation and sediment interaction;
– Incorporate the variation of resistance with the passage of the flood;
– Incorporate an embedded model to deal with alluvial friction e.g. ripples, dunes, antidunes;
– Embed the CES-AES in a dynamical model which automatically updates the channel bed based on regime theory;
– Incorporate the impact of climate change, for example, for long-term vegetation growth patterns.

Finally, it should be noted that whilst the research community continue to press the boundaries of science and computational methods, in practice there is considerable value in minimising model complexity, commensurate with producing realistic and acceptable output. As noted earlier, 1-D models, with key 3-D flow processes embedded, can be a cost-effective and efficient way of solving many river engineering problems. The CES-AES software is one such model that offers novel opportunities for learning about basic open channel flow processes as well as for gaining experience in practical problem solving in, for example, extension of rating curves, flood risk management, as well as in certain areas of ecology and geomorphology. Thus we expect that the CES-AES will find a rôle in education and be a vehicle for transferring knowledge from the research domain into professional use. In conclusion, we are convinced that there is a long-term need for 1-D modelling approaches in practice, as Antoine Chezy pointed out so many years ago.

Appendix 1

The finite element approximations for the CES equations

Final CES conveyance equations

$$gHS_o - \psi \frac{f}{8}\frac{q^2}{H^2} + \frac{\partial}{\partial y}\left(\lambda\left(\frac{f}{8}\right)^{1/2} Hq \frac{\partial}{\partial y}\left(\frac{q}{H}\right)\right) - \left(\frac{(1.015-\sigma)}{0.015}\Gamma + \frac{(\sigma-1.0)}{0.015} C_{uv} \frac{\partial}{\partial y}\left(\frac{q^2}{H}\right)\right) = 0$$

$\sigma = 1.0$ inbank; $1.0 \leq \sigma \leq 1.015$ overbank

$$gHS_o - \psi \frac{f}{8}\frac{q^2}{H^2} + \frac{\partial}{\partial y}\left(\lambda\left(\frac{f}{8}\right)^{1/2} Hq \frac{\partial}{\partial y}\left(\frac{q}{H}\right)\right) - C_{uv} \frac{\partial}{\partial y}\left(\frac{q^2}{H}\right) = 0$$

$\sigma > 1.0$ inbank; $\sigma > 1.015$ overbank (3.35a&b)

In the above equation set, the Γ term is a multiple of the hydrostatic pressure term, and it is therefore not dealt with explicitly here. Thus, consider the terms in Equation 3.35b,

$$\underset{(1)}{gHS_o} - \underset{(2)}{\psi \frac{f}{8}\frac{q^2}{H^2}} + \underset{(3)}{\frac{\partial}{\partial y}\left(\lambda\left(\frac{f}{8}\right)^{1/2} Hq \frac{\partial}{\partial y}\left(\frac{q}{H}\right)\right)} - \underset{(4)}{C_{uv} \frac{\partial}{\partial y}\left(\frac{q^2}{H}\right)} = 0$$

The finite element approximations are given by:

Term 1

$$\sum_{\text{elements}} \int_{y_i}^{y_{i+1}} gS_o \left(\sum H_i \phi_i\right) \phi_i \, dy \qquad \text{[Vector Term 1a]}$$

Term 2

$$\sum_{\text{elements}} \int_{y_i}^{y_{i+1}} -\frac{\left(\sum \beta_i \phi_i\right)\left(\sum f_i^n \phi_i\right)\left(\sum q_i^n \phi_i\right)^2}{8\left(\sum H_i \phi_i\right)^2} \phi_i \, dy \qquad \text{[Vector Term 2a]}$$

$$\sum_{elements} \int_{y_i}^{y_{i+1}} -\frac{\left(\sum \beta_i \phi_i\right)\left(\sum f_i^n \phi_i\right)\left(\sum q_i^n \phi_i\right)\left(\sum \Delta q_i^n \phi_i\right)}{4\left(\sum H_i \phi_i\right)^2} \, \phi_j \, dy \qquad \text{[Matrix Term 2b]}$$

$$\sum_{elements} \int_{y_i}^{y_{i+1}} -\frac{\left(\sum \beta_i \phi_i\right)\left(\sum q_i^n \phi_i\right)^2 \left(\sum \frac{\partial f}{\partial q}\big|_i \phi_i\right)\left(\sum \Delta q_i^n \phi_i\right)}{8\left(\sum H_i \phi_i\right)^2} \, \phi_j \, dy \qquad \text{[Matrix Term 2c]}$$

Term 3

$$\sum_{elements} \int_{y_i}^{y_{i+1}} -\frac{\lambda}{\sqrt{8}}\left(\sum q_i^n \phi_i\right)\left(\sum f_i^n \phi_i\right)^{1/2}$$

$$\times \left[\left(\sum q_i^n \frac{\partial \phi_i}{\partial y}\right) - \frac{\left(\sum q_i^n \phi_i\right)\left(\sum H_i \frac{\partial \phi_i}{\partial y}\right)}{\left(\sum H_i \phi_i\right)}\right] \frac{\partial \phi_j}{\partial y} \, dy \qquad \text{[Vector Term 3a]}$$

$$\sum_{elements} \int_{y_i}^{y_{i+1}} -\frac{\lambda}{\sqrt{8}}\left(\sum q_i^n \phi_i\right)\left(\sum f_i^n \phi_i\right)^{1/2}\left(\sum \Delta q_i^n \frac{\partial \phi_i}{\partial y}\right)\frac{\partial \phi_j}{\partial y} \, dy \qquad \text{[Matrix Term 3b]}$$

$$\sum_{elements} \int_{y_i}^{y_{i+1}} -\frac{\lambda}{\sqrt{8}}\left[\frac{\left(\sum q_i^n \phi_i\right)\left(\sum f_i^n \phi_i\right)^{1/2}\left(\sum \Delta q_i^n \phi_i\right)\left(\sum H_i \frac{\partial \phi_i}{\partial y}\right)}{\left(\sum H_i \phi_i\right)}\right]\frac{\partial \phi_j}{\partial y} \, dy \qquad \text{[Matrix Term 3c]}$$

$$\sum_{elements} \int_{y_i}^{y_{i+1}} -\frac{\lambda}{\sqrt{8}}\left(\sum \Delta q_i^n \phi_i\right)\left(\sum f_i^n \phi_i\right)^{1/2}$$

$$\times \left[\left(\sum q_i^n \frac{\partial \phi_i}{\partial y}\right) - \frac{\left(\sum q_i^n \phi_i\right)\left(\sum H_i \frac{\partial \phi_i}{\partial y}\right)}{\left(\sum H_i \phi_i\right)}\right] \frac{\partial \phi_j}{\partial y} \, dy \qquad \text{[Matrix Term 3d]}$$

$$\sum_{\text{elements}} \int_{y_i}^{y_{i+1}} -\frac{\lambda}{2\sqrt{8}} \frac{\left(\sum \Delta q_i^n \phi_i\right)\left(\sum q_i^n \phi_i\right)\left(\sum \frac{\partial f}{\partial q}\big|_i \phi_i\right)}{\left(\sum f_i^n \phi_i\right)^{1/2}}$$

$$\times \left[\left(\sum q_i^n \frac{\partial \phi_i}{\partial y}\right) - \frac{\left(\sum q_i^n \phi_i\right)\left(\sum H_i \frac{\partial \phi_i}{\partial y}\right)}{\left(\sum H_i \phi_i\right)}\right] \frac{\partial \phi_j}{\partial y} \, dy \qquad \text{[Matrix Term 3e]}$$

Term 4

$$\sum_{\text{elements}} \int_{y_i}^{y_{i+1}} -\frac{2C_{uv}\left(\sum q_i^n \phi_i\right)}{\left(\sum H_i \phi_i\right)} \left(\sum q_i^n \frac{\partial \phi_i}{\partial y}\right) \phi_j \, dy \qquad \text{[Vector Term 4a]}$$

$$\sum_{\text{elements}} \int_{y_i}^{y_{i+1}} -\frac{2C_{uv}\left(\sum q_i^n \phi_i\right)^2}{\left(\sum H_i \phi_i\right)^2} \left(\sum H_i^n \frac{\partial \phi_i}{\partial y}\right) \phi_j \, dy \qquad \text{[Vector Term 4b]}$$

$$\sum_{\text{elements}} \int_{y_i}^{y_{i+1}} -\frac{2C_{uv}\left(\sum \Delta q_i^n \phi_i\right)}{\left(\sum H_i \phi_i\right)} \left(\sum q_i^n \frac{\partial \phi_i}{\partial y}\right) \phi_j \, dy \qquad \text{[Matrix Term 4c]}$$

$$\sum_{\text{elements}} \int_{y_i}^{y_{i+1}} -\frac{2C_{uv}\left(\sum q_i^n \phi_i\right)}{\left(\sum H_i \phi_i\right)} \left(\sum \Delta q_i^n \frac{\partial \phi_i}{\partial y}\right) \phi_j \, dy \qquad \text{[Matrix Term 4d]}$$

$$\sum_{\text{elements}} \int_{y_i}^{y_{i+1}} \frac{2C_{uv}\left(\sum q_i^n \phi_i\right)\left(\sum \Delta q_i^n \phi_i\right)}{\left(\sum H_i \phi_i\right)^2} \left(\sum H_i^n \frac{\partial \phi_i}{\partial y}\right) \phi_j \, dy \qquad \text{[Matrix Term 4e]}$$

Summary of hydraulic equations used in the AES

Transition lengths

$$\frac{L_{(C,E)}}{D_N} = b_0 + b_1Fr + b_2J_D + b_3Fr^2 + b_4J_D^2 + b_5FrJ_D$$

Coefficient	Contraction $k_1\ (m^{-1})$	k_2	Expansion $k_1\ (m^{-1})$	k_2
b_0	0.0190	−8.9	0.0712	−26.2
b_1	−0.213	50.8	−0.294	197.8
b_2	0.190	26.6	0.0569	15.2
b_3	0.565	−36.2	−0.872	−22.2
b_4	−0.170	−20.4	0.0170	9.9
b_5	0.418	−39.1	0.394	−2.8

Transition energy losses

$$h_t = LS_f + C\left|\alpha_2\frac{U_2^2}{2g} - \alpha_1\frac{U_1^2}{2g}\right|$$

Transition energy loss coefficients

Subcritical flow	Supercritical flow
$C_C = 0.1$	$C_C = 0$
$C_E = \min\left\{0.3, 0.1\dfrac{B}{B*} + 0.1\right\}$	$C_E = 0$

Pressure flow

$$Q = C_d b_N Z (2g\Delta h)^{1/2}$$

Overtopping flow

$$Q = CB_{RD} H^{3/2}$$

Cross-section survey data

Table 3.1 River Trent, Yoxall.

Y	Z
0	2.59
1	2.4
2	2.42
5	2.35
8	2.36
11	2.36
14	2.34
17	2.4
20	2.42
23	2.43
26	2.41
29	2.4
32	2.43
33.5	2.44
36	2.59
39	2.24
42	2.29
45	2.17
48	1.77
51	0.33
54	0.14
57	0.02
60	0.05
63	0
66	0.04
69	0.23
72	0.71
74	1.61
76	1.7
78	1.94
79	2.04
79.7	2.25
79.9	2.59

Table 3.2 River Colorado.

Y	Z
−60	3.8
7.5	3.6
10	1.2
14	1.36
17	1.3
20	1.2
23	1.1
26	1
29	0.9
32	0.9
35	0.8
38	0.7
41	0.6
44	0.56
47	0.56
50	0.4
53	0.34
56	0.34
59	0.3
62	0.2
65	0
68	0.2
71	1.1
73.6	3.6
150	3.8

Table 3.3 River Main, Bridge End Bridge.

Y	Z
0	4.54
5.3	1.41
13.5	0.92
14.4	0
26.6	0
27.6	0.97
35.7	1.38
40.8	4.89

Table 3.4 River Severn, Montford Bridge.

Y	Z
7	7.74
13	6.94
18	6.62
23	6.46
28	6.32
31	6.21
33	5.87
36	5.46
38	4.91
39	4.15
40	3.63
41	3.22
42	1.84
43	1.36
44	1.05
45	0.29
47	0.17
50	0.06
53	0.01
56	0.02
59	0
62	0.07
63	0.53
64	1.63
65	3.13
66	3.68
67	4.05
68	4.37
69	4.93
70	5.73
72	5.91
76	6
80	6.1
84	6.16
88	6.26
92	6.37
96	6.46
100	6.54
104	6.59
108	6.74
112	6.8
116	6.87
120	6.9
124	6.9
128	6.99
133	7.02
136	7.74

Table 3.5 River La Suela.

Y	Z
4	2.51
7.3	2.14
8.1	1.64
9	0.38
10.5	0
15.6	0.02
28.9	0.2
30.4	1.22
33.7	1.9
36.6	3.07

Table 3.6 River Waiwakaiho.

Y	Z
0	5
1	4.5
2	3.8
3	3.7
4	3.6
5	3.3
6	2.9
7	2.8
8	2.3
9	2
10	1.7
11	1.6
12	1.2
13	1.1
14	1.1
15	1
16	0.6
18	0.6
19	0.55
20	0.6
21	0.7
22	0.6
23	0.45
24	0.5
25	0.45
26	0.45
27	0.3
28	0
29	0.3
30	0.5
31	0.55
32	1.1
33	3.1
34	3.6
35	4.1
36	4.6
37	5

Table 3.7 River Tomebamba.

Y	Z
1	4.3
5	1.53
5.4	1.33
7	0.39
8	0
10	0.17
12	0.23
14	0.13
16	0.24
18	0.42
20	0.64
22	0.75
24	0.83
26	0.86
28	1.1
29	1.16
30	1.33
31	1.53
45	4.3

Table 3.8 River Cole.

Y	Z
0.0	1.60
2.0	1.05
5.0	1.00
7.0	0.30
7.5	0.05
8.0	0.03
9.0	0.01
10.0	0.00
10.5	0.01
11.0	0.02
12.5	0.03
13.0	0.30
15.0	1.40
20.0	1.50
22.0	1.60
23.0	1.70

Table 3.9 River Avon.

Y	Z
0.24	0.80
0.25	0.56
0.70	0.11
1.15	0.07

(Continued)

Table 3.9 (Continued).

Y	Z
1.60	0.04
2.05	0.02
2.50	0.04
2.95	0.02
3.40	0.05
3.85	0.05
4.30	0.06
4.75	0.08
5.20	0.06
5.65	0.09
6.10	0.04
6.55	0.05
7.00	0.09
7.45	0.07
7.90	0.02
8.35	0.00
8.80	0.01
9.25	0.04
9.40	0.56
9.41	0.80

Table 3.10 Holme Bridge.

Y	Z
−102.704	158.229
−96.285	157.797
−83.447	157.333
−77.028	157.375
−64.190	157.279
−19.257	157.039
0.000	157.192
5.638	157.143
10.815	157.126
12.180	156.836
13.713	155.649
14.978	154.274
17.546	153.644
18.629	153.624
19.367	153.400
19.800	153.275
20.392	153.026
21.842	152.880
23.503	152.577
24.860	152.260
26.410	152.561
27.271	152.570
28.540	152.389
29.498	152.652
30.586	152.877

(Continued)

Table 3.10 (Continued).

Y	Z
31.840	152.959
33.570	153.098
35.269	152.878
36.428	152.972
36.429	153.275
36.429	153.646
36.431	154.556
36.695	154.589
37.722	155.329
37.955	156.485
38.432	156.529
39.069	157.168
67.870	157.452
76.076	158.986

Table 3.11 River Main culvert.

Y	Z
−50.001	38.126
−50.000	36.642
−10.750	36.196
−8.310	35.599
0.000	35.046
0.750	33.584
0.900	33.583
1.340	33.570
1.820	33.551
2.160	33.543
2.510	33.553
2.870	33.547
3.350	33.553
3.790	33.548
4.220	33.577
4.700	33.552

(Continued)

Table 3.11 (Continued).

Y	Z
5.630	33.524
5.990	33.490
6.220	33.446
6.920	33.369
7.220	33.355
8.440	33.271
8.780	33.266
9.330	33.202
9.690	33.179
10.460	33.164
10.780	33.185
11.510	33.273
12.150	33.426
12.310	33.508
12.880	33.732
13.340	33.863
13.740	33.862
14.490	34.550
23.130	38.158
49.240	38.126

Table 3.12 Trial culvert
design – Section 1
(downstream).

Y	Z
−4.000	11.000
−3.000	10.750
−2.250	10.750
−0.750	9.250
0.750	9.250
2.250	10.750
3.000	10.750
4.000	11.000

Table 3.13 Trial culvert design – Section chainage, upstream to
downstream.

Label	Distance (m)	Comment	Bed slope
4	0	Upstream	0.005
3	150	Culvert inlet	0.005
2	162	Culvert outlet	0.005
1	312	Downstream	0.005

Table 3.14 River Dane.

Y	Z
0	17
149.5	16.19
157	16.19
160	15.28
161	14.25
162	13.51
163	13.45
164	12.47
169	12.04
173	12.29
175	13.44
177	14.76
179	15.1
183	16.76
190.5	16.76

Table 3.15 River Blackwater, Coleford Bridge.

Y	Z
0	1.71
1.7	1.25
5.2	1.15
11.2	1
12.2	0.9
12.7	0.7
12.9	0.5
13.4	0.37
13.8	0
16	0
17.4	0.1
17.9	0.3
18.15	0.9
19.15	1.38
19.4	1.88
20	2.25

Table 3.16 River Torridge.

Y	Z
0	19.716
0.91	19.688
5	19.102
10	18.425
15	17.885

(Continued)

Table 3.16 (Continued).

Y	Z
20	17.64
25	17.451
30	17.333
34	17.19
37	16.985
40	16.86
43	16.698
44.5	16.549
46.5	17.518
48.5	17.136
51	17.218
54	17.321
57	17.522
60	17.953
62.5	18.348
63.5	18.348
65	18.254
68	17.962
70	17.518
73	17.504
76	17.553
79	17.559
82	17.509
85	17.472
88	17.46
91.6	17.197
92.5	16.409
93.6	14.521
117	14.426
118.1	14.472
119	15.867
119.99	16.556
120	19.716

Table 3.17 River Blackwater, Ower.

Y	Z
0	2.174
2.5	1.89
5.5	1.89
10.5	1.96
13	1.74
13.5	2
13.6	0
19.4	0
19.5	2.01

(Continued)

Table 3.17 (Continued).

Y	Z
21	1.5
24	1.82
27	1.84
30	1.78
33	1.86
36	1.9
39	1.92
42	1.79
45	1.61
48	1.51
51	1.62
54	1.74
57	2.03
60	2.54

Table 3.18 River Heathcote.

Y	Z
0	1.95
1	1.7
2	0.55
2.5	0.1
3	0.1
5	0.1
6	0
7	0.1
8	0.15
8.5	0.25
9	0.9
9.5	1.85
10	1.95

Table 3.19 River Cuenca.

Y	Z
0	2.3
3	0.6
6	0.3
9	0.2
12	0.5
15	0.1
18	0
21	0.5
24	0.4

(Continued)

Table 3.19 (Continued).

Y	Z
27	0.6
30	0.9
33	1.76
36	1.6
39	2.3

Table 3.20 River Severn, Llandrinio.

Y	Z
0	8
2	7.37
4.992424	5.9
5.998903	5.62
6.99655	5.37
7.998398	4.37
8.996046	3.87
10.00252	3.05
10.9955	2.27
12.00198	1.87
14.00143	0.4
16.00555	0
17.99615	0
19.9956	0.87
21.00208	1.35
21.99506	2.12
23.00154	3.37
23.99918	3.63
25.00103	4.3
25.99868	5.04
27.00516	6.02
27.99813	5.97
29.99758	6.27
32.0017	6.37
34.00115	6.47
38.00471	6.57
41.99476	6.87
45.99833	6.87
49.99723	6.87
54.00079	6.87
60.00379	7.37
62	8

Table 3.21 River Frome (no gravel).

Y	Z
0.3	3
0.3	2.04
0.5	1.94
0.7	1.805
0.9	1.16
1.1	1.05
1.3	1.02
1.5	1.02
1.7	0.98
1.9	0.91
2.1	0.88
2.3	0.83
2.5	0.72
2.7	0.5
2.9	0.45
3.1	0.38
3.3	0.31
3.5	0.24
3.7	0.19
3.9	0.14
4.1	0.08
4.3	0.07
4.5	0.03
4.7	0.02
4.9	0.03
5.1	0.04
5.3	0.06
5.5	0.09
5.7	0
5.9	0.12
6.1	0.15
6.3	0.19
6.5	0.26
6.7	0.32
6.9	0.4
7.1	0.47
7.3	0.58
7.5	0.69
7.7	0.7
7.9	0.72
8.1	0.74
8.3	0.74
8.5	0.72
8.7	0.68
8.9	0.72
9.1	0.74
9.3	0.83

(Continued)

Table 3.21 (Continued).

Y	Z
9.5	0.81
9.7	0.81
9.9	0.82
10.1	0.88
10.3	1.05
10.31	2.47
10.32	3

Table 3.22 River Frome (gravel).

Y	Z
0.299	4
0.3	2.04
0.5	1.94
0.7	1.805
0.9	1.16
1.1	1.05
1.3	1.02
1.5	1.02
1.7	0.98
1.9	0.91
2.1	0.88
2.3	0.87
9.9	0.87
10.1	0.88
10.3	1.05
10.31	2.47
10.32	4

Table 3.23 River Rhone, Section PK7-85.

Y	Z
0	9.5
0.005	9.119
1.009	8.989
2.014	8.511
3.018	8.301
4.023	8.333
5.028	8.376
6.032	8.393
7.037	8.353
8.041	8.319
9.046	8.323
10.051	8.27

(Continued)

Table 3.23 (Continued).

Y	Z
11.055	8.174
12.06	7.994
13.065	7.771
14.069	7.618
15.074	7.549
16.078	7.438
17.083	7.399
18.088	7.405
19.092	7.413
20.097	7.397
21.101	7.352
22.106	7.31
23.111	7.283
24.115	7.221
25.12	7.184
26.124	7.092
27.129	6.909
28.134	6.818
29.138	6.985
30.143	7.053
31.148	7.02
32.152	6.954
33.157	6.994
34.161	7.074
35.166	7.041
36.171	6.98
37.175	6.791
38.18	6.632
39.184	6.46
40.189	6.324
41.194	6.119
42.198	5.81
43.203	5.541
44.207	5.175
45.212	4.506
46.217	4.169
47.221	3.769
48.226	3.244
49.23	3.08
50.235	2.969
51.24	2.78
52.244	2.589
53.249	2.406
54.254	2.258
55.258	2.097
56.263	1.959
57.267	1.825

(Continued)

Table 3.23 (Continued).

Y	Z
58.272	1.678
59.277	1.533
60.281	1.41
61.286	1.3
62.29	1.177
63.295	1.058
64.3	0.944
65.304	0.855
66.309	0.783
67.313	0.706
68.318	0.641
69.323	0.6
70.327	0.548
71.332	0.478
72.336	0.414
73.341	0.359
74.346	0.317
75.35	0.29
76.355	0.251
77.359	0.223
78.364	0.202
79.369	0.174
80.373	0.138
81.378	0.114
82.383	0.126
83.387	0.135
84.392	0.135
85.396	0.127
86.401	0.113
87.406	0.096
88.41	0.084
89.415	0.072
90.419	0.025
91.424	0.004
92.429	0.022
93.433	0.028
94.438	0.02
95.442	0.01
96.447	0.007
97.452	0.015
98.456	0.027
99.461	0.023
100.465	0.018
101.47	0.027
102.475	0.039
103.479	0.036
104.484	0.027

(Continued)

Table 3.23 (Continued).		Table 3.23 (Continued).	
Y	Z	Y	Z
105.488	0.035	152.705	3.474
106.493	0.053	153.71	3.989
107.498	0.056	154.714	4.728
108.502	0.043	155.719	5.176
109.507	0.03	156.723	5.823
110.512	0.025	157.728	6.196
111.516	0.026	158.733	6.367
112.521	0.033	159.737	6.261
113.525	0.044	160.742	6.046
114.53	0.059	161.747	5.875
115.535	0.079	162.751	5.639
116.539	0.102	163.756	5.393
117.544	0.129	164.76	5.313
118.548	0.158	165.765	5.417
119.553	0.191	166.77	5.709
120.558	0.225	167.774	6.123
121.562	0.262	168.779	6.481
122.567	0.3	169.783	6.539
123.571	0.34	170.788	6.492
124.576	0.38	171.793	6.408
125.581	0.421	172.797	6.287
126.585	0.462	173.802	6.072
127.59	0.503	174.806	5.774
128.594	0.543	175.811	5.262
129.599	0.582	176.816	4.691
130.604	0.62	177.82	4.247
131.608	0.656	178.825	3.942
132.613	0.691	179.829	3.763
133.618	0.723	180.834	3.71
134.622	0.752	181.839	3.72
135.627	0.777	182.843	3.754
136.631	0.803	183.848	3.795
137.636	0.84	184.853	3.833
138.641	0.891	185.857	3.871
139.645	0.968	186.862	3.943
140.65	1.055	187.866	4.063
141.654	1.135	188.871	4.185
142.659	1.295	189.876	4.307
143.664	1.302	190.88	4.411
144.668	1.503	191.885	4.497
145.673	1.875	192.889	4.552
146.677	2.034	193.894	4.57
147.682	2.265	194.899	4.582
148.687	2.354	195.903	4.596
149.691	2.787	196.908	4.603
150.696	2.868	197.912	4.598
151.7	3.166	198.917	4.6

(Continued)

(Continued)

Table 3.23 (Continued).

Y	Z
199.922	4.616
200.926	4.654
201.931	4.704
202.935	4.754
203.94	4.874
204.945	4.944
205.949	4.972
206.954	4.836
207.959	4.67
208.963	4.602
209.968	4.55
210.972	4.621
211.977	4.73
212.982	4.94
213.986	5.326
214.991	5.326
215.995	5.326
217	5.326
217	9.5

Table 3.24 River Rhone, Section PK7-1.

Y	Z
0	10
0	7.415
1	7.415
2	7.415
3	7.415
4	7.173
5	6.697
6	6.196
7	5.119
8	4.335
9	3.743
10	3.215
11	2.863
12	2.676
13	2.505
14	2.368
15	2.105
16	1.916
17	1.807
18	1.567
19	1.346
20	1.159
21	1.052

(Continued)

Table 3.24 River Rhone, Section PK7-1.

Y	Z
22	0.988
23	0.98
24	1.002
25	1.02
26	1.007
27	0.985
28	0.957
29	0.913
30	0.863
31	0.817
32	0.779
33	0.663
34	0.634
35	0.606
36	0.578
37	0.539
38	0.478
39	0.432
40	0.402
41	0.361
42	0.316
43	0.284
44	0.263
45	0.187
46	0.109
47	0.075
48	0.073
49	0.07
50	0.06
51	0.038
52	0.024
53	0.016
54	0.013
55	0.011
56	0.006
57	0.001
58	−0.003
59	0
60	0.009
61	0.019
62	0.021
63	0.017
64	0.018
65	0.024
66	0.01
67	0.024
68	0.056
69	0.069

(Continued)

Table 3.24 (Continued).

Y	Z
70	0.075
71	0.091
72	0.111
73	0.118
74	0.111
75	0.113
76	0.129
77	0.144
78	0.155
79	0.162
80	0.166
81	0.179
82	0.2
83	0.226
84	0.239
85	0.271
86	0.316
87	0.344
88	0.374
89	0.401
90	0.412
91	0.421
92	0.434
93	0.454
94	0.466
95	0.511
96	0.561
97	0.583
98	0.598
99	0.648
100	0.672
101	0.688
102	0.736
103	0.77
104	0.799
105	0.848
106	0.902
107	0.944
108	0.964
109	0.989
110	1.027
111	1.052
112	1.085
113	1.1
114	1.12
115	1.136
116	1.151
117	1.182

(Continued)

Table 3.24 (Continued).

Y	Z
118	1.25
119	1.356
120	1.457
121	1.586
122	1.732
123	1.969
124	2.242
125	2.653
126	3.011
127	3.487
128	3.956
129	4.82
130	5.434
131	5.56
132	5.768
133	5.908
134	6.008
135	6.16
136	6.284
137	6.386
138	6.545
139	6.806
140	7.029
141	7.133
142	7.164
143	6.971
144	6.658
145	6.449
146	6.302
147	6.214
148	6.235
149	6.299
150	6.374
151	6.362
152	6.37
153	6.434
154	6.545
155	6.683
156	6.843
157	7.038
158	7.155
159	7.198
160	7.212
161	7.325
162	7.368
163	7.37
164	7.51
165	7.565

(Continued)

Table 3.24 (Continued).

Y	Z
166	7.441
167	7.169
168	7.116
169	7.05
170	6.987
171	6.982
172	6.995
173	6.992
174	6.9
175	6.89
176	6.797
177	6.916
178	7.034
179	7.041
180	6.921
181	7.067
182	7.161
183	7.304
184	7.567
185	7.866
186	7.665
187	7.685
188	7.748
189	7.911
190	8.079
191	8.079
191	10

Table 3.25 River Yangtze.

Y	Z
4.6	207.16
40	201.44
90	197.56
240	195.14
380	185.57
470	185.57
560	188.22
661	207.46

Appendix 4

Derivation of the governing depth-averaged equation used in the SKM

By considering the elementary control volume $dxdydz$ in Figure A4.1 for uniform flow, where x, y, z are the streamwise, lateral and vertical directions respectively, θ is the bed slope angle ($S_0 = \sin\theta$), and τ_{ij} is the shear stress in the j direction on the plane perpendicular to the i direction.

Then if advection terms (secondary flows) for uniform flow are ignored, the gravitational body force acting on the mass $\rho(dxdydz)$ in the streamwise direction must be balanced by the surface forces acting on the two sides $(dxdy)$ and the upper and lower surfaces $(dxdz)$, giving

$$\rho g \sin\theta (dxdydz) + \frac{\partial \tau_{zx}}{\partial z} dz(dxdy) + \frac{\partial \tau_{yx}}{\partial y} dy(dxdz) = 0 \tag{A4.1}$$

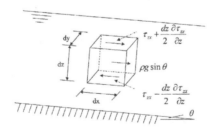

Figure A4.1 Stresses acting on an element.

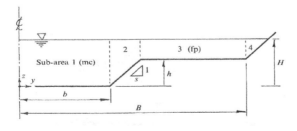

Figure A4.2 Notation for trapezoidal channel compound channel (mc = main channel and fp = floodplain).

Integrating (A4.1) over the water depth, H, with $S_0 = \sin\theta$, assuming no wind or other stresses on the water surface $(z = z_s)$, and allowing for the bed elevation to vary laterally $(z = z_{0(y)})$ then the depth-integrated form of (A4.1) is

$$\rho g S_0 H (dxdy) - \tau_{zb}(dxdy) + \left(\int_{z_0(y)}^{z_s} \frac{\partial \tau_{yx}}{\partial y} dz \right)(dxdy) = 0 \tag{A4.2}$$

Using Leibnitz's rule for the third term,

$$\int_{z_0}^{z_s} \frac{\partial \tau_{yx}}{\partial y} dz = \frac{\partial}{\partial y} \int_{z_0}^{z_s} \tau_{yx} dz + \tau_{yx}(z_0)\frac{\partial z_0}{\partial y} - \tau_{yx}(z_s)\frac{\partial z_s}{\partial y} \tag{A4.3}$$

and introducing the depth-averaged lateral shear stress, $\overline{\tau}_{yx}$,

$$\overline{\tau}_{yx} = \frac{1}{H} \int_{z_0}^{z_s} \tau_{yx} dz$$

where z_s = constant water surface level laterally and z_0 = y/s on the side slope, then (A4.2) becomes

$$\rho g S_0 H - \tau_{zb} - \frac{\tau_{yb}}{s} + \frac{\partial\left(H\overline{\tau}_{yx}\right)}{\partial y} = 0 \tag{A4.4}$$

where τ_{yb} and τ_{zb} are the shear stresses perpendicular to the y & z directions, and are related to the boundary shear stress that resists the flow at the bed, τ_b. These are shown in Figure A.4.3 for a near bed element in region 2 of Figure A.4.2, close to a sloping side wall. Thus

$$\tau_{yb}(dxdz) + \tau_{zb}(dxdy) = \tau_b dx(dy^2 + dz^2)^{1/2} \tag{A4.5}$$

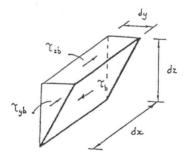

Figure A4.3 Relationship between shear stresses τ_{yb} and τ_{zb} and the boundary shear stress, τ_b, on a side slope.

Noting that $dz/dy = 1/s$, then dividing by $dxdy$ gives

$$\frac{\tau_{yb}}{s} + \tau_{zb} = \tau_b \left[1 + \left(\frac{1}{s} \right)^2 \right]^{1/2} \tag{A4.6}$$

Therefore, using the shear stress τ_b at the boundary, (A4.4) becomes

$$\rho g S_0 H - \frac{\sqrt{1+s^2}}{s} \tau_b + \frac{\partial \left(H \overline{\tau}_{yx} \right)}{\partial y} = 0 \tag{A4.7}$$

For a constant depth domain ($s \to \infty$) with H = constant, $\tau_{zb} \to \tau_b$ as $s \to \infty$, and (A4.7) becomes

$$\rho g H S_0 - \tau_b + \frac{\partial \left(H \overline{\tau}_{yx} \right)}{\partial y} = 0 \tag{A4.8}$$

It also should be noted that $\tau_{yb} + s\tau_{zb} = \tau_b \left(1 + s^2 \right)^{1/2}$ so that $\tau_{yb} \to \tau_b$ as $s \to 0$ (vertical wall).

Equation (A4.7) for a linear side slope domain and equation (A4.8) for a constant depth domain can be solved analytically by assuming a quadratic frictional resistance law based on the Darcy-Weisbach friction factor, f, the Boussinesq eddy viscosity, ε_{yx}, and a dimensionless eddy viscosity model

$$f = \frac{8\tau_b}{\rho U^2}; \quad \overline{\tau}_{yx} = \rho \varepsilon_{yx} \frac{\partial U}{\partial y}; \quad \varepsilon_{yx} = \lambda U_* H; \quad U_* = \left(\frac{\tau_b}{\rho} \right)^{1/2}; \quad \varepsilon_{yx} = \lambda \left(\frac{f}{8} \right)^{1/2} UH \tag{A4.9}$$

Using the appropriate depth-averaged forms for (A4.9) with the depth-averaged velocity, U_d,

$$U_d = \frac{1}{H} \int_0^H U \, dz; \quad \tau_b = \left(\frac{f}{8} \right) \rho U_d^2; \quad \overline{\tau}_{yx} = \rho \overline{\varepsilon}_{yx} \frac{\partial U_d}{\partial y}; \quad \overline{\varepsilon}_{yx} = \lambda U_* H \tag{A4.10 \& (2.59)}$$

Then (A4.7) can be expressed in terms of the boundary shear stress, τ_b as

$$\rho g H S_0 - \frac{\sqrt{1+s^2}}{s} \tau_b + \frac{\partial}{\partial y} \left\{ \frac{\lambda H^2}{2} \left(\frac{8}{f} \right)^{1/2} \frac{\partial \tau_b}{\partial y} \right\} = 0 \tag{A4.11}$$

or alternatively the depth-mean velocity, U_d, can be used instead, giving

$$\rho g H S_0 - \frac{\rho f U_d^2}{8} \frac{\sqrt{1+s^2}}{s} + \frac{\partial}{\partial y} \left\{ \rho \lambda H^2 \left(\frac{f}{8} \right)^{1/2} U_d \frac{\partial U_d}{\partial y} \right\} = 0 \tag{A4.12}$$

Equations (A4.11) & (A4.12) can be solved analytically (See Appendix 5), provided the appropriate boundary conditions are specified with f and λ constant in the solving domain. Since for flow in a compound trapezoidal channel, the boundary conditions are easier to specify in terms of velocity than boundary shear stress, it is preferable to solve (A4.12).

For a constant depth domain, i.e. H = constant, it may be shown that U_d is given in (A4.12) by

$$U_d = \left[A_1 e^{\gamma y} + A_2 e^{-\gamma y} + k_1 \right]^{1/2} \qquad \text{(A4.13) \& (A5.18)}$$

where

$$k_1 = \frac{8gS_0 H}{f} \quad \text{and} \quad \gamma = \left(\frac{2}{\lambda} \right)^{1/2} \left(\frac{f}{8} \right)^{1/4} \frac{1}{H} \qquad \text{(A4.14)}$$

For a linear side slope domain, i.e. H = a function of y, then it may be shown that U_d is given in (A4.12) by

$$U_d = \left[A_3 \xi^{\alpha_1} + A_4 \xi^{-(\alpha_1+1)} + \omega \xi \right]^{1/2} \qquad \text{(A4.15) \& (A5.47)}$$

where

$$\alpha_1 = -\frac{1}{2} + \frac{1}{2} \left\{ 1 + \frac{s(1+s^2)^{1/2}}{\lambda} (8f)^{1/2} \right\}^{1/2} \qquad \text{(A4.16)}$$

$$\alpha_2 = -\frac{1}{2} - \frac{1}{2} \left\{ 1 + \frac{s(1+s^2)^{1/2}}{\lambda} (8f)^{1/2} \right\}^{1/2} = -(\alpha_1 + 1) \qquad \text{(A4.17)}$$

and

$$\omega = \frac{gS_0}{\dfrac{(1+s^2)^{1/2}}{s} \dfrac{f}{8} - \dfrac{\lambda}{s^2} \left(\dfrac{f}{8} \right)^{1/2}} \qquad \text{(A4.18)}$$

where ξ is the local depth, given for a sloping element in a trapezoidal side by

$$\xi = H - \frac{(y-b)}{s} \qquad \text{(A4.19)}$$

The two solutions, (A4.13) & (A4.15), are derived and explained in full in Appendix 5 and in Section 6.4.

The addition of secondary flows now needs to be considered as they introduce some advective terms on the right-hand side of (A.4.1). From the Reynolds-Averaged Navier-Stokes (RANS) equations, introduced in section 2.3 as (2.57), and explained in more detail in section 6.2.2, the more appropriate form of (A.4.1) when secondary flows are present is

$$\rho g \sin \theta + \frac{\partial \tau_{zx}}{\partial z} + \frac{\partial \tau_{yx}}{\partial y} = \rho \left[\frac{\partial}{\partial y}(UV) + \frac{\partial}{\partial z}(UW) \right] \qquad \text{(A4.20) \& (2.57)}$$

Since SKM assumes any vertical components of velocity, W, are zero at the bed and the water surface, then when (A4.20) is integrated over the depth, H, the LHS follows exactly the previous analysis for the $\Gamma = 0$ case, through to (A4.7), and (A4.20) becomes

$$\rho g H S_0 - \tau_b \left[1 + \left(\frac{1}{s} \right)^2 \right]^{1/2} + \frac{\partial \left(H \bar{\tau}_{yx} \right)}{\partial y} = \frac{\partial}{\partial y} \left[H (\rho U V)_d \right] \qquad \text{(A4.21) \& (2.58)}$$

The experimental evidence presented in Figure 6.24 and explained in detail in section 6.2.4 suggests that the gradient term on the RHS is more or less constant in different regions of the flow, allowing it to be represented as a constant, represented by Γ, where

$$\frac{\partial}{\partial y} \left[H (\rho U V)_d \right] = \Gamma \qquad \text{(6.9), (2.60) \& (3.23)}$$

The remaining terms in (A4.7) stay the same, and the general SKM governing equation in terms of U_d is now expressed by (A4.22) rather than by (A4.12) in which Γ was zero.

$$\rho g H S_0 - \frac{f}{8} U_d^2 \left(1 + \frac{1}{s^2} \right)^{1/2} + \frac{\partial}{\partial y} \left\{ \rho \lambda H^2 \left(\frac{f}{8} \right)^{1/2} U_d \frac{\partial U_d}{\partial y} \right\} = \frac{\partial}{\partial y} \left[H (\rho U V)_d \right] \quad \text{(A4.22) \& (2.58)}$$

$$U_d = \frac{1}{H} \int_0^H U \, dz; \quad \tau_b = \left(\frac{f}{8} \right) \rho U_d^2; \quad \bar{\tau}_{yx} = \rho \bar{\varepsilon}_{yx} \frac{\partial U_d}{\partial y}; \quad \bar{\varepsilon}_{yx} = \lambda U_* H \quad \text{(A4.10) \& (2.59)}$$

The corresponding analytical solutions to (A4.22), which now include the effects of secondary flows, making the RHS non-zero, are given by (A4.23) & (A4.26) as follows:
For the constant depth domain, i.e. $H = $ constant, then

$$U_d = \left[A_1 e^{\gamma y} + A_2 e^{-\gamma y} + k_2 \right]^{1/2} \qquad \text{(A4.23)}$$

where

$$k_2 = \frac{8 g S_0 H (1 - \beta)}{f} \qquad \text{(A4.24)}$$

and

$$\beta = \frac{\Gamma}{\rho g H S_0} \qquad \text{(A4.25)}$$

For a linear side slope domain, i.e. $H = $ a function of y, then

$$U_d = \left[A_3 \xi^{\alpha_1} + A_4 \xi^{-(\alpha_1 + 1)} + \omega \xi + \eta \right]^{1/2} \qquad \text{(A4.26)}$$

where the constants α_1 and α_2, are as defined by (A4.16) & (A4.17), ω by (A4.18) as before and

$$\eta = -\frac{\Gamma}{\frac{\rho f}{8} \left(1 + \frac{1}{s^2} \right)^{1/2}} \qquad \text{(A4.27)}$$

Analytical solutions to the governing equation used in the SKM

The derivation of the SKM governing equation has been dealt with in Appendix 4, starting from the RANS equation for steady flow in the form presented as (2.57) in section 2.

$$\rho\left[\frac{\partial}{\partial y}(UV) + \frac{\partial}{\partial z}(UW)\right] = \rho g S_0 + \frac{\partial \tau_{yx}}{\partial y} + \frac{\partial \tau_{zx}}{\partial z} \qquad\text{(2.57) \& (6.43)}$$

Depth-averaging (2.57) gives

$$\rho g H S_0 - \rho\frac{f}{8}U_d^2\left(1 + \frac{1}{s^2}\right)^{1/2} + \frac{\partial}{\partial y}\left\{\rho\lambda H^2\left(\frac{f}{8}\right)^{1/2}U_d\frac{\partial U_d}{\partial y}\right\} = \frac{\partial}{\partial y}\left[H(\rho UV)_d\right] \qquad\text{(2.58)}$$

where

$$U_d = \frac{1}{H}\int_0^H U\,dz; \quad \tau_b = \left(\frac{f}{8}\right)\rho U_d^2; \quad \overline{\tau}_{yx} = \rho\overline{\varepsilon}_{yx}\frac{\partial U_d}{\partial y}; \quad \overline{\varepsilon}_{yx} = \lambda U_* H \qquad\text{(2.59)}$$

By letting Γ = secondary flow term on the RHS of equation (2.58), then

$$\frac{\partial}{\partial y}\left[H(\rho UV)_d\right] = \Gamma \qquad\text{(2.60)}$$

And (2.58) becomes

$$\rho g H S_0 - \rho\frac{f}{8}U_d^2\left(1 + \frac{1}{s^2}\right)^{1/2} + \frac{\partial}{\partial y}\left\{\rho\lambda H^2\left(\frac{f}{8}\right)^{1/2}U_d\frac{\partial U_d}{\partial y}\right\} = \Gamma \qquad\text{(2.58a)}$$

The solution of this governing equation for both laboratory and practical cases is undertaken by dividing the cross-section concerned into a number of segments transversely to the flow; in each segment the ground level is either horizontal or varies linearly with distance, y, along the section line. Thus there are two generic situations to solve Equation (2.58a); these are for a horizontal bed and for a bed level that slopes linearly with the transverse coordinate y. A solution for the whole cross-section is then constructed by joining the solutions in the individual segments together with appropriate compatibility conditions at the ends of each section as discussed in Chapters 6 & 7.

The algebraic manipulations in the various cases considered below simplify the governing equation either

1. to a second order ordinary differential equation with constant coefficients for a horizontal bed; which has exponentially varying solutions, or,
2. to a standard second order ordinary differential equation with coefficients that are functions of the independent variable for the linearly sloping bed; this standard equation is discussed in engineering mathematics texts and is called the Euler equation or the Cauchy-Euler equation or the Euler-Cauchy equation.

In each case the dependent flow variable is changed to be the square of the depth average velocity U_d and for the horizontal bed cases the independent variable remains as the transverse coordinate y. However, in the sloping bed cases the independent variable is changed from the transverse coordinate to the local depth. Applying coordinate transformation to the derivatives in Equation (2.58a) then leads to the Euler equation, whose analytic solution is well-established as being expressed as multiples of two real-number powers of the new independent variable. Finally, in each case the coefficients in the solutions are written in terms of the physical parameters of Equation (2.58a).

I FLOW IN A CONSTANT DEPTH DOMAIN WITH A HORIZONTAL BED

Let $s = \infty$, so that $1/s = 0$, and $H = $ constant.

I.I Case I when there is no secondary flow (i.e. $\Gamma = 0$) and with f & λ constant

The governing equation is

$$\rho g H S_0 - \rho \frac{f}{8} U_d^2 + \frac{\partial}{\partial y} \left\{ \rho \lambda H^2 \left(\frac{f}{8} \right)^{1/2} U_d \frac{\partial U_d}{\partial y} \right\} = 0 \qquad (A5.1)$$

which may be written as

$$D_1 - D_2 U_d^2 + \frac{\partial}{\partial y} \left\{ D_3 2 U_d \frac{\partial U_d}{\partial y} \right\} = 0 \qquad (A5.2)$$

where

$$D_1 = \rho g H S_0; \quad D_2 = \rho \frac{f}{8}; \quad D_3 = \frac{\rho \lambda H^2}{2} \left(\frac{f}{8} \right)^{1/2} \qquad (A5.3)$$

Equation (A5.2) may also be written as

$$D_1 - D_2 U_d^2 + D_3 \frac{\partial}{\partial y}\left\{\frac{\partial U_d^2}{\partial y}\right\} = 0$$

(A5.4)

Letting

$$F = U_d^2$$

$$D_1 - D_2 F + D_3 \frac{\partial^2 F}{\partial y^2} = 0$$

(A5.5)

or

$$D_1 - D_2 F + D_3 F'' = 0$$

(A5.6)

where F' denotes level of differentiation w.r.t. y
Hence

$$F'' - \frac{D_2}{D_3} F = -\frac{D_1}{D_3}$$

(A5.7)

Equation (A5.7) is a second order linear partial differential equation (pde), which may be rewritten as

$$F'' - \gamma^2 F = a$$

(A5.8)

The complementary function (CF) is found in exponential form by putting $a = 0$, giving

$$F = A_1 e^{\gamma y} + A_2 e^{-\gamma y}$$

(A5.9)

Note that an equivalent form is in terms of $\sinh(\gamma y)$ and $\cosh(\gamma y)$. This form is convenient for the solution away from a vertical wall at $y = 0$ as only the $\sinh(\gamma y)$ remains on applying $U_d = 0$ at $y = 0$.
Since

$$F' = A_1 \gamma e^{\gamma y} - A_2 \gamma e^{-\gamma y}$$

and

$$F'' = A_1 \gamma^2 e^{\gamma y} + A_2 \gamma^2 e^{-\gamma y} = \gamma^2 \left(A_1 e^{\gamma y} + A_2 e^{-\gamma y}\right) = \gamma^2 F$$

where γ^2 is defined by equations (A5.3) and (A5.10)

$$\gamma^2 = \frac{D_2}{D_3}$$

(A5.10)

Substituting for the various terms in D_1 and D_3 from (A5.3) gives

$$\gamma^2 = \frac{D_2}{D_3} = \frac{\rho f}{8} \frac{2}{\rho \lambda H^2} \left(\frac{8}{f}\right)^{1/2} = \left(\frac{2}{\lambda}\right)\left(\frac{f}{8}\right)^{1/2} \frac{1}{H^2}$$

$$\gamma = \left(\frac{2}{\lambda}\right)^{1/2} \left(\frac{f}{8}\right)^{1/4} \frac{1}{H} \tag{A5.11}$$

The particular integral (PI) is found from (A5.8) by assuming

$$F = -\frac{a}{\gamma^2} \tag{A5.12}$$

where

$$a = -\frac{D_1}{D_3} = -\frac{\rho g H S_0}{\rho \lambda H^2} 2\left(\frac{8}{f}\right)^{1/2}$$

$$a = -\frac{2gS_0}{\lambda H}\left(\frac{8}{f}\right)^{1/2} \tag{A5.13}$$

$$F = -\frac{a}{\gamma^2} = \frac{2gS_0}{\lambda H}\left(\frac{8}{f}\right)^{1/2} \frac{\lambda}{2}\left(\frac{8}{f}\right)^{1/2} H^2$$

$$F = \frac{8gS_0 H}{f} \tag{A5.14}$$

The full solution to equation (A5.7) is therefore

$$F = A_1 e^{\gamma y} + A_2 e^{-\gamma y} + \frac{8gS_0 H}{f} \tag{A5.15}$$

In terms of depth-averaged velocity, using (A5.4), means that the solution to equation (A5.1) is given by

$$U_d = \left[A_1 e^{\gamma y} + A_2 e^{-\gamma y} + \frac{8gS_0 H}{f}\right]^{1/2} \tag{A5.16}$$

Letting

$$k_1 = \frac{8gS_0 H}{f} \tag{A5.17}$$

Hence

$$U_d = \left[A_1 e^{\gamma y} + A_2 e^{-\gamma y} + k_1\right]^{1/2} \tag{A5.18}$$

where γ and k_1 are constants, given by (A5.11) and (A5.17) respectively. This implies that the two physical constants f and λ must be known or specified for each panel to which (A5.18) is applied. The two coefficients, A_1 and A_2, are determined from the boundary conditions for each panel in a specific application.

1.2 Case 2 when there is secondary flow present (i.e. $\Gamma \neq 0$) and with f & λ constant

The governing equation is

$$\rho g H S_0 - \rho \frac{f}{8} U_d^2 \left(1 + \frac{1}{s^2}\right)^{1/2} + \frac{\partial}{\partial y}\left\{\rho \lambda H^2 \left(\frac{f}{8}\right)^{1/2} U_d \frac{\partial U_d}{\partial y}\right\} = \Gamma \qquad \text{(A5.19)}$$

The solution obtained for a constant depth domain in 1.1, Case 1, may be easily adapted, since (A5.6) now becomes

$$\left(D_1 - \Gamma\right) - D_2 F + D_3 F'' = 0 \qquad \text{(A5.20)}$$

Hence

$$F'' - \frac{D_2}{D_3} F = -\frac{\left(D_1 - \Gamma\right)}{D_3} \qquad \text{(A5.21)}$$

$$F'' - \gamma^2 F = a_1 \qquad \text{(A5.22)}$$

Thus the only difference from (A5.8) is in the coefficient a_1 on the RHS. The complementary function (CF) is therefore the same as before.

The particular integral (PI) is obtained from

$$F = -\frac{a_1}{\gamma^2} \qquad \text{(A5.23)}$$

where

$$a_1 = -\frac{\left(D_1 - \Gamma\right)}{D_3} = -\frac{\left(\rho g H S_0 - \Gamma\right)}{\rho \lambda H^2} 2\left(\frac{8}{f}\right)^{1/2} \qquad \text{(A5.24)}$$

$$F = -\frac{a_1}{\gamma^2} = \frac{\left(\rho g S_0 H - \Gamma\right)}{\rho \lambda H^2} 2\left(\frac{8}{f}\right)^{1/2} \frac{\lambda}{2}\left(\frac{8}{f}\right)^{1/2} H^2$$

$$F = \frac{8\left(\rho g S_0 H - \Gamma\right)}{\rho f}$$

or

$$F = \frac{8 g S_0 H \left(1 - \beta\right)}{f} \qquad \text{(A5.25)}$$

where

$$\beta = \frac{\Gamma}{\rho g H S_0} \tag{A5.26}$$

The full solution to equation (A5.20) is therefore

$$F = A_1 e^{\gamma y} + A_2 e^{-\gamma y} + \frac{8 g S_0 H (1 - \beta)}{f} \tag{A5.27}$$

In terms of depth-averaged velocity, using (A5.4), means that the solution to equation (A5.19) may be expressed as

$$U_d = \left[A_1 e^{\gamma y} + A_2 e^{-\gamma y} + \frac{8 g S_0 H (1 - \beta)}{f} \right]^{1/2} \tag{A5.28}$$

Letting

$$k_2 = \frac{8 g S_0 H (1 - \beta)}{f} \tag{A5.29}$$

$$U_d = \left[A_1 e^{\gamma y} + A_2 e^{-\gamma y} + k_2 \right]^{1/2} \tag{A5.30}$$

where γ and k_2 are constants, given again by (A5.11) and (A5.29) respectively. It should be noted that the only difference in (A5.30) from (A5.18) is in the constant k_2, which involves the secondary flow constant, Γ. Since (A5.29) involves both f and Γ, this implies that the three physical constants f, λ and Γ must be known or specified for each panel to which (A5.30) is applied. Once again, the two coefficients, A_1 and A_2, need to be determined from the boundary conditions for each panel in a specific application.

2   FLOW IN A VARIABLE DEPTH DOMAIN WITH A LINEAR SIDE SLOPE

2.1   Case 3 when there is no secondary flow (i.e. $\Gamma = 0$) and with f & λ constant

Let s = side slope in transverse direction (1: s; vertical: horizontal), as shown in Figure 6.2, such that the local depth, ξ, is given by

$$\xi = H - \frac{(y - b)}{s} \tag{A5.31}$$

And introducing ψ, which has been defined in more general terms by (3.6), but given here in a simpler form as in practice S_0 is usually much less than S_y, as

$$\psi = \left(1 + \frac{1}{s^2}\right)^{1/2} \tag{A5.32}$$

Then with no secondary flow the governing equation is

$$\rho g \xi S_0 - \rho \frac{f}{8} U_d^2 \left(1 + \frac{1}{s^2}\right)^{1/2} + \frac{\partial}{\partial y}\left\{\rho \lambda \xi^2 \left(\frac{f}{8}\right)^{1/2} U_d \frac{\partial U_d}{\partial y}\right\} = 0 \tag{A5.33}$$

$$\rho g \xi S_0 - \rho \frac{f}{8} \psi U_d^2 + \frac{\rho \lambda}{2}\left(\frac{f}{8}\right)^{1/2} \frac{\partial}{\partial y}\left\{\xi^2 2 U_d \frac{\partial U_d}{\partial y}\right\} = 0$$

$$\rho g \xi S_0 - \rho \frac{f}{8} \psi U_d^2 + \frac{\rho \lambda}{2}\left(\frac{f}{8}\right)^{1/2} \frac{\partial}{\partial y}\left\{\xi^2 \frac{\partial U_d^2}{\partial y}\right\} = 0 \tag{A5.34}$$

Letting $F = U_d^2$ as before, then

$$\rho g S_0 \xi - \rho \frac{f}{8} \psi F + \frac{\rho \lambda}{2}\left(\frac{f}{8}\right)^{1/2} \frac{\partial}{\partial y}\left\{\xi^2 \frac{\partial F}{\partial y}\right\} = 0$$

$$\rho g S_0 \xi - \rho \frac{f}{8} \psi F + \frac{\rho \lambda}{2}\left(\frac{f}{8}\right)^{1/2}\left[\frac{\partial F}{\partial y} 2\xi \frac{\partial \xi}{\partial y} + \xi^2 \frac{\partial^2 F}{\partial y^2}\right] = 0 \tag{A5.35}$$

But since

$$\frac{\partial \xi}{\partial y} = -\frac{1}{s} \tag{A5.36}$$

Then (A5.35) becomes, in terms of differentiation w.r.t. ξ

$$\rho g S_0 \xi - \rho \frac{f}{8} \psi F + \frac{\rho \lambda}{2}\left(\frac{f}{8}\right)^{1/2}\left[\frac{\partial F}{\partial \xi}\frac{\partial \xi}{\partial y} 2\xi \frac{\partial \xi}{\partial y} + \xi^2 \frac{\partial^2 F}{\partial \xi^2}\left(\frac{\partial \xi}{\partial y}\right)^2\right] = 0$$

$$\rho g S_0 \xi - \rho \frac{f}{8} \psi F + \frac{\rho \lambda}{2 s^2}\left(\frac{f}{8}\right)^{1/2}\left[2\xi \frac{\partial F}{\partial \xi} + \xi^2 \frac{\partial^2 F}{\partial \xi^2}\right] = 0 \tag{A5.37}$$

Dividing by ρ, and letting a constant, l, be defined as

$$l = \frac{\lambda}{2 s^2}\left(\frac{f}{8}\right)^{1/2} \tag{A5.38}$$

Then since F' denotes level of differentiation w.r.t. ξ, (A5.37) may be written as

$$g S_0 \xi - \frac{f}{8} \psi F + 2 l \xi F' + l \xi^2 F'' = 0$$

Rearranging and dividing by l, gives the Euler differential equation:

$$\xi^2 F'' + 2\xi F' - \frac{f}{8l}\psi F + \frac{gS_0}{l}\xi = 0 \qquad (A5.39)$$

which may be written as

$$\xi^2 F'' + 2\xi F' - mF + n\xi = 0 \qquad (A5.40)$$

where

$$m = \frac{f\psi}{8l} \qquad (A5.41)$$

$$n = \frac{gS_0}{l} \qquad (A5.42)$$

The complementary function (CF) is given by putting $n = 0$, and is of the form

$$F = \xi^\alpha \qquad (A5.43)$$

Inserting this into (A5.40), with $n = 0$, gives

$$\xi^2 \alpha(\alpha-1)\xi^{\alpha-2} + 2\xi\alpha\xi^{\alpha-1} - m\psi\,\xi^\alpha = 0$$
$$\alpha(\alpha-1) + 2\alpha - m = 0$$
$$\alpha^2 + \alpha - m = 0$$
$$\alpha = \frac{-1 \pm \sqrt{1 + 4m}}{2}$$

Hence, using (A5.41) with (A5.32) & (A5.38)

$$\alpha = -\frac{1}{2} \pm \frac{1}{2}\left\{1 + 4\frac{f}{8}\left(1 + \frac{1}{s^2}\right)^{1/2}\frac{2s^2}{\lambda}\left(\frac{8}{f}\right)^{1/2}\right\}^{1/2}$$

$$\alpha_1 = -\frac{1}{2} + \frac{1}{2}\left\{1 + \frac{s(1+s^2)^{1/2}}{\lambda}(8f)^{1/2}\right\}^{1/2} \qquad (A5.44)$$

and

$$\alpha_2 = -\frac{1}{2} - \frac{1}{2}\left\{1 + \frac{s(1+s^2)^{1/2}}{\lambda}(8f)^{1/2}\right\}^{1/2} \qquad (A5.45)$$

Hence

$$\alpha_2 = -\alpha_1 - 1 = -(\alpha_1 + 1) \tag{A5.46}$$

The particular integral (PI) can be found from

$$F = \frac{-n}{2 - m} \xi \tag{A5.47}$$

$$F = \frac{-gS_0}{l\left(2 - \dfrac{f\psi}{8l}\right)} \xi = \frac{gS_0}{\left(\dfrac{f\psi}{8l} - 2l\right)} \xi$$

$$F = \frac{gS_0}{\dfrac{f}{8}\left(1 + \dfrac{1}{s^2}\right)^{1/2} - 2\dfrac{\lambda}{2s^2}\left(\dfrac{f}{8}\right)^{1/2}} \xi$$

$$F = \omega \xi \tag{A5.48}$$

where

$$\omega = \frac{gS_0}{\dfrac{(1 + s^2)^{1/2}}{s}\dfrac{f}{8} - \dfrac{\lambda}{s^2}\left(\dfrac{f}{8}\right)^{1/2}} \tag{A5.49}$$

The full solution to equation (A5.33) is therefore given by

$$F = A_3 \xi^{\alpha_1} + A_4 \xi^{\alpha_2} + \omega \xi$$

Using (A5.46)

$$F = A_3 \xi^{\alpha_1} + A_4 \xi^{-(\alpha_1 + 1)} + \omega \xi$$

Using

$$F = U_d^2 \tag{A5.4}$$

Thus

$$U_d = \left[A_3 \xi^{\alpha_1} + A_4 \xi^{-(\alpha_1 + 1)} + \omega \xi \right]^{1/2} \tag{A5.50}$$

where the local depth is ξ, as indicated for a side sloping element of a trapezoidal channel by (A5.31), and the two constants, α and ω, given by (A5.44) and (A5.49). This implies that the two physical constants, f and λ must be known or specified for each panel to which (A5.50) is applied. The two coefficients, A_3 and A_4, need to be determined from the boundary conditions for each panel in a specific application.

2.2   Case 4 when there is secondary flow present (i.e. $\Gamma \neq 0$) and with f & λ constant

With secondary flow, the RHS of (A5.33) is now not zero but Γ, and so the governing equation is

$$(\rho g \xi S_0 - \Gamma) - \rho \frac{f}{8} U_d^2 \left(1 + \frac{1}{s^2}\right)^{1/2} + \frac{\partial}{\partial y} \left\{ \rho \lambda \xi^2 \left(\frac{f}{8}\right)^{1/2} U_d \frac{\partial U_d}{\partial y} \right\} = 0 \qquad (A5.51)$$

As shown previously, through (A5.33)–(A5.40), the governing equation above, (A5.51), may be expressed in the equivalent form to (A5.40) as

$$\xi^2 F'' + 2\xi F' - m F + n\xi = \frac{\Gamma}{\rho l} \qquad (A5.52)$$

where F' denotes level of differentiation w.r.t. ξ. The complementary function (CF) will therefore be the same as before.

The particular integral (PI) is best obtained by making a change of variable from ξ to t so that equation (A5.52) becomes a pde w.r.t. t rather than ξ. With secondary flow, the analysis from (A5.33) to (A5.37) is repeated, so that the equivalent equations are

$$\rho g \xi S_0 - \rho \frac{f}{8} U_d^2 \left(1 + \frac{1}{s^2}\right)^{1/2} + \frac{\partial}{\partial y} \left\{ \rho \lambda \xi^2 \left(\frac{f}{8}\right)^{1/2} U_d \frac{\partial U_d}{\partial y} \right\} = \Gamma \qquad (A5.33a)$$

$$\rho g S_0 \xi - \rho \frac{f}{8} \psi F + \frac{\rho \lambda}{2s^2} \left(\frac{f}{8}\right)^{1/2} \left[2\xi \frac{\partial F}{\partial \xi} + \xi^2 \frac{\partial^2 F}{\partial \xi^2} \right] = \Gamma \qquad (A5.37a)$$

Dividing by ρ and introducing l, as before, using (A5.38), then (A5.37a) becomes (A5.52)

$$\xi^2 F'' + 2\xi F' - m F + n\xi = \frac{\Gamma}{\rho l} \qquad (A5.52)$$

Now, let

$$\xi = e^t \qquad (A5.53)$$

so that

$$\frac{d\xi}{dt} = e^t = \xi$$

and

$$\frac{d}{d\xi} = \frac{dt}{d\xi}\frac{d}{dt} = \frac{1}{\xi}\frac{d}{dt}$$

and

$$\frac{d^2}{d\xi^2} = \frac{d}{d\xi}\left(\frac{1}{\xi}\frac{d}{dt}\right) = -\frac{1}{\xi^2}\frac{d}{dt} + \frac{1}{\xi}\frac{d}{d\xi}\frac{d}{dt}\frac{dt}{dt} = -\frac{1}{\xi^2}\frac{d}{dt} + \frac{1}{\xi^2}\frac{dt}{dt^2}$$

Substituting these into (A5.52) gives

$$\xi^2\left[-\frac{1}{\xi^2}\frac{d}{dt} + \frac{1}{\xi^2}\frac{dt}{dt^2}\right] + 2\xi\left[\frac{1}{\xi}\frac{d}{dt}\right] - \frac{f}{8l}\psi F + \frac{gS_0}{l}\xi = \frac{\Gamma}{\rho l}$$

$$\frac{d^2F}{dt^2} + \frac{dF}{dt} - mF + n\xi = \frac{\Gamma}{\rho l} \tag{A5.54}$$

$$F'' + F' - mF + n\xi = \frac{\Gamma}{\rho l} \tag{A5.55}$$

Note that here F' now denotes a level of differentiation w.r.t. t, and (A5.55) now replaces (A5.40) as the primary equation to be solved.

The particular integral (PI) is obtained from

$$F = \frac{-n}{2-m}\xi - \frac{\Gamma}{lm} \tag{A5.56}$$

This differs from (A5.47) only by the addition of a second term on the RHS. Using (A5.38), (A5.41) and (A5.42), (A5.56) becomes

$$F = -\frac{gS_0}{l\left(2 - \frac{f\psi}{8l}\right)}\xi - \frac{\Gamma}{lm} = \frac{gS_0}{\left(\frac{f\psi}{8} - 2l\right)}\xi - \frac{\Gamma}{lm}$$

$$F = \frac{gS_0}{\frac{f}{8}\left(1 + \frac{1}{s^2}\right)^{1/2} - 2\frac{\lambda}{2s^2}\left(\frac{f}{8}\right)^{1/2}}\xi - \frac{\Gamma}{lm}$$

$$F = \frac{gS_0}{\frac{(1+s^2)^{1/2}}{s}\frac{f}{8} - \frac{\lambda}{s^2}\left(\frac{f}{8}\right)^{1/2}}\xi - \frac{\Gamma}{lm} \tag{A5.57}$$

Letting

$$\eta = -\frac{\Gamma}{lm} \tag{A5.58}$$

which from (A5.38) & (A5.41) is defined by

$$\eta = -\frac{\Gamma}{\frac{\rho f}{8}\left(1 + \frac{1}{s^2}\right)^{1/2}} \tag{A5.59}$$

Then the PI then simplifies to

$$F = \omega\xi + \eta \tag{A5.60}$$

where ω and η are defined by (A5.49) and (A5.59) respectively. The full solution to equation (A5.51) is therefore given by

$$F = A_3\xi^{\alpha_1} + A_4\xi^{\alpha_2} + \omega\xi + \eta \tag{A5.61}$$

Using (A5.46) this becomes

$$F = A_3\xi^{\alpha_1} + A_4\xi^{-(\alpha_1+1)} + \omega\xi + \eta \tag{A5.62}$$

Using

$$F = U_d^2 \tag{A5.4}$$

Thus

$$U_d = \left[A_3\xi^{\alpha_1} + A_4\xi^{-(\alpha_1+1)} + \omega\xi + \eta \right]^{1/2} \tag{A5.63}$$

where the local depth is ξ, as indicated for a side sloping element of a trapezoidal channel by (A5.31), and the three constants, α, ω and η by (A5.44), (A5.49) and (A5.59). This implies that the three physical constants, f, λ and Γ must be known or specified for each panel to which (A5.63) is applied. The two coefficients, A_3 and A_4, need to be determined from the boundary conditions for each panel in a specific application.

3   FLOW WITH ADDED RESISTANCE DUE TO TREES ON A FLOODPLAIN

3.1   Case 5 Flow in a constant depth domain with a horizontal bed

Let $s = \infty$, so that $1/s = 0$, and $H =$ constant. Assume that the trees are uniformly spaced trees on the floodplain and that the drag force, F_d, per unit volume is given by the formula:

$$F_d = \frac{1}{2}\rho C_d A_p U^2 \tag{A5.64}$$

where $C_d =$ drag coefficient and $A_p =$ total projected area of all trees per unit volume (1/m).

Adding this term to SKM (2.58a) for a constant water depth region

$$\rho g H S_0 - \rho \frac{f}{8} U_d^2 \left(1 + \frac{1}{s^2} \right)^{1/2} + \frac{\partial}{\partial y}\left\{ \rho\lambda H^2 \left(\frac{f}{8} \right)^{1/2} U_d \frac{\partial U_d}{\partial y} \right\} - \frac{1}{2}\rho C_d A_p U_d^2 H = \Gamma \tag{A5.65}$$

$$\rho g H S_0 - \left(\rho \frac{f}{8} \left(1 + \frac{1}{s^2} \right)^{1/2} + \frac{1}{2} \rho C_d A_p H \right) U_d^2 + \frac{\partial}{\partial y} \left\{ \rho \lambda H^2 \left(\frac{f}{8} \right)^{1/2} U_d \frac{\partial U_d}{\partial y} \right\} = \Gamma$$

$$E_1 - E_2 U_d^2 + E_3 \frac{\partial}{\partial y} \left\{ \frac{\partial U_d^2}{\partial y} \right\} = \Gamma' \tag{A5.66}$$

where

$$\Gamma' = \Gamma / \rho$$

$$E_1 = g H S_0$$

$$E_2 = \frac{f}{8} + \frac{1}{2} C_d A_p H$$

$$E_3 = \frac{\lambda H^2}{2} \left(\frac{f}{8} \right)^{1/2}$$

Repeat the process of solving the differential equation
The complementary function (CF) is

$$U_d^2 = A e^{\lambda y} \tag{A5.67}$$

Substituting this CF into (A5.66) gives

$$-E_2 + E_3 \lambda^2 = 0 \tag{A5.68}$$

$$\lambda^2 = \frac{E_2}{E_3}$$

$$\lambda = \pm \sqrt{\frac{E_2}{E_3}} \tag{A5.69}$$

The particular integral (PI) is the form

$$U_d^2 = A \tag{A5.70}$$

$$E_1 - E_2 A = \Gamma'$$

$$A = \frac{E_1 - \Gamma'}{E_2}$$

Similarly, the analytical solution to the new additional drag force in (2.58) becomes

$$F = A_1 e^{\gamma y} + A_2 e^{-\gamma y} + \frac{g H S_0 (1 - \beta)}{\dfrac{f}{8} + \dfrac{1}{2} C_d A_p H} \tag{A5.71}$$

With

$$\gamma = \left| \frac{\dfrac{f}{8} + \dfrac{1}{2} C_d A_p H}{\dfrac{\lambda H^2}{2} \left(\dfrac{f}{8}\right)^{1/2}} \right|^{1/2}$$

(A5.72)

Since from (A5.4), $F = U_d^2$, then the solution is given by

$$F = \left[A_1 e^{\gamma y} + A_2 e^{-\gamma y} + \frac{g H S_0 (1 - \beta)}{\dfrac{f}{8} + \dfrac{1}{2} C_d A_p H} \right]^{1/2}$$

(A5.73)

where the modified constants should be used.

Fortran program for solving the governing depth-averaged equation used in the SKM

Fortran source code of the program, together with two sets of input data files, are given below to solve for the velocity and bed shear stress distributions in the FCF channel, without rods on the floodplains, as an example The Fortran compiler can be installed from the following website *http://www.silverfrost.com/32/ftn95/ftn95_personal_edition.aspx* or freely downloaded from *http://www.cs.yorku.ca/~roumani/fortran/ftn.htm*. Once the compiler is installed, use the command prompt to run the Fortran program below. This program uses a simple finite difference technique to solve the SKM. Two input data files are required to run the program. The details of how to make the two files are explained after the Fortran program below.

```
!
        PROGRAM TDRMn
!
!
        REAL WL,SO,DY,Y(0:300),Z(0:300),YS1(0:300),YS2(0:300),FF(0:300)
        REAL DEV(0:300),YY(0:1000),ZZ(0:1000),HH(0:1000),SS(0:1000),DH
        REAL    AL(0:1000),AD(0:1000),AR(0:1000),U(0:1000),B(0:1000)
        REAL FFY(0:1000),DEVY(0:1000)
        INTEGER  NOS,SUB,M
!
        CALL INPUT(WL,SO,DY,NOS,Y,Z,SUB,YS1,YS2,FF,DEV)
        CALL CALCUL(WL,NOS,Y,Z,SUB,YS1,YS2,FF,DEV,YY,ZZ,HH,SS,&
          DH,FFY,DEVY,M)
!
        CALL COEF(SO,SS,DH,HH,FFY,DEVY,M,AL,AD,AR,U,B)
        CALL TRID(AL,AD,AR,U,B,M)
        CALL VELSHE(DY,M,YY,HH,FFY,U,devy)
        STOP
        END
!
!    inputting crosssection geometry and friction data files, and
!      water depth in the main channel and bed slope.
!
!
        SUBROUTINE INPUT(WL,SO,DY,NOS,Y,Z,SUB,YS1,YS2,FF,DEV)
!
        REAL WL,SO,DY,Y(0:300),Z(0:300),YS1(0:300),YS2(0:300),FF(0:300)
        REAL DEV(0:300)
        INTEGER  NOS,SUB
        CHARACTER*20 CROSS,GEOME,RESULT
```

```
!
          WRITE(*,120)
          WRITE(*,130)
                   READ(*,180) CROSS
          WRITE(*,140)
                   READ(*,190) GEOME
          WRITE(*,150)
                   READ(*,*) WL
          WRITE(*,160)
                   READ(*,*) SO
                   DY=1000.
          WRITE(*,170)
                   READ(*,190) result
          OPEN(UNIT=3,STATUS='new',FORM='FORMATTED',FILE=result)
!
          OPEN(UNIT=1,STATUS='OLD',FORM='FORMATTED',FILE=CROSS)
          READ(1,*) NOS
          DO 100 I=1,NOS
          READ(1,*) Y(I),Z(I)
!         PRINT*,Y(I),Z(I)
100       CONTINUE
!
          OPEN(UNIT=2,STATUS='OLD',FORM='FORMATTED',FILE=GEOME)
          READ(2,*) SUB
          DO 110 I=1,SUB
                   READ(2,*) YS1(I),YS2(I),FF(I),DEV(I)
!         PRINT*, YS1(I),YS2(I),FF(I),DEV(I)
110       CONTINUE
!
120       FORMAT(15X,'TWO DIMENSIONAL RIVER MODEL',&
                     /,15X,'--------------------------')

130       FORMAT(//,5X,'INPUT CROSS-SECTION DATA FILE (1-999 m):',T50)
140       FORMAT(//,5X,'INPUT SUB-AREAS FF & LAM DATA FILE  :',T50)
150       FORMAT(//,5X,'WATER LEVEL (m) :',T50)
160       FORMAT(//,5X,'LONGITUDINAL BED SLOPE :',T50)
170       FORMAT(//,5X,'reslut file :',T50)
180       FORMAT(A20)
190       FORMAT(A20)
!
          RETURN
          END
!
!
!

          SUBROUTINE CALCUL(WL,NOS,Y,Z,SUB,YS1,YS2,FF,DEV,YY,ZZ,HH,SS,&
          DH,FFY,DEVY,M)

!
          REAL WL,Y(0:300),Z(0:300),YS1(0:300),YS2(0:300),FF(0:300)
          REAL YN(0:1000),ZN(0:1000),SN(0:1000),S(0:1000),DH,YY(0:1000)
          REAL ZZ(0:1000),HH(0:1000),SS(0:1000),DEVY(0:1000),FFY(0:1000)
          REAL DEV(0:300)
          INTEGER  NOS,SUB,ST(0:1000),L,M
!
! Making uniform mesh dy of cross section, in this case, dy=DH=0.05m
! and the total width of the channel is less than 99m width.
```

```
!
        IF (Y(NOS).LE.99.0) THEN
        DH=0.05
        ELSEIF (Y(NOS).GT.99.0.AND.Y(NOS).LT.499.0) THEN
        DH=1.0
        ELSEIF (Y(NOS).GE.499.0.AND.Y(NOS).LE.999.0) THEN
        DH=1
        ENDIF
!
        L=0
        DO 210 I=1,NOS-1
          L=L+1
          S(I)=(Z(I)-Z(I+1))/(Y(I)-Y(I+1))
          ST(I)=ABS(Y(I)-Y(I+1))/DH
          YN(L)=Y(I)
          ZN(L)=Z(I)
          SN(L)=S(I)
        DO 200 J=2,ST(I)
          L=L+1
          YN(L)=YN(L-1)+DH
          ZN(L)=ZN(L-1)+DH*S(I)
          SN(L)=S(I)
!        PRINT*, L,YN(L),ZN(L),SN(L)

200       CONTINUE
210       CONTINUE
!
! The end of cross section
!
        L=L+1
        YN(L)=YN(L-1)+DH
        ZN(L)=ZN(L-1)+DH*S(NOS-1)
        SN(L)=SN(L-1)
!
! Determine water depth at each dy
!
        M=0
        DO 240 I=1,L
!
          IF ((WL-ZN(I)).LT.0.) THEN
              GOTO 240
!
          ELSEIF ((WL-ZN(I)).GE.0.) THEN
              M=M+1
              YY(M)=YN(I)
              ZZ(M)=ZN(I)
              HH(M)=WL-ZN(I)
              SS(M)=SN(I)
!        PRINT*, YY(M),ZZ(M),HH(M),SS(M)
!
              DO 220 II=1,SUB
                IF (YN(I).GE.YS1(II).AND.YN(I).LE.YS2(II)) THEN
                  DEVY(M)=DEV(II)
!
!Two cases 1 is inputting friction factor 2 is Manning coeff
! input friction factor
!                 FFY(M)=FF(II)
```

```
!
!Input Manning coefficient as FF

         if(hh(m).ne.0.0) ffy(m)=8.*9.807*ff(ii)**2/(hh(m)**(0.3333))
                   ENDIF
220              CONTINUE
              ENDIF
240      CONTINUE
!

         RETURN
         END
!
!Making matrix entries
!
         SUBROUTINE COEF(SO,SS,DH,HH,FFY,DEVY,M,AL,AD,AR,U,B)
!
         REAL SO,SS(0:1000),DH,HH(0:1000),FFY(0:1000),DEVY(0:1000)
         REAL    AD(0:1000),AR(0:1000),U(0:1000),B(0:1000),A1(0:1000)
         REAL    A3(0:1000),A4(0:1000),A5(0:1000),A6(0:1000),AL(0:1000)
         REAL A2(0:1000)
         INTEGER M

!
            DO 300 I=1,M
!

         A1(I)=4.*HH(I)**2*DEVY(I)*SQRT(FFY(I)/8.)/2.
         A2(I)=DEVY(I)*SQRT(FFY(I)/8.)*(HH(I+1)**2-HH(I-1)**2)/2.
         A3(I)=HH(I)**2*SQRT(FFY(I)/8.)*(DEVY(I+1)-DEVY(I-1))/2.
         A4(I)=HH(I)**2*DEVY(I)*(SQRT(FFY(I+1)/8.)-SQRT(FFY(I-1)/8.))/2.
!
! drag force formula FD=1/2*Cd*N*d*H N =No per unit area
!    A5(I)=4*DH*DH*FFY(I)/8.*SQRT(1.+SS(I)*SS(I))+FD
!
         A5(I)=4*DH*DH*FFY(I)/8.*SQRT(1.+SS(I)*SS(I))
         A6(I)=4.*DEVY(I)*HH(I)*HH(I)*SQRT(FFY(I)/8.)
!
          AL(I)=A1(I)+A2(I)+A3(I)+A4(I)
          AD(I)=-A5(I)-A6(I)
          AR(I)=A1(I)-A2(I)-A3(I)-A4(I)
          B(I)=-4.*9.807*HH(I)*SO*DH*DH
!
300      CONTINUE
!
         U(1)=0
         U(M)=0
!
         RETURN
         END
!
!Solve the Matrix
!
         SUBROUTINE TRID(SUB,DIAG,SUP,X,B,M)
!
         REAL SUB(M),DIAG(M),SUP(M),B(M),X(M)
!
         IF (M.GT.2) GOTO 400
         X(2)=B(2)/DIAG(2)
         RETURN
```

```
!
400       DO 410 I=3,M-1
            RATIO=-SUB(I)/DIAG(I-1)
            DIAG(I)=DIAG(I)+RATIO*SUP(I-1)
            B(I)=B(I)+RATIO*B(I-1)
410       CONTINUE
!
          X(M-1)=B(M-1)/DIAG(M-1)
          I=M-1
          DO 420 NP=3,M-1
            I=I-1
            X(I)=(B(I)-SUP(I)*X(I+1))/DIAG(I)
420       CONTINUE
!
          RETURN
          END
!
!  Print out results of velocity and bed shear stress after solving
!  matrix
!
          SUBROUTINE VELSHE(DY,M,YY,HH,FFY,U,devy)
!
          REAL FFY(0:1000),U(0:1000),SH(0:1000),YY(0:1000),HH(0:1000)
          REAL  UU(0:1000),  t(0:1000),devy(0:1000),DY,Q
!           CHARACTER*20 RESULT
!
!           WRITE(*,510)
!          print*,' result file'
!               READ(*,550) RESULT
!               OPEN(UNIT=3,STATUS='NEW',FILE=RESULT)
          WRITE(3,520)
          Q=0
          DO 500 I=1,M
          UU(I)=SQRT(U(I))
          SH(I)=UU(I)*UU(I)*DY*FFY(I)/8.
          Q=Q+(YY(I+1)-YY(I))*(SQRT(U(I+1))+SQRT(U(I)))*(HH(I+1)+HH(I))/4.
          if(i.gt.1.and.i.lt.m) then
          ddy=yy(i+1)-yy(i-1)
          T(I)=0.5*dy*devy(i)*hh(i)*hh(i)*(U(I+1)-U(I-1))/ddy*FFY(I)/8.
          endif

          WRITE(3,530) YY(I),UU(I),SH(I),T(i)
!         WRITE(*,*) YY(I),UU(I),FFY(I),Q
500       CONTINUE
          WRITE(3,540) Q
510       FORMAT(//,5X,'OUTPUT FILE NAME :',T50)
520       FORMAT(10X,'DISTAN!E        VELOCITY       BED SHEAR FORCE st st')
530       FORMAT(5X,F12.5,3X,F12.6,4X,(2F12.6))
540       FORMAT(//,10X,'DIS!HARGE :',F15.5,'m**3')
550       FORMAT(A20)
!
          RETURN
          END
```

Two input data are required into the program when it executes. There are FCFCROSS and FCF_FF2, but the names can be changeable to appropriate files names.

The first one is FCFCROSS

```
8
0    0.35
0.2   0.15
3.2   0.15
3.35 0.0
4.85 0.0
5.0   0.15
8.0   0.15
8.2   0.35
```

FCFCROSS file consists of
8: total numbers of data in the file.

1 row: $y = 0$ m, at the top of floodplain bank and $z = 0.35$ m, a height of the top of floodplain bank.
2 row: $y = 0.2$ m at the edge of floodplain bank and $z = 0.15$ m, a height of floodplain.
3 row: $y = 3.2$ m at the edge of floodplain and $z = 0.15$ m a height floodplain.
4 row: $y = 3.35$ m at the corner of main channel and $z = 0$ m at the bed
5 row are $y = 4.85$ m at the other corner of the main channel and $z = 0$ at the bed.
6 row: $y = 5$ m at the edge of floodplain and $z = 0.15$ m a height floodplain.
7 row: $y = 8.0$ m at the edge of floodplain bank and $z = 0.15$ m, a height of floodplain.
8 row: $y = 8.2$ m, at the top of floodplain bank and $z = 0.35$ m, a height of the top of floodplain bank.

The second is FCF_FF2

```
7
0.     0.2   0.01   0.15
0.2    3.2   0.01   0.15
3.2    3.35 0.01   0.15
3.35 4.85 0.01   0.15
4.85 5.0   0.01   0.15
5.0    8.0   0.01   0.15
8.0    8.2   0.01   0.15
```

FCF_FF2 consists of
1st column: y coordinates
2nd column: z coordinates
3rd column: friction factor, f.
4th column: non-dimensional eddy viscosity, λ.
1st row: subsection 1,
2nd row: subsection 2,
3rd row: subsection 3 etc.

References

Abbot, M.B.: *Hydroinformatics*, Avebury Technical, Aldershot, (1991).

Abbot, M.B. and Basco, D.R.: *Computational fluid dynamics: an introduction for engineers*, Longman Scientific & Technical, J. Wiley, (1989), pp. 1–425.

Abril, J.B.: Numerical modelling of turbulent flow and sediment transport by the finite element method, *MPhil. Thesis, Department of Civil Engineering, The University of Birmingham, England*, (1995).

Abril, J.B.: Numerical modelling of turbulent flow, sediment transport and flood routing by the finite element method, *PhD Thesis, Department of Civil Engineering, The University of Birmingham*, UK, (1997), pp. 1–334.

Abril, J.B.: Updated RFMFEM finite element model based on the SKM for depth-averaged river flow simulation, *EPSRC Technical Report* GR/R54880/01, The University of Birmingham (2001).

Abril, J.B. and Knight, D.W.: Sediment transport simulation of the Paute river using a depth-averaged flow model, *River Flow 2002, Proc. Int. Conf. on Fluvial Hydraulics*, Louvain-la-Neuve, Belgium, Sept., Vol. 2, Balkema, (2002), pp. 895–901.

Abril, J.B. and Knight, D.W.: Stage-discharge prediction for rivers in flood applying a depth-averaged model, *Journal of Hydraulic Research, IAHR*, 42, No. 6, (2004), pp. 616–629.

Abril, J.B. and Knight, D.W.: Simultaneous prediction of rating curves and junction split flow relationships in a natural river network, *RiverFlow 2012*, Proc. Int Conf. on Fluvial Hydraulics, International Association for Hydro-Environment Engineering and Research, IAHR, Costa Rica, September, (2012), pp. 1–8.

Ackers, P.: Resistance of fluids flowing in channel and pipes, *Hydraulics Research Paper No. 1*, HMSO, London, (1958), pp. 1–39.

Ackers, P., White, W.R., Perkins, J.A. and Harrison, A.J.: *Weirs and flumes for flow measurement*, J. Wiley, (1978), pp. 1–327.

Ackers, P.: Sediment transport: the Ackers and White theory revised, *Report SR 237, HR Wallingford*, UK, (1990).

Ackers, P.: Hydraulic design of straight compound channels, *SR Report SR 281*, HR Wallingford, October, Vols. 1 & 2, (1991), pp. 1–131 & 1–139.

Ackers, P.: 1992 Gerald Lacey Memorial Lecture – Canal and river regime in theory and practice: 1929–92, *Proc. Instn Civ. Engrs Wat., Marit. & Energy*, 96, Sept., Paper No. 10019, (1992a), pp. 167–178.

Ackers, P.: Hydraulic design of two stage channels, *Proc. Instn Civ. Engrs Wat., Marit. & Energy*, 96, Dec., Paper No. 9988, (1992b), pp. 247–257.

Ackers, P.: Stage-discharge functions for two-stage channels: The impact of new research, *Journal Instn Water & Environmental Management*, Vol. 7, No. 1, February, (1993a), pp. 52–61.

Ackers, P.: Flow formulae for straight two-stage channels, *Journal of Hydraulic Research*, IAHR, Vol. 31, No. 4, (1993b), pp. 509–531.

Ackers, P.: Sediment transport in open channels: Ackers and White update, *Proc. Instn. Civil Engineers, London, Water, Maritime & Energy*, 101 Water Board Technical Note 619, (1993c), pp. 247–249.

Ackers, P. and White, W.R.: Sediment transport – new approach and analysis, *Proc. Hydraulic Eng. Div.*, ASCE, Vol. 99, No. 11, (1973), pp. 2041–2060.

Alavian, V. and Chu, V.H.: Turbulent Exchange Flow in Shallow Compound Channel, *Proceedings of the 21st Congress of the International Association for Hydraulic Research*, 19–23 August, Melbourne, Australia, (1985), pp. 446–451.

Alhamid, A.A.I.: Boundary shear stress and velocity distributions in differentially roughened trapezoidal channels, *PhD thesis, The University of Birmingham*, England, UK, (1991).

Allen, J. and Ullah, M.I.: The flow of water through smooth open channel of narrow rectangular T-shaped cross section, *Proc. Inst. Civil Engrs.*, February, (1967), pp. 325–349.

Anastasiadou-Partheniou, L. and Samuels, P.G.: Automatic calibration of computational river models, *Proceedings of ICE Water and Maritime Engineering*, Vol. 130, Issue 3, September, (1998), pp. 154–162.

Anderson, M.G., Walling, D.E.S. and Bates, P.D.: *Floodplain processes*, J. Wiley, (1996), pp. 1–658.

Argawal, V.C., Garde, R.J. and Ranga Raju, K.G.: Resistance to flow and sediment transport in curved alluvial channels, *Fourth Congress, Asian and Pacific Division*, IAHR, Thailand, 11–13 September, (1984), pp. 207–218.

Armanini, A., Righetti, M. and Grisenti, P.: Direct measurement of vegetation resistance in prototype, Journal of Hydraulic Research Vol. 43, No. 5, (2005), pp. 481–487.

Armitage, P.D. and Cannan, C.E.: Annual changes in the summer patterns of mesohabitat distribution and associated faunal assemblages. Hydrological Processes, 14, (2000), pp. 3161–3179.

Ashworth, P.J., Bennett, S.J., Best, J.L. and McLelland, S.J.: *Coherent flow structures in open channels*, J. Wiley, (1996), pp. 1–733.

Atabay, S.: Stage-discharge, resistance and sediment transport relationships for flow in straight compound channels, *PhD thesis, Department of Civil Engineering, The University of Birmingham*, (2001), pp. 1-.

Atabay, S.: Accuracy of the ISIS bridge methods for prediction of afflux at high flows, *Water and Environment Journal*, CIWEM, 22, No. 1, (2008), pp. 1–10.

Atabay, S. and Knight, D.W.: Bridge afflux experiments in compound channels, *Report supplied to jba*, Department of Civil Engineering, The University of Birmingham (2002).

Atabay, S., Knight, D.W. and Seckin, G.: Influence of a mobile bed on the boundary shear in a compound channel, *River Flow 2004, Proc. 2nd Int. Conf. on Fluvial Hydraulics, 23–25 June, Napoli, Italy* [Eds M. Greco, A. Carravetta and R.D. Morte], Vol. 1, (2004), pp. 337–345.

Atabay, S., Knight, D.W. and Seckin, G.: Effects of overbank flow on fluvial sediment transport rates, *Water Management*, Proceedings of the Instn. of Civil Engineers, London, 158, WM1, March, (2005), pp. 25–34.

Atabay, S. and Knight, D.W.: 1-D modelling of conveyance, boundary shear and sediment transport in overbank flow, *Journal of Hydraulic Research*, IAHR, Vol. 44, No. 6, (2006), pp. 739–754.

Ayyoubzadeh, S.A.: Hydraulic aspects of straight-compound channel flow and bed load sediment transport, *PhD thesis, Department of Civil Engineering, The University of Birmingham*, (1997), pp. 1–380.

Baattruo-Pedersen, A. and Riis, T.: Impacts of different weed cutting practices on macrophyte species diversity and composition in a Danish stream, *River Research Applications*, 20, (2004), pp. 103–114.

Babaeyan-Koopaei, K., Ervine, D.A., Carling, P.A. and Cao, Z.: Velocity and turbulence measurements for two overbank flow events in River Severn, *Jnl. of Hydraulic Engineering*, Vol. 198, No. 10, (2002), pp. 891–900.

Bagnold, R.A.: An approach to the sediment transport problem from general physics, Professional Paper 422-I, *US Geological Survey*, Washington, (1966).

Baptist, M.J.: Modelling floodplain biogeomorphology, *PhD Thesis, Technical University, Delft*, (2005), pp. 1–195.

Barnes, H.H.: Roughness characteristics of natural channels, *US Geological Survey*, Water-Supply Paper No. 1849, Washington DC, (1967), pp. 1–214.

Barr, D.I.H.: Two additional methods of direct solution of the Colebrook-White function, *Proc. Inst. of Civil Engineers*, Vol. 59, (1975), pp. 827.

Barr, D.I.H.: Osborne Reynolds' turbulent flow resistance formulae, *Proc. Institution of Civil Engineers*, Part 2, Vol. 67, (1979), pp. 743–750.

Batchelor, G.K., Moffatt, H.K. and Worster, M.G.: *Perspectives in fluid dynamics: a collective introduction to current research*, Cambridge University Press, (2000), pp. 1–631.

Bazin, H.: French translates to "A new formula for the calculation of discharge in open channels", Memoire No. 41, *Annales des Ponts et Chaussees*, Vol. 14, No. 7, (1897), pp. 20–70.

Bedient, P.B. and Huber, W.C.: *Hydrology and floodplain analysis*, Addison Wesley, Reading, Massachussetts, USA, (1988), pp. 1–650.

Benn, J.R. and Bonner, V.R.: Scoping study into hydraulic performance of bridges and other structures, including the effects of blockage, at high flows, *Report for the Environment Agency*, JBA, December, (2001).

Benn, J.R., Mantz, P., Lamb, R., Riddell, J. and Nalluri, C.: Afflux at Bridges and Culverts: Review of current knowledge and practice, *Environment Agency R&D Technical Report W5A-061/TR1*, (2004), ISBN 1-8443-2291-2, 111 pp.

Berz, G.: Flood disasters: lessons from the past – worries for the future, *Proc. Instn. of Civil Engineers, Water, Maritime and Energy Division*, London, Vol. 142, Issue 1, March, (2000), pp. 3–8.

Bettess, R. and White, W.R.: Extremal hypotheses applied to river regime, In *Sediment transport in gravel bed rivers* [Eds C.R. Thorne, J.C. Bathurst and R.D. Hey], J. Wiley & Sons, (1987), pp. 767–789.

Bettess, R.: Sediment transport and alluvial resistance in rivers, *Report by HR Wallingford to MAFF/EA*, MAFF/Defra project FD0112 and W5i 609, (2001).

Bettess, R., White, W.R. and Reeve, C.E.: On the width of regime channels, *Int. conf. on River regime*, Edited by White, W.R., J. Wiley & Son, (1988), 149–161.

Beven, K.: A manifesto for the equifinality thesis, *Journal of Hydrology*, 320, (2006), pp. 18–36.

Beven, K. and Freer, J.: Equifinality, data assimilation, and uncertainty estimation in mechanistic modelling of complex environmental systems using the GLUE methodology, *Journal of Hydrology*, 249, (2001), pp. 11–29.

Biery, P.F. and Delleur, J.W.: Hydraulics of single span arch bridge constrictions, *Journal of the Hydraulics Division*, American Society of Civil Engineers, New York, February, Vol. 88, No. 2, March, (1962), pp. 75–108.

Biglari, B. and Sturm, T.W.: Numerical modeling of flow around bridge abutments in compound channel, *Journal of Hydraulic Engineering*, American Society of Civil Engineers, New York, February, Vol. 124, No. 2, (1998), pp. 156–164.

Blalock, M.E. and Sturm, T.W.: Minimum specific energy in compound open channel flow, *Journal of the Hydraulics Division*, ASCE, Vol. 107, No. HY6, (1981), pp. 699–717.

Blasius, P.R.H.: Das Aehnlichkeitsgesetz bei Reibungsvorgangen [in *Flüssigkeiten, Forschungsheft 131*], Vereins Deutscher Ingenieure, Berlin, (1913), pp. 1–41.

Blench, T.: *Mobile bed fluviology*, University of Alberta Press, (1969), pp. 1–222.

Bos, M.G., Replogle, J.A. and Clemmens, A.J.: *Flow measuring flumes for open channels*, J. Wiley, (1984), pp. 1–321.

Bousmar, D.: Flow modelling in compound channels: momentum transfer between main channel and prismatic and non-prismatic floodplains, *PhD Thesis, Universite Catholique de Louvain*, (2002).

Bousmar, D. and Zech, Y.: Momentum transfer for practical flow computation in compound channels, *Journal of Hydraulic Engineering*, ASCE, Vol. 125, No. 7, July, (1999), pp. 696–706.

Bousmar, D. and Zech, Y.: Velocity distribution in non-prismatic compound channels, *Proc. ICE*, Water Management, Vol. 157, (2004), pp. 99–108.

Bousmar, D., Wilkin, N., Jacquemart, J.H. and Zech, Y.: Overbank flow in symmetrically narrowing floodplains, *Journal of Hydraulic Engineering*, ASCE, Vol. 130, No. 4, April, (2004), pp. 305–312.

Bradley, J.N.: Hydraulics of bridge waterways, *Hydraulic Design Series, Edition 2*, Department of Transport, Federal Highway Administration, Washington, DC, (1978).

Bronstert, A.: Overview of current perspectives on climate change, Chapter 3 in *River Basin Modelling for Flood Risk Mitigation* [Eds D.W. Knight and A.Y. Shamseldin], Taylor & Francis, (2006a), pp. 59–75.

Bronstert, A.; The effects of climate change on flooding, Chapter 4 in *River Basin Modelling for Flood Risk Mitigation* [Eds D.W. Knight and A.Y. Shamseldin], Taylor & Francis, (2006b), pp. 77–791.

Brown, F.A.: Sediment transport in river channels at high stage, *PhD thesis, Department of Civil Engineering, The University of Birmingham*, (1997).

Brown, P.M.: Afflux at British bridges, *Report No. 60*, HR Wallingford, October, (1985).

Brown, P.M.: Afflux at arch bridges, *Report No. 182*, HR Wallingford, December, (1988).

Brundett, E. and Baines, W.D.: The production and diffusion of vorticity in duct flow, *Jnl. Fluid Mech.*, Vol. 19, (1964), pp. 375–394.

Buisson, R.S.K., Wade, P.M., Cathcart, R.L., Hemmings, S.M., Manning, C.J. and Mayer, L.: *The Drainage Channel Biodiversity Manual: Integrating Wildlife and Flood Risk Management*, Association of Drainage Authorities and Natural England, Peterborough, (2008), pp. 1–189.

Burnham, M.W. and Davis, D.W.: Effects of data errors on computed steady-flow, *Journal of Hydraulic Engineering*, ASCE, 116(7), (1990), pp. 914–929.

Cao, S.: Regime theory and a geometric model for stable alluvial channels, *PhD thesis, Department of Civil Engineering, The University of Birmingham*, (1995), pp. 1–350.

Cao, S. and Knight, D.W.: Regime theory of alluvial channels based upon the concepts of stream power and probability, *Proc. Instn. of Civil Engineers, Water, Maritime and Energy Division*, London, Vol. 118, Issue 3, Sept., Paper No. 11029, (1996), pp. 160–167.

Cao, S. and Knight, D.W.: Design for hydraulic geometry of alluvial channels, *Journal of Hydraulic Engineering*, ASCE, Vol. 124, No. 5, May, (1998), pp. 484–492.

CAPM: Aquatic Weed Control Operation – Best Practice Guidelines, *Centre for Aquatic Plant Management* Report W111, (1997).

Casellarin, A., di Baldassarre, G., Bates, P.D. and Brath, A.: Optimal cross-sectional spacing in Preissmann scheme 1-D hydrodynamic models, *Journal of Hydraulic Engineering*, ASCE, Vol. 135, February, (2009), pp. 96–105.

Cebeci, T. and Bradshaw, P.: *Momentum transfer in boundary layers*, Hemisphere Publishing Co. McGraw-Hill, (1977), pp. 1–407.

Chadwick, A., Morfett, J. and Borthwick, M.: *Hydraulics in Civil and Environmental Engineering*, Spon Press, Taylor & Francis Group, (2004), pp. 1–644.

Chang, H.H.: Energy expenditure in curved open channels, *Journal of Hydraulic Engineering*, ASCE, Vol. 109, No. 7, (1983), pp. 1012–1022.

Chang, H.H.: Regular meander path model, *Journal of Hydraulic Engineering*, ASCE, Vol. 110, No. 10, October, (1984), pp. 1398–1411.

Chang, H.H.; *Fluvial processes in river engineering*, J. Wiley, (1988).

Chanson, H.: *The Hydraulics of Open Channel Flow: An Introduction*, Arnold, (1999), pp. 1–495.

Charbeneau, R.J., Henderson, A.D. and Sherman, L.C.: Hydraulic Performance Curves for Highway Culverts, *Journal of Hydraulic. Engineering*, ASCE, 132(5), (2006), pp. 474–481.

Chezy, A.: See 'Antoine Chezy, histoire d'une formule d'hydraulique' by G. Mouret, *Annales des Ponts et Chaussees*, Vol. II, (1921) and 'On the origin of the Chezy formula' by C. Herschel, *Journal Association of Engineering Societies*, Vol. 18, (1768), pp. 363–368.

Chlebek, J.: Modelling of simple prismatic channels with varying roughness using the SKM and a study of flows in smooth non-prismatic channels with skewed floodplains, *PhD thesis, Department of Civil Engineering, The University of Birmingham*, UK, (2009), pp. 1–527.

Chlebek, J. and Knight, D.W.: A new perspective on sidewall correction procedures, based on SKM modelling, *RiverFlow 2006*, September, Lisbon, Portugal, Vol. 1, [Eds Ferreira, Alves, Leal and Cardoso], Taylor & Francis, (2006), pp. 135–144.

Chlebek, J. and Knight, D.W.: Observations on flow in channels with skewed floodplains, *RiverFlow 2008*, [Eds M.S. Altinakar, M.A. Kokpinar, I. Aydin, S. Cokgar and S. Kirkgoz], Cesme, Turkey, Vol. 1, (2008), pp. 519–527.

Chow, V.T.: *Open Channel Hydraulics*, McGraw-Hill, (1959), pp. 1–680.

CIRIA: *Culvert design guide*, [Ed Ramsbottom, D., Day, R. and Rickard, C.], Construction and Industry Research and Information Association, Report 168, London, (1997), pp. 1–189.

CIWEM: River Engineering – Part II, Structures and Coastal Defence Works, *Water Practice Manual 8*, [Ed T.W. Brandon], the Chartered Institution of Water and Environmental Management, London, (1989), pp. 1–332.

Colebrook, C.F.: Turbulent flow in pipes, with particular reference to the transition region between the smooth and rough pipe laws, *Journal of the Institution of Civil Engineers*, 11, (1939), pp. 133–156.

Colebrook, C.F. and White, C.M.: Experiments with fluid friction in roughened pipes, *Proc. Roy. Soc.*, A, Vol. 161, (1937), pp. 367–381.

Colson, B.E., Ming, C.O. and Arcement, G.J.: Backwater at bridges and densely wooded flood plains, Bogue Chitto near Johnston Station, Mississippi, *USGS Hydrologic Atlas Series No. 591*, (1978).

Cowan, W.L.: Estimating hydraulic roughness coefficients, *Journal of Agricultural Engineering*, Vol. 37, (1956), pp. 473–475.

Cunge, J.A.: On the subject of a flood propagation computation method (Muskingum method), *Journal of Hydraulic Research*, IAHR, Vol. 7, No. 2, (1969), pp. 205–230.

Cunge, J.A., Holly, F.M. and Verwey, A.: *Practical aspects of computational river hydraulics*, Pitman Publishing Ltd, 1–420 (reprints available from Institute of Hydraulic Research, The University of Iowa, IA 52242-1585, USA, (1980), pp. 1–420.

Darby, S. (lead writer), Alonso, C., Bettess, R., Borah, D., Diplas, P., Julien, P., Knight, D.W., Li, L., Pizzuto, P., Quick, M., Simon, A., Stevens, M., Thorne, C.R. and Wang, S.: River width adjustment. Part II: Modelling, *Journal of Hydraulic Engineering*, ASCE, Task Committee on Hydraulics, Bank Mechanics and the Modelling of River Width Adjustment, Sept., 124(9), (1998), pp. 903–917.

Darcy, H.: Sur des recherches experimetales relatives au mouvement des eaux dans les tuyaux, *Computs rendus des séances de l'Academie des Sciences*, Mallet-Bachelier, Paris, (1857), pp. 1–268 [and atlas "Experimental Research Relating to the Movement of Water in Pipes"].

Davies, A.J.: *The finite element methods – a first approach*, Clarendon Press, Oxford, (1980).

Dawson, F.H.: The seasonal effects of aquatic plant growth on the flow of water in a stream, EWRS 5th Symposium on Aquatic Weeds, (1978), pp. 71–78.

de Bruijn, K.M., Klijn, F., Mc Gahey, C., Mens, M.P.J. and Wolfert, H.: Long-term strategies for flood risk management, Scenario definition and strategic alternative design, *FLOODsite report T14-08-01*, Delft Hydraulics, Delft, the Netherlands, (2008).

Defalque, A., Wark, J.B. and Walmsley, N.: Data density in 1-D river model, *Report SR 353, HR Wallingford*, March, (1993), pp. 1–78.

Defra/EA: Reducing uncertainty in river flood conveyance: scoping study, R&D Technical Report, prepared by Evans, E.P., Pender, G., Samuels, P.G. and Escarameia, M. for the *Environment Agency* Project W5 A-057, HR Wallingford, UK, (2001), pp. 1–127.

Defra/EA: Risk, Performance and Uncertainty in Flood and Coastal Defence: A Review, *R&D Report FD2302/TR1, Environment Agency R&D Dissemination Centre*, c/o WRc, Swindon, SN5 8YF, (2002), pp. 1–124.

Defra/EA: Reducing uncertainty in river flood conveyance, Interim Report 2: Review of Methods for Estimating Conveyance, Project W5 A-057, prepared by HR Wallingford for the *Environment Agency*, UK, (2003a), pp. 1–88.

Defra/EA: Reducing uncertainty in river flood conveyance, Roughness Review, Project W5 A-057, prepared by HR Wallingford for the *Environment Agency*, UK, (2003b), pp. 1–218.

Defra/EA: The investigation and specification of flow measurement structure design features that aid the migration of fish without significantly compromising flow measurement accuracy, with the potential to influence the production of suitable British Standards, R&D Technical Report to EA prepared by HRW, Project W6-084, *Environment Agency*, UK, (2003c).

Defra/EA: Reducing uncertainty in river flood conveyance, Interim Report 3: Testing of Conveyance Methods in 1D River Models, Project W5 A-057, prepared by HR Wallingford for the *Environment Agency*, UK, (2004a), pp. 1–237.

Defra/EA: Reducing uncertainty in river flood conveyance, Conveyance Manual, Project W5 A-057, prepared by HR Wallingford for the *Environment Agency*, UK, (2004b), pp. 1–115.

Defra/EA: Recommendations for maintaining the scientific relevance of the Conveyance and Afflux Estimation System, prepared by HR Wallingford (C. Mc Gahey and P. Samuels) for *Environment Agency*, R&D Project SC070032, April, (2009), pp. 1–115.

Diehl, T.H.: Potential Drift Accumulation at Bridges, *US Department of Transportation*, Federal Highway Administration Research and Development, Turner-Fairbank Highway Research Center, Virginia, Publication No. FHWA-RD-97-028, (1997).

Drazin, P.G. and Riley, N.: *The Navier-Stokes equations: a classification of flows and exact solutions*, London Mathematical Society, Lecture Note Series, 334, Cambridge University Press, (2006), pp. 1–196.

Du Boys, P.F.D.: Le Rhone et le rivier a lit affouillable, *Annales des Ponts et Chausses*, Vol. 18, No. 5, (1879).

Dupuis, V., Proust, S., Berni, C. and Paquier, A.: Mixing layer development in compound channel flows with submerged and emergent rigid vegetation over the floodplains. *Experiments in Fluids*, Vol. 58, No. 30, (2017). DOI: 10.1007/s00348-017-2319-9.

EA.: Environmental options specifications for flood maintenance, *Environment Agency* Report, March, (1997), pp. 1–40.

EA.: Environmental Guidelines for Vegetation Management in Channel and on Banks, *Environment Agency*, Technical Report W135 (1998a).

EA.: Management of vegetation on raised embankments, *Environment Agency*, Technical Report W133 (1998b).

EA.: *Scoping Study for Reducing Uncertainty in River Flood Conveyance*, (Available from Environment Agency R&D Dissemination Centre, WRc, Frankland Road, Swindon, Wilts SN5 8YF), (2001).

EA: Environmental options for flood defence maintenance works, *Environment Agency* Report, Peterborough, November, (2003), pp. 1–46.

EA: River Habitat Survey in Britain and Wales – Field Survey Guidance Manual, published by the *Environment Agency*, (2003), pp. 1.1–5.13.

EA: *Hydraulic Performance of River Bridges and Other Structures at High Flows*, Phase 1 Scoping Report, R&D Project: W5 A-065, Environment Agency, Rio House, Waterside Drive, Aztec West, Almondsbury, Bristol BS32 4UD, (2004).

EA: Environment Agency's Hi-Flows site, Retrieved 5 September 2005 from the World Wide Web: http://www.environment-agency.gov.uk/hiflowsuk/, (2005).

Einstein, H.A.: Formulas for the transportation of bedload, *Trans. of ASCE*, Vol. 107, (1942), pp. 561–577.

Einstein, H.A.: Der Hydraulische oder Profil-Radius, *Schweizerische Bauzeitung*, Zurich, 103, (1934), pp. 89–91.

Einstein, H.A. and Banks, R.B.: Fluid resistance of composite roughness, *Trans. American Geophysical Union*, Vol. 31, No. 4, (1950), pp. 603–610.

Einstein, H.A. and Li, H.: Secondary currents in straight channels, *Trans. American Geophysical Union*, Vol. 39, (1958), pp. 1085–1088.

Elder, J.W.: The dispersion of marked fluid in a turbulent shear flow, *Journal Fluid Mechanics*, Vol. 5, No. 4, (1959), pp. 544–560.

Elliott, S.C.A. and Sellin, R.H.J.: SERC Flood Channel facility: skewed flow experiments, *Journal of Hydraulic Research*, IAHR, Vol. 28, No. 2, (1990), pp. 197–214.

Engelund, F. and Hansen, H.: Investigations of Flow in Alluvial Streams, *Journal of Hydraulics Division*, Vol. 92, HY2, (1966), pp. 315–326.

Ervine, D.A., Babaeyan-Koopaei, K. and Sellin, R.H.J.: Two-dimensional solution for straight and meandering overbank flows, *Journal of Hydraulic Engineering*, ASCE, Vol. 126, No. 9, September, (2000), pp. 653–669.

Ervine, D.A. and Baird, J.I.: Rating curves for rivers with overbank flows, *Proc. ICE Part 2: Research and Development*, Vol. 73, June, (1982), pp. 465–472.

Ervine, D.A. and Ellis, J.: Experimental and computational aspects of overbank floodplain flow, *Trans. of the Royal Society of Edinburgh: Earth Sciences*, Vol. 78, (1987), pp. 315–475.

Ervine, D.A. and MacLeod, A.B.: Modelling a river channel with distant floodbanks, *Proc. ICE Jnl. for Water, Maritime and Energy*, Vol. 136, March (1999), pp. 21–33.

Escarameia, M., May, R.W.P., Jiming, M. and Hollinrake, P.G.: Performance of sluice gates for free and submerged flows, *Draft Report SR 344*, unpublished, March, HR Wallingford, (1993).

ESDU: Losses caused by friction in straight pipes with systematic roughness elements, *Engineering Sciences Data Unit* (ESDU), September, London, (1979), pp. 1–40.

European Commission: Directive 2000/60/EC of the European Parliament and of the Council, *Establishing a framework for Community action in the field of water policy*, published in the Official Journal of the European Union, L 327, 22 December, (2000), pp. 1–73.

European Commission: Directive 2007/60/EC of the European Parliament and of the Council, *On the assessment and management of flood risks*, published in the Official Journal of the European Union, L 288/27, 6 November, (2007).

Evans, E.M., Ashley, R., Hall, J., Penning-Rowsell, E., Saul, A., Sayers, P., Thorne, C. and Watkinson, A. Foresight. Future flooding. *Scientific Summary: Volume I – Future risks and their drivers*, Office of Science and Technology, London, UK (2004).

FEH: *Flood Estimation Handbook*, Vols. 1–5, Centre for Ecology and Hydrology, Wallingford, Oxon, UK, [ISBN 0-948540-94], (1995).

FHWA: *Hydraulic design of highway culverts*, US Department of Transportation, Federal Highway Administration, Washington DC, (2001).

Fisher, K.: Handbook for assessment of hydraulic performance of environmental channels, Report SR 490, *HR Wallingford*, UK, (2001), pp. 1–180.

Fleming, G.: *Flood risk management*, Thomas Telford, London, (2002), pp. 1–250.

Flintham, T.P. and Carling, P.A.: The prediction of mean bed and wall boundary shear in uniforem and compositely rough channels, *Proc. Int. Conf. on River Regime* (Ed W.R. White), Wiley, Chichester, (1988), pp. 267–287.

Franke, P.G. and Valetin, F.: The determination of discharge below gates in case of variable tailwater conditions, *Journal of Hydraulic Research* Vol. 7, No. 4, (1969), pp. 433–447.

Fukuoka, S.: *Floodplain risk management*, Balkema, (1998), pp. 1–357.

Fukuoka, S. and Fujita, K.: Prediction of flow resistance in compound channels and its application to design of river courses, *Jnl. Hydr., Coast. and Envir. Eng.*, JSCE, Vol. 411, No. 12, (1989), pp. 63–72 (in Japanese).

Ganguillet, E. and Kutter, W.R.: Versuch zur augstellung einer neuen allgeneinen furmel für die gleichformige bewegung des wassers in canalen und flussen, *Zeitschrift des Osterrichischen Ingenieur – und Architecten-Vereines* 21, (1), 6–25 and (2), (1869), pp. 46–59. Translated into English by: Hering, R. and Trautwine, J.C. (1888 and 1891). *A General Formula for the Uniform Flow of Water in Rivers and Other Channels*, New York, NY: John Wiley & Sons.

Garcia-Navarro, P. and Playan, E.: *Numerical Modelling of Hydrodynamics for Water Resources*, Taylor & Francis (2007), pp. 1–389.

Garde, R.J. and Ranga-Raju, K.G.: *Mechanics of sediment transportation and alluvial stream problems*, Wiley Eastern Ltd, (1977), pp. 1–483.

Garton, J.E. and Green, J.E.P.: Vegetation lined channel design procedures, *Transaction ASAE*, Vol. 26, No. 2, (1983), pp. 436–439.

Gibson A.H.: *Hydraulics and its applications*, Constable and Company, London, (1908), pp. 1–813.

Gibson A.H.: On the depression of the filament of the maximum velocity in a stream flowing through an open channel, *Proc. Royal Society of London, Series a*, Vol. 82, (1909), pp. 149–159.

Gordon, N.D., Mc Mahon, M.D. and Finlayson, B.L.: *Stream hydrology: an introduction for ecologists*, John Wiley and Sons, Chichester, (1992).

Gouldby, B.P. and Kingston, G.: Uncertainty and sensitivity method for flood risk analysis, *FLOODsite research report, HR Wallingford*, (2007).

Gouldby, B.P. and Samuels, P.G.: Language of risk, project definitions. *FLOODsite report T32-04-01*, (2005, updated 2009), pp. 1–56.

Graf, W.H.: *Hydraulics of Sediment Transport*, McGraw-Hill, New York, (1971), pp. 1–513.

Grass, A.J., Stuart, R.J. and Mansour-Tehrani, M.: Vortical structures and coherent motion in turbulent flow over smooth and rough boundaries, *Trans. Physical Sciences and Eng.*, Vol. 336, No. 1640, Turbulent flow structure near walls, Part 1, (1991), pp. 35–65.

Green, M., Dale, M., Gebre, H. and Knight, D.W.: The impact of climate change on the threshold of flooding for an urban UK river, *Defra/EA Annual Conference on River Engineering*, Telford, (2009) (in press).

Greenhill, R.: An Investigation into compound meandering channel flow, *PhD Thesis, Department of Civil Engineering, University of Bristol*, UK, (1992).

Guan, Y., Altinakar, M.S. and Krishnappan, B.G.: Modelling of lateral flow distribution in compound channels, *Proceedings River Flow 2002*, compiled by Y. Zech and D. Bousmar, Louvain, Belgium, Vol. 1, (2002), pp. 169–175.

Gunawan, B., Sterling, M., Tang, X. and Knight, D.W.: Measuring and modelling flow structures in a small river, *Riverflow 2010*, Proc. Int. Conf. on Fluvial Hydraulics, [Eds A. Dittrich, K. Koll, J. Aberle and P. Geisenhainer], Braunschweig, Germany, Sept. 8–10, Bundesanstalt fur Wasserbau (BAW), Karlsruhe, Germany, Vol. I, (2010), 179–186.

Haderlie, G.M. and Tullis, B.P.: Hydraulics of Multi-barrel culverts under Inlet Control. *J. Irrigation and Drainage Eng.*, ASCE, 134(4), (2008), pp. 507–514.

Hall, J. and Solomatine, D.: A framework for uncertainty analysis in flood risk management decisions, *Int. Journal of River Basin Management*, Vol. 6, No. 2, (2008), pp. 85–98.

Hall, J., Harvey, H., Solomatine, D. and Lal Shrestha, D.: Development of framework for the influence and impact of uncertainty, *FLOODsite Task 20 Report T20-08-04*, (2008), available at: www.floodsite.net, pp. 1–168.

Hamill, L.: Improved flow through bridge waterways by entrance rounding, *Proc. Instn Civil Engrs Mun. Engineering, 121*, March, (1997), pp. 7–21.

Hamill, L.: *Bridge hydraulics*, Spon, (1999), pp. 1–367.

Harris, I.: A Comparison of the Macrophyte Cover and Macroinvertebrate Fauna at three Sites on the river Kennet in the mid 1970s and late 1990s. The Science of the Total Environment, 282–283, (2002), pp. 121–142.

Hearne, J., Johnson, I. and Armitage, P.: Determination of ecologically acceptable flows in rivers with seasonal changes in the density of macrophyte. Regulated Rivers – Research & Management, 9, (1994), pp. 117–184.

Heatlie, F.: Turbulent flow around bluff bodies at the floodplain edge, *PhD thesis, Loughborough University*, (2011).

HEC.: Flow transitions in bridge backwater analysis, River Analysis System, Research Document 42, *US Army Corps of Engineers*, Hydrologic Eng. Center, Davis, (1995), pp. 1–67.

HEC.: Bridge hydraulic analysis with HEC-RAS, Technical Paper No. 151, *US Army Corps of Engineers*, Hydrologic Eng. Center, Davis, (1996), pp. 1–17.

HEC.: *River Analysis System, Version 3.1.2*, US Army Corps of Engineers Hydrologic Engineering Center, Davis, CA, USA, (2004).

HEC-RAS.: River Analysis System, User's Manual, Version 4, *US Army Corps of Engineers*, Hydrologic Eng. Center, Davis, March, (2008), pp. 1–747, available at: http://www.hec.usace.army.mil/software/hec-ras/hecras-document.html.

Henderson, F.M.: *Open Channel Flow*, Macmillan, (1966), pp. 1–522.

Hicks, D.M. and Mason, P.D.: Roughness characteristics of New Zealand Rivers, *National Institute of Water & Atmospheric Research Ltd*, September, (1998), pp. 1–329.

Hine, D. and Hall, J.W.: Analysing the robustness of engineering decisions to hydraulic model and hydrological uncertainties, in Harmonising the Demands of Art and Nature in Hydraulics: *Proc. 32nd Congress of IAHR, Venice*, 1–6 July (2007).

Hinze, J.O.: Experimental investigation on secondary currents in turbulent flow through a straight conduit, *Applied Science Research*, Vol. 28, (1973), pp. 453–465.

Hoffmans, G.J.C.M. and Verheij, H.J.: *Scour Manual*, A.A. Balkema, Rotterdam, (1997).

Horton, R.E.: Separate roughness coefficients for channel bottoms and sides, *Engineering News-Rec.*, 111, (1933), pp. 652–653.

Hotchkiss, R.H., Thiele, E.A., Nelson J. and Thompson P.L.: Culvert Hydraulics: Comparison of current computer models and recommended improvements, Transportation Research Record: *Journal of the Transportation Research Board*, No. 2060, (2008), pp. 141–149.

Huthoff, F., Roos, P.C., Augustijn, D.C.M. and Hulscher, S.J.M.H.: Interacting divided channel method for compound channel flow, *Journal of Hydraulic Engineering*, American Society of Civil Engineers, New York, August, Vol. 134, No. 8, (2008), pp. 1158–1165.

ICE: *Learning to live with rivers*, Final report of the Institution of Civil Engineers' Presidential Commission to review the technical aspects of flood risk management in England and Wales, [by G. Fleming, Frost, L., Huntingdon, S., Knight, D.W., Law, F.M. and Rickard, C.], November, (2001), pp. 1–83.

Ikeda, S.: Self-formed straight channels in sandy beds, *Journal of Hydraulics Division*, ASCE, Vol. 107, No. 4, (1981), pp. 389–406.

Ikeda, S.: Role of lateral eddies in sediment transport and channel formation, *Proc. Seventh Intl. Symposium on River Sedimentation*, [Eds A.W. Jayawardena, Lee, J.H.W. and Wang, Z.Y.] Hong Kong, December 1998, Balkema, (1999), pp. 195–203.

Ikeda, S., Kawamura, K., Toda, Y. and Kasuya, I.: Quasi-three dimensional computation and laboratory tests on flow in curved open channels, *Proc. of River Flow 2002*, Editors Bousmar & Zech, Balkema, Louvain La-Neuve, Belgium, Vol. 1, (2002), pp. 233–245.

Ikeda, S. and Kuga, T.: Laboratory study on large horizontal vortices in compound open channel flow, *Annual Jnl. of hydraulic Eng.*, JSCE, Vol. 45, (1997), pp. 493–498.

Ikeda, S. and McEwan, I.K.: *Flow and sediment transport in compound channels: the experiences of Japanese and UK research*, IAHR Monograph Series, International Association of Hydraulic Engineering Research, Madrid, Spain, (2009), pp. 1–333.

Imamoto, H. and Ishigaki, T.: Secondary flow in compound open channel flow, *Proc. HydroComp 89*, Dubrovnik, Yugoslavia, 13–16th June, (1989), pp. 235–243.

IPCC.: *Intergovernmental Panel on Climate Change, Climate Change 2001, Impacts, Adaptation and Vulnerability*, Cambridge University Press (2001).

ISIS V2.3: Online help manual, http://www.wallingfordsoftware.com/products/isis/, *HR Wallingford/Halcrow*, (2004).

ISO.: *Measurement of liquid flow in open channels. Part 1: Establishment and operation of a gauging station*, ISO 1100-1, International Standards Organization, Geneva, Switzerland, (1996), pp. 1–20.

ISO 18320.: *Measurement of liquid flow in open channels. Part 2: Determination of the stage-discharge relation*, ISO 1100-2:2010, International Standards Organization, Geneva, Switzerland, (2018), pp. 1–49.

ISO.: *Liquid flow measurement in open channels – Round-nose horizontal broad-crested weirs*, ISO 4374, International Standards Organization, Geneva, Switzerland, (1990), pp. 1–18.

ISO.: *Hydrometric determinations – Flow measurement in open channels using structures – Flat-V weirs*, ISO 4377, International Standards Organization, Geneva, Switzerland, (2002), pp. 1–35.

ISO.: *Hydrometry Open channel flow measurement using thin-plate weirs*, ISO 1438, International Standards Organization, Geneva, Switzerland, (2008), pp. 1–59.

Jalonen, J. and Jarvela, J.: Modeling the flow resistance of woody vegetation using physically based properties of the foliage and stem, *Proc. of 37th World Congress, International Association for Hydraulic and Engineering Research* (IAHR), Chengdu, China, September, Water Resources Research, (2014), Vol. 50, No. 1, (2013), pp. 229–245.

James, C.S.: Evaluation of methods for predicting bend loss in meandering channels, *Journal of Hydraulic Research*, ASCE, Vol. 120, No. 2, February, (1994), pp. 245–253.

James, C.S. and Wark, J.B.: Conveyance estimation for meandering channels, Report SR329, *HR Wallingford*, UK, (1992), pp. 1–91.

Janes, M., Fisher, K., Mant, J. and de Smith, L.: River Rehabilitation Guidance for Eastern England Rivers, prepared by *The River Restoration Centre* for the EA, November, (2005).

Kaatz, K.J. and James, W.P.: Analysis of alternatives for computing backwater at bridges, *Journal of Hydraulic Engineering*, American Society of Civil Engineers, New York, April, Vol. 123, No. 9, (1997), pp. 784–792

Karcz I.: Secondary currents and the configuration of a natural stream bed, *Journal of Geophysical Research*, Vol. 71, (1966), pp. 3109–3116.

Kawahara: Drag coefficient distribution in LES of vegetated open channel flows, *River Flow 2016*, (2016), pp. 2240–2245, 20160714.

Kells, J.A.: Hydraulic performance of damaged-end corrugated steel pipe culverts, Canadian. Journal of Civil Engineering., 35, (2008), pp. 918–924.

Keulegan, G.H.: Laws of turbulent flow in open channels, *Journal National Bureau of Standards*, Washington, D.C., Research Paper 1151, Vol. 21, (1938), pp. 707–741,

Khatibi, R.H., Williams, J.R. and Wormleaton, P.R.: Identification problem of open-channel friction factors, *Journal of Hydraulic Engineering*, American Society of Civil Engineers, New York, December, Vol. 123, No. 12, (1997), pp. 1078–1088.

Kindsvater, C.E. and Carter, R.W.: Tranquil flow through open-channel constrictions, *Transactions of the American Society of Civil Engineers*, New York, 120, (1955), pp. 955–992.

Kindsvater, C.E., Carter, R.W. and Tracy, H.J.: Computation of peak discharge through constrictions, *Circular 284, US Geological Survey*, Washington DC, (1953).

Kitigawa, A.: Past and recent developments of flood control in the Tone River, [in *Floodplain Risk Management*, S. Fukuoka (Ed)], Balkema, (1998), pp. 61–74.

Kline, S.J., Reynolds, W.C., Schraub, F.A. and Runstadler, P.W.: The structure of turbulence in boundary layers, *Journal of Fluid Mechanics*, Vol. 30, (1967), pp. 741–773.

Kline, S.J., Morkovin, M.V., Sovran, G., Cockrell, D.J., Coles, D.E. and Hirst, E.A.: Proc. AFOSR-IFP – Stanford Conference 1968, *Computation of turbulent boundary layers, Vol. I & II*, Stanford Univ. Press, (1969).

Knight, D.W.: Some field measurements concerned with the behaviour of resistance coefficients in a tidal channel, *Estuarine, Coastal and Shelf Science*, Academic Press, London, No. 12, (1981), pp. 303–322.

Knight, D.W.: Issues and directions in river mechanics – closure of sessions 2, 3 and 5, In *'Issues and directions in hydraulics'*, [Ed T. Nakato and R. Ettema], An Iowa Hydraulics Colloquium in honour of Professor John F Kennedy, Iowa Institute of Hydraulic Research, Iowa, USA, Balkema, (1996), pp. 435–462.

Knight, D.W.: Flow and sediment transport in two-stage channels, *Proc. 2nd IAHR Symposium on River, Coastal and Estuarine Morphodynamics*, RCEM2001, Japan, (2001), pp. 1–20.

Knight, D.W.: Introduction to flooding and river basin modelling, Chapter 1 in *River Basin Modelling for Flood Risk Mitigation* [Eds D.W. Knight and A.Y. Shamseldin], Taylor & Francis, (2006a), pp. 1–19.

Knight, D.W.: River flood hydraulics: theoretical issues and stage-discharge relationships, Chapter 17 in *River Basin Modelling for Flood Risk Mitigation* [Eds D.W. Knight and A.Y. Shamseldin], Taylor & Francis, (2006b), pp. 301–334.

Knight, D.W.: River flood hydraulics: calibration issues in one-dimensional flood routing models, Chapter 18, in *River Basin Modelling for Flood Risk Mitigation* [Eds D.W. Knight and A.Y. Shamseldin], Taylor & Francis, (2006c), pp. 335–385.

Knight, D.W.: Modelling overbank flows in rivers – data, concepts, models and calibration, Chapter 1 in *Numerical Modelling of Hydrodynamics for Water Resources* [Eds P. Garcia-Navarro and E. Playan], Taylor & Francis, (2008a), pp. 3–23.

Knight, D.W.: A review of some laboratory and field data for checking models of rivers with inbank and overbank flows, *Advances in Hydro-Science and Engineering, Vol. VIII, Proc. 8th International Conference on Hydro-Science and Engineering*, ICHE, [Eds S.S.Y. Wang, M. Kawahara, K.P. Holtz, T. Tsujimoto and Y. Toda], Nagoya, Japan, Sept., Abstract book pp. 201–202 and full paper in CD pp. 812–821, (2008b).

Knight, D.W., 1997, Performance characteristics of bridgeflow kerb channel units for the drainage of bridge pavement surfaces, *Final report on hydraulic performance tests for CIS Ltd.*, The University of Birmingham, March, 1–160.

Knight, D.W.: Hydraulic Problems in Flooding: from Data to Theory and from Theory to Practice, in *Experimental and Computational Solutions of Hydraulic Problems* [Ed P. Rowinski], 32nd International School of Hydraulics, Institute of Geophysics, Polish Academy of Sciences, Lochow, Poland, May 2012, GeoPlanet: Earth and Planetary Series, Springer, (2013), pp. 19–52.

Knight, D.W.: River hydraulics – a view from midstream, *Journal of Hydraulic Research*, IAHR, International Association for Hydro-Environmental Engineering and Research, Taylor & Francis, Vol. 51, No. 1, (2013), pp. 2–18 (and Discussion in Vol. 52, No. 1, (2014)).

Knight, D.W. and Abril, B.: Refined calibration of a depth-averaged model for turbulent flow in a compound channel, *Proc. Instn. of Civil Engineers, Water, Maritime and Energy Division*, London, Vol. 118, Issue 3, Sept., Paper No. 11017, (1996), pp. 151–159.

Knight, D.W. and Cao, S.: Boundary shear in the vicinity of river banks, *ASCE National Conference on Hydraulic Engineering*, Buffalo, New York, August, Vol. 2, (1994), pp. 954–958.

Knight, D.W. and Brown, F.A.: Resistance studies of overbank flow in rivers with sediment using the Flood Channel Facility, *Journal of Hydraulic Research*, IAHR, Vol. 39, No. 3, (2001), pp. 283–301.

Knight, D.W. and Demetriou, J.D.: Flood plain and main channel flow interaction, *Journal of Hydraulic Engineering*, ASCE, Vol. 109, No. 8, August, (1983), pp. 1073–1092.

Knight, D.W., Demetriou, J.D. and Hamed, M.E.: Boundary shear in smooth rectangular channels, *Journal of Hydraulic Engineering*, ASCE, Vol. 110, No. 4, April, (1984), pp. 405–422.

Knight, D.W. and Lai, C.J.: Turbulent flow in compound channels and ducts, *Proc. 2nd Int. Symposium on Refined Flow Modelling and Turbulence Measurements*, Iowa, USA, Sept., Hemisphere Publishing Co., (1985), pp. I21-1 to I21-10.

Knight, D.W. and Macdonald, J.A.: Open channel flow with varying bed roughness, *Journal of the Hydraulics Division*, ASCE, Vol. 105, No. HY9, Proc. Paper 14839, Sept., (1979), pp. 1267–1183.

Knight, D.W. and Patel, H.S.: Boundary shear stress distributions in rectangular duct flow, *Proc. 2nd Int. Symposium on Refined Flow Modelling and Turbulence Measurements*, Iowa, USA, Sept., Hemisphere Publishing Co., (1985), I22-1 to I22-10.

Knight, D.W. and Samuels, P.G.: Localised flow modelling of bridges, sluices and hydraulic structures under flood flow conditions, *Proc. 34th MAFF Conference of River and Coastal Engineers*, Keele University, June, (1999), pp. 1.2.1–1.2.13.

Knight, D.W. and Samuels, P.G.: Examples of recent floods in Europe, *Journal of Disaster Research*, Fuji Technology Press, Tokyo, Japan, Vol. 2, No. 3, (2007), pp. 190–199.

Knight, D.W. and Sellin, R.H.J.: The SERC Flood Channel Facility, *Journal of the Institution of Water and Environmental Management*, Vol. 1, No. 2, October, (1987), pp. 198–204.

Knight D.W. and Shamseldin, A.: *River Basin Modelling for Flood Risk Mitigation* [Eds D.W. Knight and A.Y. Shamseldin], Taylor & Francis, (2006), pp. 1–607.

Knight, D.W. and Shiono, K.: Turbulence measurements in a shear layer region of a compound channel, *Journal of Hydraulic Research*, IAHR, Vol. 28, No. 2, 175–196, (Discussion in IAHR Journal, 1991, Vol. 29, No. 2, (1990), pp. 259–276).

Knight, D.W. and Shiono, K.: River channel and floodplain hydraulics, in *Floodplain Processes*, (Eds Anderson, Walling and Bates), Chapter 5, J. Wiley, (1996), pp. 139–181.

Knight, D.W. and Tang, X.: Zonal discharges and boundary shear in prismatic channels, *Journal of Engineering and Computational Mechanics*, Proceedings of the Instn. of Civil Engineers, London, Vol. 161, EM2, June, (2008), pp. 59–68.

Knight, D.W. and Yu, G.: A geometric model for self formed channels in uniform sand, *26th IAHR Congress*, London, International Association for Hydraulic Research, September, HYDRA 2000, Vol. 1, Thomas Telford, (1995), pp. 354–359.

Knight, D.W., Al-Hamid, A.A.I. and Yuen, K.W.H.: Boundary shear in differentially roughened trapezoidal channels, In *Hydraulic and Environmental Modelling: Estuarine and River Waters* (Eds R.A. Falconer, K. Shiono and R.G.S. Matthew), Ashgate Press, (1992), pp. 3–14.

Knight, D.W., Omran, M. and Abril, J.B.: Boundary conditions between panels in depth-averaged flow models revisited, *Proceedings of the 2nd International Conference on Fluvial Hydraulics: River Flow 2004*, Naples, 24–26 June, Vol. 1, (2004), pp. 371–380.

Knight, D.W., Omran, M. and Tang, X.: Modelling depth-averaged velocity and boundary shear in trapezoidal channels with secondary flows, *Journal of Hydraulic Engineering*, February, ASCE, Vol. 133, No. 1, January, (2007), pp. 39–47.

Knight, D.W., Shiono, K. and Pirt, J.: Prediction of depth mean velocity and discharge in natural rivers with overbank flow, *Proc. Int. Conf. on Hydraulic and Environmental Modelling of Coastal, Estuarine and River Waters*, (Ed. R.A. Falconer, P. Goodwin, R.G.S. Matthew), Gower Technical, University of Bradford, September, Paper 38, (1989), pp. 419–428.

Knight, D.W., Yuen, K.W.H. and Alhamid, A.A.I.: Boundary shear stress distributions in open channel flow, in *Physical Mechanisms of Mixing and Transport in the Environment*, (Eds K. Beven, P. Chatwin and J. Millbank), J. Wiley, Chapter 4, (1994), pp. 51–87.

Knight, D.W., Aya, S., Ikeda, S., Nezu, I. and Shiono, K.: "Flow Structure", Chapter 2 in Ikeda and McEwan, "Flow and sediment transport in compound channels: the experiences of Japanese and UK research" (2009), pp. 5–114.

Knight, D.W., Brown, F.A., Valentine, E.M., Nalluri, C., Bathurst, J.C., Benson, I.A., Myers, W.R.C., Lyness, J.F. and Cassells, J.B.: The response of straight mobile bed channels to inbank and overbank flows, *Proc. Instn. of Civil Engineers, Water, Maritime and Energy Division*, London, Vol. 136, Dec., (1999), pp. 211–224.

Knight, D.W., Tang, X., Sterling, M., Shiono, K. and Mc Gahey, C.: Solving open channel flow problems with a simple lateral distribution model, *Riverflow 2010*, Proceedings of the Int. Conf. on Fluvial Hydraulics, [Eds A. Dittrich, K. Koll, J. Aberle and P. Geisenhainer], Braunschweig, Germany, Sept. 8–10, Bundesanstalt fur Wasserbau (BAW), Karlsruhe, Germany, Keynote address, Vol. I, (2010), pp. 41–48.

Kolkman, P.A.: Discharge relationships and component head losses for hydraulic structures, in *Discharge characteristics, Hydraulic Design Considerations, Hydraulic Structures Design Manual 8*, 55–151, edited by Miller, D. S, International Association for Hydraulic Research, Balkema, (1994).

Kothyari, U.C., Hayashi, K. and Hashimoto, H.: Drag coefficient of unsubmerged rigid vegetation stems in open channel flows, *Journal of Hydraulic Research*, IAHR, 47(6): (2009), pp. 691–699.

Kouwen, N. and Li, R.M.: Biomechanics of vegetative channel linings, *Journal of Hydraulic Engineering*, ASCE, Vol. 106, (1980), pp. 1085–1103.

Krishnamurthy, M. and Christensen, B.A.: Equivalent roughness for shallow channels, *Journal of Hydraulics Division*, ASCE, Vol. 98, HY12, (1972), pp. 2257–2262.

Lai, C.J.: Flow resistance, discharge capacity and momentum transfer in smooth compound closed ducts, *PhD Thesis, Department of Civil Engineering, The University of Birmingham, UK*, (1987), pp. 1–170.

Lamb, R., Mantz, P., Atabay, S., Benn, J. and Pepper, A.: Recent advances in modelling flood water levels at bridges and culverts, *Proceedings of 41st Defra Flood and Coastal Risk Management Conference*, 4–6 July, York, (2006), pp. 3a.1.1–3a.1.9.

Lambert, M.F. and Myers, W.R.C.: Estimating the discharge capacity in straight compound channels, *Proc. Instn. of Civil Engineers, Water, Maritime and Energy*, Paper 11530, Vol. 130, June, (1978), pp. 84–94.

Larock, B.E.: Gravity-affected flow from planar sluice gates, *Journal of the Hydraulics Division*, Proceedings of the American Society of Civil Engineers, HY 4, July, (1969), pp. 1211–1226.

Latapie, A.: Testing of a New Method for Conveyance Estimation in Natural Rivers, MSc Thesis, *Heriot-Watt University*, Edinburgh, (2003), pp. 1–81.

Laurensen, E.M.: Friction slope averaging in backwater calculations, *Journal of Hydraulic Engineering*, ASCE, Vol. 112, No. 12, December, (1985), pp. 1151–1163.

Leggett, D.J. and Elliott, C.: Improving the Implementation and Adoption of Flood and Coastal Defence R&D Results, *R&D Technical Report W5G-003*, Environment Agency, Rio House, Waterside Drive, Aztec West, Almondsbury, Bristol BS32 4UD, (2002).

Leopold, L.B., Bagnold, R.A., Wolman, M.G. and Brush, L.M.: Flow resistance in sinuous or irregular channels, *USGS Professional Paper* 282-D, Washington DC, (1960), pp. 111–134.

Liao, H. and Knight, D.W.: Analytic depth-averaged velocity and boundary shear stress distributions for V-shaped channel with vertical side walls, in Sustainable Water Management and River Development [Eds Lin Luo and Shaw Lei Yu], *Proc. 5th Int. Conf. on Urban Watershed Management and Mountain River Protection and Development*, Chengdu, China, April 3–5, Vol. I, Sichuan University Press, (2007), pp. 289–297.

Liao, H. and Knight, D.W.: Analytic stage-discharge formulas for flow in straight prismatic channels, *Journal of Hydraulic Engineering*, ASCE, Vol. 133, No. 10, October, (2007a), pp. 1111–1122.

Liao, H. and Knight, D.W. Analytic stage-discharge formulae for flow in straight trapezoidal open channels, *Advances in Water Resources*, Elsevier, Vol. 30, Issue 11, November, (2007b), pp. 2283–2295.

Lighthill, M.J.: Introduction – boundary layer theory, *Laminar Boundary Layers*, Chapter 2, [Editor I. Rosenhead], Clarendon, (1963).

Lighthill, M.J.: Turbulence, [in *Osborne Reynolds and engineering science today*, Chapter 2, Eds McDowell D.M. and Jackson J.D.], Manchester University Press, England, (1970), pp. 83–146.

Liu, C., Shan, Y., Liu, X. and Yang, K.: Method for assessing discharge in meandering compound channels, *Water Management*, Proc Institution of Civil Engineers, London, ICE, WM1, February (2016), pp. 17–29.

Liu, H.K.: Mechanics of sediment ripple formation, *Journal of Hydraulics Division*, Vol. 83, HY2, April, (1957).

Lotter, G.K.: Considerations on hydraulic design of channels with different roughness of walls, Trans. All-Union Scientific Research Institute of Hydraulic Eng., Leningrad, Vol. 9, (1933), pp. 238–241.

Manning, R.: On the flow of water in open channels and pipes, *Trans. Inst. Civil Engineers of Ireland*, Dublin, 20, (1891), pp. 161–207.

Mantz, P.A. and Benn, J.R.: Computation of afflux ratings and water surface profiles, *Water Management*, Proc. Instn. of Civil Engineers, London, 162, WM1, February, (2009), pp. 41–55.

Mantz, P.A., Lamb, R. and Benn, J.R.: Afflux Estimation System Hydraulic Reference, *Environment Agency Science Report SC030218/SR*, (2007), pp. 1–142.

Matthai, H.F.: Measurement of peak discharge at width contractions by indirect method, *Techniques of Water Resources Investigations of the United States geological Survey*, A4, Book 3, Applications of Hydraulics, US Govt. Printing Office, Washington DC, (1967).

May, R.W.P., Ackers, J.C. and Kirby, A.M.: Manual on scour at bridges and other hydraulic structures, *CIRIA Report C551*, Construction and Industry Research and Information Association, London, (2002), pp. 1–225.

Mc Gahey, C.: A practical approach to estimating the flow capacity of rivers, PhD Thesis, Faculty of Technology, *The Open University (and British Library)*, May, (2006), pp. 1–371.

Mc Gahey, C. and Samuels, P.G.: A practical approach to uncertainty in conveyance estimation, *Proc. of the 39th Defra Flood and Coastal Management Conf.*, The University of York, 29th June–1st July, (2004), pp. 10.3.1–10.3.12.

Mc Gahey, C., Ramsbottom, D., Panzeri, M. and Millington, R.: National flood hazard mapping for Scotland – an innovative approach, *Proceedings of the 40th Defra Flood and Coastal Management Conference*, The University of York, 5th–7th July, (2005), pp. 2.4.1–2.4.11.

Mc Gahey, C., Samuels, P.G. and Knight, D.W.: A practical approach to estimating the flow capacity of rivers – application and analysis, *RiverFlow 2006*, September, Lisbon, Portugal, Vol. 1, (2006), pp. 303–312.

Mc Gahey, C., Samuels, P.G., Knight, D.W. and O'Hare, M.T.: Estimating river flow capacity in practice, *Journal of Flood Risk Management*, CIWEM, Vol. 1, (2008), pp. 23–33.

Mc Gahey, C., Knight, D.W. and Samuels, P.G.: Advice, methods and tools for estimating channel roughness, *Water Management*, Proceedings of the Instn. of Civil Engineers, London, Vol. 162, Issue WM6, December, (2009), pp. 353–362.

Melville, B.W. and Coleman, S.E.: *Bridge scour*, Water Resources publications, LLC, Colorado, USA, (2000), pp. 1–550.

MIKE11 V3.11: General reference manual, 1st edition, Danish Hydraulic Institute, (1995).

Miller, D.S.: *Discharge Characteristics*, Hydraulic Structures Design Manual No. 8, IAHR, Balkema, (1994), pp. 1–249.

Miller, J.B.: *Floods: people at risk, strategies for prevention*, United Nations, New York, (1997).

Mockmore, C.A.: Flow around bends in stable channels, *Transactions of ASCE*, Vol. 109, (1944), pp. 593–618.

Mohammadi, M.: Gradually varied and spatially varied flow in bridge deck units, *PhD Thesis, The University of Birmingham*, (1998), pp. 1–348.

Mohammadi, M. and Knight, D.W.: Threshold for sediment motion in V-shaped channels, 28th IAHR Congress, Graz, Austria, August, Session B5, (1999), pp. 1–6.

Montes, J.S.: *Hydraulics of Open Channel Flow*, ASCE Press, New York, USA, [ISBN 0-7844-0357-0], (1998), pp. 1–712.

Moody, L.F.: Friction factors for pipe flow, *Transactions of the American Society of Mechanical Engineers*, ASME, Vol. 66, (1944), pp. 671–684.

Morvan, H., Knight, D.W., Wright, N.G., Tang, X. and Crossley, M.: The concept of roughness in fluvial hydraulics and its formulation in 1-D, 2-D & 3-D numerical simulation models, *Journal of Hydraulic Research*, IAHR, Vol. 46, No. 2, (2008), pp. 191–208.

Muto, Y.: Turbulent flow in two-stage meandering channels, PhD Thesis, *Department of Civil Engineering, Bradford University*, UK, (1997).

Myers, W.R.C.: Momentum transfer in a compound channel, *Journal of Hydraulic Research*, IAHR, Vol. 16, No. 2, (1978), pp. 139–150.

Myers, W.R.C. and Lyness, J.J.: Flow resistance in rivers with floodplains, final report on research grant GR/D/45437, *University of Ulster*, (1989).

Myers, W.R.C. and Brennan, E.K.: Flow resistance in compound channels, *Journal of Hydraulic Research*, IAHR, Vol. 28, No. 2, (1990), pp. 141–155.

Myers, W.R.C., Knight, D.W., Lyness, J.F., Cassells, J. and Brown, F.A.: Resistance coefficients for inbank and overbank flows, *Proc. Instn. of Civil Engineers, Water, Maritime and Energy Division*, London, Vol. 136, June, (1999), pp. 105–115.

Myers, W.R.C. and Lyness, J.F.: Discharge ratios in smooth and rough compound channels, *Journal of the Hydraulic Engineering, American Society of Civil Engineers*, Vol. 123, Issue 3, March, (1997), pp. 182–188.

Naden, P.S., Rameshwaran, P. and Vienot, P.: Modelling the influence of instream macrophytes on velocity and turbulence, *Proceedings 5th International Symposium on Ecohydraulics – Aquatic Habitats Analysis and Restoration*, Editors D.G. de Jalon Lastra and P.V. Martinez, Madrid, Spain, (2004), pp. 1118–1122.

Naden, P.S., Rameshwaran, P., Mountford, O. and Robertson, C.: The influence of macrophyte growth, typical of eutrophic conditions, on river flow, velocities, and turbulence production, *Hydrological Processes*, 20, (2006), pp. 3915–3938.

Nagler, F.A.: Obstructions of bridge piers to the flow of water, *Transactions of the American Society of Civil Engineers*, 82, (1918), pp. 334–395.

Nakato, T. and Ettema, R.: *Issues and directions in hydraulics*, Balkema, (1996), pp. 1–495.

National Rivers Authority: *Design of straight and meandering channels*, NRA R&D Report 13, Environment Agency, Rio House, Waterside Drive, Aztec West, Almondsbury, Bristol BS32 4UD, (1994).

Navratil, O., Albert, M.B. and Gresillon, J.M.: Using a 1D steady flow model to compare field determination methods of bank-full stage, *Proceedings of River Flow 2004* [Eds Greco, Carravetta and Della Morte], Naples, June, Balkema, (2004), pp. 155–161.

Neill, C.R.: (1973), *Guide to bridge hydraulics*, University of Toronto Press for Roads and Transportation Association of Canada, Toronto, (1973), pp. 1–191.

Nepf, H.: Drag, Turbulence and Diffusion in Flow Through Emergent Vegetation, *Water Resources Research* 35, (1999), pp. 479–489.

Newton, I.: *Methodus fluxionum et serierum infinitarum*, (1644–1671).

Nezu, I. and Nakagawa, H.: *Turbulence in Open-Channel Flows*, IAHR Monograph Series, A.A. Balkema, Rotterdam, (1993), pp. 1–281.

Nezu, I. and Nakayama, T.: 祢津家久, 中川博次. 一様開水路および閉管路の三次元乱流構造に関する研究. 土木学会論文集. 1986 (369):89–98.

Nezu, I. and Nakayama, T.: Space-time correlation structures of horizontal coherent vortices in compound open-channel flows by using particle-tracking velocimetry, *Journal of Hydraulic Research*, Vol. 35, No. 2, (1997), pp. 191–208.

Nezu, I., Onitsuka, K. and Iketani, K.: Coherent horizontal vortices in compound open-channel flows, *Hydraulic Modeling* (ed. V.P. Singh, et al.), Water Resources Publications, Colorado, USA, (1999), pp. 17–32.

Nezu, I. and Rodi, W.: Experimental study on secondary currents in open channel flow, *Proc. of 21st IAHR Congress*, Melbourne, 2, (1985), pp. 115–119.

Nikora, V.: Hydrodynamics of rough-bed turbulent flows: spatial averaging perspective, *River Flow 2008* [Eds M.S. Altinakar, M.A. Kokpinar, I. Aydin, S. Cokgar and S. Kirkgoz], Kubaba Services, Turkey, (2008), pp. 11–19.

Nikuradse, J.: Translates to: Laws of flow in rough pipes, *Verein deutscher Ingenieure*, Forschubgsheft, No. 361, Berlin, (1933).

NIRA: National flood map for Northern Ireland, Proposed Methodology Report, prepared by RPS-KMM & HR Wallingford for *Northern Ireland Rivers Agency*, June, (2005), pp. 1–37.

Novak, P., Moffat, A.I.B., Nalluri, C. and Narayanan, R.: *Hydraulic structures*, E&FN Spon, (2007), pp. 1–700.

O'Hare, M.T., Stillman, R.A., Mc Donnell, J. and Wood, L.: The effects of mute swan grazing on a keystone macrophyte, Freshwater Biology Published article online: 23 Aug, (2007), doi: 10.1111/j.1365-2427.2007.01841.x.

O'Hare, M.T., Scarlett, P., Henville, P., Ryaba, T., Cailes, C. and Newman, J.: Variability in Manning's n estimates for vegetated rivers – Core Site Study: Intra- and Inter-Annual Variability, *An Aquatic Plant Management Group Report*, Centre for Ecology & Hydrology, UK, (2008), pp. 1–33.

Okada, S. and Fukuoka, S.: Land-form features in compound meandering channels and classification diagram of flood flows based on sinuosity and relative depth, *River Flow 2002* [Eds D. Bousmar and Y. Zech], Balkema, (2002), pp. 205–212.

Okada, S. and Fukuoka, S.: Plan shape features in compound meandering channels and classification diagram of flood flows based on sinuosity and relative depth, *Journal of Hydroscience and Hydraulic Engineering*, Vol. 231, No. 1, May, (2003), pp. 41–52.

Omran, M.: Modelling stage-discharge curves, velocity and boundary shear stress distributions in natural and artificial channels using a depth-averaged approach, *PhD thesis, Department of Civil Engineering, The University of Birmingham*, (2005), pp.

Omran, M. and Knight, D.W.: Modelling the distribution of boundary shear stress in open channel flow, *RiverFlow 2006*, September, Lisbon, Portugal, Vol. 1, [Eds Ferreira, Alves, Leal and Cardoso], Taylor & Francis, (2006), pp. 397–404.

Omran, M., Knight, D.W., Beaman, F. and Morvan, H.: Modelling equivalent secondary current cells in rectangular channels, *RiverFlow 2008*, [Eds M.S. Altinakar, M.A. Kokpinar, I. Aydin, S. Cokgar and S. Kirkgoz], Cesme, Turkey, Vol. 1, (2008), pp. 75–82.

Oshikawa, H., Hashimoto, A., Tsukahara, K. and Komatsu, T.: Impacts of recent climate change on flood disaster and preventive measures, *Journal of Disaster Research*, Vol. 23, No. 2, (2008), pp. 131–141.

Ozaki and Ishigaki, T., Shiono, K., Ferreira, E., Chandler, J.H. and Wackrow, R.: Turbulent characteristics in open channel flow for one line trees, Proceedings of 2013, IAHR Congress, Chengdu, China., (2013), pp.

Palmer, V.J.: A method for designing vegetated waterways, *Agriculture Engineering*, Vol. 26, No. 12, (1945), pp. 516–520.

Patel, V.C.: Calibration of the Preston tube and limitations on its use in pressure gradients, *Journal of Fluid Mechanics*, Vol. 23, Part 1, (1965), pp. 185–208.

Pavlovskii, N.: On a design for uniform movement in channels with non-homogenous walls, *All-Union Scientific Research Institute of Hydraulic Engineering*, Leningrad, Vol. 3, (1931), pp. 157–164.

Perkins, H.J.: The formation of streamwise vorticity in turbulent flow, *Journal of Fluid Mechanics*, Vol. 44, Part 4, (1970), pp. 721–740.

Petryk, S. and Bosmajian, G.: Analysis of flow through vegetation, *Journal of Hydraulic Division*, ASCE, Vol. 101, No. 7, (1975), pp. 871–884.

Pitt, M.: *Learning lessons form the 2007 floods*, Defra/Environment Agency, June, (2008), pp. 1–462.

Powell, R.W.: Resistance to flow in rough channels, *Eos Transactions AGU*, Vol. 31, (1950), pp. 575–582.

Prandtl, L.: Bericht über Untersuchungen zur ausgebildeten Turbulenz, *Z. Angew. Math, Meth.*, Vol. 5, (1925), pp. 136–139 and Proc. 2nd Int. Congr. Applied Mech., Zurich, 1926, pp. 62–75; also Coll. Works II, pp. 736–751.

Prandtl, L.: *The Essentials of Fluid Dynamics*, Blackie & Son Ltd, 1953, pp. 111–131.

Press, W.H., Teukolsky, S.A., Vetterling, W.T. and Flannery, B.P.: *Numerical recipes in Fortran 90 – the art of parallel scientific computing*, Vol. 2, Cambridge University Press, (1996).

Preston, J.H.: The determination of turbulent skin friction by means of Pitot tubes, *J. Roy. Aeronautical Soc.*, 58, (1954), pp. 109–121.

Preissmann, A.: Propagation of translatory waves in channels and rivers, *Proceedings of 1st Congress of French Association for Computation*, AFCAL, Grenoble, France, (1961), pp. 43–442.

Rajaratnam, N.: Free flow immediately below sluice gates, *Journal of the Hydraulics Division*, ASCE, Vol. 3, No. 4, April, (1977), pp. 345–351.

Rajaratnam, N. and Humpries, J.A.: Free flow upstream of vertical sluice gates, *Journal of Hydraulic Research*, Vol. 20, No. 5, (1982), pp. 427–437.

Rajaratnam, N. and Muralidhar, D.: Boundary shear distribution in rectangular open channels, *La Houille Blanche*, No. 6, (1969), pp. 603–609.

Rajaratnam, N. and Subramanya, K.: Practical problems of sluice-gate flow, *Water Power*, March, (1969), pp. 112–115.

Raju, K.G.R., Asawa G.l., Rana O.P.S. and Pillai, A.S.N.: Rational assessment of blockage effect in channel flow past smooth circular cylinders, *Journal of Hydraulic Research*, IAHR, Vol. 21, No. 4, (1983), pp.

Rameshwaran, P. and Shiono, K.: Quasi two dimensional model for straight overbank flows through emergent vegetation on floodplains, *Journal of Hydraulic Research. International Association for Hydraulic Engineering Research*, Vol. 45, No. 3, (2007), pp. 302–315.

Rameshwaran, R. and Willetts, B.B.: Conveyance prediction for meandering two-stage channel flows, *Proceedings ICE Journal for Water, Maritime and Energy*, Paper 11765, Vol. 136, Sept., (1999), pp. 153–166.

Ramette, M.: Hydraulique et morphologie des. rivières: quelques principes d'etude et applications, Compagnie Natinale du Rhone, *Formatione Continue*, France, (1992).

Ramsbottom, D., Day, R. and Rickard, C.: *Culvert Design guide*, CIRIA Report 168, Construction and Industry Research and Information Association, London, (1997), pp. 1–189.

Ramsbottom, D. and Whitlow, C.: Extension of rating curves at gauging stations: best practice guidance manual, Prepared for the *Environment Agency*, R&D Manual W6-061/M, October, (2003), pp. 1–247.

Raudkivi, A.J.: *Loose Boundary Hydraulics*, A.A. Balkema, Rotterdam, Netherlands, (1998), pp. 1–538.

Raven, P.J., Holmes H.T.H., Dawson, F.H., Fox, P.J.A., Everard, M., Fozzard, I. and Rouen, K.J.: River habitat quality: the physical character of rivers and streams in the UK and the Isle of Man, River Habitat Report No. 2, *Environment Agency*, (1998), pp. 1–90.

Ree, W.O. and Palmer, V.J.: Flow of water in channels protected by vegetative linings, Bulletin No. 967, Soil Conservation Service, *US Department of Agriculture*, Washington, (1949), pp. 1–115.

ReFEH: revised *Flood Estimation Handbook*, Vols. 1–5, Centre for Ecology and Hydrology, Wallingford, Oxon, UK, (2007).

Rehbock, T.: (1922). See Theodor Rehbock, (1864–1950), in Hydraulics and Hydraulics Research: A Historical Review, [Ed Gunther Garbrecht], International Association for Hydraulic Research, Balkema, (1987), pp. 281–291.

Reinius, E.: Steady uniform flow in open channels, Bulletin 60, Division of Hydraulics, *Royal Institute of Technology*, Stockholm, Sweden, (1961).

Reynolds, A.J.: *Turbulent flows in engineering*, Wiley, (1974), pp. 1–462.

Rezaei, B.: Flow in channels with non-prismatic floodplains, *PhD Thesis, Department of Civil Engineering, The University of Birmingham*, (2006).

Rezaei, B. and Knight, D.W.: Overbank flow in compound channels with non-prismatic floodplains, *Journal of Hydraulic Engineering*, ASCE, 137, No. 8, August, (2011), pp. 815–824.

Rhodes, D.G.: An experimental investigation of the mean flow structure in wide ducts of simple rectangular and trapezoidal compound cross section, examining in particular zones of high lateral shear, *PhD Thesis, The University of Birmingham*, (1991), pp. 1–355. (and Volumes 2–4 of supplementary graphs).

Rhodes, D.G. and Knight, D.W.: Lateral shear in wide compound duct, *Journal of Hydraulic Engineering*, ASCE, Vol. 121, No. 11, (1995), pp. 829–832.

Rhodes, D.G. and Knight, D.W.: Anomalous measurements in a compound duct, *ASCE Symposium on Fundamentals and Advancements in Hydraulic Measurements and Experimentation*, Buffalo, New York, August, Vol. 2, (1994), pp. 311–320.

Roberts, J.D., Free-Hewish, R.J. and Knight, D.W. Modelling of hydraulic drag forces on submersible bridge decks, *Proc. 4th Road Engineering Association of Asia and Australasia (REAA)*, Jakarta, Indonesia, August, (1983), pp. 1–11.

Rodi, W.: Turbulence models and their application in hydraulics, *State of the Art Paper, Int. Association for Hydraulic Research, IAHR*, Delft, Netherlands, (1980), pp. 1–104.

Rouse, H.: *Elementary Mechanics of Fluids*, Wiley, New York, (1946), pp. 1–376.

Rozovskii, I.L., Flow of water in bends or open channels, *The Academy of Science of the Ukraine SSR*, translated from Russian by the Israel Program for Scientific Translations, Jerusalem, Israel, (1957).

RS V5.7: Online Help Manual, InfoWorks RS software, Copyright *Wallingford Software*, (2004).

Samuels, P.G.: Modelling of river and flood plain flow using the finite element method, *PhD Thesis*, Department of Mathematics, University of Reading, HR Wallingford Report SR61, November, (1985).

Samuels, P.G.: Lateral shear layers in compound channels, *Proceedings International Conference on Fluvial Hydraulics*, Budapest (also published at *HR Wallingford*, UK), (1988).

Samuels, P.G.: Some analytical aspects of depth averaged flow models, *Proc. Intl. Conf. Hydraulic and Environmental Modelling of Coastal, Estuarine and River Waters*, Bradford, England, 19–21st September, (1989a).

Samuels, P.G.: Backwater length in rivers, *Proc. Instn Civ. Engrs*, Part 2, Vol. 87, December, (1989b), pp. 571–582.

Samuels, P.G.: Cross-section location in 1-D models, *Proc. Int. Conf. on River Flood Hydraulics*, (Ed. W.R. White), J. Wiley & Sons, Wallingford, September, Paper K1, (1990), pp. 339–348.

Samuels, P.G.: Uncertainty in Flood Level Prediction, XXVI Biannual Congress of the IAHR, HYDRA2000, London, Sept. 1995, *Volume 1 of Proceedings published by Thomas Telford*, (1995).

Samuels, P.G.: Risk and Uncertainty in Flood Management, Chapter 25 in River Basin *Modelling for Flood Risk Mitigation* [Eds D.W. Knight and A.Y. Shamseldin], Taylor & Francis, (2006), pp. 481–517.

Samuels, P.G., Dijkman, J. and Klijn, F.: An Analysis of the Current Practice of Policies on River Flood Risk Management in Different Countries, *Irrigation and Drainage*, Vol. 55, S141–156, (2006).

Sanjou and Nezu, I.: Turbulence structure and concentration Exchange property in compound channel flows with emergent trees on the floodplain edge, *Int. J. River Basin Management*, 9(34), (2011), pp. 181–193.

Sayers, P.B., Hall, J.W. and Meadowcroft, I.C.: Towards risk-based flood hazard management in the UK. *Proceedings of ICE – Civil Engineering*, Vol. 150, Special Issue 1, (2002), pp. 36–42.

Sayre, W.W. and Albertson, M.L.: Roughness spacing in rigid open channels, *Journal of Hydraulic Division*, ASCE, Vol. 87, HY3, (1961), pp. 121–150.

Schlichting, H.: Aerodynamische Probleme des Hochstauftriebes, *Lecture at Third Int. Congr. Aero. Sci.* (ICAS), Stockholm, Sweden, (1962), pp. 1–14 (1965).

Schlichting, H.: *Boundary layer theory*, McGraw-Hill, (1979), pp. 1–817.

Schmidt, A.R. and Yen, B.C.: Theoretical development of stage-discharge ratings for subcritical open-channel flows, *Journal of Hydraulic Engineering*, American Society of Civil Engineers, New York, Vol. 134, No. 9, September, (2008), pp. 1245–1256.

Seckin, G., Yurtal, R. and Haktanir, T.: Contraction and expansion losses through bridge constrictions, *Journal of Hydraulic Engineering*, American Society of Civil Engineers, New York, Vol. 124, No. 5, May, (1998), pp. 546–549.

Seckin, G., Haktanir, T. and Knight, D.W.: A simple method for estimating flood flow around bridges, *Water Management*, Proceedings of the Instn. of Civil Engineers, London, 160, Issue WM4, December, (2007), pp. 195–202.

Seckin, G., Knight, D.W., Atabay, S. and Seckin, N.: Effect of the Froude number on the assessment of the bridge afflux, *Water Management*, Proceedings of the Instn. of Civil Engineers, London, Vol. 161, Issue WM2, April, (2008a), pp. 97–104.

Seckin, G., Knight, D.W., Atabay, S. and Seckin, N.: Improving bridge afflux prediction for overbank flows, *Water Management*, 161(5), (2008b), pp. 253–260.

Seed, D.J., Samuels, P.G. and Ramsbottom, D.M.: Quality assurance in computational river modelling, *First Interim report SR 374*, HR Wallingford, (1993).

Sellin, R.H.J.: A laboratory investigation into the interaction between the flow in the channel of a river and that over its floodplain, *La Houille Blanche*, Vol. 19, No. 7, (1964), pp. 793–801.

Sellin, R.H.J.: Hydraulic performance of skewed two-stage flood channel, *Journal of Hydraulic Research*, IAHR, Vol. 33, No. 1, (1995), pp. 43–64.

Sellin, R.H.J. and van Beesten, D.P.: Berm vegetation and its effect on flow resistance in a two-stage river channel: an analysis of field data, *Proc. of River Flow 2002*, Belgium, Zwets & Zeitlinger, ISBN 90-5809-509-6, Vol. 1, (2002), pp. 319–327.

Sellin, R.H.J. and van Beesten, D.P.: Conveyance of a managed vegetated two-stage river channel, *Proc. ICE Journal of Water Management*, The Institution of Civil Engineers, Vol. 157, March, No. WM1, (2004), pp. 21–33.

Sellin, R.H.J., Ervine, A.E. and Willetts, B.B.: Behaviour of meandering two-stage channels, *Water Maritime and Energy*, Proc. Institution of Civil Engineers, Vol. 101(2), (1993), pp. 99–111.

SEPA: Second generation flood map for Scotland – methodology report, Report EX5098, prepared by HR Wallingford for the *Scottish Environment Protection Agency*, January, (2006), pp. 1–110.

Sharifi, S., Knight, D.W. and Sterling, M.: A novel application of a multi-objective evolutionary algorithm in open channel flow modelling, *Journal of Hydroinformatics*, IWA Publishing, 11.1, (2009), pp. 31–50.

Shaw, E.M.: *Hydrology in practice*, Chapman and Hall, London, (1994), pp. 1–569.

Shen, H.W.: *River mechanics*, Vols. I & II, Colorado State University, Fort Collins, (1971), pp. 1–897.

Shields, F.D., Coulton, K.G. and Nepf, H.: Representation of vegetation in two-dimensional hydrodynamic models, *Journal of Hydraulic Engineering*, American Society of Civil Engineers, New York, Vol. 143, No. 8, August (2017), pp. ?–?.

Shields, A.: Anwendung Aenlich Keitsmechanik und der Trubulenz-schung auf Die Geschiebebewegung, *Mitteilungen de Preussischen Veruchsanstalt fur Wasserbau und Schiffbau*, Berlin, Germany, (1936).

Shiono, K. and Knight, D.W.: Two-dimensional analytical solution for a compound channel, *Proc. 3rd Int. Symposium on Refined Flow Modelling and Turbulence Measurements*, Tokyo, Japan, July, (1988), pp. 503–510.

Shiono, K. and Knight, D.W.: Mathematical models of flow in two or multi stage straight channels, *Proceedings International Conference on River Flood Hydraulics*, (Ed. W.R. White), Wiley & Sons Ltd., Paper G1, (1990), pp. 229–238.

Shiono, K. and Knight, D.W.: Turbulent open channel flows with variable depth across the channel, *Journal of Fluid Mechanics*, Vol. 222, (1991), pp. 617–646 (and Vol. 231, October, p. 693).

Shiono, K. and Muto, Y.: Complex flow mechanisms in compound meandering channels with overbank flow, *Jnl. of Fluid Mechanics*, Vol. 376, (1998), pp. 221–261.

Shiono, K., Al-Romaih, J.S. and Knight, D.W.: Stage-discharge assessment in compound meandering channels, *Journa of Hydraulic Engineering*, ASCE, Vol. 125, No. 1, Jan., (1999a), pp. 66–77.

Shiono, K., Muto, Y., Knight, D.W. and Hyde, A.F.L.: Energy losses due to secondary flow and turbulence in meandering channels with overbank flow, *Journal of Hydraulic Research*, IAHR, Vol. 37, No. 5, (1999b), pp. 641–664.

Shiono, K., Ishigaki, T., Kawanaka, R. and Heatlie, F.: Influence of one line vegetation on stage-discharge rating curves in compound channel, *33rd IAHR Congress: Water Engineering for a Sustainable Environment*, Vancouver, Canada, (2009), pp. 1475–1482.

Shiono, K. and Rameshwaran, P.: Mathematical modelling of bed shear stress and depth averaged velocity for emergent vegetation on floodplain in compound channel, *E-proceedings of the 36th IAHR World Congress 28 June–3 July, 2015, The Hague, the Netherlands*, (2015), pp.

Shiono, K., Takeda, M., Yang, K., Sugihara, Y. and Ishigaki, T.: Modelling of vegetated rivers for inbank and overbank flows, River Flow 2012, *Proceedings of the International Conference on Fluvial Hydraulics*, Vol. 1, (2012), pp. 263–269.

Simons, D.B. and Senturk, F.: Sediment transport technology, Water Resources Publications, Colorado, USA, (1992).

Shucksmith, J.D., Boxall, J.B. and Guymer, I.: Bulk flow resistance in vegetated channels: analysis of momentum balance approaches based on data obtained in aging live vegetation, *Journal of Hydraulic Engineering*, ASCE, Vol. 137, No. 12, (2011), pp. 1624–1635.

Spooner, J. and Shiono, K.: Modelling of meandering channels for overbank flow, *Water, Maritime and Energy*, Proc. ICE, Volume 156, September, (2003), pp. 225–233.

Statzner, B., Lamouroux, N., Nikora, V. and Sagnes, P.: The debate about drag and reconfiguration of freshwater macrophytes: comparing results obtained by three recently discussed approaches, *Journal of Freshwater Biology*, Vol. 51, (2006), pp. 2173–2183.

Stearns, F.P.: On the current meter, together with a reason why the maximum velocity of water flowing in open channels is below the surface, *Trans. of ASCE*, Vol. 12, No. 216, (1883), pp. 331–338.

Stokes, G.G.: On the theories of internal friction of fluid motion and of the equilibrium and motion of elastic solids, *Cambridge Philosophical Society*, Vol. 8, (1845), pp. 287–305.

Strickler, A.: Translates to: Some contributions to the problem of the velocity formula and roughness factors for rivers, canals and closed conduits, *Mitteilungen des Eidgenossischen Amtes fur Wasserwirtschaft*, Bern Switzerland, No. 16, (1923).

Sturm, T.W.: *Open Channel Hydraulics*, McGraw-Hill Series in Water Resources and Environmental Engineering, New York, [ISBN 0-07-062445-3], (2001), pp. 1–493.

Sun, X.: Flow characteristics in compound channels with and without vegetation, *PhD thesis, Loughborough University*, UK; (2007).

Sun, X. and Shiono, K.: Flow resistance of One-line Emergent Vegetation along the floodplain edge of a Compound Open Channel, Advances in Water Resources, Vol. 32, (2009), pp. 430–438.

Sun, X., Shiono, K., Fu, X., Yang, K. and Huang, T.L.: Application of Shiono and Knight Method to compound open channel flow with one-line emergent vegetation, *Advanced Materials Research*, Vol. 663, (2013), pp. 930–935.

Tagg, and Samuels, P.G.: Modelling flow resistance in tidal rivers, *Proc. Int. Conf. on Hydraulic and Environmental Modelling of Coastal, Estuarine and River Waters*, Bradford, September, (1989), pp. 441–452.

Tanaka, N., Takenaka, H., Yagisawa, J. and Morinaga, T.: Estimation of drag coefficient of a real tree considering the vertical stand structure of trunk, branches and leaves, *Int. J. of River Basin Management*, Vol. 9, Nos. 3–4, (2011), pp. 221–230.

Tang, X. and Knight, D.W.: Experimental study of stage-discharge relationships and sediment transport rates in a compound channel, *Proc. 29th IAHR Congress*, Hydraulics of Rivers, Theme D, Vol. II, September, Beijing, China, Tsinghua University Press, (2001), pp. 69–76.

Tang, X., Knight, D.W. and Samuels, P.G.: Volume conservation in variable parameter Muskingum-Cunge method, *Journal of Hydraulic Engineering*, ASCE, Vol. 125, No. 6, June, (1999a), pp. 610–620.

Tang, X., Knight, D.W. and Samuels, P.G.: Wave speed-discharge relationship from cross-section survey, *Proc. Instn. of Civil Engineers, Water and Maritime Engineering*, London, 148, June, Issue 2, (2001), pp. 81–96.

Tang, X., Knight, D.W. and Sterling, M.: Analytical model for streamwise velocity in vegetated channels, *Journal of Engineering and Computational Mechanics*, Proceedings of the Instn. of Civil Engineers, London, Vol. 164 EM1, (2011), pp. 91–102.

Tang, X., Knight, D.W. and Samuels, P.G.: Variable parameter Muskingum-Cunge method for flood routing in a compound channel, *Journal of Hydraulic Research*, IAHR, Vol. 37, No. 5, (1999b), pp. 591–614.

Tennekes, H. and Lumley, J.L.: *A first course in turbulence*, Massachusetts Institute of Technology Press, USA, (1972), pp. 1–300.

Theodorsen, T.: Mechanism of turbulence, *2nd Midwestern Conf. on Fluid Mechanics*, Ohio State University, (1952), pp. 1–18.

Thijsse, J.T.: Formulae for the friction head loss along conduit walls under turbulent flow, *Proc., 3rd IAHR Congress*, Grenoble, France, Vol. 3, No. 4, (1949), pp. 1–11.

Thorne, C.R., Hey, R.D. and Newson, M.D.: *Applied fluvial geomorphology for river engineering and management*, J. Wiley, (1997), pp. 1–376.

Thorne, C., Knight, D.W. et al.: River width adjustment. Part I: Processes and mechanisms, *Journal of Hydraulic Engineering*, ASCE, Task Committee on Hydraulics, Bank Mechanics and the Modelling of River Width Adjustment, Sept., Vol. 124, No. 9, (1998), pp. 881–902.

Thorne, C.R., (chair), Alonso, C., Bettess, R., Borah, D., Darby, S., Diplas, P., Julien, P., Knight, D.W., Li., L., Pizzuto, P., Quick, M., Simon, A., Stevens, M. and Wang, S.: River width adjustment. Part II: Processes and mechanisms, *Journal of Hydraulic Engineering*,

ASCE, Task Committee on Hydraulics, Bank Mechanics, and Modelling of River Width Adjustment, 124(9), (1998), pp. 881–902.

Toebes, G.H. and Sooky, A.A.: Hydraulics of meandering rivers with floodplains, *Journal of Waterways and Harbours Division*, ASCE, Vol. 93, (1967), pp. 213–236.

Tominaga, A., Nezu, I., Ezaki, K. and Nakagawa, H.: Turbulent structure in straight open channel flows, *Journal of Hydraulic Research*, IAHR, 27(1), (1989), pp. 149–173.

Tominaga, A. and Nezu, I.: Turbulent structure in compound open-channel flows, *Journal of Hydraulic Engineering*, ASCE, Jan, 117(1), (1991), pp. 21–41.

Tominaga, A. and Knight, D.W.: Numerical evaluation of secondary flow effects on lateral momentum transfer in overbank flows, *River Flow 2004, Proc. 2nd Int. Conf. on Fluvial Hydraulics, 23–25 June, Napoli, Italy* [Eds M. Greco, A. Carravetta and R.D. Morte], Vol. 1, (2004), pp. 353–361.

Townend, I.H., Eliott, C. and Sayers, P.B.: Controlled dissemination and uptake of research, *Proceedings of 42nd Defra Flood and Coastal Risk Management Conference*, 3–5 July, New York, (2007), pp. 5a.1.1–5a.1.10.

Tsikata, J.M., Katopodis, C. and Tachie, M.F.: Experimental study of turbulent flow near model trashracks, *Journal of Hydraulic Research*, IAHR, Vol. 47, No. 2 (2009), pp. 275–280. doi: 10.3826/jhr.2009.3381.

Tullis, B.P., Anderson, D.S. and Robinson, S.C.: Entrance loss coefficients and inlet control head-discharge relationships for buried-invert culverts, *Journal of Irrigation and Drainage Eng.*, ASCE, 134(6), (2008), pp. 831–839.

USBPR.: Hydraulics of Bridge Waterways, US Dept. of Transportation, FHWA, *Hydraulic Design Series No. 1*, published electronically in 1978, (1978).

USBR, Progress report on results of studies on design of stable channels, *Hydraulic Laboratory Report No. Hyd-352, United States Bureau of Reclamation*, Engineering Laboratories Branch and Office of Chief Designing Engineer, Design and Construction Division, Denver, Colorado, June (1952), pp.1–34.

USGS.: Backwater at bridges and densely wooded flood plains, *Hydrologic Atlas Series Nos. 590–611*, (1978).

Vanonoi, V.A.: Transportation of suspended sediment by water, *Trans. of ASCE*, Vol. 111, (1946), pp. 67–133.

van Prooijen, B.C.: Transverse momentum exchange in compound channel flows, *Proc. XXX IAHR Conf.*, John F Kennedy Student Paper Competition, Thessaloniki, Greece, 2003.

van Prooijen, B.M.: Shallow mixing layers, *PhD Thesis, Department of Civil Engineering*, Delft University of Technology Report 04-1, ISSN 0169-6548, (2004).

van Prooijen, B.C., Battjes, J.A., Uijttewaal, W.S.J.: Momentum exchange in straight uniform compound channel flow, *Journal of Hydraulic Engineering*, ASCE, 131(3), (2005), pp. 175–183.

van Rijn, L.C.: The prediction of bedforms and alluvial roughness, *Proc. Euromech 156, Mechanics of sediment transport* [Eds B. Sumer and A. Muller], 12–14 July, Istanbul, Turkey, Balkema, (1982), pp. 133–135.

van Rijn, L.C.: Sediment transport, Part III: bedforms and alluvial roughness, *Journal of Hydraulic Engineering*, Vol. 110, HY12, December, (1984), pp. 1733–1754.

Versteeg, H.K. and Malalaskera, W.: *An introduction to computational fluid dynamics: the finite volume method*, Longman Scientific & Technical, (1995), pp. 1–257.

Vidal, J.P., Moisan, S., Faure, J.B. and Dartus, D.: River model calibration, from guidelines to operational support tools, *Environmental Modelling & Software* (available on-line via Science Direct), Volume 22, Issue 11, November, (2007), pp. 1628–1640.

Vreugdenhil, C.B.: *Computational hydraulics: an introduction*, Springer-Verlag, Berlin, (1989), pp. 1–182.

Wallis, S.G. and Knight, D.W.: Calibration studies concerning a one-dimensional numerical tidal model with particular reference to resistance coefficients, *Estuarine, Coastal and Shelf Science*, Academic Press, London, No. 19, (1984), pp. 541–562.

Wang, S.: On the reliability and accuracy of computational models for Hydroscience and Engineering investigations, *US-China Workshop on Advanced Computational Modelling in Hydroscience and Engineering*, Oxford, Mississippi, USA, 19, September, (2005).

Wark, J.B.: Discharge assessment in straight and meandering compound channels, *PhD Thesis, Department of Civil Engineering, University of Glasgow*, UK, (1993).

Westlake, D.F. and Dawson, F.H.: Thirty years of weed cutting on a chalk-stream. EWRS 6th Symposium on Aquatic Weeds (1982).

Weisbach, J.: Lehrbuch der Ingenieur, *und Maschinen-Mechanik*, Braunschweig, (1845).

White, W.R.: Sediment transport in channels – a general function, SR 102, *HR Wallingford*, UK, (1972).

White, W.R., Bettess, R. and Paris, E. Analytical approach to river regime, *Journal of the Hydraulics Division*, ASCE, Vol. 108, HY10, October, (1982), pp. 1179–1193.

White, W.R., Paris, E. and Bettess, R.: The frictional characteristics of alluvial streams: a new approach, *Proc. Inst. of Civil Engineers, London*, Vol. 69, No. 1, Sept., (1980), pp. 737–750.

White, W.R., Paris, E. and Bettess, R.: Tables for design of stable alluvial channels, Report No. 208, *Hydraulics Research Station*, Wallingford, (1981).

Whitehead, E., Brown, P. and Hollinrake, P.: The hydraulic roughness of vegetated channels, Report SR305, *HR Wallingford*, UK, (1992), pp. 1–38.

Whitlow, C.D. and Knight, D.W.: An investigation of the effect of different discretisations in river models and a comparison of non-conservative and conservative formulation of the de St. Venant equations, In *Hydraulic and Environmental Modelling: Estuarine and River Waters* (Eds R.A. Falconer, K. Shiono and R.G.S. Matthew), Ashgate Press, (1992), pp. 115–126.

Wormleaton, P.R., 1988, Determination of discharge in compound channels using the dynamic equation for lateral velocity distribution, *Proc. Int. Conf. on Fluvial Hydraulics*, Belgrade, Hungary, June, pp.

Wormleaton, P.R.: Floodplain secondary circulation as a mechanism for flow and shear stress redistribution in straight compound channels, In *Coherent Flow Structures in Open Channels* [Eds Ashworth, Bennett, Best and McLelland], Chapter 28, J. Wiley, (1996), pp. 581–608.

Wormleaton, P.R. and Merrett, D.J.: An improved method of calculation for steady uniform flow in prismatic main channel/floodplain sections, *Journal of Hydraulic Research*, IAHR, Vol. 28, No. 2, (1990), pp. 157–174.

Wormleaton, P.R., Allen, J. and Hadjipanos, P.: Discharge assessment in compound channel flow, *Journal of the Hydraulics Division*, Proc. ASCE, Vol. 108, HY9, Sept., (1982), pp. 975–993.

Wormleaton, P.R., Allen, J. and Hadjipanos, P.: Flow distribution in compound channels, *Journal of the Hydraulics Division*, Proc. ASCE, Vol. 111, HY2, February, (1985), pp. 357–361.

Wormleaton, P.R., Hey, R., Sellin, R.H.J., Bryant, T.B., Loveless, J. and Catmur, S.E.: Behaviour of Meandering Overbank Channels with Graded Sand Beds, *Journal of Hydraulic Engineering*, American Society of Civil Engineers, 131(8), (2005), pp. 665–681.

Wormleaton, P.R., Sellin, R.H.J., Bryant, T.B., Loveless, J., Hey, R. and Catmur, S.E.: Flow Structures in a Two-Stage Channel with Mobile Bed, *Journal of Hydraulic Research*, IAHR, 42(2), (2004), pp. 145–162.

Wright, J.F., Gunn, R.J.M., Winder, J.M., Wiggers, R., Vowles K., Clarke, R.T. and Harris, I.: A Comparison of the Macrophyte Cover and Macroinvertebrate Fauna at three Sites on the river Kennet in the mid 1970s and late 1990s, *The Science of the Total Environment*, 282–283, (2002), pp. 121–142.

Wright, N.G., Crossley, A.J., Morvan, H.P. and Stoesser, T.: Detailed validation of CFD for flows in straight channels, *Proc. of the 2nd Intl. Conf. on Fluvial Hydraulics*, River Flow 2004, Naples, 24–26 June, Vol. 2, (2004), pp. 1041–1048.

WSPRO.: Bridge waterways analysis: Research report, by Shearman, J.O. and Kirby, W.H., Schneider, V.R. and Flippo, H.N., *Report No. FHWA/RD-86/108, NTIS*, VA, USA, (1986).

Wu, B. and Guo, J.J.: Limit ratio for choking phenomenon, *2nd Int. Conf. On Flooding and Flood Defences*, Beijing, China, September, (2002), pp. 1–7.

Wu, B. and Molinas, A.: Choked flows through short transitions, *Journal of Hydraulic Engineering*, American Society of Civil Engineers, New York, Vol. 127, No. 8, August, (2001), pp. 657–662.

Wunder, S., Lahmann, B. and Nestmann, F.: Determination of the drag coefficients of emergent and just submerged willows, *International Journal of River Basin Management*, Vol. 9, Nos. 3–4, (2013), pp. 231–236.

Xin, S. and Shiono, K.: Flow resistance of one-line emergent vegetation along the floodplain edge of a river channel, *Advances in Water Resources, 32*, (2009), pp. 430–438.

Yalin, M.S. and Ferreira da Silva, A.M.: *Fluvial processes*, IAHR Monograph, (2001), pp. 1–197.

Yang, C.T.: *Sediment transport, theory and practice*, McGraw-Hill, (1987), pp. 1–396.

Yang, K., Nie, R., Liu, X. and Cao, S.: Modeling Depth-Averaged Velocity and Boundary Shear Stress in Rectangular Compound Channels with Secondary Flows, *Journal of Hydraulic Engineering*, ASCE, Vol. 139, No. 1, January, 2013, pp. 76–83.

Yarnell, D.L.: Bridge piers as channel obstructions, *Technical Bulletin No. 442, US Dept. of Agriculture*, Washington, D.C., November, (1934).

Yen, B.C.: *Channel flow resistance: centennial of Manning's formula*, Colorado, USA, Water Resources Publications, (1991), pp. 1–453.

Yen, C.L. and Overton, D.E.: Shape effects on resistance in floodplain channels, *Journal of the Hydraulics Division*, ASCE, Vol. 99, No. 1, (1973), pp. 219–238.

Yu, G. and Knight, D.W.: Geometry of self-formed straight threshold channels in uniform material, *Proc. Instn. of Civil Engineers, Water, Maritime and Energy Division*, London, Vol. 130, March, Paper No. 11398, (1998), pp. 31–41.

Yuen, K.W.H.: A study of boundary shear stress, flow resistance and momentum transfer in open channels with simple and compound trapezoidal cross section, *PhD thesis, Department of Civil Engineering, The University of Birmingham*, (1989).

Yuen, K.W.H. and Knight, D.W.: Critical flow in a two stage channel, *Proc. Int. Conf. on River Flood Hydraulics*, (Ed. W.R. White), Wallingford, September, J. Wiley & Sons, Paper G4, (1990), pp. 267–276.

Zegzhda, A.P.: Teoriia podobiia I metodika rascheta gidrotekhnicheskikh modelei (Translates as: Theory of Similarity and Methods of Design of Models for Hydraulic Eng.), *Gosstroiizdat*, Leningrad, (1938).

Zienkiewick, O.C.: The finite element method, 3rd Edition, *McGraw-Hill Book Co. Ltd.*, UK, (1977).

Author index

Subject index